SADT™

SADT™
Structured Analysis and Design Technique

David A. Marca

Intelligent Systems Technologies Group
Digital Equipment Corporation
Hudson, Massachusetts

Clement L. McGowan

Computer and Intelligent Systems Laboratory
GTE Laboratories Incorporated
Waltham, Massachusetts

with a Foreword by

Douglas T. Ross

SofTech, Inc.
Waltham, Massachusetts

McGraw-Hill Book Company

New York St. Louis San Francisco Auckland Bogotá
Hamburg London Madrid Milan Mexico
Montreal New Delhi Panama Paris São Paulo
Singapore Sydney Tokyo Toronto

Library of Congress Cataloging-in-Publication Data

Marca, David.
SADT : structured analysis and design technique / David A. Marca and Clement L. McGowan
with a foreword by Douglas T. Ross.
p. cm.
Bibliography: p.
Includes index.

ISBN 0-07-040235-3

1. Electronic data processing—Structured techniques. 2. SADT (Computer system) 3.
System design. I. McGowan, Clement L. II. Title.
QA76.9.S84M36 1987
004.2′1—dc19 87-15307

1 2 3 4 5 6 7 8 9 0 SEM/BKP 8 9 3 2 1 0 9 8 7

ISBN 0-07-040235-3

The editors for this book were Theron Shreve and Georgia Kornbluth, the designer was Elliot Epstein, and the production supervisor was Dianne Walber. It was set in Times Roman by ComCom.

Printed by Semline, Inc., and bound by The Book Press.

The material in this book expresses only the views of its authors, and is in no way a statement of Digital Equipment Corporation or GTE Laboratories Incorporated.

The pronouns "he" and "his" and terms such as "foreman" have been used in a purely generic sense in this book to accommodate the text to the limitations of the English language and avoid awkward grammatical constructions.

To all who developed, extended, taught, and used the SADT methodology over these past 14 years

Contents

Foreword

It is very good that this, the first book on SADT, has finally been written, and it is nice to be invited to supply a foreword for it. The authors, both good friends and colleagues of mine, are well prepared to guide you, the reader, along the paths they and others have found successful in the practice and teaching of the methodology in diverse settings. Like all good teachers, they put their own stamp and emphasis on the material that they cover, but you will find that most of what I mention here as a general overview is fully and compatibly addressed in the book itself.

SADT is not so much an invention as a discovery on my part. I first properly used the central "SA box" notation of what I much later came to call "Structured Analysis" in the first figure of the first interim report of the APT project at MIT 30 years ago. It simply was a concise expression of the important thought I had to convey about what today would be called a "hierarchic, layered, modular system." Each layer was a complete system (box), supporting and controlled by the system (box) above it. (APT is a standard language, like Fortran, Cobol, or Lisp, but is used for automatic programming of machine tools, and the original system architecture still is in use today.) The concept of *decomposition,* another central feature of SADT, still was hidden, however. Before it arrived on the scene, the second major proper use of the SA box showed up similarly in the key figure of my prize-winning 1967 paper, "The AED Approach to Computer-Aided Design." The AED project followed APT at MIT (in the 1960s) and pioneered most of the design and tool-building techniques of what now is the field of *software engineering* as well as CAD proper. The figure shows the "system of systems for building systems" for what today is called "compiler compiler" technology for creating special user-oriented languages. Rather than being stacked in layers, table-controlled general processors (for spelling, grammar, code generation, and execution), each a complete system (box), were chained together, output to input, to make the compiler. The tables, themselves, were generated similarly. So the SA box was fully recognized and in practical use, even though not yet named.

Significantly, it was between these two instances of the graphic use of the graphic SA box that the importance of what now is known as "top-down hierarchic decomposition" was emphasized—but not graphically—in my 1960 "Statement of Objectives" technical report that launched the AED project. There I stressed that CAD had to be *object-oriented,* and that objects, including the systems we would make to deal with them, should be defined and described *outside-in,* to whatever depth was needed to expose their details.

We used and came to understand these crucial concepts in practical, working form, literally over the next dozen years, for the final preparation stage for structured analysis came in a 6-week, intensive, confidential consulting project in 1972 (after we had left MIT to found SofTech in 1969) in which I was asked for the general design of a complete *factory of the future* for a commercial client. A dozen separate user languages and their corresponding integrated systems were involved, along with a shared database system with essentially no downtime and triple-redundant, recoverable data as well as operating functions. All this had to operate on a hierarchy of technical and business computers, control systems, and inventory and handling machinery for raw materials, parts, and tools, in addition to the working machine tools and people and orders and. . . . Above all, the design had to be *credible.* It had to justify the go-ahead decision, and then be strong enough to control the project itself.

In a frenzy of activity I drew together all the years of tested ideas and experience in an ad hoc design methodology, emphasizing interfaces between modular systems and their use as "fire-wall protection" for the required ultra-reliability—and I completed the project on schedule. Along with recent chats I had had with S. Hori, who used almost the same SA box notation in his form of "cell modeling" of human-directed activities, this was the background that allowed another frenzied 6-week activity to, in 1973, create the first *Author Guide* and use it in the original training of analysts in "the Architecture Method" used in the Air Force Computer-Aided Manufacturing (AFCAM) Project. In the following year I coined the name "Structured Analysis" for that methodology, which finally brought together the SA box and SA decomposition in one graphic "language for blueprinting systems," and the rest is history.

After all these years, how do I view SADT? More and more I am interested in telling people *why* it works so well. An overview comprising my most recent attempt to express those thoughts is a suitable close to this foreword. Refer to the overview below, occasionally, as you progress through the rest of the book. You will find more and more insight, as you learn from practice.

The world and everything in it, including our thoughts about it, can be viewed as a system of interacting systems of systems. A system has *boundary, behavior,* and *substance,* each of which is defined by the mutual interaction of that system with other systems, with which it composes still other systems. Thus a delivery truck has electrical, mechanical, and hydraulic systems: Concretely, it has driving control, delivery schedule, and cost-of-ownership systems. Abstractly, a science has subject, theories, laboratories, curriculum, experiments, budgets, and students. A novel has setting, characters, plot, and style. History (past, present, future—possible or experienced) can include them all.

How can such a variety of systems be treated rigorously? By adopting and using directly the language *most natural* to each as a subject—whether it be a "natural" language, such as English, the technical jargon of an artificial formal language, mathematical formulas, or even pictures of shapes and forms. *How can the imprecision and ambiguity of such a range of expressions be countered?* In the same way that *any* natural language does—by imposing on those expressions dominant forms of punctuation which has no meaning of its own but constrains, structures, and contextualizes the interpretation of those expressions so that only the intended meaning can filter through.

How can such a range of disparate subject matters be accommodated in one universal punctuation scheme? By providing a concise, precise, human- and machine-readable *written format* that itself *models* the boundary, behavior, and substance of any chosen subject, expressed in its natural language. *What does it mean "to model"? Can one model model all?* Here is the definition:

M *models* A if M *answers questions* about A.

One modeling *scheme* can be used to model any chosen subject. Every model is limited in the answers it provides, but there is no limit on how or what models model, just as there is no limit on natural thought. *So what is the universal unit of universal punctuation for unrestricted rigorous structured analysis?* The SA box:

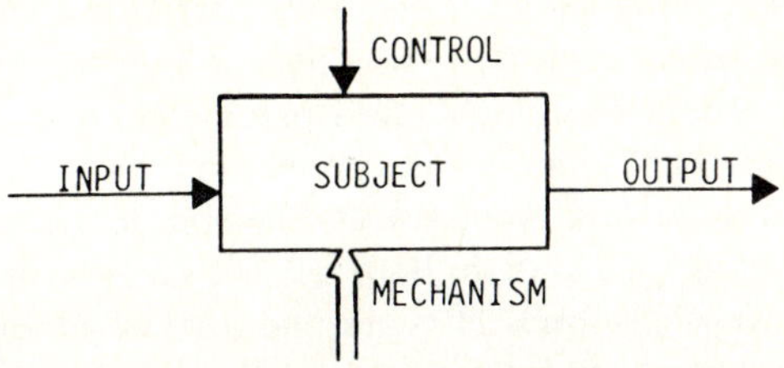

Under control, input is transformed into output by the mechanism.

Outputs from one box may be inputs or controls (or mechanisms) for other boxes. [Boxes are *named* and arrows are *labeled* using (their) natural language.] Arrows may *branch* and *join,* and any box may be *decomposed,* as a *whole,* into its constituent *parts* on a detail *diagram.* Inputs, controls, and outputs define the *interfaces* between boxes, and mechanisms allow constrained *mixing* of subjects, when appropriate. Box and diagram boundaries *must match,* and the resulting hierarchic, interconnected collection of diagrams is a *model.* By the definition of "to decompose," every grammatical model has a rigorously defined *meaning.*

How is complexity controlled, so that the meaning can be grasped? A diagram is restricted to three to six boxes for gradual exposition of detail. Huge models are shunned in favor of many small, interrelated models, whose meanings mutually support each other's roles in structuring a complex subject understandably. *How is understandability ensured? How practical is the approach?* Only well-disciplined thought can yield well-structured models. SA authors are taught explicit methods, using the feedback of the visible structuring of the SA box-and-arrow graphic language, to organize the creation of good models. Then, to ensure that the models actually express the intended thoughts, SA readers are taught explicit methods for interpreting models correctly. The SA *Author/Reader Cycle* provides methods by which concise comments on understandability can be fed back to the author as ongoing quality assurance for the intended readership. A well-defined SA *Librarian* role supports team efforts, but is simple enough to be useful to the single author, as well. The simplicity of the mechanics, from the thought discipline to the use of the language to the shuffling of papers, ensures that the ability to communicate is made more productive and effective without artificial burden. Experience builds rapidly, with active readers.

Is SA intended to replace other forms of communication? Not at all—it complements them. The box-to-diagram shared boundary is called a "node." The *Node Index* of diagram titles for a model is precisely the same as the standard, well-structured *topic outline* of ordinary good documentation. The primary *reading rule* for reading a model is: read the diagram first, and only then read the matching SA text for that diagram. A compact SA *Reference Language* allows reference to any aspect of the diagram, so that paths and collective relationships can be pointed out to ensure the message is conveyed. This brief, well-organized text can be of great assistance in organizing a standard presentation, matching the outline. For broad overview purposes, the intermediate boundaries imposed during analysis by the three-to-six-boxes rule may be inappropriate for synthesis purposes. Rigorous methods are taught for compressing model portions into large, multibox schematic diagrams, which then can further be transformed into standard illustrations—but with the important knowledge that they are fully justified and supported by the full analysis of the model and match the resulting condensed topic outline as table of contents for a standard presentation.

Often the dynamics of a system is important, while at other times a knowledge base must be shared. How does SA relate to simulation and databases? The *orientation* of a model (its context, viewpoint, and purpose) may be directed so that the resulting structured thoughts feed parameters to available simulation, database, or structured programming design methodologies. There are two main styles of SA modeling: *activity* models stress the *happenings* of the system; *data* models stress the *things* of a system. Both use the same box-and-arrow graphic language, in dual ways (the roles of boxes and arrows interchange). The most complete modeling links complementary models of each type together, for the same subject, using the *SA tie* process. *Activation rules* may be associated with continuous or discrete data flows in the models, and *Data Dictionary*–type constraints may be applied to things and happenings. The fundamental meaning of an activity model portion can be derived algorithmically, as a loop-with-exits *structured program,* with the addition of a simple marking to indicate when input and control data is consumed.

What is the track record of SA? With very few exceptions, SA has successfully been applied, either as SofTech's Structured Analysis and Design Technique (SADT) or as the activity-only, government version IDEF0 by thousands of people on hundreds of projects in many areas, since its introduction in 1973. Most applications have been concerned with human-machine-computer systems in business, manufacturing, defense, communications, and project design and control. It is hoped that the availability of this book will extend these areas into humanities and science areas, as well. SA is applicable to any interesting subject.

Douglas T. Ross
SofTech and MIT
November 1986

Preface

The use of expert systems, fourth-generation languages, and computer-aided manufacturing systems is ever-increasing. The success of such systems depends critically upon our ability to precede their implementation and deployment by defining the overall problem addressed, specifying which system functions should be automated, identifying human-machine interface points, and determining how the system and its environment interconnect. In other words, the *system engineering* function is critical to building quality systems. System engineering is a discipline for specifying subsystems, components, and how they interconnect; for identifying the constraints under which a system must operate; and for deciding upon an effective combination of people, machines, and software to realize a system. SADT is one of the best-known and most widely used system engineering methods.

SADT—an abbreviation for Structured Analysis and Design Technique—is a graphic notation and an approach to system description. Douglas T. Ross introduced it almost two decades ago. Since then, system analysts at SofTech, Inc., have refined and used SADT on a wide variety of problems. Telephonics software, system maintenance and diagnostics, long-range and strategic planning, computer-aided manufacturing, computer-aided design, computer system configuration, personnel training, embedded software for defense systems, financial management, and inventory control are some of the areas of successful SADT usage. The wide range of problem domains to which SADT has been applied testifies to its generality and power. The Integrated Computer-Aided Manufacturing Program of the U.S. Department of Defense recognized the utility of SADT by standardizing and making public a subset of it called IDEF0. Such standardization, combined with growing automated support and this book, means that SADT is now more readily available and easier to use.

Under the name IDEF0, SADT has been used by thousands of people in hundreds of military and industrial organizations. In the commercial world, SADT is typically used for requirements definition. As such, it competes with the dataflow-oriented methods of E. Yourdon's Structured Design, T. DeMarco's Structured Analysis, and C. Gane and T. Sarson's Structured Systems Analysis, and with the data-structuring methods of M. Jackson's Systems Development and J.D. Warnier and K. Orr's Structured Systems Development. Unlike these other structured analysis techniques, which have their origins in software design, SADT was originally developed to describe a system and its environment before defining requirements—software or otherwise. In other words, by focusing on the task of describing systems in general, the developers of SADT created a graphic language and a set of analysis procedures that can be used to understand a system before its implementation is envisioned. As a result, SADT is typically used at the earliest points in the system development process, commonly termed the "system life cycle," and is sometimes followed by the above-mentioned methods.

During the 10 years we, the authors of this book, have both used and taught SADT, we have felt the need for a book covering its concepts and showing how it is most commonly practiced today. Until now, public information on SADT has been scattered in technical papers and in nonclassified U.S. government "deliverables" typically available to those in the fields of software engineering and computer science. The major goal of this book is, therefore, to make SADT available to a much wider audience than past publications—in fact, to anyone involved in any kind of system science discipline such as medicine, ecology, cognitive science, and economics. To accomplish this, we present SADT here in a straightforward, expository fashion, focusing on how a single SADT "model" of a system is developed. We also show in a step-by-step fashion how an SADT model is built, so that, by the time you finish this book, you will be able to understand SADT models and know how they are made. Naturally, our presentation reflects our experiences with using, and our perpectives about, SADT. Accordingly, the book concludes with a variety of examples, taken from actual analysis projects, that illustrate different styles of SADT usage and the wide range of SADT's applicability.

The book focuses discussion by presenting the general system concepts inherent in SADT, explaining how SADT is commonly practiced, describing how the SADT modeling process is managed, discussing current automated support for SADT, and presenting real-life examples. Part I, Concepts of Activity Modeling, covers the fundamental system-modeling concepts SADT is built upon. Part II, Authoring Activity Models and Diagrams, shows how the graphic language of SADT is used to break down a system description into a hierarchic set of diagrams. Part III, The SADT Review Process, discusses the SADT Author/Reader Cycle and how it is used to improve the quality of a model through a special peer review process. Many pratitioners believe that this is the best practical process for assuring the quality and correctness of their work. Thus, we have presented the Author/Reader Cycle as a stand-alone guide that can be applied to any technical effort. Part IV, Finishing and Managing a Modeling Effort, discusses the ways to monitor and manage a team of SADT analysts, and includes a set of heuristics which can help determine when a model should be stopped. Part VI, Applications of SADT, presents several complete models in a variety of problem domains.

We have also written this book to present SADT as it is most commonly practiced, using workshops that are designed to teach you how to make a single SADT Activity Model. These hands-on exercises will give you the knowledge and practice you will need to start using SADT. The 25 workshops of Part V, Workshops—Developing an Activity Model and a Specification, are organized to correspond closely to the chapters in this book, and are packaged so that each covers a separate and complete step in building a model. The workshops can also be used in a classroom setting. Early workshops emphasize group activities, so the senior people of a class can become role models for the less experienced. Later workshops emphasize individual activities, so everyone in the class achieves the same level of SADT modeling experience. This mixing of group and individual participation establishes a classroom setting of nonthreatening learning—a key to introducing and teaching any new methodology—and is a major reason why we have been very successful in using these workshops over the past decade to teach 1-week SADT courses to classes of about 20 people.

By presenting SADT in this way, we are addressing an audience of practitioners in the "systems" area. Also, this book should be of particular interest to those who are responsible for commissioning, reading, or writing system specifications. We therefore view this book as a practitioner's primer—an SADT "starter's kit" and reference guide. Our strategy in writing this book has been to teach SADT from many viewpoints: concepts explain the underpinnings of SADT, practice shows how to best use the methodology, and examples show the results of SADT practice in industry today. We feel strongly that collecting this knowledge into a single document most effectively conveys SADT to the widest possible audience.

Acknowledgments

We would, above all, like to thank Douglas T. Ross and Clare Feldmann of SofTech, Inc., for their help in providing material and guidance during the initial stages of this work. We would also like to thank SofTech, Inc., for allowing us to use their extensive SADT library of articles, papers, deliverables, and models.

A special thanks goes to Dan Thornhill for his detailed review of the manuscript and his many helpful suggestions. Thanks also go to Clare Feldmann, Dan Thornhill, Ken Schoman, and John Brackett for their reviews and for their examples during many systems analysis projects, which gave us insights into the SADT methodology that can only be learned by seeing SADT in action.

Also, we are grateful to Peter Freeman who always held on to the idea that an SADT book should be published.

Introduction: The Motivation behind SADT

SADT was developed in the late 1960s, during the structured programming revolution. While many were struggling with software issues, a few were concerned with the harder problem of building large-scale systems made up of people and machines as well as software, like those in the areas of telephonics, manufacturing, and weapons command and control. Traditional large-scale system development practitioners at that time were just beginning to see the need for more discipline. Thus, the engineering community began to formalize the process of system development into a sequence of phases such as:

- Analysis — Determine what the system will do.
- Design — Define subsystems and their interfaces.
- Implementation — Create the subsystems independently.
- Integration — Combine the subsystems into a single unit.
- Testing — Verify the system workings.
- Installation — Make the system operational.
- Operation — Use the system.

This sequence was always repeated, because the system never perfectly met the perceived needs of its users, and because users' needs changed. But there still were problems with this management-directed model of system development. The maintenance costs, the costs incurred after a system was delivered, started to greatly exceed the development costs and were rising at an alarming rate because of the poor quality of the initially built system.

Some believed that the nature of errors during development was the key to rising maintenance costs. Studies showed that a high percentage of the errors in a system were made during analysis or design, many fewer errors were made during implementation and testing, and the cost in time and dollars to identify and correct errors increased as development progressed. For instance, an uncorrected analysis error could cost 2 times as much to correct during design, 10 times as much to correct during testing, and 100 times as much to correct during system operation and maintenance. Errors made during analysis or design took around twice as long to find and around 5 times as long to correct as errors made in subsequent phases. Also, analysis and design errors were often found by the users, causing much dissatisfaction.

Experience brought out several problems with the more traditional approaches to system development. There was no uniform approach. Users' involvement in the process was not controlled. Checking for consistency was lacking or sporadic.

The results of one development task did not integrate well with others. The process was difficult to quantitatively and qualitatively measure. It was said that while system builders were using techniques like structured programming and top-down design, they were solving either unstated problems, poorly stated problems, or even well-stated problems that were misunderstood. The problems in system development were also becoming less traceable to either the hardware or to programming, and the most crucial errors were being made early in the system development process. Often, these errors were due to incomplete functional specifications and to inconsistencies between the specification and the design.

Engineers knew that systems were growing in complexity and were often poorly defined. System complexity and growth are facts of life—a premise that must be accepted as a given. But faulty system definition is not a given; it is the result of inadequate system development methods. A thesis soon emerged: Improved analysis methods are the key to cost-effective, maintainable, and reliable systems. Better methods, tailored specifically for use very early in the process, were required to attack the key problems of traditional large-scale system development. The practicing of SADT grew out of this belief. By using SADT during the first steps of system development, the problem under consideration could be much better understood. This would reduce both development and maintenance costs, as well as improving the reliability of the system. SADT was a way to reduce costly errors by structuring the early products of system development, by improving communication between users and developers, and by smoothing the transition from analysis to design.

Douglas T. Ross expressed some of his PLEX theories in a system description language and methodology which he called "Structured Analysis and Design Technique" and which is commonly known as "SADT." Initial work on SADT began in 1969. Its first major application was during a project for a large aerospace effort in 1973, when it underwent some revision by those at SofTech, Inc. In 1974 SADT was further refined and transferred to one of Europe's largest telephonics companies. The market introduction of SADT occurred in 1975 after a year of productization. By 1981, SADT had been used by over 50 companies, on over 200 projects, involving over 2000 people, and spanning dozens of problem domains including telephonics, aerospace manufacturing, command and control, inventory control, and data processing. With its extensive use today in Europe, the Far East, and U.S. aerospace manufacturing (under the name IDEF0), those figures have grown substantially. SADT is therefore notable among today's system description methodolgies for its wide range of usage.

Why has SADT been so widely applied? First, SADT is the only methodology capable of easily representing systems characteristics like control, feedback, and mechanism. This is because SADT was initially developed from a more general systems engineering perspective, as opposed to all the other structured methods that had their origins in software design. Second, SADT complemented the existing system development concepts and standards of the time, had explicit procedures supporting group work, and had the advantage of being applicable to the earliest stages of system development. Also, extensive use showed that it could be integrated with other structured techniques and that SADT pictures acted like blueprints, tying together different techniques used to describe specific parts of a system with different levels of detail.

Thus, the inadequate system specification practices of the times motivated the creation of the SADT graphic language, and its heavy use matured SADT into a complete methodology, capable of improving the quality of the first work products made during system development. So, SADT started out as a language to describe the workings of systems in general, and as it became more heavily used, its procedures for system description were expanded and refined. In opening chapters of this book, we discuss the system description concepts underlying SADT, its graphic language, and its procedures for describing systems. In later chapters, we look over the shoulder of someone using SADT to see how the methodology is used to describe a system and how a specification can be developed from that description. At the end, we present several system descriptions from a variety of problem domains as testimony to SADT's wide range of applicability.

Part

I

Concepts of Activity Modeling

This part of the book summarizes the SADT methodology, describes the conceptual underpinnings upon which SADT is based, and explains why the graphic language and modeling process of SADT can be used to develop rich system descriptions. The subset of the complete SADT methodology which concentrates on system functionality is called "activity modeling" and is the focus of this book. The conceptual aspects of SADT activity modeling have been collected in this part to simplify your search when you have questions of a theoretical nature about SADT.

Chapter 1 introduces systems in general, discusses where SADT falls in the spectrum of system description methods, and introduces the basic concepts upon which SADT was developed. Chapter 2 defines the syntax and usage rules required to draw a single SADT diagram. Chapter 3 defines the rules required to tie together SADT diagrams into a model, even though that model may be rapidly evolving. Chapter 4 explains how the SADT modeling process, a critical part of the entire SADT methodology, facilitates validation of models. Chapter 5 goes into more detail on SADT diagrams and what system concepts they convey. Chapter 6 explores some of the detailed concepts underlying how SADT models are built.

These chapters illustrate the concepts and practice of SADT through a Prototype Machine Shop problem, which is used consistently in Parts I to IV of this book. Portions of the Prototype Machine Shop model illustrate SADT concepts and show how SADT is typically used. While Part I does define the concepts behind SADT, these concepts alone are not sufficient for making accurate, readable, and usable SADT models. The additional processes of analysis, synthesis, peer review, and project management (all of which are covered in the other parts of this book) are also vital components of the SADT methodology.

Chapter

1

Systems and Models

We are about to embark on a study of system modeling. By the word "system" we mean a set of interacting components with relationships among them. The world we live in can be viewed as a complex interconnected set of natural and constructed systems. These systems range from complex (e.g., the planetary bodies in a solar system) to medium complexity (e.g., the space shuttle) to extremely complex (e.g., molecular interactions in living organisms). Science has developed many disciplines to explain and understand systems at either end of this vast spectrum of complexity. For example, mechanics can explain the gravitational attraction between two planets, and chemistry can predict molecular interactions in a glass of boiling water. Constructed systems, however, often fall roughly in the middle of this spectrum. For example, our worldwide telephone network has many thousands of switches, yet the interaction among these switches cannot come close to the number of molecular interactions in even a very small glass of water. From a general systems point of view, these systems are often thought of as having "medium" complexity.

By the word "modeling" we mean the act of developing an accurate description of a system. What is especially difficult is to describe systems of medium complexity, such as telephone network switching, air traffic control, submarine guidance, automobile assembly, the space shuttle, and the operations of a refining plant. From a human perspective, these systems are hard to describe because they are large enough to make impossible (in a practical sense) the enumeration of their components and interactions, and yet they are small enough to make impossible their simplification using generalizing assumptions (e.g., as physics does). Our inability to easily describe, and hence understand, these kinds of systems makes their specification, development, and maintenance time-consuming and expensive, and increases the risk of their failure. As technology grows, accurate system description becomes more vital.

The methodology called "SADT"—an abbreviation for Structured Analysis and Design Technique—was developed specifically to help people describe and understand constructed systems that fall into the spectrum of "medium" complexity. SADT was initially developed and field-tested from 1969 to 1973. The methodology was strongly influenced by PLEX, ideas from Hori's Human-Directed Activity Cell model, general systems theory, software engineering, and even cybernetics. It received extensive use starting in early 1973 on "large-system" problems such as real-time telephonic communications design, computer-aided manufacturing (CAM), command and control software, and military readiness. Since then, it has been successfully applied to describe a large number of complicated constructed systems in a wide variety of domains (e.g., banking, oil refining, manufacturing process planning, weapon systems guidance, logistics management, technology planning, software engineering). The reason for this success is that SADT is a complete methodology for developing system descriptions, centered on the concepts of system modeling.

1.1 SADT Models

An SADT system description is called a "model." An SADT model employs both natural and graphic languages to convey meaning about a particular system. The natural language comes from the people describing the system, and the graphic language comes from SADT. As we will see, the graphic language of SADT provides structure and precise semantics for the natural language contained in a model. The SADT graphic language organizes natural language in a particular and unique way, and it is because of this that SADT can describe systems formerly beyond our ability to explain well.

From an SADT perspective, a model focuses on either system activities or system things. Historically, those SADT models that focus on system activities are called "activity models," and those that focus on system things are called "data models." Activity models present system activities in a successively detailed manner, and they define the relationship among those activities through the things of the system. Data models are the duals of activity models, and thus present a continually detailed description of system things interrelated by system activities. The complete SADT methodology also includes the construction of multiple models to properly describe a complex system. This book focuses on how to make a single activity model, and therefore the topics of SADT data modeling and multiple SADT models are beyond the scope of this book.

Parts I to IV of this book present those SADT concepts, techniques, and processes that relate to developing activity models. As an illustration to the technical discussion, we will give examples of how an actual activity model is built. The system to be described, taken from the aerospace manufacturing domain, is a machine shop that makes parts for experimental aircraft, commonly called a "prototype machine shop." The SADT model we will develop is intended to describe how the machine shop operates so that a training manual for new personnel can be written. Appendix A contains a complete description of the problem and a summary of the work done by the shop.

1.2 A Model Answers Questions

An SADT model is a complete, concise, and consistent description of a system which is developed for a particular reason. That reason, called the "purpose" of the model, is derived from the way SADT formally defines a model:

> M is a model of a system S if M can be used to answer questions about S with an accuracy of A.

The purpose of a model is, by definition, answering a set of questions. These questions remain in the back of one's mind during analysis, and thus they guide and influence the development of a model. This is because the model itself must be able to answer the question set to a predetermined level of accuracy. If the model cannot answer all questions in the question set and cannot answer them to the required level of accuracy, then we say the model has not yet fulfilled its original purpose. By defining models in this way, SADT forms the basis for a practical modeling discipline.

The implications and interpretations of this definition have had a major impact on how SADT has been used in practice. Commonly, the question set for a model is developed very early in an SADT modeling effort, and is usually summarized by one or a few sentences. For example, Figure 1-1 shows the work of a person using SADT to define the purpose of the Prototype Machine Shop model. Notice that the person has, after reading the problem statement and process summary, listed a set of questions and summarized that list with a single sentence. This sentence becomes the stated purpose of the model, and the question set remains as detailed backup to that sentence. When the model is complete, the questions listed will be answered by the information contained in the model.

What would be a suitable level of accuracy for the Prototype Machine Shop model? Since the model will be used to write a training manual, a reasonable level of accuracy is reached when each machine shop activity defined in the model can be written about with one paragraph of text. The essence of accuracy is therefore something "do-able" and measurable. Other systems analysis methods (i.e., alternate ways of describing a system) fail to mention this critical element of a model's overall purpose. By understanding *how well* the questions must be answered, the person developing the model will know when to stop modeling (i.e., when the model is good enough to suit its purpose).

1.3 A Model Has a Single Subject

A model is an understanding of a system. The system itself, therefore, is the subject of the model. However, the system being modeled never exists in isolation: it is always intertwined with its environment. Also, it is often hard to tell where a system ends and where its environment begins. For this reason SADT stresses the need for precisely defining the boundary of a system. Therefore, an SADT model always bounds its subject. That is, a model states exactly what the subject is and is not by describing what makes up the system and by implying what lies beyond the system. By bounding a subject, an SADT model helps focus attention on just the system being described and avoids introducing extraneous subjects. These are the reasons for our claim that SADT models have a single subject.

1.4 A Model Has One Viewpoint

Closely attached to the definition of a model is the perspective from which it is made. Since the quality of a system description is dramatically reduced if it remains unfocused, SADT requires that a model be developed from one and only one particular perspective. This perspective is called the "viewpoint" of the model. For exam-

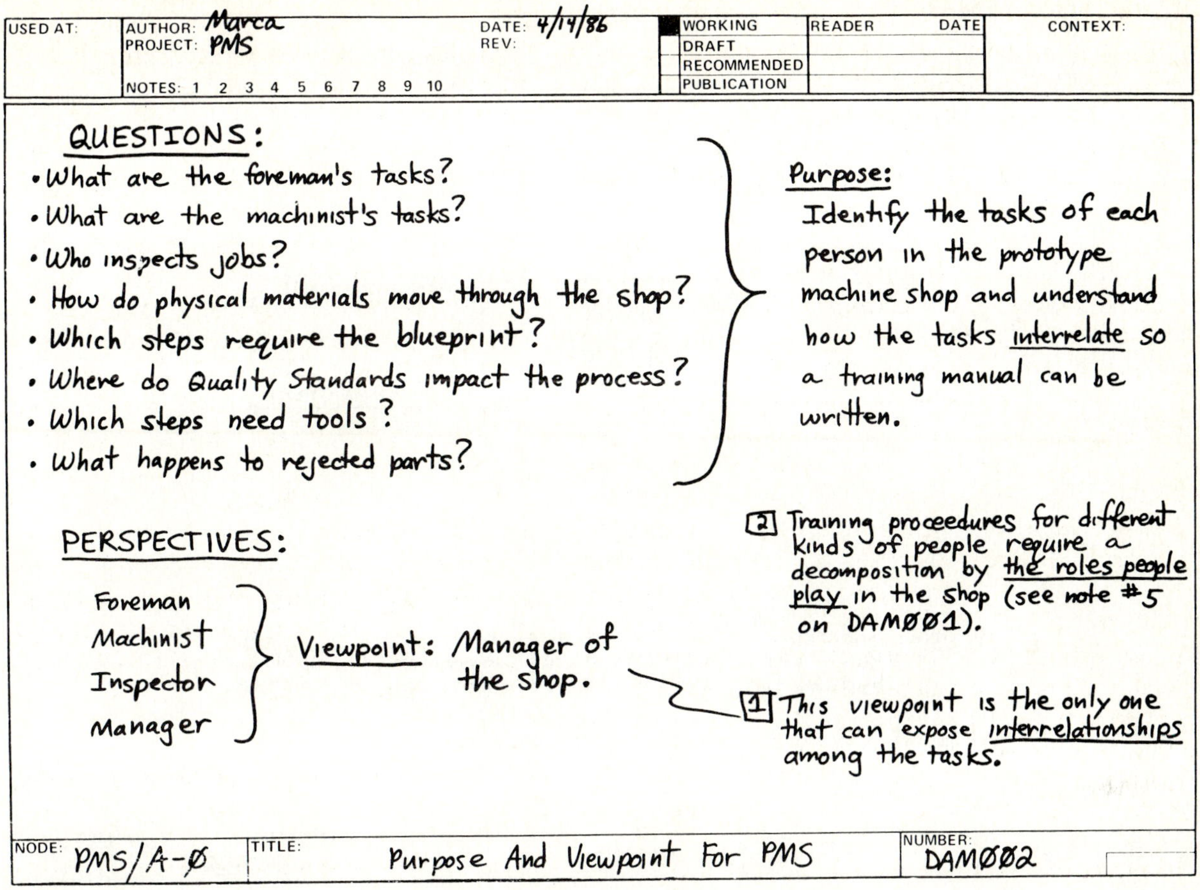

FIGURE 1-1 Developing the Purpose and Viewpoint for a Model

ple, Figure 1-1 shows how the author of the Prototype Machine Shop model enumerated possible perspectives (e.g., machinist, inspector) from which to describe the machine shop.

"Viewpoint" is best thought of as a place, person, or thing one can stand in to view the system in operation. From this fixed perspective, a consistent system description can be developed so the model will not drift around the topic and incorrectly mix together unrelated descriptions. For example, by not picking a particular viewpoint, the Prototype Machine Shop model may very well mix a discussion of managing the machines on the shop floor with a discussion of how a part is actually machined. If that occurs, the reader of the model may find it difficult to discern the separate responsibilities of shop personnel.

Sometimes, only one viewpoint out of many will yield a description that will suit the purpose of the model. For example, either the *foreman,* the *machinist,* or the *inspector* viewpoint could be used to develop a consistent description of the machine shop, but no one of these alone will yield a model that could be used to write a training manual for all the personnel. Only the *manager* of the shop is in the correct position to view all the technical tasks done by everyone in the machine shop. The *manager* viewpoint is also the perspective from which to see how the tasks of individuals interrelate, as pointed out by the note on Figure 1-1. The *manager* viewpoint allows the builder of the model to describe the role each person plays in making special parts and how tasks are coordinated.

1.5 Models Are Coordinated Sets of Diagrams

Once a subject, a purpose, and a viewpoint are selected, the iterative portion of the SADT modeling process is started. The subject defines what to include in, and exclude from, the model. The viewpoint guides the SADT model builder in selecting the right things to say about the subject and staying focused when saying them. The purpose becomes the criterion for determining when to stop the model. The end result of this process is a collection of carefully coordinated descriptions, starting from a very high-level description of the entire system and ending with detailed descriptions of system operation.

Each of these carefully coordinated descriptions is called a "diagram." SADT models collect and organize

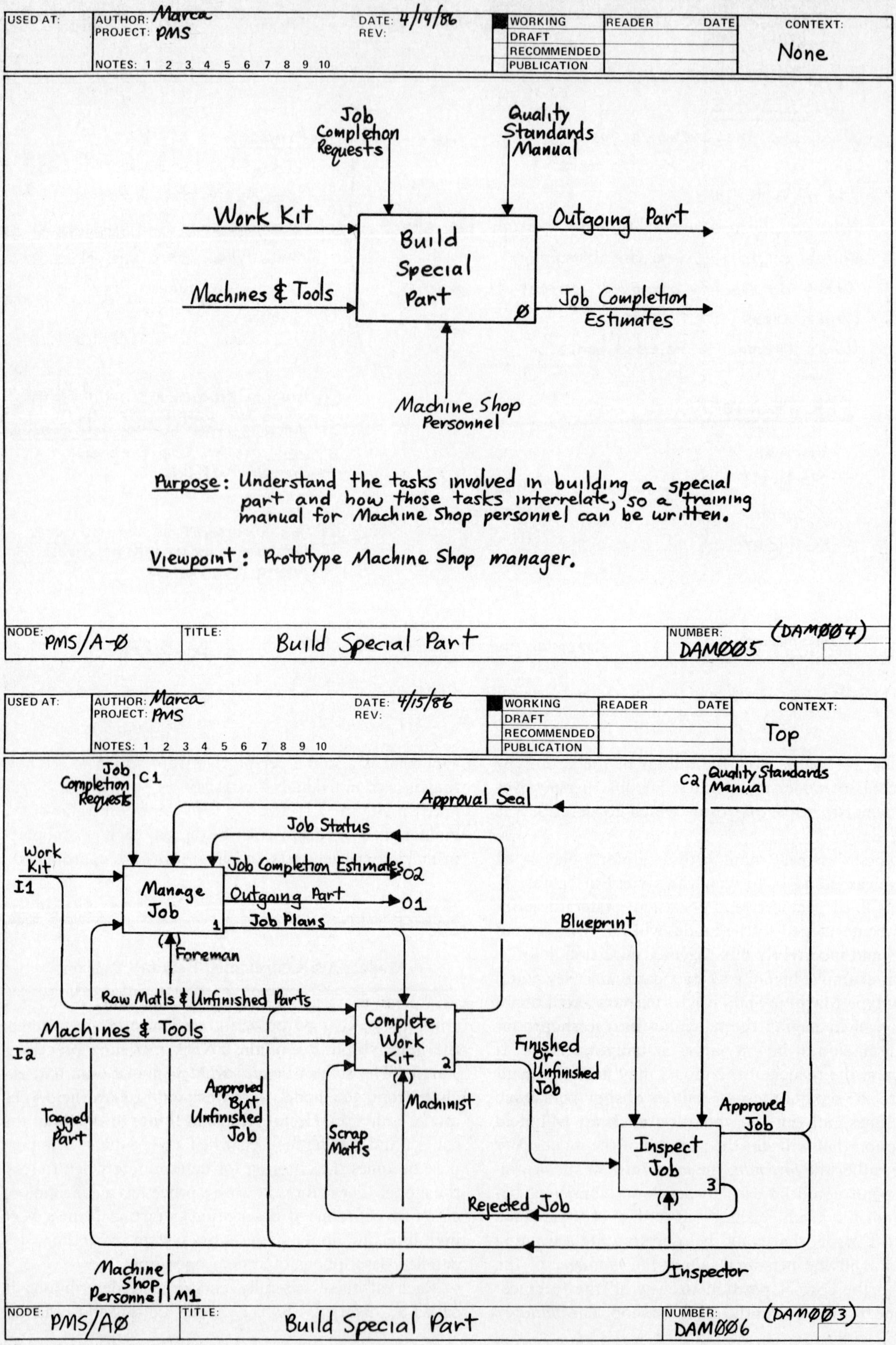

FIGURE 1-2 Two Interconnected SADT Diagrams

diagrams into a hierarchic structure, in which diagrams at the top of the model are less detailed than those at the bottom. In other words, an SADT model can be thought of as a tree-shaped collection of diagrams—the top diagram being most general and the bottom diagrams being most detailed. For example, Figure 1-2 presents two diagrams in the Prototype Machine Shop model. The top diagram, at the top of the model, describes the machine shop as an activity which basically converts incoming work into machined parts under some kind of quality control. The bottom diagram details the top one by identifying the three major activities of the machine shop: managing job completion, completing work kits, and inspecting ongoing work. Thus, the overall activity described in the top diagram is detailed by the three activities in the bottom diagram. This is an example of how SADT organizes a system description by shaping it into a hierarchy of ever-increasing detail.

Figure 1-2 also shows the three activities in the bottom diagram having an influence on each other, indicated by arrows which represent machine shop things. If you look carefully at the diagram, you will notice that some of the arrows go to the edge of it. Look more carefully and you will notice that the names of these arrows match the names of the arrows in the top diagram. This is an example of how SADT connects the diagrams in a model to each other through the things of the system. Also, this connection scheme requires that system things be named and accounted for in a consistent way in order for two diagrams to be considered connected. For example, a total of seven arrows touch the box in the top diagram, and every one of those arrows can be found coming from or leading to the edge of the bottom diagram.

1.6 Summary

The complexity involved with describing many constructed systems is caused by the fact that our systems are hard to enumerate and cannot be simplified through generalizing assumptions. SADT was developed specifically to describe such complex systems through modeling. An SADT model is a system description that has a single subject, a purpose, and a viewpoint. The purpose is a set of questions to be answered by the model. The viewpoint is the perspective from which the system is being described. Purpose and viewpoint are deep concepts in SADT. We chose to give them a very brief introduction in this chapter, postponing more detailed discussion until Chapters 5 and 6. An SADT model organizes its description into a hierarchy of interconnected diagrams. The top of this tree structure is the most general description of the system, and the bottom of the model comprises the most detailed descriptions.

Suggested Readings

Brackett, J., and C. McGowan: "Applying SADT to Large System Problems," SofTech Technical Paper TP059, January 1977.

Connor, M.: "Structured Analysis and Design Technique—SADT," Auerbach portfolio 32-04-02, 1979.

Freeman, P.: "Requirements Analysis and Specification: The First Step," Advances in Computer Technology—1980, August 1980.

Hori, S.: "Human-Directed Activity Cell Model," CAM-I Long Range Planning Final Report, CAM-I, Inc., 1972.

Miller, J.: *Living Systems,* McGraw-Hill, New York, 1978.

Ross, D.: "PLEX1: Sameness and the Need for Rigor," SofTech Deliverable no. 9031-1.1, December 1975.

———: "PLEX2: Sameness and Type," SofTech Deliverable no. 9031-2.0, December 1975.

———: "Reflections on Requirements," *IEEE Transactions on Software Engineering,* vol. SE-3, no. 1, January 1977.

———: "Doug Ross Talks about Structured Analysis," *IEEE Computer,* July 1985.

——— and K. Schoman: "Structured Analysis for Requirements Definition," *IEEE Transactions on Software Engineering,* vol. SE-3, no. 1, January 1977.

SofTech, Inc.: "Introduction to IDEF0," SofTech Deliverable no. 7500-14, September 1979.

Weinberg, G.: *An Introduction to General Systems Thinking,* John Wiley, New York, 1975.

Chapter

2

Diagram Syntax and Usage

The diagram is the basic work unit in making a model. The person who makes diagrams and models is commonly called an "analyst" or, in SADT terminology, an "author." Diagrams have their own syntax rules, separate from the syntax rules for models. It is important to understand these rules because there is specific meaning to the graphic notation. As we will see, SADT graphics are capable of defining distinct system activities and showing how these activities influence each other. Since this book covers only SADT activity diagramming, this chapter will discuss only the syntax rules for SADT activity diagrams.

2.1 Diagrams Have Boxes and Arrows

An SADT diagram contains boxes and arrows. The boxes represent activities of the system being modeled. Arrows connect boxes together and represent interfaces or interconnections between the boxes (Figure 2-1). A diagram is named with a title, placed in the bottom center portion of the diagram form. Each diagram also has standard identification information written on it: the author of this diagram, what project the work is a part of, the date of diagram creation or last revision, and diagram status. All identification information is placed in the top portion of the diagram form.

2.2 Boxes Represent Activities

The boxes on an SADT diagram are rectangles. A box represents a function or an active part of a system, so boxes are named with verbs or verb phrases. For example, the names of boxes on the ***complete work kit*** diagram are ***evaluate job progress, pick tools, setup machining place,*** and ***machine and assemble*** (Figure 2-1). Also, SADT requires that no fewer than three and no more than six boxes appear on any one diagram. These limits keep the complexity of diagrams and models from going beyond the ability of people to read, understand and use them. In other words, SADT diagrams and models are comprehensible.

Unlike all the other graphic-based structured analysis methods, each side of an SADT box has a specific and special meaning. The left side of a box is reserved for inputs, the top side is reserved for controls, the right side is reserved for outputs, and the bottom side is reserved for mechanisms. This notation represents certain system principles: inputs are transformed into outputs, controls constrain or dictate under what conditions transformations occur, and mechanisms describe how the function is accomplished. For example, box 4 of the ***complete work kit*** diagram could be interpreted as follows:

> ***Parts, raw materials,*** and ***scrap materials*** are ***machined and assembled*** into ***machining results*** according to the ***next job order step,*** using a ***ready machine place.***

2.3 Boxes Have Dominance

SADT boxes are never placed randomly on a Diagram Form. Instead, they are placed according to their relative order of importance as judged by the author of the diagram. SADT calls this relative ordering "dominance." Dominance can be thought of as the influence one box has over other boxes in a diagram. For example, the most dominant box of a diagram could be the first in a required sequence of activities, or it could be a planning or monitoring activity that influences all the other activities, such as ***evaluate job progress*** in Figure 2-1. Usually, the most dominant box is placed in the upper left-hand corner of the diagram, the least dominant box in the lower right-hand

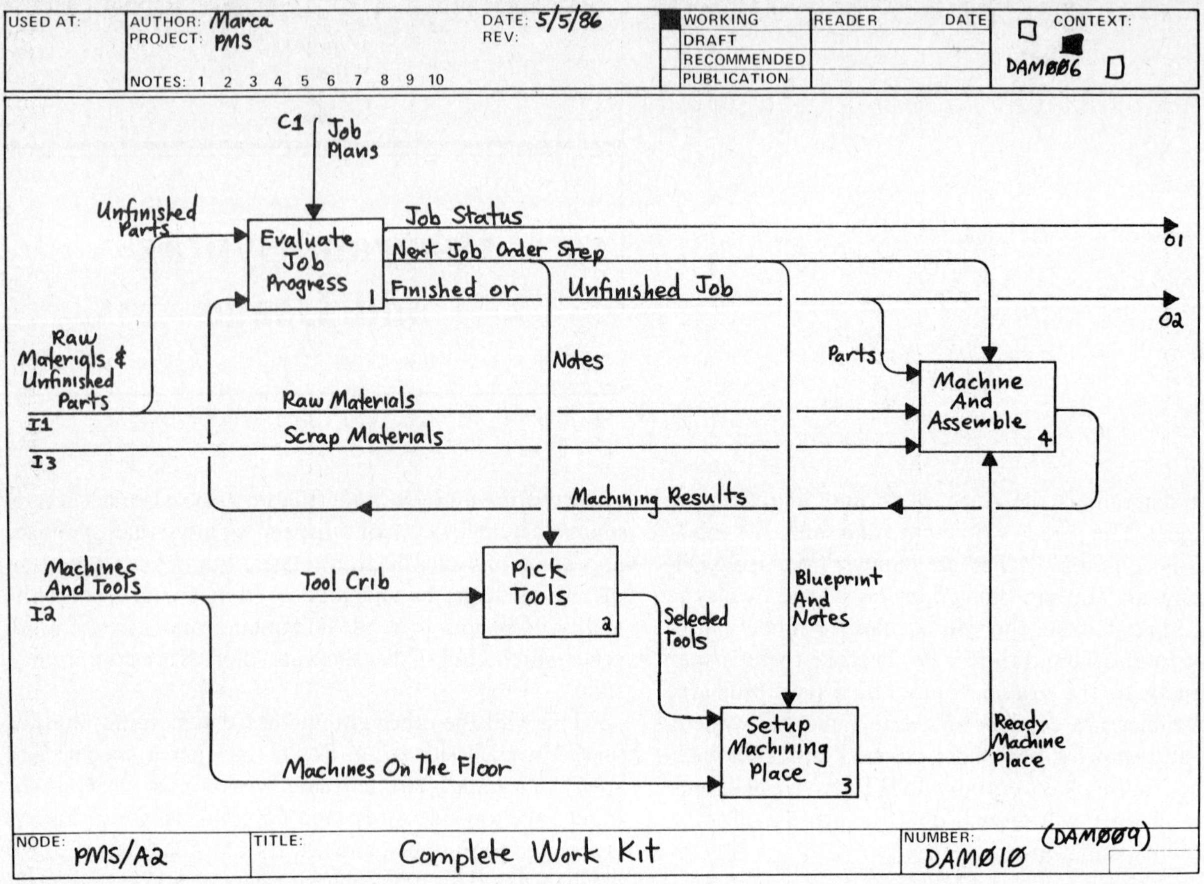

FIGURE 2-1 A Typical SADT Diagram

corner. The result is a "staircase" pattern, similar to the one shown for boxes 1 to 3 in Figure 2-1.

The positions of the boxes on the paper reflect the result of evaluating dominance by the author. Thus it is the topology of the diagram which truly defines which functions have more influence over others. To reinforce this, an SADT analyst can number each box according to its dominance over the other boxes. Dominance can therefore be indicated by a digit, placed in the lower right-hand corner of each rectangle: 1 can indicate the most dominant, 2 can indicate the next most dominant, and so on. For example, Figure 2-1 shows ***evaluate job progress*** influencing all other machining steps through ***next job order step,*** and so that box was numbered 1. Also, since ***setup machining place*** must happen before ***machine and assemble,*** they were numbered 3 and 4, respectively.

In SADT, boxes must get numbered. Box numbers provide unique identifiers for system activities and automatically organize those activities into a model hierarchy. By using box numbers to approximate the influence each box has over the other, an analyst can organize a model around functional dominance. This makes the hierarchical ordering of the activities in a model consistent with the level of influence of each activity on the rest of the system. Therefore, we strongly recommend numbering boxes according to their dominance whenever possible.

2.4 Arrows Represent Things

The arrows of an SADT diagram are single lines with arrowheads at their ends. For SADT activity diagrams, an arrow represents a collection of things. We stress use of the general word "things" here, because SADT arrows can, for example, represent plans, computer data, machines, and information. For instance, the arrows of the ***complete work kit*** diagram in Figure 2-1 represent written procedures (e.g., ***next job order step***), physical materials (e.g., ***raw materials and unfinished parts***), hand tools (e.g., ***tool crib***), mechanical drawings (e.g., ***blueprint***), work areas (e.g., ***ready machine place***), and managerial information (e.g., ***job status***). However, in the field of systems analysis, it is common to hear the word "data" used to mean "thing." This is a vestige from the early days when systems analysis was typically done for software systems.

Since SADT arrows represent things, they are labeled with nouns or noun phrases adjacent to some part of the arrow's line. We strongly recommend that arrow labels be placed as close as possible to the arrow line within the limits of preserving diagram readability. This avoids confusion about which arrow is named by a label, and also eliminates the need for extra graphic connectors (i.e., squiggles—see Chapter 19). For example, notice that all arrow labels of the *complete work kit* diagram were written adjacent to their respective arrows. We recommend adopting this style of arrow labeling to keep your diagrams uncluttered and simple to read.

2.5 Arrows Represent Interconnections among Boxes

There are four possible relationships that things may have with activities: input, control, output, and mechanism. Each of these relationships is represented by an arrow connected to one particular side of a box. By convention, the left side of a box is reserved for input arrows, the top is for control arrows, the right side is for output arrows, and the bottom is for mechanism arrows. Thus, the sides of a box classify, in a purely graphic way, the things represented by the arrows that touch the box.

Input arrows represent those things used and transformed by activities. For example, *raw materials* are transformed by the *machine and assemble* activity when the part is being made. Control arrows represent the things that constrain activities. Usually, control arrows stand for information that directs what activities do. For example, *next job order step* determines which tools to select, which shop floor machine is required, and how the tools and machine are used to make a part. Output arrows represent those things into which inputs are transformed. For example, *machine and assemble* transforms *raw* and *scrap materials* into a *machining result,* which eventually becomes a part. Mechanism arrows represent, at least in part, how activities (i.e., the functions of the system) are realized. For example, *setup machining place* organizes hand *tools* and *machines on the floor* into an efficient work area for the next job step. This work area, called a *ready machine place,* denotes where the machinist produces the part during the *machine and assemble* activity. Thus mechanisms represent the physical aspects of an activity (e.g., storage places, people, organizations, devices).

Therefore, an SADT diagram is made up of boxes interconnected by arrows which define how boxes influence each other. This influence may take the form of either sending an output to another activity for further transformation, or providing the control information that governs what another activity must actually do. For example, *machine and assemble* influences *evaluate job progress* by giving it *machining results* for evaluation, and *evaluate job progress* influences what *machine and assemble* must do next through *next job order step.* In other words, there is a very strong constraining relationship from *evaluate job progress* to *machine and assemble,* and there is also a weaker input-output relationship from *machine and assemble* to *evaluate job progress.* Thus, SADT diagrams are not flowcharts, nor are they just dataflow diagrams. They are constraint diagrams that describe both input-to-output transformations and the constraint rules on those transforms. In this way arrows document the interfaces between system activities and between the system and its environment.

In SADT, only five kinds of box interconnections are required to describe these relationships: control, input, control-feedback, input-feedback, and output-mechanism. Control and input connections are the simplest, because they represent direct influences that are intuitively obvious and very common. A control connection happens when the output of one box directly constrains a box of lesser dominance. For example, *evaluate job progress* constrains *pick tools* according to those detailed *notes* written for the *next job order step.* An input connection happens when the output of one box becomes the input to a box of lesser dominance. For example, the output of *evaluate job progress* called *finished or unfinished job* becomes the input to the *machine and assemble* activity as the *next job order step* is executed.

Both control-feedback and input-feedback connections are more complicated, because they represent iteration or recursion. That is, the outputs from one activity influence the future operations of other functions that subsequently influence the original activity. A control-feedback connection happens when the output of one box constrains a box of greater dominance. For example consider the *build special part* diagram of Figure 2-2. The *manage job* activity constrains the *inspect job* activity by the *blueprint* which documents tolerances. Also, the *approval seal,* the output of *inspect job,* constrains *manage job* because it is the *approval seal* which indicates the part has been finished. So, the *approval seal* constrains a future activation of *manage job.* This is therefore drawn by a backward-looping arrow. An input-feedback connection happens when the output of one box becomes the input of another box with more dominance. For instance, those jobs which are rejected by the *inspect job* activity are sent to future activations of *complete work kit* as *scrap materials.* (This is a good example of how systems often have feedbacks in them to make efficient use of waste.)

Output-mechanism connections are interesting and somewhat unusual. They represent the situation in which the output of one activity becomes the means by which another activity gets done. For example, Figure 2-1 shows the *setup machining place* activity producing an output of *ready machine place,* which is the mechanism to *machine and assemble.* This means that a *ready machine place* must be present in order to start machining. In this case, the mechanism arrow denotes a strong sequencing relationship: preparations must be complete before work can start.

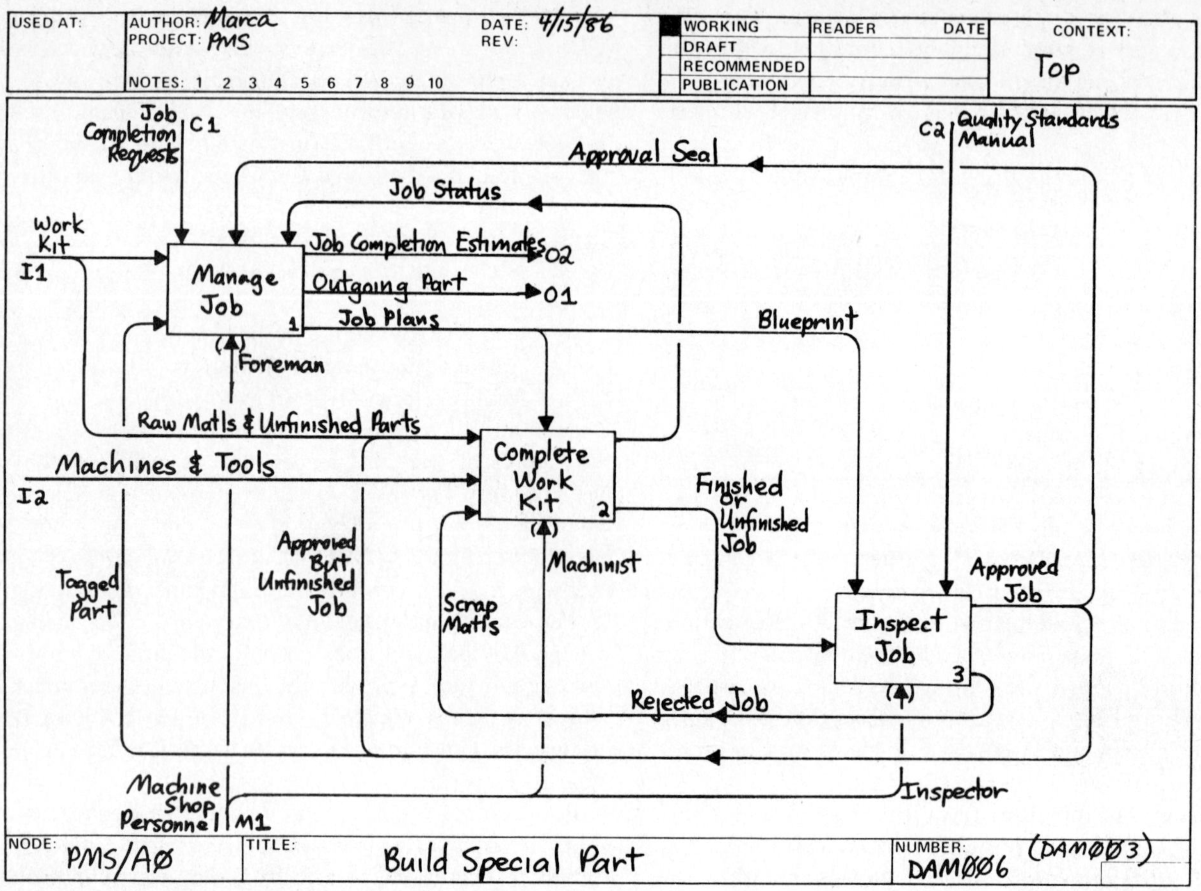

FIGURE 2-2 A Diagram Exhibiting Feedbacks

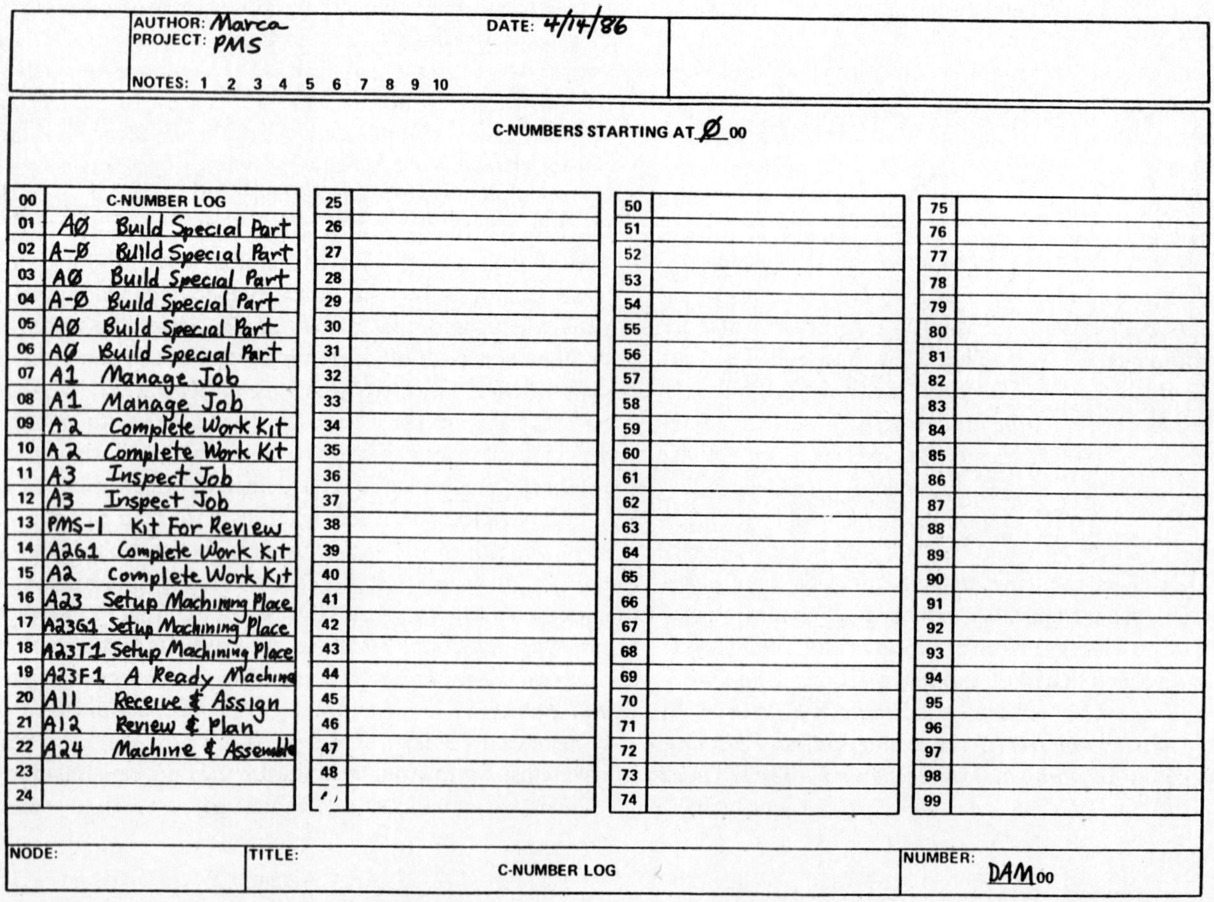

AUTHOR: Marca
PROJECT: PMS
DATE: 4/14/86
NOTES: 1 2 3 4 5 6 7 8 9 10

C-NUMBERS STARTING AT Ø 00

00	C-NUMBER LOG
01	AØ Build Special Part
02	A-Ø Build Special Part
03	AØ Build Special Part
04	A-Ø Build Special Part
05	AØ Build Special Part
06	AØ Build Special Part
07	A1 Manage Job
08	A1 Manage Job
09	A2 Complete Work Kit
10	A2 Complete Work Kit
11	A3 Inspect Job
12	A3 Inspect Job
13	PMS-1 Kit For Review
14	A261 Complete Work Kit
15	A2 Complete Work Kit
16	A23 Setup Machining Place
17	A2361 Setup Machining Place
18	A23T1 Setup Machining Place
19	A23F1 A Ready Machine
20	A11 Receive & Assign
21	A12 Review & Plan
22	A24 Machine & Assemble

NODE:
TITLE: C-NUMBER LOG
NUMBER: DAM00

FIGURE 2-3 An SADT C-Number Log

Output-mechanism connections therefore often deal with resource allocation issues (e.g., required tools, trained people, physical space, equipment, funding, materials).

2.6 Arrows Are Collections of Things

An SADT arrow rarely represents one thing by itself. Generally, it stands for a collection of things. For example, the arrow named ***work kit*** really contains a ***job order,*** a ***blueprint,*** a ***time card,*** and some ***raw materials*** and ***unfinished parts.*** Because arrows are collections of things, they can have multiple tails (sources) and multiple heads (destinations). Therefore, arrows have the ability to branch apart and join together in arbitrarily complex ways. For example, all or part of an arrow can go to or come from one or more boxes, like ***approved job*** in Figure 2-2. Notice how different components of the ***approved job*** arrow go to the other boxes on the diagram (e.g., ***approval seal*** controls ***manage job*** while ***approved but unfinished job*** is input to ***complete work kit***). Here is a case where arrow branches reveal the contents of a collection. Also in Figure 2-2, notice how ***work kit*** and ***approved but unfinished job*** combine together to form ***raw materials & unfinished parts,*** an input of ***complete work kit.*** This is a case in which two arrows join to form a larger collection. SADT conventions for drawing and labeling arrow branches and joins were designed to explain how arrows represent the coming together and splitting apart of collections of things.

2.6.1 Arrow Branches

SADT arrow branches, drawn as lines that move away from each other, indicate that all or part of an arrow's contents can appear in each arrow branch. An arrow is always labeled before its branch in order to name the collection. Also, each branch of the arrow can be labeled or unlabeled, according to the following rules:

- Branches which are not labeled are assumed to contain all the things indicated by the label before the branch (i.e., all things go to the branches).
- Branches which are labeled after the branch point are assumed to contain either all or some of the things indicated by the aggregate label before the branch (i.e., each branch label makes explicit what it carries).

For example, in the ***build special part*** diagram, the arrow labeled ***approved job*** comprises several things and branches in several directions. The ***approval seal*** influences ***manage job, approved but unfinished parts*** go to ***complete work kit*** for another machining pass, and the ***tagged part*** goes to ***manage job*** for final checkoff and delivery.

2.6.2 Arrow Joins

SADT arrow joins, drawn as lines that come together, indicate that all the contents of each branch go into making the aggregate arrow after the join. An arrow is always labeled after its join to indicate the new collection formed by the combination. Also, each branch before the join can be labeled or unlabeled, according to the following rules:

- Branches which are not labeled are assumed to contain all the things indicated by the aggregate label after the join (i.e., all things comes from both branches).
- Branches which are labeled before the join are assumed to contain either all or some of the things named by the aggregate label after the join (i.e., the branch's label makes explicit what it carries).

For example, both the ***raw materials and unfinished parts*** of a ***work kit*** come together with ***approved but unfinished job*** to form the major input to the ***complete work kit*** activity. ***Raw materials and unfinished parts*** is a name that describes both those things, so the arrow label after the join gets that name.

2.7 Tracking Versions of Diagrams with C-Numbers

During the development of an SADT model, a diagram with its boxes and arrows is usually redrawn several times, creating different versions of that diagram. SADT uses a diagram configuration control scheme, built upon chronological numbers or "C-Numbers," to distinguish different versions of the same diagram from each other. A C-Number code is constructed from the author's initials and unique sequence number. These codes are put in the lower right-hand corner of the SADT form. For example, notice that DAM010 is the unique C-Number for the complete work kit diagram of Figure 2-1. If a diagram replaces an older version, the author puts the previous C-Number in parentheses, so there is a backward link to prior work. For example, diagram DAM010 supersedes its previous version, DAM009.

Each author on an SADT project keeps a log of every diagram he or she produces, numbering each with the next higher integer. The SADT C-Number Log Form is used to do this. For example, Figure 2-3 gives the C-Number Log for the Prototype Machine Shop modeling effort for this book. Notice how the form controls the next number to be used so that the author records only the node number and title for each new diagram being made.

2.8 Summary

SADT diagrams, which are made up of from three to six boxes interconnected by arrows, go through several versions during the building of a model. C-Numbers are used to distinguish the different versions of a particular diagram. The boxes on a diagram represent system functions or activities, and the arrows represent the many and varied things of a system. Boxes are generally drawn on a diagram according to their dominance, which is their relative importance to each other. The arrows that interconnect boxes represent collections of things and can split apart and join together in a variety of complex ways. But even though arrows may split or join together, they must at all times conserve the things they represent.

Suggested Readings

Connor, M.: "Structured Analysis and Design Technique—SADT," Auerbach portfolio 32-04-02, 1979.

Ross, D., and J. Bracket: "An Approach to Structured Analysis," *Computer Decisions,* September 1976.

Ross, D.: "Structured Analysis (SA): A Language for Communicating Ideas," *IEEE Transactions on Software Engineering,* vol. 3, no. 1, January 1977.

——— and K. Schoman: "Structured Analysis for Requirements Specification," *IEEE Transactions on Software Engineering,* vol. 3, no. 1, January 1977.

SofTech, Inc.: "IDEF0 Author's Guide to Creating Activity Diagrams," SofTech Deliverable no. 7500-13, September 1979.

———: "Integrated Computer-Aided Manufacturing (ICAM) Report: Functional Modeling Manual (IDEF0)," contract no. F33612-78-C-5158, SofTech, Inc., 1981.

Chapter

3

Model Syntax and Usage

A single SADT diagram is complex in its own right, for it has between three and six boxes interconnected by many arrows. Several such diagrams are required to adequately describe a system. When diagrams are collected and connected together, they become an SADT model. SADT models have syntax rules quite distinct from, but complementary to, the syntax rules for diagrams. SADT model syntax allows an analyst to define the boundary of a model, connect diagrams into a unified whole, and ensure "plug-to-plug" compatibility between diagrams. In fact, no other structured analysis method is as rigorous as SADT in linking together diagrams into carefully organized collections called models.

3.1 A System Is Represented by a Single Box

An SADT model is a hierarchically organized set of diagrams. In general, diagrams are made up of from three to six boxes, each of which can potentially be detailed by another diagram. Each box can be thought of as a single, well-defined subject. The division of such a subject (i.e., the name of a box) into its structured pieces (i.e., the boxes and arrows that make up a diagram) is called "decomposition."

Decomposition forms boundaries, and in SADT, each box is considered to be the formal boundary around a portion of the entire system being described. In other words, a box and the arrows that touch it define the precise boundary of the diagram which decomposes that box. That diagram, called the "child" diagram, describes everything about the box and its associated arrows, and describes nothing outside that boundary. The box being decomposed is called the "parent" box, and its containing diagram is likewise called the "parent" diagram. An SADT diagram is, therefore, a decomposition of a bounded subject.

This principle of bounding a subject happens at every level in the model. At the very top, a single box and several arrows are used to define the boundary of the entire model. A single box describes the overall task performed by the system. The arrows that touch the box describe the major controls, inputs, and outputs of the system. The diagram that consists of this single box and its arrows defines the boundary around the system and is called the "Context Diagram" for the model. The box therefore represents the system boundary: everything inside the box is a part of the system being described; everything outside the box constitutes the system's environment.

For example, Figure 3-1 shows the top of the Prototype Machine Shop model. The box, entitled ***build special part,*** describes the overall activity of the machine shop and is numbered 0. (The topmost box of a model is always numbered 0.) This box, therefore, stands for the entire prototype machine shop. The arrows ***job completion requests*** and ***quality standards manual*** constrain how the prototype machine shop transforms ***work kits*** and ***machines and tools*** into ***outgoing parts*** and ***job completion estimates.*** They define the interface between the prototype machine shop and the rest of its aerospace manufacturing company.

3.2 Identifying Decompositions with Node Numbers

SADT models evolve through a process of top-down structured decomposition. The process of structured decomposition first breaks the one-box boundary of the model into a single diagram having from three to six boxes, then breaks one or more of these boxes into another diagram with another three to six boxes, and so on. The title for each diagram is taken verbatim from the box it decomposes. The result of this process is a model whose

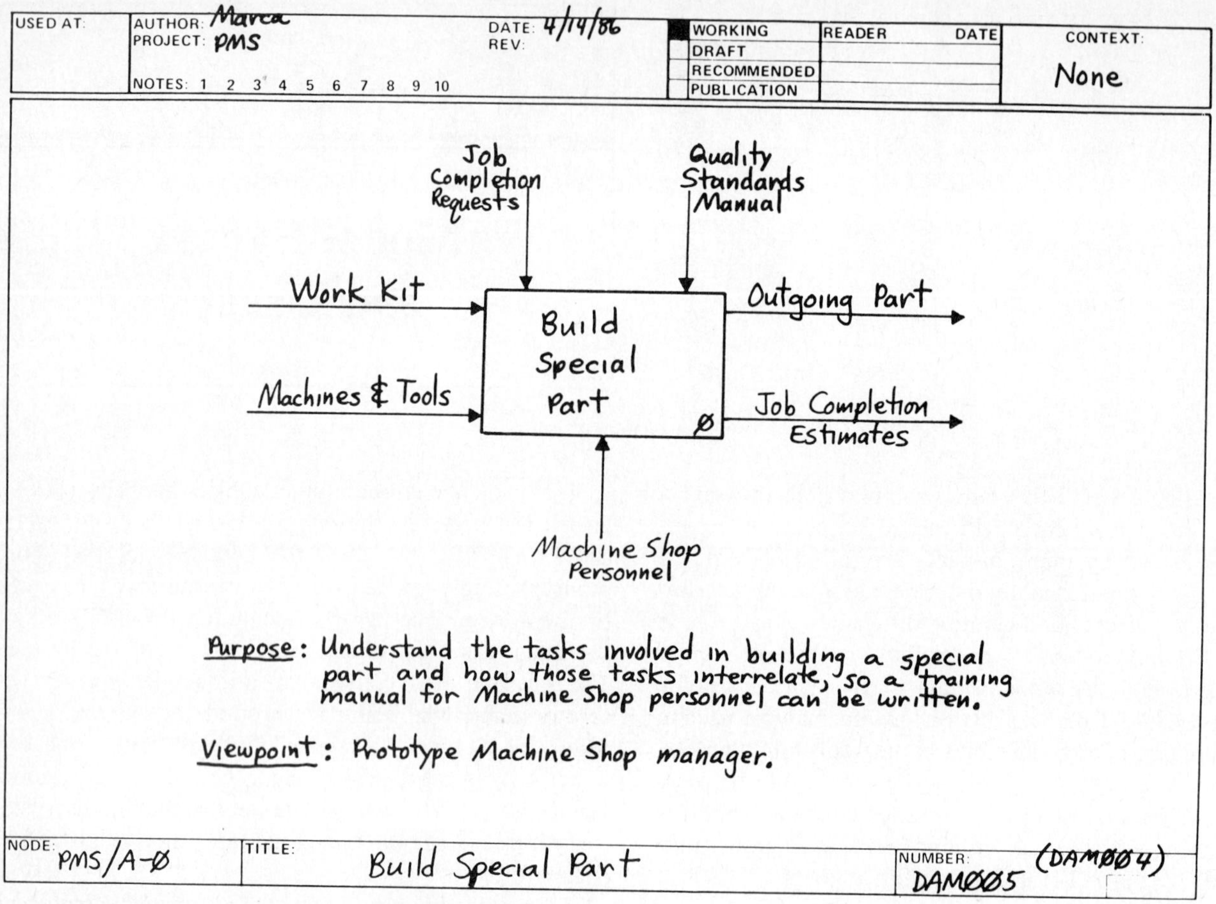

FIGURE 3-1 The Context Diagram for a Model

top diagram describes a system in general "black-box" terms and whose bottom diagrams describe very detailed aspects and operations of the system.

Each diagram is therefore an integral piece of an entire model. SADT identifies each diagram in a particular model by what is called a "node number." The node number for the context diagram has the form: model name or abbreviation, slash, the capital letter A (for activity diagram), a hyphen, and zero. For example, the node number for the context diagram of the Prototype Machine Shop model is PMS/A-0. The node number for the diagram that decomposes the boundary is the same node number without the hyphen (e.g., PMS/A0). All other node numbers are formed by taking the node number of the parent diagram and appending to it the number of the box that is being decomposed. For example, Figure 3-2 shows two diagrams in the Prototype Machine Shop model. The node number of the first diagram is PMS/A0, and the node number of the second diagram is PMS/A1. The PMS/A1 diagram decomposes box 1 on the PMS/A0 diagram. (It is customary to drop the leading zero when forming node numbers. This is why PMS/A1 is written instead of PMS/A01.)

3.3 Linking Decompositions with C-Numbers

Just as C-Numbers are used to identify unique versions of a diagram, they are also used to link together diagrams both downward and upward in the model hierarchy. Normally, the C-Number of the diagram that decomposes a box is first placed underneath that box on the parent diagram. This forms the "downward" link from the parent diagram to the child diagram. For example, notice in Figure 3-2 that DAM008, the C-Number for the ***manage job*** diagram, is placed below box 1 on the ***build special part*** diagram. This indicates that the ***manage job*** function has been decomposed.

Once the downward link is made, the parent diagram is summarized and referenced on the child diagram. In the Context Area of the SADT form (upper right-hand corner), the author draws one tiny square for each box on the parent diagram, shades in the box that the diagram decomposes, and places the C-Number of the parent diagram near the shaded box. This forms the "upward" link to the parent diagram. Such a method of linking together

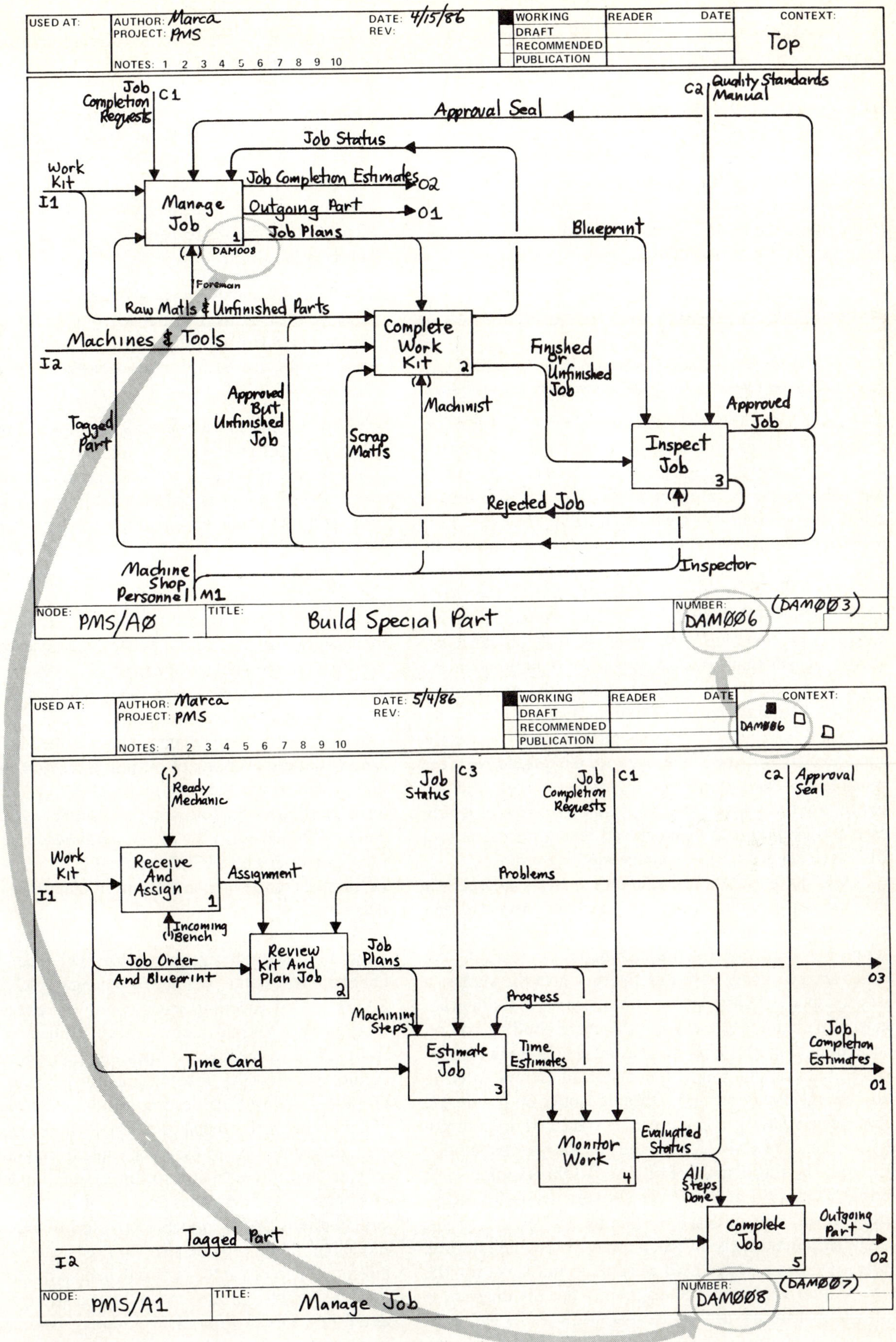

FIGURE 3-2 Linking Parent and Child Diagrams Together

diagrams by unique version numbers ensures that the intended version of a diagram becomes part of the model in a reliable way. In other words, the C-Number notation accurately controls the introduction of new diagrams into the model hierarchy. For example, notice in Figure 3-2 that the Context Area of the form for ***manage job*** has three boxes, one for each box of the ***build special part*** diagram. Notice too that the first box is shaded, indicating that the ***manage job*** box is the one being decomposed by this diagram and that diagram DAM006 contains the boundary for this decomposition.

3.4 ICOM Codes Ensure Diagrams Plug Together

Any good structured analysis methodology that lets one produce individual diagrams must have a way to ensure that all diagrams correctly connect together to make a consistent model. SADT diagrams have external arrows—arrows that appear to come from and go to the edge of the paper. These arrows are the interface between the diagram and the rest of the model. SADT requires all external arrows on a diagram to agree with the arrows that form the boundary of that diagram. In other words, a diagram must be "plug-to-plug" compatible with its parent. Usually this means that external arrows match in number and name (but not absolute position) with the arrows that touch the boundary box on the parent diagram. For example, box 1 of the ***build special part*** diagram has eight boundary arrows: the inputs ***work kit*** and ***tagged part;*** the controls ***job completion requests, approval seal,*** and ***job status;*** and the outputs ***job completion estimates, outgoing part,*** and ***job plans.*** You can find all these external arrows with their names on the ***manage job*** diagram.

SADT has a notation that allows analysts to precisely identify and validate arrow connections between diagrams. It is an encoding scheme called "ICOM," which stands for *i*nput, *c*ontrol, *o*utput, and *m*echanism. ICOM codes are extremely powerful, for they give the analyst a way to quickly verify that the external arrows of a diagram match the boundary arrows of the corresponding box on the parent diagram. They also ensure consistent decomposition, because one must account for all arrows entering and leaving a diagram. For example, notice on Figure 3-2 that ***job completion requests*** can be traced from its origin (C1 on box 0 of PMS/A-0) at the boundary of the model, through the top diagram PMS/A0, to the ***monitor work*** box (C3 on box 4 of PMS/A1). (We discuss useful exceptions to this rule in detail in Chapter 5.)

When a diagram is started, the arrows touching the box being decomposed are used as sources and sinks for the arrows that appear on the diagram. After the diagram is drawn, its external arrows are matched with its parent to ensure consistency. One way to do this is to think of assigning ICOM codes to the external arrows of the diagram in the following way:

- First, imagine drawing the new diagram inside the box being decomposed. Run the external arrows to almost the edge of the diagram. Now visually connect each external arrow of the diagram to the corresponding boundary arrow of the box. (We sometimes used dashed lines in Figure 3-3 to represent this visual connection process.)
- Second, give a code for each visual connection. Use I for connections between input arrows, C for connections between control arrows, O for connections between output arrows, and M for connections between mechanism arrows.
- Third, append to each letter a digit indicating the absolute position where each boundary arrow touches the parent box relative to all other boundary arrows of its type—for inputs and outputs, count top to bottom, and for controls and mechanisms, count left to right (refer to Figure 3-3). Now write each code at the tip of each external arrow.

Figure 3-3 gives an example of a subject boundary (i.e., a box and its arrows) and the diagram that decomposes it. Notice that the subject boundary was drawn in bold to emphasize how external arrows are tied to their respective boundary arrows, and also notice that, for this example only, we annotated the diagram with dashed lines to represent visual connections of output arrows to their proper boundary arrows. (The other matches were quite easy to visually locate.) By following the above scheme, ICOM codes of I1, I2, C1, C2, O1, O2, and M1 were produced for Figure 3-3. Thus, ICOM encoding is done with respect to the sides of the box on the parent diagram that is being decomposed.

By following this ICOM encoding scheme, one creates a set of implicit off-page connectors which can be quickly changed when the boundary changes. (Compare the SADT ICOM encoding scheme with the alternate of just labeling the external arrows in an explicit manner, say from A to E, respectively.) These implicit page connectors facilitate the SADT reading and review practice which checks for consistent decomposition. ICOM codes also simplify the work needed to manually make local changes on a diagram, and tie together explicit diagram versions so that they correctly fit into the existing model. Thus, in our opinion, ICOM codes are one of the most important contributions SADT makes to the technology of graphic modeling. They provide the required rigor while giving authors the freedom to work independently, draw legibly, and choose appropriate terminology at successive levels of detail in a model without invalidating prior work.

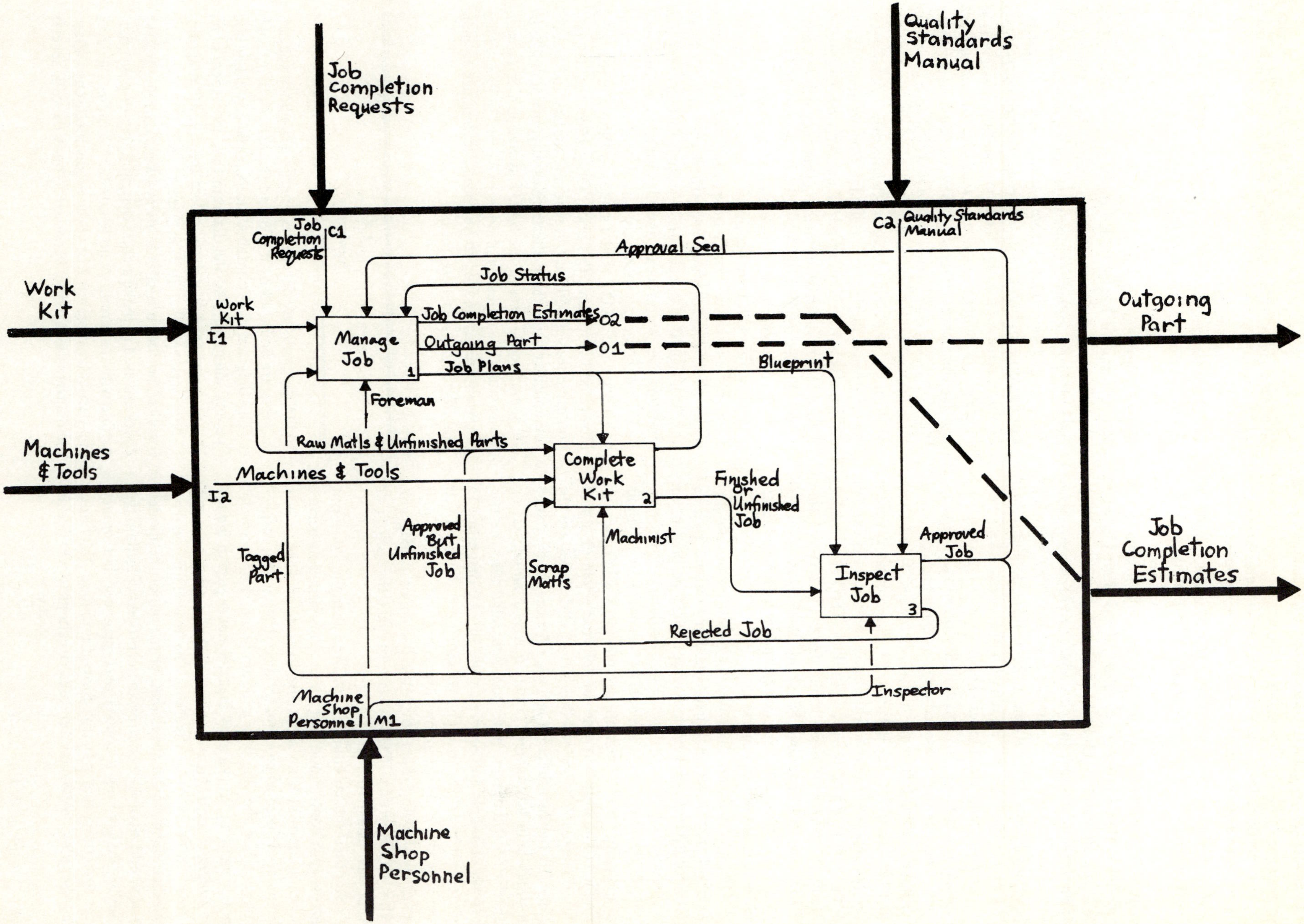

FIGURE 3-3 Codifying Connections between Diagrams the SADT Way

3.5
Notations for Less Common Arrow Interfaces

Node numbers, C-Numbers, and ICOM codes handle the vast majority of model intraconnection situations. However, several special situations can arise between the parent and the child diagrams in which a judicious use of model syntax improves a system description: (1) when external arrows branch and join, (2) when arrows change from input to control and vice versa, and (3) when arrows "tunnel." We will give examples of each of these situations so you will be able to recognize them and understand their meaning. But we will do so with the caveat that these representational techniques should be used only at very special times to clarify and simplify the system description. They should be looked upon as conveniences, not as crutches for bad systems analysis practices. In all these special cases, the conservation of data across diagram boundaries is preserved. That is, all input data is used in some way to generate all output data. The key to understanding these situations comes from the concept that SADT arrows represent hierarchic collections of data. (Refer to Chapter 5 for more explanation of arrow hierarchies and arrows in general.)

One special situation involves the splitting or joining of external arrows between diagrams. For example, two external output arrows on a diagram could be part of an aggregate output arrow on the box boundary. What happens in this case is that the analyst has not shown the join graphically on the diagram, but has instead left the join implicit. The only way one can tell that a join or a split has occurred is to notice that the ICOM codes are the same for the two different arrows. (We have shown such a situation in Workshop 7, where ***budget & money,*** C1 on the FF/A-0 diagram, have been separated into the two arrows of ***budget*** and ***money*** on the FF/A0 diagram.) In almost all cases, we strongly recommend making explicit the fact that external arrows are joined together or split apart by drawing them so on the decomposed diagram. This avoids letting ICOM's state the fact that a join or split has occurred.

Another special situation involves changing an input arrow to a control and vice versa. This situation occurs when a control (or an input arrow) that touches the box boundary is used as an input (or a control) arrow, respectively, when the diagram is decomposed. For example, notice again in Workshop 7 that the ***money*** part of ***money & budget,*** a control arrow on FF/A-0, is used as an input arrow on the FF/A0 diagram. The reason the analyst bundled them was to stress that, at the top of the model, ***budget*** and ***money*** are controls to the process of feeding the family. ***Budget*** was a control over planning, but ***money*** was something that was transformed into ***food items*** during shopping. We have rarely seen situations in which this technique is required. We recommend you think very hard about alternate ways of drawing the diagram before using this technique.

Two other special situations arise when arrows "tunnel" between diagrams. A tunneled arrow is either (1) an external arrow that did not appear on the parent diagram (i.e., it has a hidden source) or (2) an arrow which touches a box but does not appear on the diagram that decomposes that box (i.e., it has a hidden destination). Tunneled arrows that come from a hidden source have their tails parenthesized, indicating that the arrows came from some other part of the model or from outside the model directly. For example, in Figure 3-2, the ***ready mechanic*** arrow, C1 of the ***receive and assign*** box on diagram PMS/A1, is tunneled and hence no such arrow touches the ***manage job*** box on the parent diagram PMS/A0. Tunneled arrows that go to a hidden destination have their heads parenthesized indicating that the arrow is going to some other part of the model or to outside the model directly, or that it will not be treated further in the model. For example, all the mechanism arrows on the ***build special part*** diagram of Figure 3-2 are tunneled, indicating that they will not show up on the decompositions of their respective boxes.

From our experience, all these situations rarely happen, and when they do, they happen for very specific reasons. Though we have seen many times when these techniques were useful, we recommend using them with extreme caution. Poorly used, they can quickly become crutches for bad modeling. So, we advocate their use only by experienced SADT analysts, and then only rarely. Instead, we suggest you always use SADT model syntax in the standard ways illustrated throughout this book. Only when the description becomes very difficult to read and understand should you consider using these alternatives. If you would like to see some special uses of the SADT syntax just discussed, please study the industrial models presented in Part VI for the ways they employed these special techniques to help simplify their system descriptions.

3.6
Summary

SADT diagrams are decompositions of a bounded subject. Subjects are bounded by a box and the arrows that touch it. The diagram that contains the boundary is called the "parent" diagram, and the diagram that decomposes one box on the parent diagram is called the "child" diagram. C-Numbers are used to link parent and child diagrams together, so that a model is always kept current. ICOM codes are used to make a child diagram plug-to-plug compatible with its parent. A node number identifies the level of a particular diagram within the model hierarchy. When the diagrams in a model become too hard to read, special techniques like tunneling can be judiciously used to simplify the system description.

Suggested Readings

Ross, D.: "Removing the Limitations of Natural Language (with Principles behind the RSA Language)," SofTech Deliverable 9061-25, July 1979.

SofTech, Inc.: "IDEF0 Author's Guide to Creating Activity Diagrams," SofTech Deliverable no. 7500-13, September 1979.

———: "Introduction to IDEF0," SofTech Deliverable no. 7500-14, September 1979.

———: "Integrated Computer-Aided Manufacturing (ICAM) Final Report: IDEF0 Functional Modeling Manual," Contract no. F33615-73-C-5158, January 1981.

Chapter

4

The Modeling Process

Though SADT owes much of its success to its graphic language, just as much credit must be given to its modeling process. The process of SADT guides an analyst through obtaining knowledge about a domain, documenting acquired knowledge by making a model, and validating a model through iterative review. Also, the process advocates a particular way of doing consistent and reliable structured decomposition—a key to good systems analysis. SADT is unique in its ability to provide both a graphic language and a process with which to develop correct and useful system descriptions.

In fact, we claim SADT is a true methodology because it integrates an iterative process for model development, configuration control notation for models, a diagram reference language, and a model activation language with its graphic system description language, as well as recommending how to manage analysis projects. The configuration control notation ensures that new diagrams fit correctly into an evolving hierarchic model. The SADT reference language, a shorthand method for referencing diagram parts, facilitates note writing when reviewing a model. The activation language allows one to declaratively specify system execution rules—an often essential last step in system description.

Shown in Figure 4-1 is the SADT modeling process, described with an SADT diagram. The diagram emphasizes that the SADT modeling process is an iterative set of steps whose product is a correct system description. What makes this process run so effectively is its organization, which is centered on the specific roles people play on SADT projects: experts are information sources, authors develop diagrams and models, a Librarian coordinates the written exchange of information, readers review and validate models, and a Technical Review Committee approves a model for usage. Presented here is an overview of this modeling process. Detailed discussion of the steps in this process is given in Chapters 5 and 6 and throughout Parts II and III.

4.1
Knowledge Is Acquired through Interviews

Interviewing (box 1 in Figure 4-1), a proven knowledge-acquisition technique, is that part of the modeling process which brings facts to light about the system under study. To get the best facts, SADT suggests interviewing multiple information sources (e.g., documents, people, and the system in operation). Regardless of the information source, SADT recommends interviewing with a distinct purpose in mind. This means defining your informational needs before selecting the right source for the next interview. During the interview, the SADT graphic language is used as a medium for taking notes, which become the grist for making diagrams. The process of interviewing is covered in detail by Chapter 7.

4.2
Documenting Acquired Knowledge

Authoring (box 2 in Figure 4-1) is the second major step in the modeling process, the step in which an SADT analyst documents his or her acquired knowledge about the problem domain by drawing one or more diagrams. The process of authoring is defined by a specific method of detailing a bounded subject. To summarize, the SADT author first analyzes the things of the system and then uses this new understanding to analyze system activities. From these analyses, a diagram is drawn by the bringing together of similar things and activities. This particular way

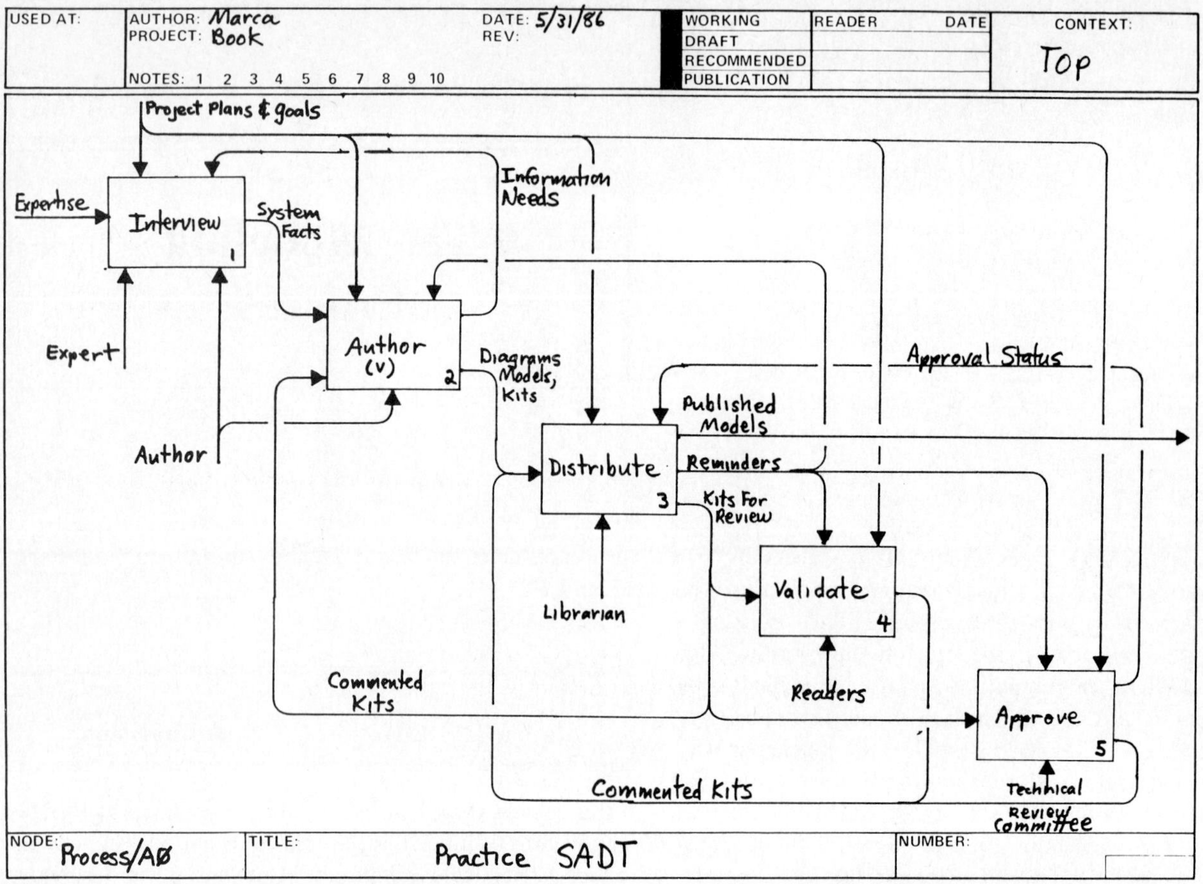

FIGURE 4-1 The SADT Modeling Process

of doing and documenting systems analysis is unique to SADT.

4.3 Models Are Validated through Iterative Review

Modeling with SADT is an engineering process. That is, models are made of reality, and these models go through a series of refinements until they accurately represent the real world. Part of SADT is concerned with the iterative process in which both the author and expert continually confer (verbally and in writing) about the validity of the evolving model. This iterative review process is called the Author/Reader Cycle.

The Author/Reader Cycle starts when an author wishes to distribute a piece of work in order to receive feedback on it. Material to be distributed is given in the form of "kits," small packets of work results to be critiqued by other people within a specified time. Written comments are placed on kits in the form of numbered comments. Commented kits are thus the feedback authors get on their work. Readers, the people who put comments on kits, are the ones who read and critique the evolving model (see box 4 in Figure 4-1). This is possible because the SADT graphic language produces diagrams and models that can be quickly and easily read. (The simplicity of the graphic language was therefore no accident, because only by reading can one gain an understanding of the system, and only from understanding can meaningful review about the validity of a model be given.)

Normally, many readers critique a particular kit at the same time, and all their comments are sent to an author by a deadline. The author then responds to every comment and consolidates the critiques. Through review of this nature, a reasonably quick convergence of ideas is possible. Thus the SADT process supports concurrent and asynchronous review of a model, the most efficient way of distributing work among a team of people. This demonstrates that the SADT modeling process is an engineering discipline, for iterative teamwork is the hallmark of engineering. This is because an SADT model is rarely built by a lone author; it is usually built by a team of people. In fact, several authors can jointly work on different parts of a model because each box in a model represents a bounded subject, suitable for independent analysis and decomposition. Thus the model itself coordinates cooperative authoring while the SADT modeling process coordinates the simultaneous review of evolving ideas. The complete details of this engineering process are given in Part III.

4.4 Coordinating the Review Process

The issue of timely response is critical to the success of a modeling effort, because out-of-date information can potentially make a system development effort fail. That is why SADT reserves a special role for the task of supervising the review process. Figure 4-1 depicts that special role being played by the Librarian, who acts as the coordinational hub in the SADT modeling process by ensuring that material is distributed in a consistent and timely manner. The Librarian distributes received kits, monitors them, and issues reminders so commented kits are returned on time to authors and so authors respond to the suggestions of readers in a timely manner. The Librarian also publishes finalized models after they are approved for use.

4.5 Models Are Used after Approval

Recall that SADT models are built for a particular purpose, and that purpose is written on the A-0 diagram of the model. In one sense, this purpose defines how the model will be used. Thus, once a correct model has been completed to a desired level of detail and has been validated, it can then be used to fulfill its intended purpose. For example, though the Prototype Machine Shop model was made to describe what different people do in the machine shop, the resulting model was always intended to be used as a guide to writing a training manual for new personnel. If the model describes accurately how people in the shop work but cannot be used to write a training manual, then it is unusable. Thus a correct model might not be a usable one.

The SADT modeling process recommends establishing a special team of people responsible for ensuring that usable and correct models are developed during an SADT analysis effort. This special team, called the "Technical Review Committee" (see box 5 on Figure 4-1), is responsible for the quality control of the models produced by SADT authors during a project. This committee reviews ongoing work with the long-range goals of the total project in mind. By doing so, it critiques models to ensure they can and will be properly used during the project in order to achieve the overall project goals.

Thus, the Technical Review Committee is often in the best position to evaluate the direction a project is taking and to suggest mid-course corrections. The committee does this through review for approval. Models that have reached the required level of detail and accuracy from a technical perspective are distributed to the members of the Technical Review Committee for their approval. The committee then critiques the model for its usability. If deemed usable by the committee, it is published. Otherwise, comments are sent to the authors indicating areas that need further work.

4.6 Summary

SADT is a methodology, because it integrates a modeling process, configuration control, supplemental languages, and project management with its graphic language. The modeling process can be broken down into interviewing experts, authoring diagrams and models, distributing written information, validating evolving models, and approving models for final use. This process works well because people play specific roles during a project, and because the Librarian ensures the timely exchange of information. Also, SADT projects are successful because a Technical Review Committee reviews models from a usability perspective.

This marks the end of discussion about the basic concepts related to SADT activity diagrams and activity models. Chapters 5 and 6 cover some advanced concepts pertaining to diagrams, models, and the process of making them, called "authoring." You may therefore want to skip to Part II at this time to see how an activity model is started. If you do, feel free to return to Chapters 5 and 6 when you want to learn about more advanced SADT concepts.

Suggested Readings

Mihram, A.: "The Modeling Process," *IEEE Transactions on Systems, Man and Cybernetics,* vol. 2, no. 5, November 1972.

Ross, D., and K. Schoman: "Structured Analysis for Requirements Definition," *IEEE Transactions on Software Engineering,* vol. 3, no. 1, January 1977.

SofTech, Inc.: "Introduction to IDEF0," SofTech Deliverable no. 7500-14, September 1979.

———: "Integrated Computer-Aided Manufacturing (ICAM) Final Report: IDEF0 Functional Modeling Manual," Contract no. F33615-73-C-5158, January 1981.

Chapter

5

Some Advanced Diagram and Model Concepts

This chapter discusses some of the more advanced concepts that make SADT a superior system description language. From the viewpoint of making activity diagrams, the quality of a diagram very much depends on how well one analyzes the objects of the system. The better analysis one does on system objects, the better the activity diagrams will describe the functionality of the system. Data decomposition must therefore initially precede and be done concurrently with activity decomposition. Since good data decomposition is the key to good activity decomposition, we will concentrate in this chapter on explaining in more detail the nature of SADT arrows in both diagram and model contexts.

5.1 Arrow Contents Are Varied

In SADT, arrows represent *things.* We stress the word "things" here because, unlike traditional dataflow diagrams of software systems, SADT diagrams have arrows that represent a wide variety of things. For example, in Figure 5-2, the ***build special part*** diagram has arrows that represent plans, machines, tools, raw materials, documents, engineering drawings, finished parts, people, verbal requests, and estimates. Good system descriptions must contain such a variety of things in order to adequately explain how a system works, both in detail and with its environment. The ability of SADT to represent a variety of things with arrows results from SADT being a description methodology for systems, as opposed to just software. In order to help both analysts and experts understand and describe how various things relate to each other, the SADT graphic language distinguishes between things transformed by a system and things that govern how the transformations are done. SADT calls these "inputs" and "controls," respectively.

5.2 Arrows Are Decomposable

That SADT arrows are decomposable is critical to SADT being a powerful system description language. Arrow branches and joins are the syntax that allows one to describe the decomposition of arrow contents. This syntax was not chosen arbitrarily. Think of a thick telephone cable. If you cut such a cable in two and looked at it, you would notice that the main cable was made up of several somewhat smaller cables, which were in turn made up of even smaller cables, and so on down to individual wires. SADT arrows can be thought of as cables that organize and carry a variety of things. This is why they are drawn as arrows that branch and join.

For example, consider the arrows named ***job plans*** and ***approved job*** in the diagram of Figure 5-1. The diagram tells us that these arrows name collections of things, because each splits into two separate arrows having different labels. These arrows tell us that ***blueprint*** is part of ***job plans*** and that an ***approved job*** is made up of a ***tagged part*** and an ***approval seal,*** or it could also be an ***unfinished job.*** This is all we can glean from the ***build special part*** diagram about these arrows, but we can always look at decompositions of the boxes on this diagram to find out in more detail what these or other arrows contain. For instance, in the ***complete work kit*** diagram of Figure 5-4, we discover that ***machines and tools*** is made up of a ***tool crib*** and ***machines on the floor.***

Thus, the arrows on a diagram that split and join reflect a hierarchy of things within those arrows. But these arrows only describe that hierarchy as it relates to the partic-

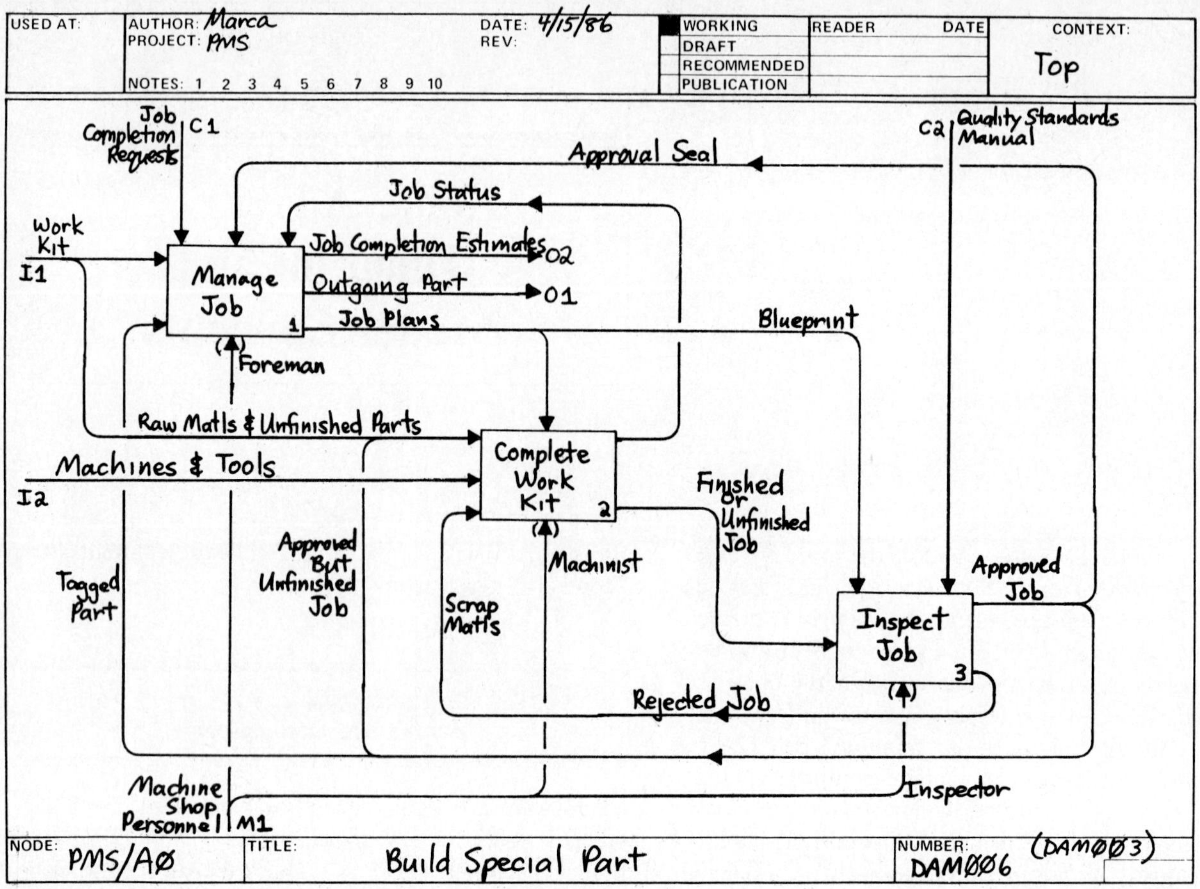

FIGURE 5-1 A Diagram Containing Arrows that Split and Join

ular system functions that appear on the diagrams themselves. In fact, one rarely uncovers the complete hierarchy of an arrow on a single diagram. Usually it takes reading through most of the model, and sometimes the details of particular arrows are never exposed because of the viewpoint of a model. That is why SADT advocates generating a supplemental description of the complete hierarchy of system things by making a Glossary Page for each diagram in a model and by compiling these Glossary Pages into a Data Dictionary. In this way, the Data Dictionary, a vital model supplement, becomes the central repository of the complete hierarchy of system things (see Chapter 18).

Sometimes an activity splits an arrow into its components, just as a prism splits white light into colors. In this case, it is important to study what the activity does to gain more knowledge about the contents of its arrows and how they relate to each other. For example, in Figure 5-4, *job plans* is a control to the *evaluate job progress* activity, one of whose outputs is *next job order step,* which in turn contains machining *notes.* This suggests that *next job order step* and *notes* are components of *job plans,* though we must look at the decomposition of *evaluate job progress* and possibly investigate the Data Dictionary to be sure of this assumption.

5.3 Arrows Can Be Tunneled

Regardless of what arrows contain, all arrows on a diagram can, potentially, be tunneled. As Chapter 3 points out, tunneling is the way to hide the source or destination of an arrow. The two kinds of tunneled arrows, those with hidden sources and those with hidden destinations, do not "pass through" the boundary defined by a box and its arrows. Tunneled arrows that appear from an unknown source have their tails parenthesized, indicating that they came from some other part of the model or from outside the model directly. For example, on diagram PMS/A1 in Figure 5-2, the *ready mechanic* arrow has a tunneled tail. Because of this, the arrow does not appear on the PMS/A0 diagram. Tunneled arrows that go to an unknown destination have their heads parenthesized, indicating that they go to some other part of the model or

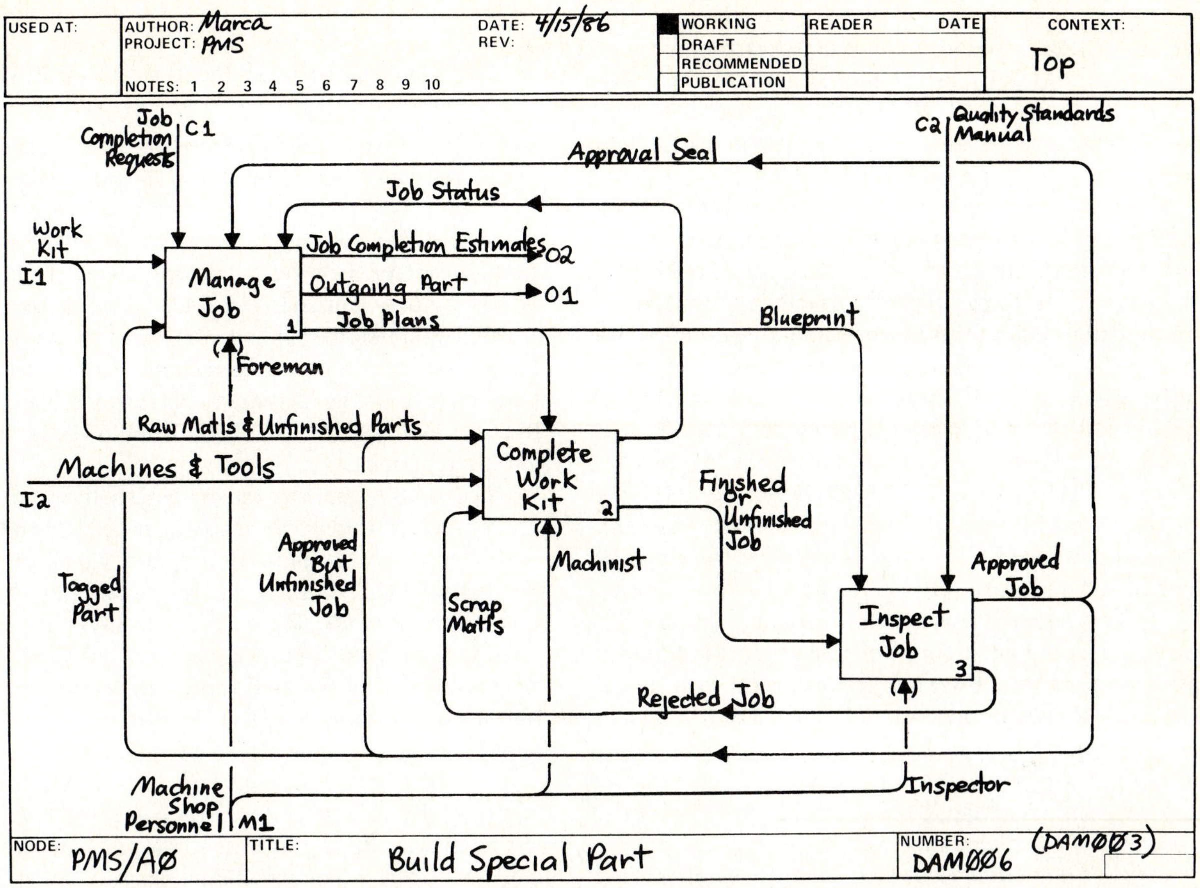

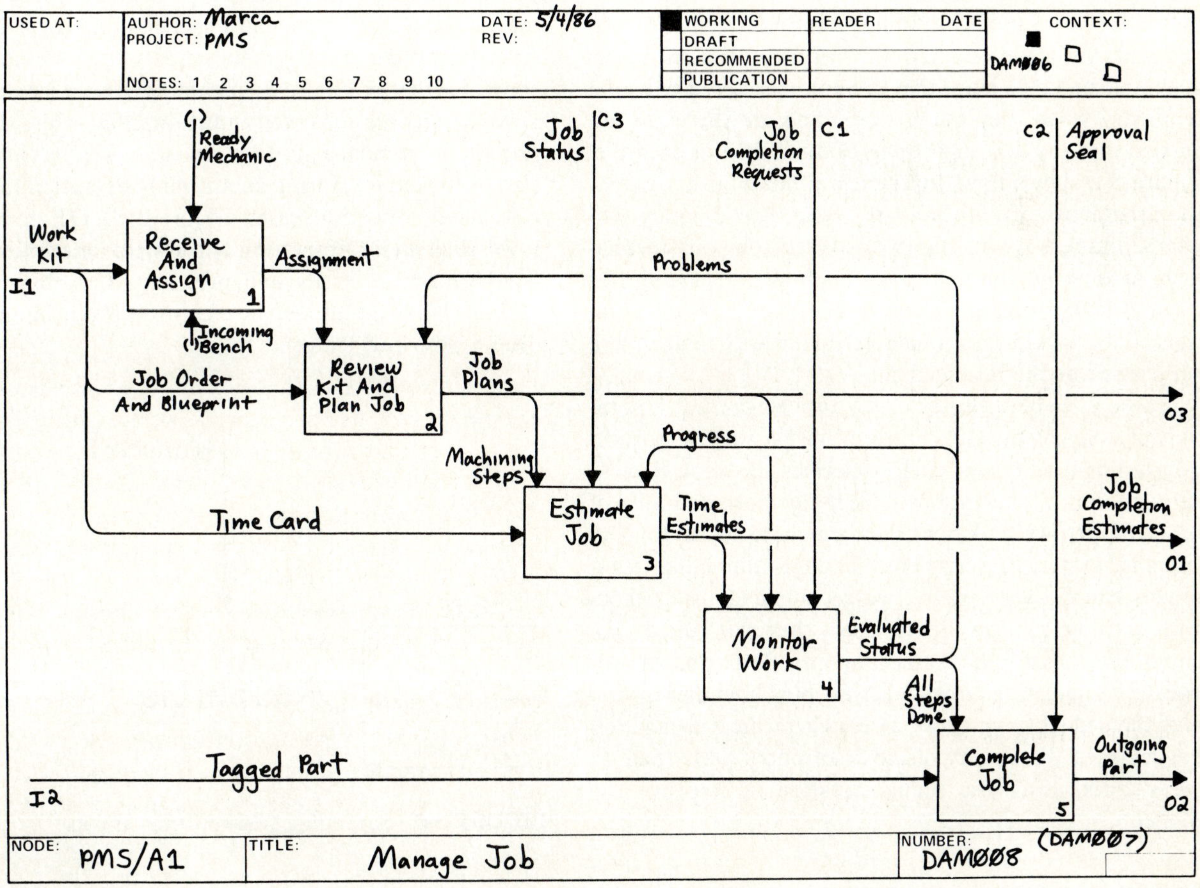

FIGURE 5-2 Suppression of Facts and Exposition of Necessary Details

to outside the model directly, or are not further treated in the model. For example, the mechanism arrow for box 2 on diagram PMS/A0 has a tunneled head, and thus does not appear on the PMS/A1 diagram. The word "tunnel" then is quite appropriate, for one can imagine a tunneled arrow going "underground."

The tunnel notation was developed after SADT was extensively used in a number of application domains. Experience showed that when describing complex systems, many arrows were required to correctly explain the system in detail. Often, these arrows could be bundled into aggregates, but sometimes important things not previously shown at higher levels in a model came to light as more detail was uncovered. Further, these details were usually not important enough to show at the higher levels of the model. To avoid cluttering the higher-level diagrams of a model with unwanted details, the tunnel notation was developed. This notation provides the ability to control the appearance of required detail without confusing the less detailed descriptions of parent diagrams. Figure 5-2 gives a good example of using tunnel notation to avoid propagating unwanted detail upward in a model: The ***ready mechanic*** arrow on diagram PMS/A1 is required at the ***manage job*** level, but propagating the arrow up to the ***build special part*** diagram would only confuse that diagram's message. Since ***ready mechanic*** is out of place on the A0 diagram, it was tunneled.

The tunnel notation also makes it easy to suppress facts appropriate only for high levels in the model. This minimizes the chance of cluttering decompositions with unnecessary information. Arrows with parenthesized heads accomplished this, since they are not considered part of a boundary when they touch a box, and are therefore not carried into any decompositions of that box. For example, Figure 5-2 shows how, by tunneling mechanism arrows, the decompositions of the ***build special part*** diagram will not be cluttered with uninformative or obvious mechanism arrows touching all boxes. Such facts would clutter those decompositions without adding any new information. They would very much be in the way, and so are suppressed at the box boundary.

These examples illustrate why tunnels are not a notation of convenience—they are instead a very important way to fine-tune how a model describes a system. Browse through Part VI, and you will see many more examples of tunneled arrows and how they were employed to help simplify system descriptions. Though tunneling techniques can be very useful, we recommend using them with extreme caution. Poorly used, they can quickly become crutches for bad modeling practices. So, we advocate propagating arrows across box boundaries first and then identifying those cases where the propagation has lessened the quality of the description. Only then do we suggest using the tunnel notation to improve the model.

5.4 The Difference between Input and Control Arrows

One of the major differences between SADT and other structured analysis methods is the distinguishing of inputs from controls. One could argue that just the inputs to and outputs of transformation functions are sufficient to describe a system. However, by emphasizing the difference between inputs and controls, SADT gives an analyst the capability to describe explicitly the constraints imposed on transformation functions. Constraints, therefore, give a truer picture of how a system operates, for they describe the facts and the rules that must be followed by transformation functions.

For example, consider the ***assemble*** activity box in Figure 5-3, which transforms a ***seat,*** a set of ***legs*** and a ***back*** into a ***chair.*** A dataflow description would stop right there, but SADT allows an analyst to add more information about ***assemble.*** Figure 5-3 also says that ***assemble*** requires a ***blueprint*** to operate correctly. The ***blueprint*** is clearly not part of the final ***chair,*** but it does play a vital role in the ***assemble*** activity. Without the ***blueprint,*** assembly of chairs can be a very unorganized activity, or at best, there is considerable room for various assembly strategies. By adding the ***blueprint*** control arrow, the analyst has made a strong statement: the ***blueprint*** is the only guide to follow when assembling chairs.

By explicitly stating that ***blueprint*** constrains ***assemble,*** the analyst leaves no room for speculation. This creates an opportunity for stronger statements. For example, the control arrow of Figure 5-3 could have been labeled ***blueprint and its special notes,*** indicating that while the ***blueprint*** is the standard guide to assembly, ***special notes*** must also be looked for during an assembly operation to handle infrequent exception cases. Without SADT control arrows, such an interpretation of a system description could not be made so easily and precisely. The ability of the graphic language to separate controls from inputs is indeed a powerful feature.

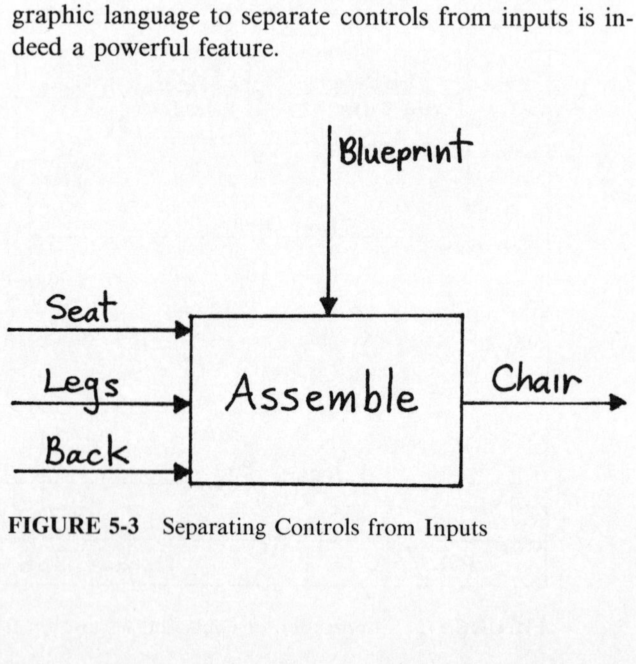

FIGURE 5-3 Separating Controls from Inputs

5.5 Mechanism Arrows Define How Activities Are Realized

A second major difference between SADT and other structured analysis methods is the importance SADT places on mechanism, a concept unique to SADT. In SADT, a system is described first from a functional perspective. But the system must ultimately be realized (i.e., it must be made operational), and that is when SADT mechanism arrows play an important role in describing the system. SADT mechanism arrows give an analyst the ability to precisely define how a particular function will operate, what storage is required by that function, who will do it, and so on. With mechanism arrows, an analyst can add the final details to a story that will leave no doubt about what happens when a function executes.

For example, consider again the ***assemble*** function described in Figure 5-3, but add a mechanism arrow to the bottom of the box. Now consider how ***assemble*** might actually be done if the mechanism arrow were labeled ***glue.*** You might have envisioned slatted pieces of wood that were first glued, then stuck together and allowed to dry. Now consider ***assemble*** when the mechanism arrow is labeled ***screwdriver.*** You may have imagined the chair parts containing wood screws and having a peg-in-hole connection scheme, which requires simply plugging the parts together and tightening the screws. The scenarios are remarkably different, and all with the change of one single word on the mechanism arrow! This example also illustrates the case in which mechanism arrows make explicit the things used in carrying out an activity.

Mechanisms can be used for many different purposes. For example, mechanisms can be used to identify who will do particular functions. As pointed out in Figure 5-2, the mechanism arrows on the ***build special part*** diagram make it explicit that the major functions of the prototype machine shop will be done by each of the three different kinds of people in the shop: ***foreman, machinist,*** and ***inspector.*** Another example of using mechanism arrows in this way can be found in the model in Chapter 28. This model shows two or more mechanism arrows touching the same box to document the fact that different people work together to complete a particular activity. In other words, many mechanism arrows touching a box can represent a coordinated activity.

Mechanisms can also be used to indicate that certain functions support the operations of other functions and must therefore always occur in a required sequence. For example, Figure 5-4 shows that ***setup machining place*** must happen before ***machine and assemble,*** because the ***machining place*** must be ready before work can be done. Here is a case in which the system by its very nature demands that a certain sequence of operations be followed. As a final example, notice that in the ***manage job*** diagram of Figure 5-2, the mechanism arrow for box 1 labeled ***incoming bench*** defines where to look for newly arrived ***work kits.*** In this situation, the analyst wanted to stress that the ***incoming bench*** is important to the ***receive and assign*** activity in the prototype machine shop. All these examples show the utility and importance of mechanism arrows in describing various aspects of system operation and realization.

5.6 Control and Dataflow Feedbacks

The concept of feedback is fundamental to systems theory in general. Feedback happens when the output of a particular function A affects the output of another function, B, whose output affects a different activation of function A. Fundamental to SADT is its ability to describe two different kinds of feedback situations: control feedback and dataflow feedback. Distinguishing these two kinds of feedback situations is vital, for control feedbacks have a more profound affect on system operation than do dataflow feedbacks. Let's see why.

Dataflow feedback between two functions happens when the output of one function becomes an input of another. For example, in the ***build special part*** diagram in Figure 5-2, the ***inspect job*** function exhibits dataflow feedback to the ***complete work kit*** function. This is an example of feedback occurring because the system is trying to make efficient use of its waste (i.e., using a ***rejected job*** as scrap to reduce the need for ***raw materials***). Another feedback between those same two functions is ***approved but unfinished job.*** This is an example of feedback occurring because iteration happens in the system to refine inputs to a desired level of quality. In this case, machining is done and inspections are made until the part has dimensions that are within the levels of tolerance predetermined by the ***blueprint.***

Control feedback happens when the outputs of two functions constrain each other. The classic chicken-and-egg scenario is an example of control feedback. The ***build special part*** diagram in Figure 5-2 shows control feedback between ***manage job*** and ***complete work kit*** through ***job status.*** In this case, ***job status*** reflects the actual progress made against the schedule defined in ***job plans.*** Based on the latest ***job status,*** the supervisor revises the ***job plans,*** which in turn affects the future work of the machinist on that job. This is an example of how efficient and productive systems tie their planning and work activities closely together through control feedback.

These examples illustrate the difference between control feedback and dataflow feedback. The mutually constraining activities in a control-feedback loop often cannot

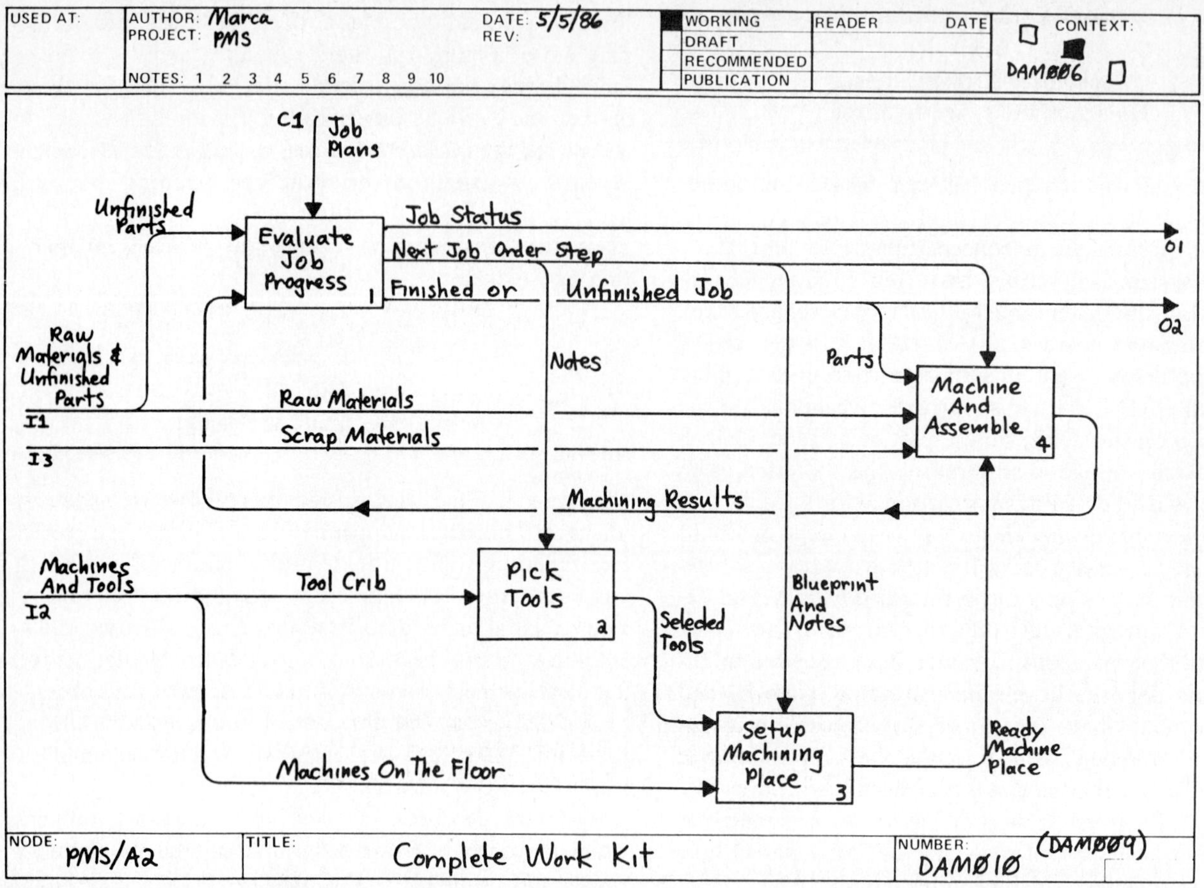

FIGURE 5-4 Specifying How Functions Support Each Other

function well without the other, and sometimes the system can fail if the feedback is not maintained. For example, making reliable ***job completion plans*** is, in part, the result of experience gained while ***machining parts,*** which is the result of planning the work. Dataflow feedback, on the other hand, does not bind together functions as tightly as control feedback. For example, not recycling ***rejected jobs*** as scrap would not reduce the effectiveness of the prototype machine shop, nor would it reduce the quality of parts. Not recycling scrap would simply make the system unnecessarily wasteful. Consequently, control feedbacks are considerably stronger than dataflow feedbacks in the way they bind functions together.

5.7 Summary

Development of a good functional decomposition relies on first doing a good decomposition of system things. Arrows are hierarchic collections of system things and describe how these things are decomposed by using branches and joins. All arrows can, potentially, be tunneled. Tunneled arrows allow one to suppress unwanted facts as well as expose necessary details at the appropriate level in a model. SADT distinguishes those things transformed by a system (i.e., inputs) from those things that govern system transformations (i.e., controls). Control arrows contain the rules and the facts that constrain the operation of a function. Input arrows contain things that get transformed. Thus, control arrows and input arrows usually represent things of a very different nature. Mechanism arrows are used to document how a particular function is realized (e.g., how it is done or who does it). Control feedbacks mutually constrain functions, while dataflow feedbacks allow for recycling and iteration. Thus, control feedbacks are considered stronger than dataflow feedbacks in the way they bind two functions together.

Suggested Readings

Miller, J.: *Living Systems,* McGraw-Hill, New York, 1978.

Ross, D.: "An Essay on Activity Diagramming," SofTech Technical Report no. 7104, November 1976.

SofTech, Inc.: "IDEF0 Author's Guide to Creating Activity Diagrams," SofTech Deliverable no. 7500-13, September 1979.

———: "Integrated Computer-Aided Manufacturing (ICAM) Final Report: IDEF0 Functional Modeling Manual," Contract no. F33615-73-C-5158, January 1981.

Chapter

6

Some Advanced Authoring Concepts

Parts II to IV discuss the basics of authoring (i.e., the process an SADT author goes through to develop a model). This chapter expands upon those basic concepts by discussing some of the important advanced concepts that make SADT a superior method for modeling systems. Since this book focuses on SADT activity models, we will cover only those advanced SADT topics which relate to the development of functional system descriptions by considering the individual author in the process of developing a single model.

6.1 SADT Models Structure Natural Language

SADT distinguishes itself as a system description method by starting with the natural language of experts and imposing a structure on that language with its graphics. Natural language allows an expert to freely express the operation of the system in familiar and comfortable terminology, through writing and speech during interviews. Then, during the diagramming process, the author takes the words of the expert and applies special punctuation around those words by using the SADT graphic language. SADT graphics work in a way similar to punctuation because they give greater precision to the words they touch and surround. By using this "advanced form of punctuation," the SADT author gives a more precise and terse description of the system without sacrificing quality, because the SADT graphics removes the ambiguity from the natural language descriptions given by the expert.

Ambiguity is removed because SADT graphics are given a standard interpretation. Formally, a single SADT box called B having associated input arrows I, control arrows C, output arrows O, and mechanism arrows M is translated into the sentence: "Activity B transforms I into O as determined by C using M." The formal translation of SADT diagrams into sentences follows this pattern exactly, though we recommend composing sentences that conform semantically to the pattern but are also interesting to read. For instance, the system description told by the model boundary diagram in Figure 6-1 can be written as:

> The prototype machine shop ***builds special parts.*** It does so by using a ***work kit*** and ***machines and tools*** and, through a process governed by the ***quality standards manual,*** creating an ***outgoing part.*** Also, occasional ***job completion requests*** require current ***job completion estimates*** to be reported.

Ambiguity is also removed by decomposing and refining the high-level diagrams, which restricts the possible ways they can be interpreted, until the low-level diagrams are detailed enough to give precise meaning to the things and activities of the system. In other words, an SADT model constrains the unfolding of its story. It hierarchically organizes the system description into pieces not unlike the chapters and sections of this book. For example, look at the narrative of the specification developed by Workshop 25. Notice that each diagram permits a specific train of thought, which is reflected in the writing of several paragraphs of text from the diagram. This happens because, though a box and its arrows are semantically equivalent to one topic sentence, there are usually several other important facts that must also be told about that box in order to meet the purpose of the model. Since SADT organizes the major topic sentences through hierarchical diagrams with boxes ordered by dominance, the associated facts are collected together and are told in an understandable fashion without repetition or conflict.

6.2 Viewpoint Influences Emphasis and Terminology

Consistent storytelling is an art practiced by SADT authors. Critical to a consistent story line is the concept of viewpoint. In SADT, a model should only be written from

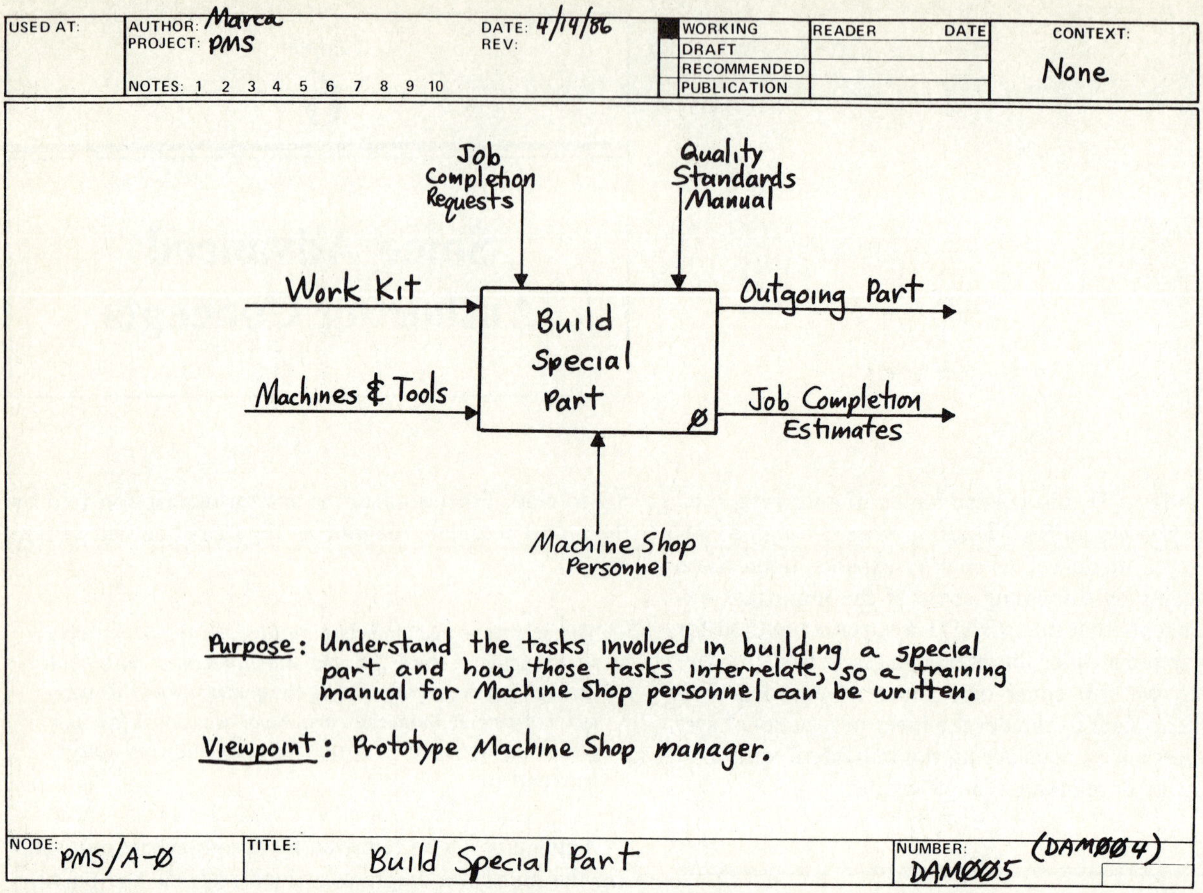

FIGURE 6-1 The Boundary Description for a Model

a single viewpoint. By selecting and adhering to a single viewpoint for a particular model, the author gains two important things. First, a single viewpoint always emphasizes certain system aspects and deemphasizes others in a consistent way. For example, by selecting the viewpoint of the prototype machine shop manager, the author initially gives equal weight to the tasks done by all machine shop personnel. As Figure 6-2 shows, managing jobs, completing work kits, and inspecting jobs are the three most important functions of the machine shop as seen by the manager.

If, on the other hand, the viewpoint of the inspector was taken, machining and management activities would have been deemphasized in the model, and the resulting A0 diagram would look similar to, though not exactly like, Figure 6-3. Clearly, selecting machines or hand tools and monitoring job progress against planned completion estimates are not important to an inspector, so these facts would be omitted from the inspector's view of the prototype machine shop and thus would never appear in his or her model. Yet other activities like evaluating the progress of a job in terms of required inspection points would still be important to emphasize. So, the diagram in Figure 6-3 is only an approximation of what the A0 diagram of the inspector's model would look like. This demonstrates why models with different viewpoints tend to contain overlapping subject matter.

Selecting a single viewpoint also ensures consistent terminology. Viewpoint is best understood as how the system looks from the perspective of a single person. Often, viewpoint is a particular role played by a person who is part of the system. Since it is common for each role to have its own special jargon, selecting a single viewpoint also means using an established set of terminology. For example, notice the words in Figure 6-3 that describe job inspection from the manager's viewpoint. Words like ***tolerance standards*** and ***appraised job*** are used. Now compare these words with those in Figure 6-4 which describe machining parts from the manager's viewpoint. Words like ***blueprint, raw materials,*** and ***tool crib*** are used. One can imagine two entirely different people talking about their work when reading these two diagrams. This illustrates the kind of terminology the author would encounter when interviewing the inspector and machinist respectively.

Therefore, choosing a viewpoint means emphasizing certain system aspects and acquiring a particular terminology. Without well-defined emphasis and terminology, consistent storytelling is impossible in a practical sense. One of the hardest jobs of the SADT author is staying

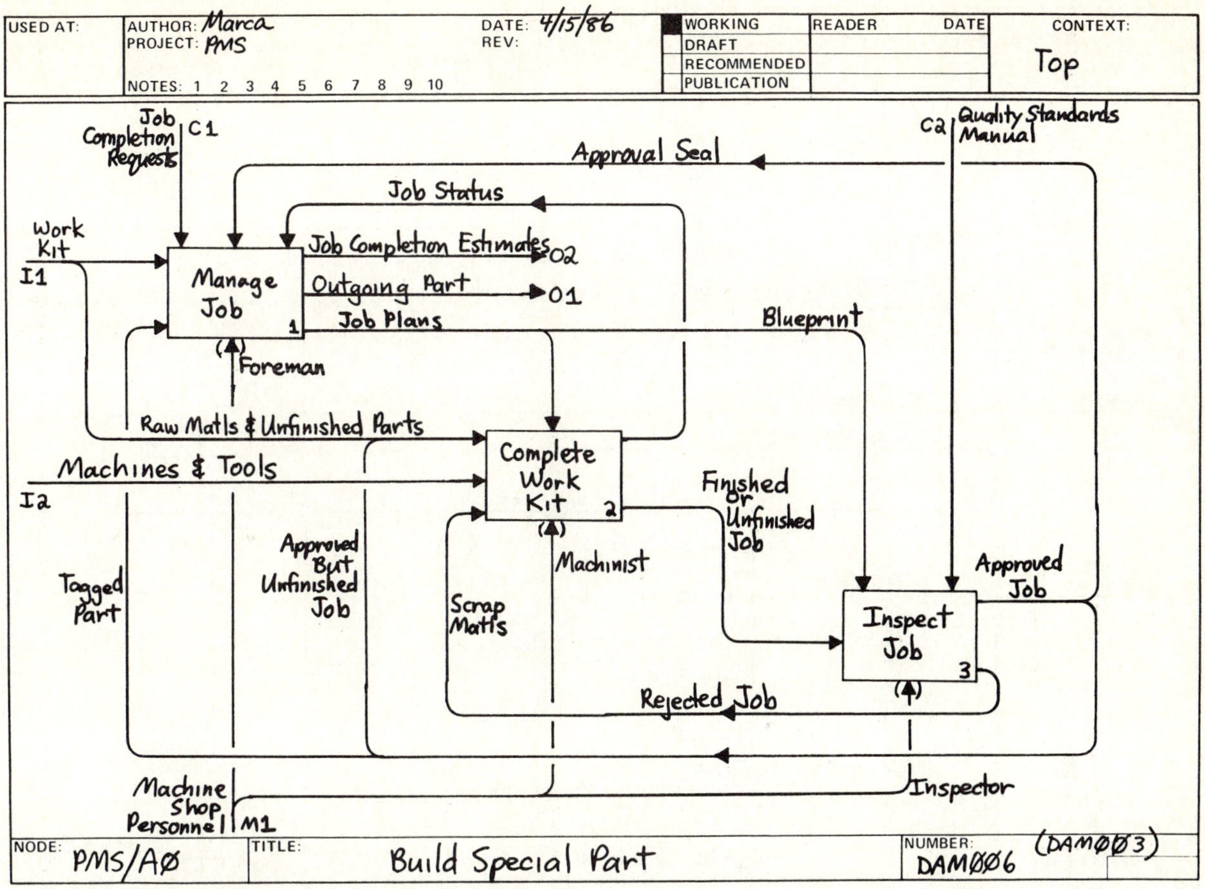

FIGURE 6-2 Giving Equal Emphasis to All Shop Tasks

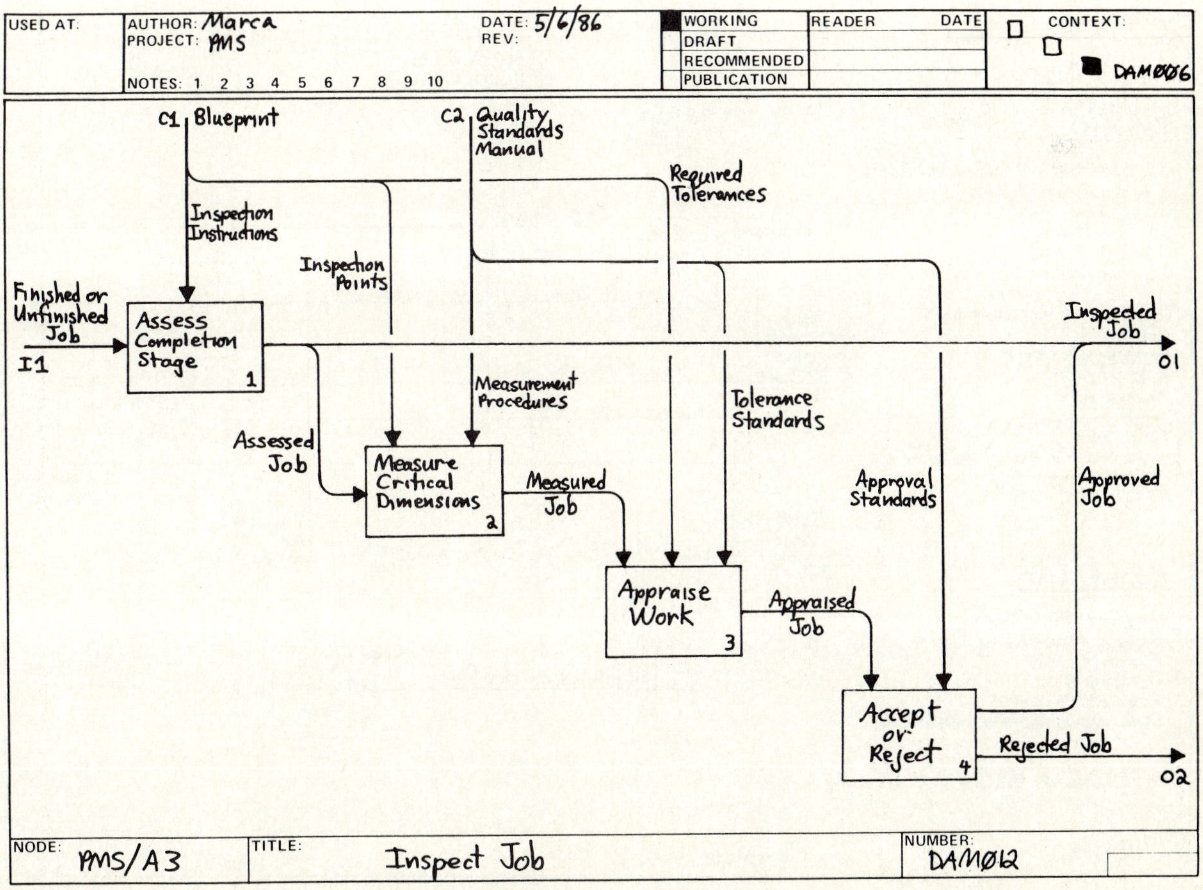

FIGURE 6-3 Some Activities Important to Inspectors

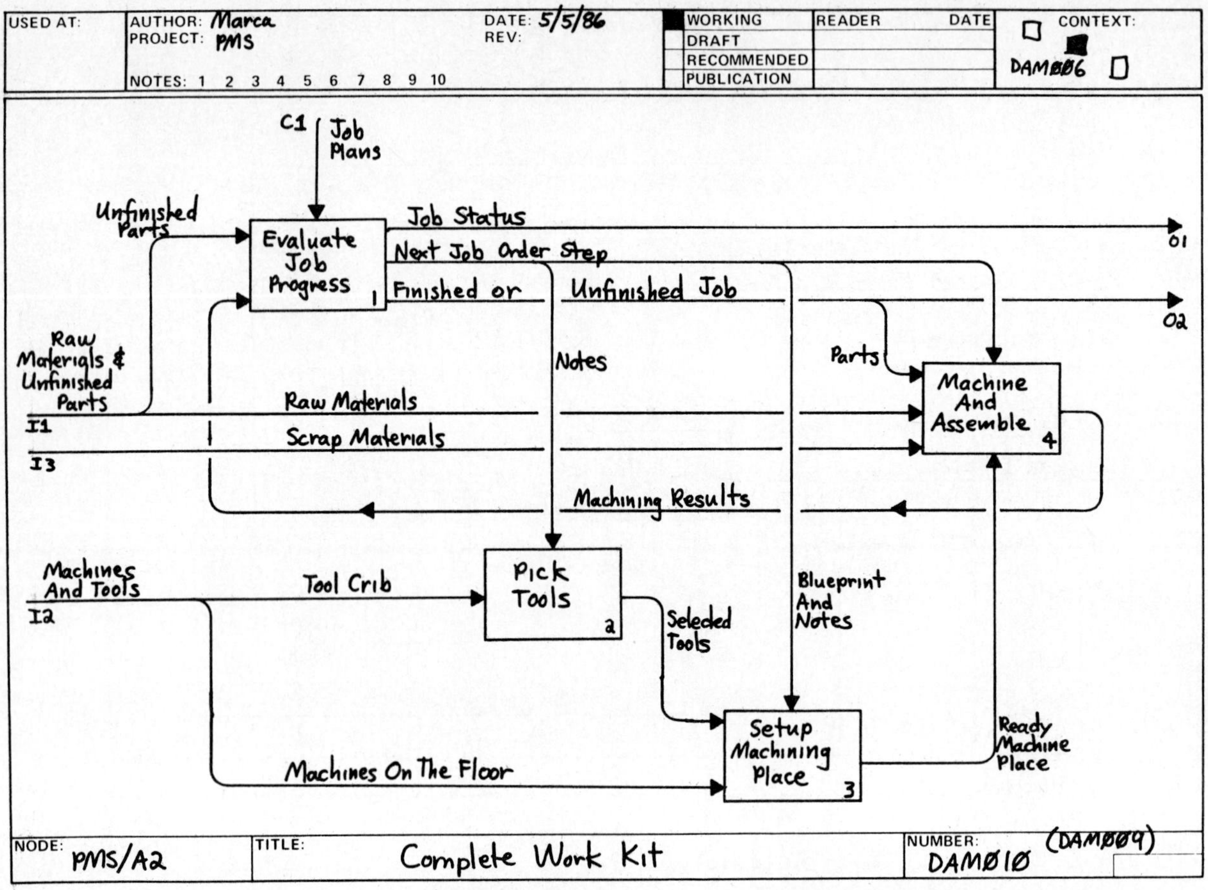

FIGURE 6-4 Some Terminology Related to Machining Parts

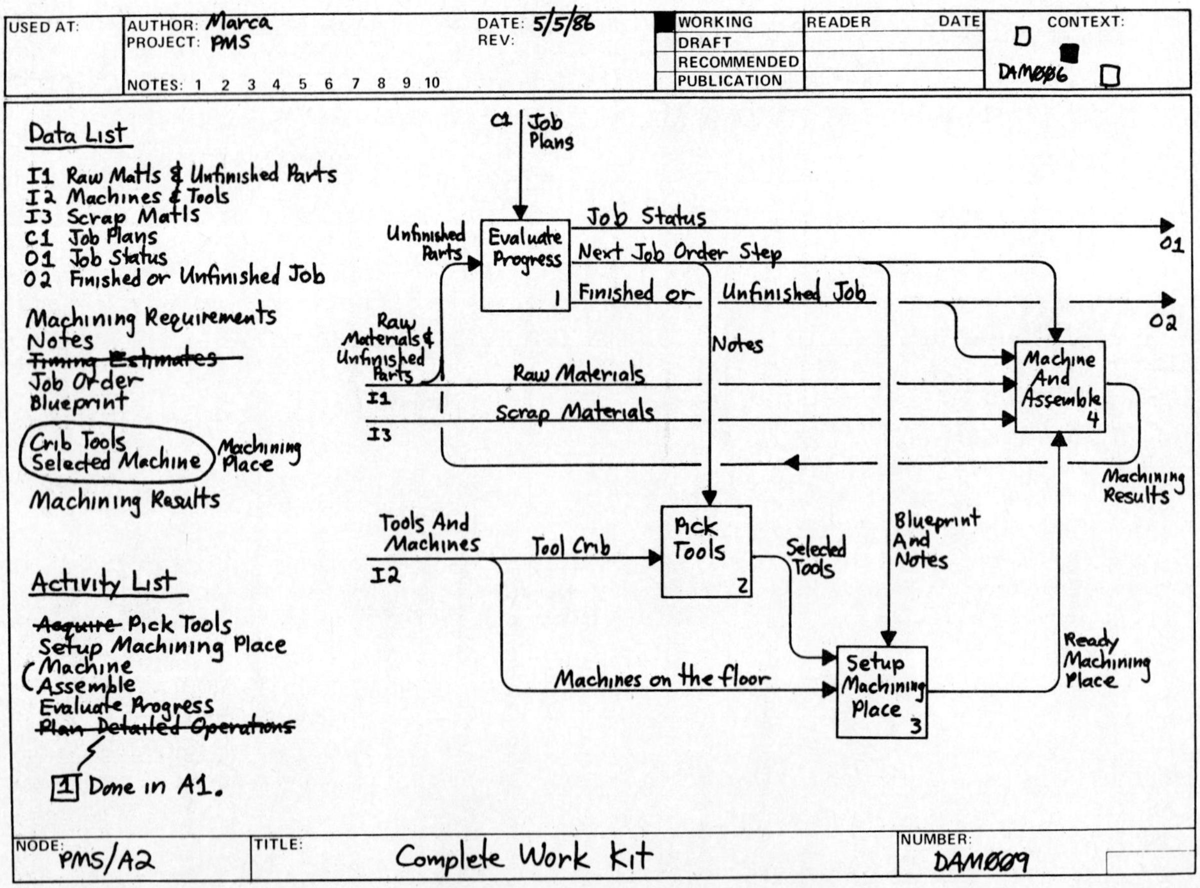

FIGURE 6-5 Examples of Analysis and Synthesis during Decomposition

within a chosen viewpoint, for it is very natural to discover many details about a system that do not fit into the current viewpoint, and the urge is strong to present such new information. For example, facts about planning a job touch upon both the ***foreman*** and the ***machinist.*** The ***foreman*** is most interested in the overall plan and how long it will take, while the ***machinist*** is most interested in the individual machining steps required to do a job. By separating these two concerns and putting them into the A1 and A2 diagrams respectively, the author has clearly and separately described the work of two very different members of the machine shop.

This strategy of separating concerns happens during an SADT modeling effort when authors select an appropriate viewpoint for the next model to be made. For example, if one wanted to build a model that only described job planning and monitoring, the ***foreman*** viewpoint could be selected. Similarly, if only machining parts were to be the subject of a model, then the ***machinist*** viewpoint would seem more appropriate. If these two models were made, they would contain overlapping subject matter. These overlapping descriptions of the prototype machine shop would each emphasize a different aspect of the shop. Viewpoint does not, therefore, limit subject matter, but instead forces the analyst to emphasizes certain aspects of a system over others.

6.3 Decomposition during Modeling

Decomposition is the process of developing a diagram which details a particular box and its associated arrows. What results is a description which seems to have "broken apart" the parent box into smaller, more detailed activities. Couple this with the fact that the word "analysis" means to break apart, and you have the original rationale behind the term "Structured Analysis." But decomposition is more than analysis: it also involves *synthesis.* True decomposition involves first breaking a subject into smaller pieces and then combining the pieces into a meaningful yet more detailed description of that subject. It is interesting to note that a model shows only the result of this interplay between analysis and synthesis.

Following the rules of SADT, the author first does analysis and synthesis on system things, documenting how the things defined in the bounded subject have been broken apart. For example, Figure 6-5 shows how the Data List starts with all the boundary arrows and their ICOM codes and ends up with their constituents. The author then scans this Data List, and possibly combines some constituents together, to identify those things that will act as controls.

Then, the author does similar analysis and synthesis for system activities but does them under the influence of the Data List. Through a process of free association and thinking about the new controls, an Activity List of more detailed system functions is derived. For example, ***notes*** and ***blueprint*** caused the author to think of the ***pick tools*** and ***setup machine place*** activities. Activities are listed until the flow of new thoughts slows down. Then these functional pieces are combined into a meaningful (i.e., balanced) set of from three to six boxes. (It is important to note here that adhering to this three-to-six rule forces the author to use abstraction and decomposition in order to gradually expose the details of a system.) After the functional pieces are identified, the Data List and Activity List are used to sketch the diagram. Both analysis and synthesis occur again, resulting in aggregates of things which are the arrows that interconnect the boxes.

This method of decomposition is important to SADT, for SADT advocates that the analysis of system things is the major influence on the analysis of system activities. When decomposition is done in this way, the resulting boxes on a diagram come into being largely because of the control arrows. Consequently, control arrows dictate the quality of and the rationale behind the resulting decomposition. For example, Figure 6-5 says that the information contained in ***notes*** is used by a machinist to select tools. If this were not true, ***notes*** might not be a control arrow and ***pick tools*** might not represent an activity.

The rule of three to six boxes per diagram is also unique to SADT. It is well known that the short-term memory capabilities of people are made up of about seven perception categories, each of which can hold about seven distict items of information. SADT takes a conservative stand, allowing an upper limit of six boxes—one box per category. The box name and its boundary arrows are the informational items that are put into a category as we read a diagram. In this way, SADT diagrams are designed not to tax the short-term memory limits of people. However, memory capabilities of people vary. Our experience indicates that diagrams of four or five boxes, with no more than five arrows touching each box, are close to optimal for the amount of information that can be quickly communicated to a wide audience.

6.4 Some Decomposition Strategies

Too often, authors do not pay enough attention to the decomposition strategy they adopt when they develop a diagram. Something that separates a senior SADT author from junior ones is constant awareness of decomposition strategies and their impact on model quality. A wide variety of decomposition strategies is available to the author. (In fact, whole books have been dedicated to the subject, and the Suggested Readings section of this chapter cites some important references.) Here is a very brief summary of some commonly used strategies for you to draw upon during model development.

Often, a functional decomposition strategy (i.e., a decomposition based on the functional relatedness of system activities) is best, because it forces an author to think hard about *what* the system does, independently of *how* the system operates. Functional decompositions also tend to deemphasize sequencing while exposing the required constraints on system activities. Thus, we strongly recommend using this strategy whenever possible. However, there are times when and situations in which a purely functional decomposition strategy might not produce the most useful model. Below are some cases in which other strategies might prove better.

Decomposition by the roles people or organizations play can be a useful first strategy for developing system descriptions that document how people coordinate with each other while they work. For example, Figure 6-2 shows how decomposition by roles describes the communication among machine shop personnel and what functions are done by a particular person. This diagram is therefore a composite picture of work in the machine shop, and each of its children is a more focused discussion of a particular kind of work (e.g., job management, machining, inspection). Sometimes the interconnections among roles is small, as in the prototype machine shop. Very often, however, interconnections among roles are extremely numerous and complicated. That is why we recommend using this strategy only to start a model of a system in the class often referred to as P3: people, paper, and procedures. This will help you gain an initial understanding of the system, from which a more functional decomposition of the entire system can be developed.

For command and control systems, decomposition according to already known and stable subsystems can be an effective strategy. It results in a set of models, one model for each subsystem or major component. A final composite model must then be built to connect these separate models together in order to describe the whole system. For example, the authors of this book jointly worked on a project that built separate models for a torpedo, a countermeasure, and the way a submarine maneuvers, and connected these models together to describe how a particular submarine defends itself. We recommend using a decomposition-by-subsystems strategy only when the major pieces of the system do not change (e.g., the chassis and engine in an automobile), because unstable subsystem boundaries will quickly invalidate the individual models as well as the composite.

Some systems act as processes that continually refine their inputs into a final product, such as an oil refinery. A decomposition strategy centered around the birth-to-death lifeline (commonly called the "life cycle") of a key system input can be effective for describing such processes. For example, the Feed Family model in Appendix C is the result of decomposing that system according to the steps required to transform bought food into eaten meals. We recommend using this strategy when the purpose of the system is to refine one of its major inputs and when you can easily determine the successive states of refinement for that input.

When all else fails, decomposition according to the physical process will always work. Emphasis on phases of activities, stages of completion, or steps to be taken are the result of this kind of decomposition strategy. For example, all the low-level diagrams (i.e., A311, A312, A313) in the training model of Chapter 31 describe the detailed sequence of steps taken to prepare military training material. Though such a strategy is useful for describing existing processes such as a manufacturing plant in operation, the result is often a very sequential system description which may not properly identify mutually constraining activities and may hide mandatory sequencing. We recommend using this strategy only if the purpose of the model is to describe the physical process itself or as a last-ditch strategy when you cannot see clearly how to proceed.

6.5 Settling on a Decomposition Strategy

It is not uncommon to try different decomposition strategies when starting a model. From our experience, a few false starts may be required to develop a quality A0 diagram. Normally, the first decomposition attempt that produces an A0 diagram gets tremendous scrutiny from the Reader Community. Often, this initial description will not, when taken to more detail, suit the purpose of the model. Do not be discouraged when this happens, so long as your first attempts are clear and concise. Remember, it is better to be clear than correct when starting a model, because the collective knowledge of the experts, the Reader Community, and other authors will help you develop a solid summary description that will, when detailed, satisfy the purpose of the model.

For example, scan through the workshops and notice how the A0 diagram of the Feed Family model was changed as its major activities were being decomposed. In this case, the interfaces and expected functionality of the major activities underwent a great amount of change, and this change had to be reflected on the A0 diagram as required by SADT syntax rules. Also, even though the initial strategy of decomposition according to the birth-to-death lifeline of food was adequate, all the diagrams in the model underwent revision to obtain a correct and consistent story for that particular scenario. (See Appendix C for the complete Feed Family model.) Thus, even if your initial decomposition strategy was a good one, expect a lot of change on the A0 diagram.

6.6
Precision Determines when Decomposition Stops

The constant but controlled evolution that the diagrams of a model undergo is an example of an engineering process. In other words, SADT is an engineering process because it allows for the successive refinement of a system description. But, as in any engineering effort, the refinement process always stops at some point. SADT models are built not only to answer a specific set of questions, but they must answer those questions to a given level of precision. So, in SADT, decomposition stops when those diagrams that form the bottom of a model are sufficiently detailed to meet the purpose of the model. In other words, decomposition is no longer required when the model is precise enough to answer all the questions summarized by the purpose.

For example, the purpose of the model that describes the prototype machine shop is to identify the tasks of working in the machine shop to a sufficient level of detail so that they can be described in a training manual. Notice that the low-level diagram given in Figure 17-1 presents a set of tasks that can be easily written about with one paragraph of text for each box. In fact, this is the precision test for that model: stop when each box can be explained by one paragraph of text. For the purposes of the training manual, one paragraph is considered enough precision. Therefore, as in other engineering disciplines, the development of an SADT model stops when precisions are within tolerances.

6.7
Summary

SADT models structure natural language into precise and terse models, and it is through the graphic language of SADT that ambiguity is removed from natural language explanations of a system. The viewpoint of a model influences what aspects of the system are emphasized and what aspects are deemphasized, and selecting a viewpoint means using a particular terminology. Decomposition is the process used to build a model, and it involves both analysis and synthesis steps. SADT advocates analyzing system things first as a prelude to analyzing system activities. Decomposition strategies are numerous and varied. Often, settling on a decomposition strategy and a quality A0 diagram requires several iterations and many changes. Decomposition stops when the model is precise enough to answer the questions that constitute its purpose.

Suggested Readings

Alexander, C.: *Notes on the Synthesis of Form,* Harvard University Press, Cambridge, Mass., 1964.

Jackson, M.: *System Development,* Prentice-Hall, Englewood Cliffs, N.J., 1983.

Orr, K.: *Structured Systems Development,* Yourdon Press, New York, 1977.

Miller, G.: "The Magical Number Seven, Plus Or Minus Two: Some Limits on Our Capacity for Information Processing," *The Psychology Review,* vol. 63, no. 2, March 1956.

Marca, D., and C. McGowan: "Static and Dynamic Data Modeling for Information System Design," *6th International Conference on Software Engineering Proceedings,* September 1982.

Parnas, D.: "On the Criteria to be Used in Decomposing Systems into Modules," *CACM,* December 1972.

Pirsig, R.: *Zen and the Art of Motorcycle Maintenance,* Bantam Books, New York, 1974.

Ross, D.: "An Essay on Activity Diagramming," SofTech Technical Report no. 7104, November 1976.

SofTech, Inc.: "IDEF0 Author's Guide to Creating Activity Diagrams," SofTech Deliverable no. 7500-13, September 1979.

———: "Integrated Computer-Aided Manufacturing (ICAM) Final Report: IDEF0 Functional Modeling Manual," Contract no. F33615-73-C-5158, January 1981.

Part

II

Authoring Activity Models and Diagrams

This part of the book shows how an SADT analyst, or author, develops a description of how a system works by drafting an SADT activity model. Discussion concentrates on how knowledge is acquired through interviews, how diagrams are created, how models are started, decomposition, and critiquing one's own work for accuracy of message and visual complexity. This part is designed to walk you through the first phase of systems analysis, in which an author sketches the functional description of a system.

Chapter 7 discusses the skills you need to collect facts from experts so you can construct a model. Chapter 8 guides you through an analysis session that produces the first attempt at a model. Chapter 9 guides you through a second analysis session that decomposes part of the model made in Chapter 8. Chapter 10 shows you how to critique your own work before sending it out for review. Chapter 11 presents the essential set of standard drawing conventions used to simplify SADT diagrams.

The skills of interviewing and doing the mental steps of analysis and synthesis are discussed and illustrated throughout these chapters. The Prototype Machine Shop model will be used to illustrate both diagram and model syntax rules and the process of drawing diagrams and building models. Thus, by going through these chapters in sequence, you will get a feel for what goes on in the mind of a systems analyst during the process of using SADT to start building a model of a system.

Chapter

7

Gathering Information

This chapter focuses on the techniques good SADT analysts use to acquire domain knowledge, the techniques used to extract from domain experts facts about the system to be described. These techniques are called "data collection" by SADT practitioners, but are more commonly known as "interviewing" and "knowledge acquisition" in the computer field. Whatever the name, good fact-gathering skills are critical to building accurate and usable models from expert knowledge. Thus, an essential part of the SADT methodology is its knowledge-acquisition techniques.

Interviewing is a fact-gathering mission, and the first interview formally marks the beginning of the actual modeling effort. To hold an interview, the analyst first locates the best source of information (e.g., a document, a particular person) and then arranges to "interview" the source. The goal of this and each interview should be to acquire only enough knowledge to either start or continue a particular portion of a model. After the first interview, the SADT model is used to determine what information must be acquired in the next interview. Thus, by following the hierarchy of an SADT model, a series of interviews can be conducted that addresses increasingly specific details about the domain.

SADT analysts use their models to focus their interviews and to optimize the time they spend with their information sources. This leads to a more efficient style of work, minimizes the time spent going back over details that were forgotten or unclear, and reduces the chances of posing the same question many times to an expert. This way, experts are more willing to talk with analysts, thus preventing one killer of analysis projects—the stopping of information flow.

7.1 Sources of Information

Experts are common sources of information. They are often the best sources because they are familiar with the current nuances and undocumented aspects of the system. Most important, experts know many facts that are not documented or are hard to explain. These facts, sometimes referred to as "causal knowledge," can only be gotten through interviews with the experts. To prepare for such interviews, we recommend investigating other information sources, like documents. Many worthwhile interviewing strategies are available to tap these sources. The other strategies we commonly use are:

- Reading documents
- Observing existing operations
- Surveying with a questionnaire
- Using your own knowledge
- Inventing a description

Documents are a good information source because they are often readily available and can be "interviewed" at a leisurely or intense pace. Reading documents is a good way to get an initial overview of a system and to formulate questions for interviews with experts. We recommend that, when possible, an analysis project should form and maintain a library of relevant documents. For SADT projects, these documents can be kept in the project library and distributed in small work packets called "kits" (see Chapter 13).

Observing the operations of the system being modeled is a good knowledge-acquisition strategy and should be done whenever possible. Through observation and possibly participation, analysts gain firsthand knowledge about day-to-day operations. Questions often arise when you see a system in operation—questions that would never come up if the analyst only read documents or talked with experts. The observation strategy is too often overlooked, and we encourage analysts to try it more often. We do, however, recommend caution; too much observation can lead to identifying too closely with the way things are currently done. This loss of objectivity can prevent you from seeing alternate ways of describing and accomplishing system functions.

Questionnaires can be used to tap large groups of experts in short periods of time. They can be used, for example, to quickly obtain many different viewpoints about the operation of a particular part of a system. They can also be used to poll experts about which parts of the system need the most improvement. In practice, questionnaires are somewhat unreliable in their ability to extract a variety of good information from the experts. We recommend using a questionnaire only after you become familiar enough with a system so you can ask specific and accurate questions. This way, the questionnaire will have a better chance of being correctly interpreted and answered.

Sometimes, senior analysts have studied many systems of a particular type (e.g., manufacturing, telephonics, accounting). Over the course of many projects, they have acquired a large amount of domain knowledge—that is, knowledge about a particular class of system. Domain knowledge is very useful because it often embodies concepts that span many particular systems—concepts that can be generally applied. If you have such domain expertise, choose when to use yourself as an information source. We recommend interviewing yourself to get the most from what you know. We also urge caution in using this strategy, because domain knowledge may lack information specific to the system under study. When you need specifics, don't rely on your own domain knowledge; interview the experts instead.

Another useful strategy is to invent a description and have it corrected by an expert. Invented descriptions can show alternative schemes of system operations—schemes never thought of by the experts. They require clout to work, however. We suggest that the author have domain expertise and that there be at least one friendly sounding board in the expert community before an invented description is attempted. This way, the description may reasonably reflect reality and will have a good chance of being critiqued in a favorable light. Invented descriptions can fail, however, if the experts are not open to new possibilities. Ideally, the author should establish strong rapport with a few key experts before using this strategy.

7.2 Types of Interviews

Regardless of the information sources used, several types of interviews are held during an SADT analysis project. The type of interview conducted varies depending on both the kind of information needed and the desired objective. The most common types of interviews are:

- Fact-finding interviews
- Problem identification interviews
- Solution discussions
- SADT Author/Reader Talks

Fact-finding interviews are held when trying to determine how a system currently operates. Problem identification interviews are useful when you are defining what is wrong with the current system. Solution discussions can be held when you want to identify how the future system must operate in order to solve deficiencies in the current system. SADT Author/Reader Talks (see Chapter 12) are informal discussions held when there is disagreement between an analyst and an expert.

A categorization of interviews is given here for two reasons: First, it is extremely valuable to understand the primary objective for holding an interview. A clear purpose helps to focus sidetracked discussions. Second, it gives an analyst the capability to sort out varied information given during an interview by the expert. Typically, experts talk about the way things work now, what is wrong, and how things should be done throughout an interview. These categories provide buckets in which to place each of these particular kinds of facts.

7.3 The Interview Process

We recommend following a standard interview process for all types of information sources. Our purpose here is to formalize the process enough to maximize the flow of information and to prevent situations that could cause the interview to fail to meet its objectives. To these ends, we recommend a three-step approach: preparation, conducting the interview, and follow-up.

7.3.1 Preparation

Preparing for an interview is critical when you get only one chance to talk with an expert (i.e., the president of a bank, the general of an army), in order to optimize the time you spend with the information source and to maintain good information flow during the actual interview. We recommend the following steps:

- Select an appropriate interviewee.
- Make an appointment.
- Establish a tentative agenda.
- Review background information.
- Coordinate with your team.

Selecting the interviewee is the first and most important step. Too often, an author does not get the needed information because the right source was not selected. Be concerned about selecting the correct level in an organization, the right kind of document, the best facility to observe, the person who has the required expertise, etc., and check your selection with other team members.

After selecting the interviewee, make an appointment as soon as possible. State the purpose for the meeting and limit the discussion to an hour or less. Information flow will slow down appreciably after an hour, so do not schedule more than can be discussed in that time. If the topic is large, find a way to break the discussion into several 1-hour meetings.

Establish an agenda immediately after making the appointment. Identify the areas of questioning, and write down specific questions, especially all questions that must get answered in order for you to proceed. Review appropriate background material while formulating the agenda. Consult relevant documents and glossaries of terminology, and use the SADT model to help bound the subject matter of your questions. Finally, do not forget to coordinate your preparations with your SADT project team. The last thing you want is to interview someone who just yesterday talked to your colleague about the very same subject you want to discuss. Proper coordination avoids wasting the expert's time and minimizes duplicate efforts by the authors of a model.

7.3.2
Conducting the Interview

The most important thing about conducting the interview is to properly start and maintain the flow of information from the expert to you. We recommend taking time to properly open the interview, gathering information by notes if possible, and closing the interview session in a smooth way. Let us discuss each of these points in more detail.

When you open the interview, be sure to introduce yourself and define the purpose of the meeting. This will help avoid misinterpretations and bound the discussion. Also, confirm the extent of note taking. Assure the expert about confidentiality and that accuracy will be ensured by a subsequent follow-up. Then, start the information flow by posing the first question. Remember that the first question usually sets the stage for the rest of the interview, so ask one that motivates.

Gather information by taking notes on everything (e.g., technical terms, relationships among parts of the system), and by limiting the time you talk. List SADT activities and data, and if you can, sketch a diagram. Keep the information coming by asking questions that qualify and validate the answers to your questions. For example, here are some questions that can help qualify or validate information:

- Can you give me an example?
- How long ago did this happen?
- Are there exceptions to this rule?
- Could you give me some numbers that support what you say?

Above all, do not contradict. Never ask leading questions or questions that require short or "yes" or "no" answers. Instead, record what is said, and ask for summaries and explanations. You will get the most out of the interview if the expert is free to say what she or he wants to say, not what you would like to hear. Watch out for the following situations:

- You have already received sufficient information.
- You are receiving large amounts of inappropriate information.
- You are becoming saturated with information.
- The expert is becoming tired.
- You are having frequent clashes with the expert.

Any of these is good cause for terminating the interview.

When you think the interview should be terminated, close out the meeting in a smooth manner. Summarize the main points, and review things that may have been overlooked or misinterpreted. Schedule another interview if more discussion is required, and get recommendations for other interviews during the close. Last, let the expert know when and how you intend to use the information provided and when you will send material for review.

7.3.3
Follow-Up

Always follow up an interview immediately after meeting with the expert. This way, the expert gets immediate feedback, and you minimize the chance of losing an important point. First review and complete your notes, and then make an SADT Glossary Page as a way of defining new concepts and terminology. Next sketch a diagram, identifying further questions to ask and areas to investigate. As soon as possible, develop a good copy of these diagrams and glossary pages, gather them into a small packet of reviewable material (a kit), and send the kit to the expert.

7.4
Interviewing Tips

Below are a few important tips to keep in mind when conducting the interview. While it may be difficult to remember and apply them all the time, review these points after an interview to see where you can improve your fact-gathering skills:

- Determine whether your information is largely fact or opinion by asking qualifying questions. Also, try to determine how the expert fits into his or her organization. This can help you better interpret and qualify responses.
- Always ask for numbers and units (e.g., time, volume, cost). Numbers and units give credence to what is being said.
- Always ask for sources and destinations of data. Find out their format, required retention time, intended use, what transformations are required, and so on. These specifics will help define what the data is and why it exists.

Here are suggestions that will help maintain the smooth flow and authenticity of information from the expert:

- Let there be times of silence while the expert thinks. Let the expert decide what to say next. Never interrupt by suggesting an answer or asking another question.
- Try not to ask leading, loaded, or self-answering questions, for these give the expert no chance to relay knowledge. And try not to ask controlling questions, for they break the information flow.
- Take notes to keep your mind on the subject matter and to prepare yourself for asking the next question. Do not become a stenographer, however, or you may lose control of the interview.

Like drawing good SADT diagrams and models, interviewing is a very important skill to master. By practicing the techniques in this chapter, you can become a more accomplished interviewer. The better interviewer you are, the easier it will be to acquire the domain knowledge essential for your SADT diagrams and models.

7.5
Summary

Interviewing is the way SADT analysts acquire knowledge about the problem domain. Experts are a common source of information, though documents, observation, and questionnaires are also handy sources. Before conducting an interview on an information source, the SADT analyst determines which type of interview is most appropriate: a fact-finding, problem identification, or solution discussion interview or an SADT Author/Reader Talk. Also before the interview, the analyst should prepare by selecting appropriate interviewees, making appointments, establishing agendas, and reviewing background information. Vital to the actual interview are a proper opening, maintaining information flow, and a proper close. Also, it is good to review your notes immediately after an interview to be sure that they include all the important points discussed.

Suggested Readings

Bravoco, R.: "Supplementary Information concerning Interview Standards and Techniques," *SofTech,* June 1976.

FitzGerald, J., and A. FitzGerald: *Fundamentals of Systems Analysis,* John Wiley, New York, 1973.

Freeman, P.: "Why Johnny Can't Analyze," *Systems Analysis and Design: A Foundation for the 1980s Workshop Proceedings,* September 1980.

Gorden, R.: *Interviewing: Strategy, Techniques and Tactics,* Dorsey Press, 1980.

Hartman, Matthes, and Proeme: *Management Information Systems Handbook,* McGraw-Hill, New York, 1968.

Kahn and Cannel: *The Dynamics of Interviewing Theory, Techniques and Cases,* John Wiley, New York, 1966.

Mumford, E.: *Designing Human Systems for New Technology,* Manchester Business School, Manchester, England, 1983.

Parkin, A.: *Systems Analysis,* Little, Brown, Boston, 1980.

SofTech, Inc.: "Practical Aspects of Data Collection, IDEF0 Architects Guide," SofTech Deliverable no. 7500-10, September 1979.

Weinberg, G.: *Rethinking Systems Analysis and Design,* Little Brown, Boston, 1982.

Chapter

8

Starting a Model

Starting an SADT model means creating A0 and A-0 diagrams appropriate for review. These two diagrams tell a complete story about the system under study at a very high level of detail. To do this, the analyst makes an initial attempt to develop a single decomposition of the system and to summarize this decomposition. The decomposition, the A0 diagram, highlights the most important activities and things of the system. The summary, the A-0 diagram, treats the system as a black box by giving it a name and defining its most important inputs, controls, outputs, and (possibly) mechanisms.

8.1 The Steps

To start a model, an SADT analyst prepares, gathers information, decomposes the subject, and summarizes the decomposition. Preparation involves selecting the purpose of the model (e.g., describe how the machine shop makes parts), the viewpoint from which the model is being presented (e.g., foreman, machinist), the kind of model being made (e.g., process flow), and the intended use of the model after it has been made and verified (e.g., train a new machinist). Preparation thus sets the stage so that information can be gathered with the least amount of effort.

Gathering information involves any combination of reading documents, observing existing operations, surveying a community of experts with a questionaire, interviewing one or more experts, using one's own knowledge, and inventing a scenario that will be later corrected. Chapter 7 describes in detail how each of these techniques is used to gather information during a modeling effort.

When decomposing the subject, one concentrates first on the data that passes to and from the whole system. Decomposing the entire system begins with listing its major data classes and its major activities. By doing so, the overall operations of the system are mentally reviewed. Normal situations, pathological situations, feedbacks, and error cases are all considered. Then, these lists are annotated to show the major classes, or aggregates of both data and function. Finally, the annotated lists are used to produce an A0 diagram, which is then summarized by the A-0 diagram.

8.2 Determining Purpose and Viewpoint

At some point very early in the development of a model, the purpose and viewpoint of the model are determined. The purpose is defined by stating the questions to be answered by the model, and the viewpoint is defined by selecting the proper perspective from which to describe the system. Sometimes purpose and viewpoint can be defined before any diagrams are drawn. For example, it is possible to identify the purpose of the Prototype Machine Shop model, because it mentioned in the problem statement: "Understand the tasks of each person in the machine shop well enough to describe them to new personnel." We strongly recommend drafting a purpose and selecting a viewpoint for a new model as soon as possible. But first, try writing down some specific questions the model will answer, to ensure your purpose is accurately stated, and try to envision the system from several alternative viewpoints before selecting one. Figure 8-1 shows how this was done for the Prototype Machine Shop problem.

It is sometimes the case that determining purpose and viewpoint at the very start of a modeling effort can be extremely difficult. When this happens we suggest first trying to generate data and activity lists and possibly drawing an A0 diagram. By doing so, you will get a feel for the system and will start to ascertain whether the A0 diagram is describing the system from a useful perspective. You may have to actually try several alternate A0 diagrams before you feel confident enough to state a useful purpose and viewpoint. We strongly recommend this technique whenever you are unsure of the proper purpose and

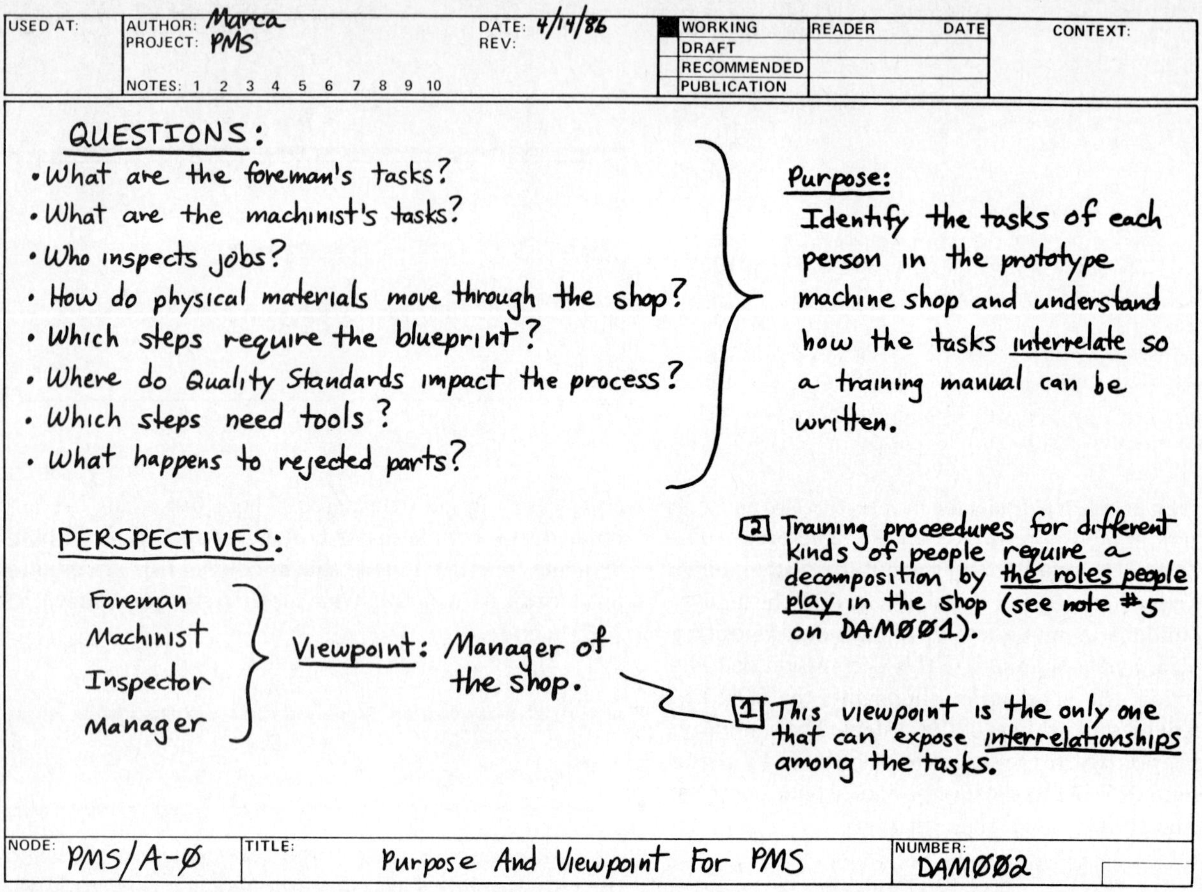

FIGURE 8-1 Determining Purpose and Viewpoint

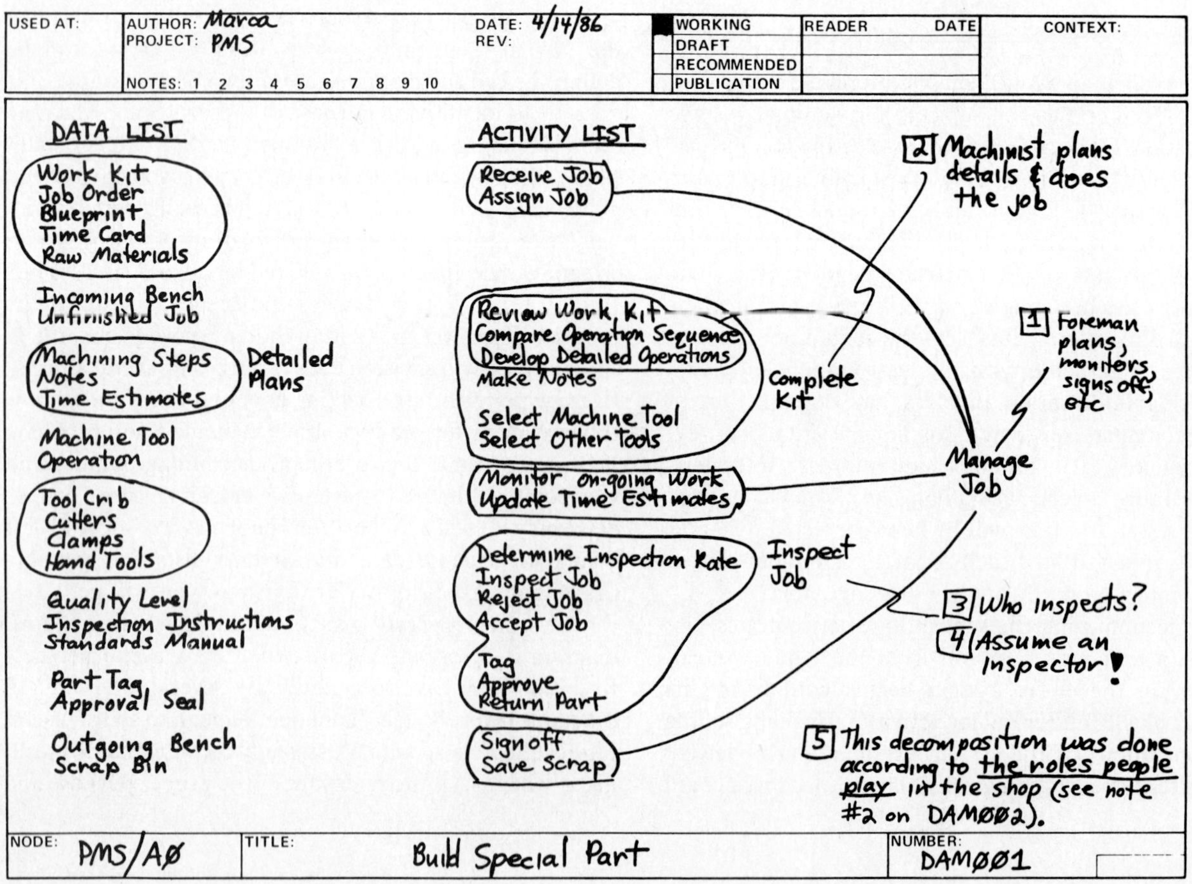

FIGURE 8-2 Making Data and Activity Lists

perspective for your work. For example, Workshops 1 and 2 in Part V demonstrate how generating initial data and activity lists can be a valuable prerequisite to determining purpose and viewpoint.

8.3 Generating a Data List

SADT calls the lists of system things generated during model development "Data Lists," "data" therefore being synonomous with "thing." We will thus use the term "Data List" in our discussion of modeling with SADT. Generating Data Lists is the starting point for developing each diagram in an SADT activity model. The rule is to list data first and functions second. Start your diagram by first thinking of all the major groups and categories of data used and generated by the system. And don't be too selective; list all reasonable possibilities. When in doubt, list what comes to your mind, for it is better to list too much than to be incomplete in the analysis. For example, notice that in Figure 8-2 many detailed things were listed, even though the analyst tried to produce a diagram that summarizes the entire shop.

Current analytical methods too often concentrate on function to the exclusion of data. By first writing a Data List, one avoids jumping immediately into a functional decomposition. Data Lists help strengthen an analysis by touching upon the activities of the system without concentrating on them, avoiding the oversights that are commonly made when one mentally has a preconceived functional breakdown, ensuring that the data gets thought about, and identifying the constraints that dictate the functional decomposition.

SADT activity diagrams represent functional boundaries and their constraints, and constraint is inherent in all systems. By exposing constraints first, the natural required structure of a system is exposed. Without constraints, an SADT activity diagram would be nothing more than a dataflow picture. Without constraint arrows, the diagrams could not tell its readers why the analyst chose a particular decomposition. Since SADT distinguishes input arrows from control arrows (the information vital to explaining decomposition), SADT diagrams clearly explain the system under study and why it is being decomposed in a particular way.

8.4 Generating an Activity List

Once a Data List is built, use it to create an Activity List. To do this, envision the activities of the system that use each class or collection of data. Remember, several separate classes of data are often used by a single function. Indicate which classes or collections of data are used by particular functions. This will often indicate data of a similar nature, which can be aggregated into metaclasses. As you get further along in your list, mentally review your initial thoughts to see if they were reasonable, because initial thoughts sometimes do not fit within the stated purpose and viewpoint of the model. On the other hand, do not automatically discount initial ideas that do not seem immediately correct. Rethinking can provide insights into the working of a system not obvious at first glance, and you may decide to reinstate an original idea after you have drawn several other diagrams.

The activities are listed on the same page with the Data List. We suggest not trying to make associations between activities when making the initial list. Instead, try to concentrate first on a particular activity and how it relates to a group of data. Try also to pick functions that seem appropriate to work on the very general data classes you just listed. For example, notice in Figure 8-2 the amount of clustering done to aggregate detailed functions into more general ones. Further, take care to not be too selective by writing down activities even if you question whether or not they are done by the system. Borderline functions (i.e., activities that could be done by the system or by its environment) are very hard to initially pin down as being in or out of a model.

Now cluster all the activities into aggregates. Aim for three to six functional groupings. Try to balance the groupings so each seems to be at the same level of complexity, each contains roughly the same "amount" of functionality, and each collects together functions with similar operations and purpose. For example, notice in Figure 8-2 that the initial list of system activities was clustered to obtain a trio of high-level functions. Clustering is not always easy. You may find it difficult to decide upon the "best" way to cluster functions together at a particular level in a model. Don't worry, because a bad clustering will show its weakness when it is decomposed. When this happens, you can always go back and try alternate clusterings.

8.5 Drawing the A0 Diagram

The data and activity lists provide the initial content of the diagram. Now, form must be given to this content in order to properly describe the system. One does this in SADT by drawing a diagram. We recommend that first-time users of SADT follow a particular order: (1) lay out the boxes, (2) draw primary constraint arrows, (3) draw external arrows, and (4) draw all remaining arrows. After a while, you will become experienced enough to deviate from this procedure and to draw the boxes and arrows

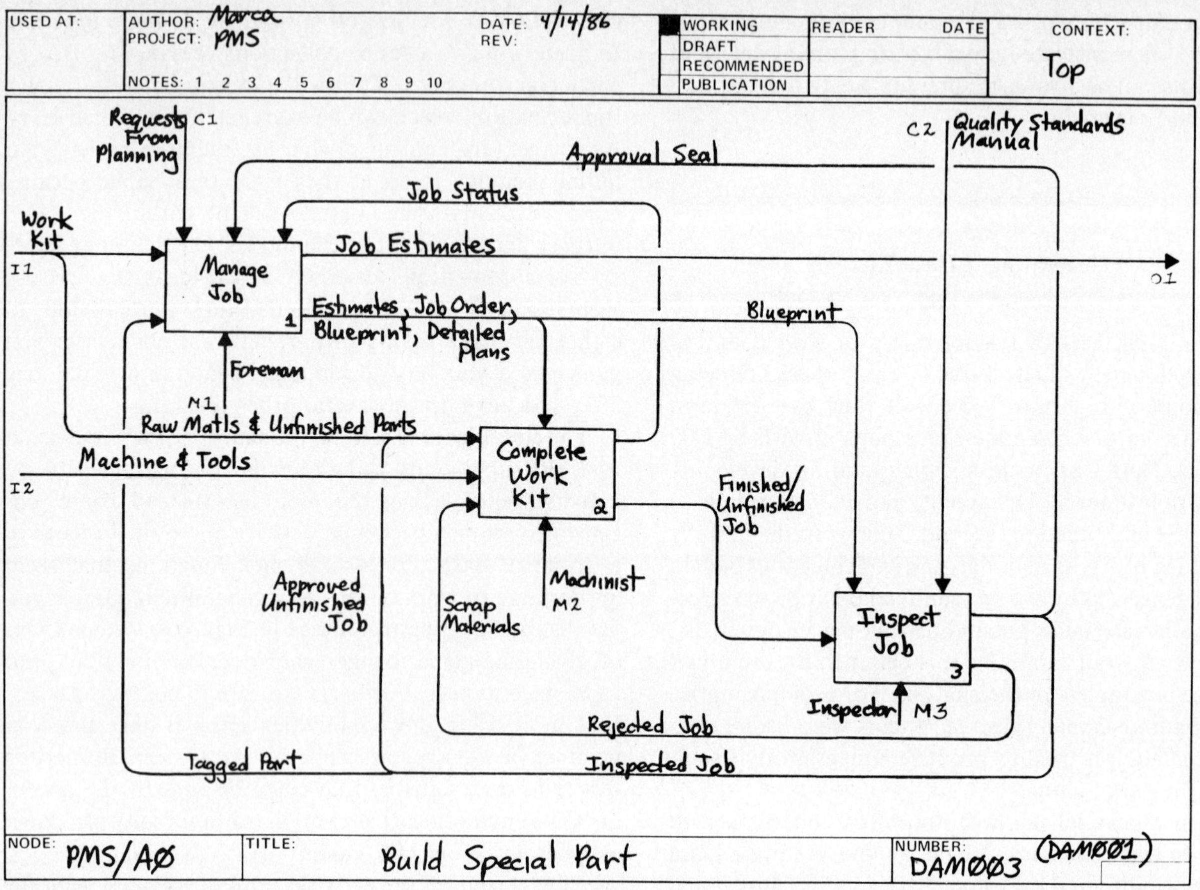

FIGURE 8-3 An A0 Diagram

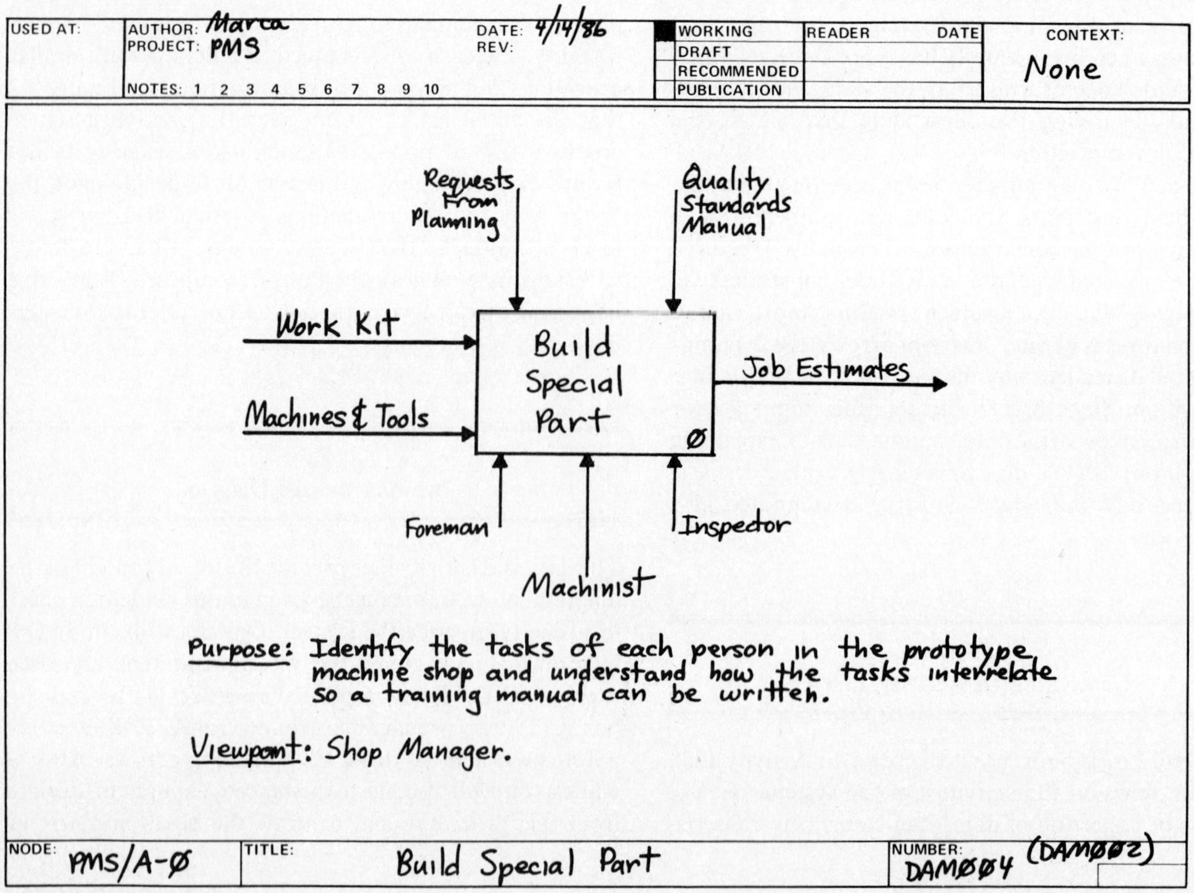

FIGURE 8-4 An A-0 Diagram

according to the message you want to convey with the diagram.

Box layout is the most important step in drawing a diagram. Boxes are placed according to their dominance (e.g., perceived importance or required sequence). The most dominant box is usually put in the upper left-hand corner and the least dominant box in the lower right-hand corner. This results in a layout in which the more dominant boxes constrain the less dominant boxes in the diagram, creating a "staircase" pattern. Dominance is critical in telling a clear story. For example, it makes little sense to talk about inspecting a job before the part is made.

The primary constraint arrows, the second most important pieces of information, are drawn next. They give the reasons why the subject of the diagram has been decomposed into the three to six system activities which are the boxes. For example, the ***quality standards manual*** is a major influence on how a partially machined part is inspected. When drawing these arrows, review each one to see if it really does impact the way the subject has been decomposed. Also, check the data list to see if other constaint arrows are missing. If so, you may want to check the way you did the decomposition.

Primary constraint arrows are always external arrows, that is, they represent data coming from the environment immediately surrounding the diagram. The next step in making the diagram is to draw in all the remaining external arrows, and giving each of them an associated ICOM code. This way, all data coming into and out of the system is accounted for in the picture. Forgetting an external arrow is an interface error, one of the most common errors made during systems analysis. It is very easy to forget data that crosses an interface when you are involved in decomposing a subject, because it is easy to focus on the details. Drawing all the external arrows first will improve the accuracy of the diagram by including all data that crosses the interface.

Finally, draw all the other arrows that detail the overall workings of the system. Draw any remaining constraints among boxes first. For example, the reviewed ***blueprint*** influences part inspection. Develop the major dataflow path second. For example, Figure 8-2 shows ***raw materials*** being worked on according to ***job plans,*** and jobs being inspected sometimes several times according to the ***blueprint.*** Consider pathalogical data flows (e.g., error cases) third, and detail all feedbacks in the data flow fourth. For example, ***rejected jobs*** are recycled as ***scrap materials.*** Draw all error feedbacks last.

One very important modeling concept should be raised at this time. Trying to draw any diagram neatly the first time is impractical. Sketching is the best way to initially give form to data and activities. While the analyst is sketching, a story begins to unfold in his or her mind. Though fuzzy at first, the story sharpens as the sketch is completed. Often, arrows are relabeled, boxes are renamed, arrows get crossed out, and boxes get moved in the process. So we recommend sketching a diagram first and then redrawing that same diagram to clarify your thinking, sharpen your story, and produce a description that can be reviewed by others.

8.6 Summarizing the A0 Diagram

Summarizing is a critical last step when starting a model. Remember, every SADT diagram has a parent diagram that contains its context, and this context is a single box with a set of input, control, and output arrows. The top diagram of the model (i.e., the A0 diagram) is no exception. Its context is the A-0 diagram—the summary for the entire model. The A-0 diagram has several purposes. First, it states the overall function of the system. For example, the box in Figure 8-4 entitled ***build special part*** clearly indicates what the shop does. Second, it provides the set of major data classes or collections the system uses and produces. For example, the ***quality standards manual*** is used to control the process of making an ***outgoing part.*** Third, the A-0 diagram shows the relationships among the major data classes by distinguishing them from each other. For example, the ***work kit*** is considered an input, something that is changed by the process, while the ***quality standards manual*** controls how the shop makes quality parts. The A-0 diagram is therefore a complete summary of the system under study.

An A-0 diagram is created by using the information already captured in the A0 diagram. A single large box is first drawn in the middle of the SADT form, and named with the title of the A0 diagram. At this point, it helps to verify that the title is a good overall description of the system. Once the box appears to be correct, every external arrow on the A0 diagram is drawn on the A-0 diagram touching the appropriate side of the box. It helps here to verify that the label for each arrow is appropriate to describe what passes between the system and its environment. After you draw the input and output arrows, step back for a moment and check the accuracy of the overall dataflow now described. After drawing the control arrows, verify that they are indeed the ones that govern the way the system transforms its inputs. Finally, write down the purpose and viewpoint of the model underneath the large box, and check them against the story told by the box and its arrows.

You will often find that this summarization, a process of mental abstraction, will clarify your system description, because, when making the summary, arrow labels are revised to more accurately name the data passed between the system and its environment. Also during summarization, arrows are often clustered together to simplify the message of the model. This shows up as arrows which split into their constituent parts on the A0 diagram. For example, the A0 diagram in Figure 8-3 says that ***raw materials & unfinished parts*** go into making up a ***work kit.***

The drawing of the A-0 diagram signals the end of starting a model. At this point, a first attempt has been made to summarize and describe the overall workings of the system and how the system fits into its environment. Even though they have limited detail, the A-0 and A0 diagrams tell a complete story because they introduce the major inputs, controls, outputs, and activities of the system. The summary provided by A-0 and A0 diagram pair is, therefore, the goal an analyst tries to achieve when starting an SADT model.

8.7 Summary

To start a model, an SADT analyst prepares, gathers information, decomposes the subject, and summarizes the decomposition. During preparation the purpose of the model is selected, the viewpoint of the model is determined, and the intended use of the model is decided upon. Preparation sets the stage so that information can be gathered with the least amount of effort. Decomposition means first creating a Data List, second creating an Activity List, and third drawing an A0 diagram. The final step involves summarizing the A0 diagram with an A-0 diagram, which contains major system inputs, outputs, and controls, as well as the stated purpose and viewpoint of the model.

Suggested Readings

DeMarco, T.: *Structured Analysis and System Specification,* Yourdon Press, New York, 1978.

DeMarco, T.: "Specification Modeling," *Guide50 Proceedings,* 1980.

Greenspan, S., and J. Mylopoulos: "Capturing More World Knowledge in the Requirements Specification," *6th International Conference on Software Engineering Proceedings,* September 1982.

Jackson, M.: *System Development,* Prentice-Hall, Englewood Cliffs, N.J., 1983.

Marca, D.: *Applying Software Engineering Principles,* Little, Brown, Boston, 1984.

Orr, K.: *Structured Systems Development,* Yourdon Press, New York, 1977.

Ross, D.: "Some Further Observations about Structured Analysis," SofTech Technical Report no. 9031-6, January 1976.

———: "An Essay on Activity Diagramming," SofTech Technical Report no. 7104, November 1976.

Rubinstein, M.: *Patterns of Problem Solving,* Prentice-Hall, Englewood Cliffs, N.J., 1975

SofTech, Inc.: "IDEF0 Author's Guide to Activity Diagrams," SofTech Deliverable no. 7500-13, September 1979.

———: "Integrated Computer-Aided Manufacturing (ICAM) Final Report: IDEF0 Functional Modeling," Contract no. F33615-78-C-5158, January 1981.

Weinberg, G.: *Rethinking Systems Analysis and Design,* Little, Brown, Boston, 1982.

Chapter

9

Continuing a Model

Starting a model involves the techniques of defining subject, purpose, and viewpoint; bounding the subject; drawing both top-level and summary diagrams; listing data and activities; clustering activities into boxes; and forming relationships among boxes using the Data List. Continuing a model builds upon these techniques, and extends the model to a finer level of detail. This requires developing a single diagram for possibly every box on the top-level diagram, developing diagrams for each box in those new diagrams, and repeating the process until the model describes its subject to a sufficient level of detail to meet its purpose. Continuing a model is thus a recursive process.

9.1 Decomposing a Bounded Subject

Decomposing a model is very similar to, but simpler than, starting a model. Why? Because an SADT analyst always decomposes a model by staying within the context defined by a box and its arrows on one of the diagrams of the model. This boundary, called a "subject boundary," is defined in two ways. First, the subject, purpose, and viewpoint of each new diagram are already defined on the A0 diagram. Second, each box that is decomposed into a new diagram is already a bounded subject. In other words, it identifies the function taking place and all the data that this function requires or generates. Drawing a diagram from this information is simpler because a data list is developed from the arrows coming into and going out of the box and because an activity list is a detailed expansion of the box name. The process of decomposing a bounded subject goes as follows:

1. Pick a box on a diagram.
2. Consider the subject defined by the box.
3. Create the new diagram.
4. Criticize the new diagram.
5. Try alternate decompositions.
6. Revise the new diagram.
7. Revise all related diagrams.

Steps 1 to 3 constitute the creative part of the process. Here is where the analyst concentrates on unveiling new information about the subject to a finer level of detail, with a goal of clear communication. Steps 4 to 7 make up the self-review part of the process, in which, after developing the new diagram, the analyst reviews what the diagram says and also reviews it in relation to its parent diagram. Changes are then made to the diagram and related diagrams accordingly, with a goal of clear communication *to others.* We will discuss steps 1 to 3 in more detail in this chapter. Chapter 10 will cover the steps involved with self-review and subsequent revision.

9.1.1 Picking a Box

Decomposition is started by first reading the A0 diagram to find the "most revealing" box. This box is the one that, when chosen and decomposed, will illuminate many aspects of the A0 diagram and will have much influence on future decompositions of other boxes on the A0 diagram. Ideally, selecting the most revealing box is a balanced decision that accounts for: dominance, functional complexity, and understanding. The "best" box may not be the one that is most difficult to understand. The "best" box to decompose first is the one which will give you the most insight into the system under study.

For example, consider Figure 9-1, the A0 diagram for the Prototype Machine Shop model. It has three boxes: ***manage job, complete work kit,*** and ***inspect job. Manage job*** appears to be the most complex since it is the most

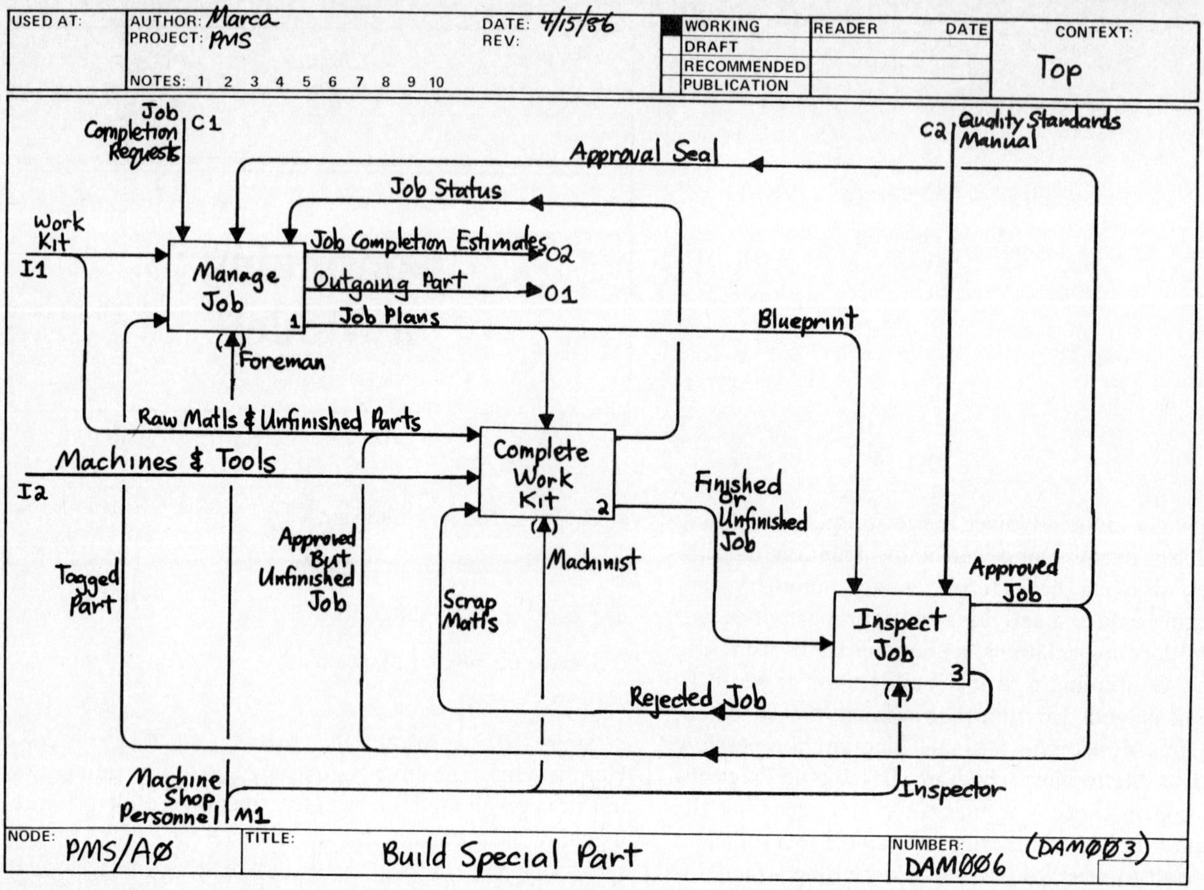

FIGURE 9-1 An A0 Diagram Ready to Decompose

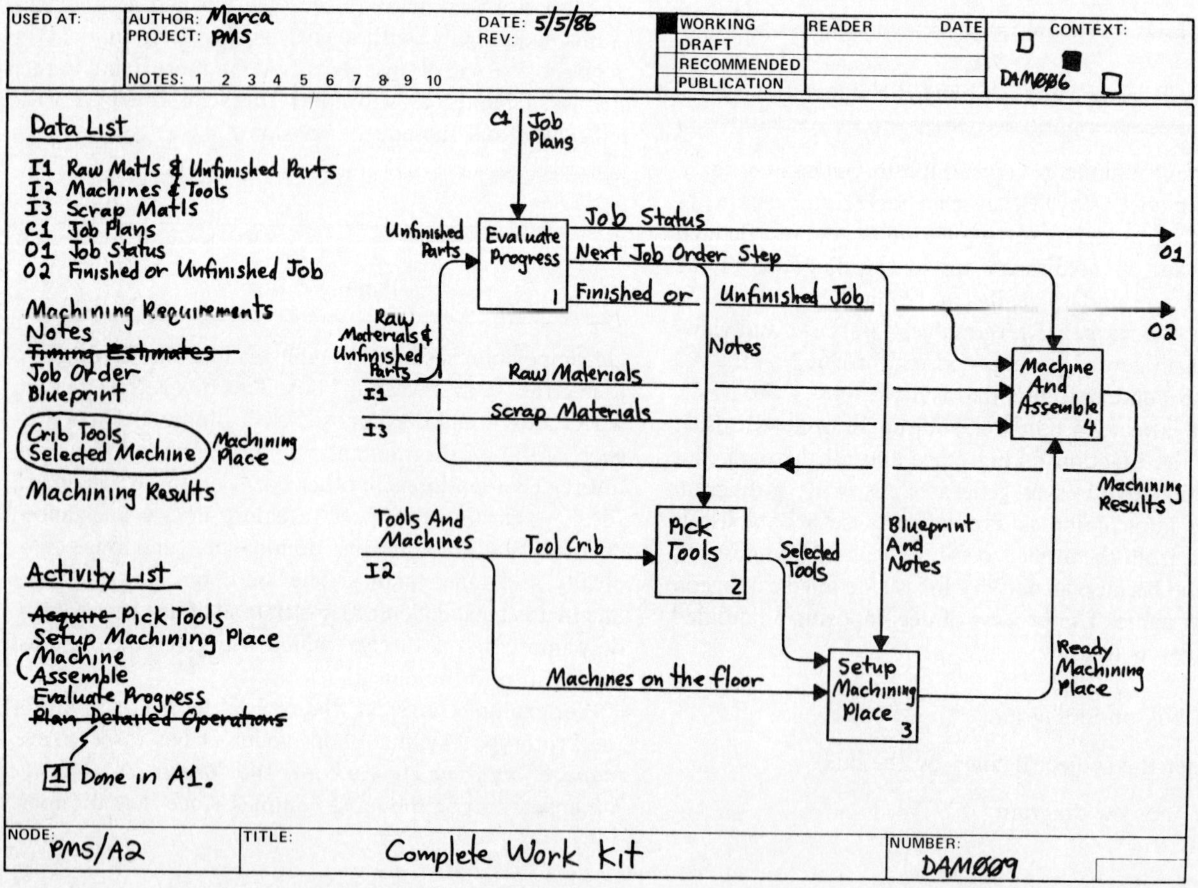

FIGURE 9-2 Sketching the Decomposition of a Box

dominant box and has the most associated arrows. But, upon close inspection, it can be seen that many of its arrows are simple feedbacks or outputs. ***Inspect job*** seems to be the simplest activity, so decomposing it might not give us much new information. Thus, these two boxes are not good candidates for a first decomposition. ***Complete work kit*** seems most interesting, because it participates in many feedback loops and has a wide variety of inputs, controls, and outputs. After understanding how the ***machinist*** completes a ***work kit,*** we expect to be in the best position to further decompose the rest of the model.

9.1.2
Considering the Subject Defined by the Box

Box 2, ***complete work kit,*** now becomes the sole subject to be decomposed. To do this, we first quickly review the summary diagram (please see the A-0 diagram in Figure 8-4) and mentally review the purpose and viewpoint of the model. By doing so, we see that we must describe ***complete work kit*** from the ***manager***'s viewpoint in such a way that training procedures can be written for all new machine shop personnel. Also we scan box 2 on the A0 diagram and its arrows in order to focus on the particulars. For example, the mechanism arrow labeled ***machinist*** indicates that we can decompose this box emphasizing just what machinists do.

Now we develop an initial Data List from all the arrows touching the box, using ICOM's to ensure no interface data gets forgotten. For example, ***job plans, machines & tools,*** and ***scrap parts*** are some of the entries of the initial data list (see the upper left-hand column in Figure 9-2). This list is now refined by decomposing those initial data items or defining new but closely related data items. For example, ***notes*** was thought of when ***job plans*** was considered further. Third, we develop an Activity List from the Data List, keeping always within the functionality described by the box on the top diagram. For example, notice that ***pick tools, setup machining place, machine, assemble,*** and ***evaluate progress*** all seem to be activities that a ***machinist*** does to complete work kits.

Remember to strive for an understandable amount of complexity by clustering activities and data: four or five activities is usually best. Too much data and too many activities often attempt to say too much. This results in a cluttered diagram. Conversely, too few activities often say very little and make a diagram marginally worthwhile. When you are sure the balance is right, review the words you wrote and see if they fall completely within the subject defined by the box and the boundary arrows appearing on the parent diagram. After that, you are ready to draw the diagram.

9.1.3
Creating the New Diagram

The new diagram is created in the same manner as the A0 and A-0 diagrams: boxes are positioned and drawn by their dominance (i.e., by the way the boxes constrain each other), primary constraint arrows are drawn second, external arrows are drawn third, and internal arrows are drawn last. We again stress here that this procedure, like the one given in Chapter 8, is a guideline for first-time SADT users. In general, SADT diagrams are drawn according to the message they are conveying. Those familiar with SADT let the content of the diagram influence the way they draw it.

For example, Figure 9-2 illustrates the result of an SADT author who has just finished sketching a diagram of a bounded subject. Here are some important points about that diagram: The ***next job order step*** arrow is a vital piece of information, for it defines what the next machining step must do. Thus, since it constrains every other activity, ***evaluate progress,*** the producer of ***next job order step,*** is the most dominant box for this diagram. ***Job plans (C1)*** is definitely required before anything can happen, because the contents of this external arrow define the actual sequence of machining steps. Thus it is a very important constraint arrow. Also, the author kept in mind the dataflow occurring in the diagram, especially dataflow feedbacks. For example, notice that ***machining results*** is an input feedback from ***machine and assemble*** to ***evaluate progress.*** This means that a ***machining result*** is an ***unfinished part*** which is evaluated after each ***job step***—exactly what happens in the shop.

9.2
Catching Interface Errors

Engineers know that a large percentage of errors happen at interfaces. For example, a cracked gasket may cause a sewer pipe connection to leak, a computer chip with one bent pin could fail when put in a circuit board, or too tight a terminal screw could cause an electrical wire to break. Similarly, errors frequently occur at the interface points in models. For SADT, the interface points are where diagrams connect to their parent. That is why it is prudent to always take care to correctly connect every decomposition to its parent using ICOM's.

The sample decomposition given in this chapter showed you one way to do this: Write down the names and associated ICOM's for all the arrows that make up the subject boundary before generating a Data List. Doing this will lessen the likelihood of forgetting a portion of the subject boundary when you are immersed in doing the decomposition. (Chapter 3 presents a second way.) After making the decomposition, refer back to the appropriate box on the parent diagram, and connect every external arrow of the new diagram to the corresponding arrow that touches the box. Doing this verifies that no required connection was overlooked. We recommend using both techniques as a way of double-checking interface connections between diagrams.

9.3 Arrow Sketching Principles and Techniques

Arrows express relationships among boxes which must be true. Arrows are not drawn to express sequences which could be followed. Instead, they express relationships among boxes independent of potential sequencing. For example, the diagram of Figure 9-2 does not restrict possible sequencing situations. In particular, ***pick tools*** may or may not happen before ***setup machining place*** depending on the nature of the ***next job order step.*** Also, once the ***ready machine place*** exists, ***machine and assemble*** can happen repeatedly until the ***next job order step*** must be done. These triggerings lead to varied scenarios, causing boxes to "activate" at different times for different purposes.

You may start to draw an arrow and question whether or not it really belongs in the diagram. Leaving the arrow out of the first sketch of a diagram risks creating an error of omission. Since SADT diagrams are always reviewed by others for mistakes and since missing arrows can never be read by anyone, leaving questionable arrows out of a diagram opens the door to forgetfulness and closes the door to insight. We recommend drawing questionable arrows on a diagram with a note asking your readers if they should be included. The comments you get back will be invaluable because they will usually give you enough new facts to decide whether or not the arrow belongs in the diagram.

SADT arrows are aggregates of things, and thus are capable of potentially "carrying" much data. For example, ***raw materials*** could be just a hunk of steel or could be sheets, rolls, and blocks of wood, plastics, and metals of various kinds. Good SADT analysts do not specify all the things in an arrow. At most, we recommend giving a definition of a new arrow using a Glossary Page and possibly specifying its contents when you decompose the boxes that touch the arrow. This way, you give enough information to readers to allow them to understand your diagram, without wasting your time by specifying too much of the model too early, which would inevitably require vast amounts of rework.

9.4 Summary

The first part of the process of decomposition involves picking a box on a particular diagram and considering the subject defined by that box and its arrows. The detailing of that box is done by making data and activity lists and then drawing the diagram. It is always wise to check ICOM codes throughout decomposition, because interface errors during model development are quite common. also, try to include questionable arrows or boxes on the diagram, and indicate your concerns with notes. This will signal to your readers that you would like help resolving those issues.

Suggested Readings

Alexander, C.: *Notes on the Synthesis of Form,* Harvard University Press, Cambridge, Mass., 1964.

Miller, G.: "The Magical Number Seven, Plus or Minus Two: Some Limits On Our Capacity for Processing Information," *The Psychology Review,* 1956.

Parnas, D.: "On the Criteria to be Used in Decomposing Systems into Modules," *CACM,* December 1972.

Ross, D.: "An Essay on Activity Diagramming," SofTech Deliverable no. 7104, November 1976.

———: "Principles of Structuring," SofTech Technical Paper no. 082, November 1978.

Rubenstein, M.: *Patterns of Problem Solving,* Prentice-Hall, Englewood Cliffs, N.J., 1975.

SofTech, Inc.: "IDEF0 Author's Guide to Creating Activity Diagrams," SofTech Deliverable no. 7500-13, September 1979.

———: "Integrated Computer-Aided Manufacturing (ICAM) Final Report: IDEF0 Functional Modeling," Contract no. F33615-78-C-5158, January 1981.

Thomas, M.: "Functional Decomposition: SADT," Infotech State of the Art Report on Structured Analysis and Design, 1978.

Chapter 10

Self-Review by the Author

Decomposition involves recursively exploding each box on a diagram into a diagram at the next level of detail. Techniques such as listing data and activities, clustering activities into boxes, and forming relationships among boxes using a Data List are used during decomposition. The SADT steps of picking a box, considering its subject, and creating the new diagram guide the analyst through this creative process. The goal of the analyst during these steps is to first clearly communicate to himself or herself.

After creating a decomposition, one critiques it before sending it to others for more extensive review. The goal of this critique step is to communicate clearly to others. Good SADT analysts separate the creative process from the self-review process when they decompose a box. They take the time to step back from their work and productively criticize it, knowing that a few moments of self-review will catch errors that frequently get in the way of receiving useful feedback from their readers. The topics of diagram creation and critique have been separated into Chapters 9 and 10, respectively, to highlight their differences and to emphasize the need for doing both during decomposition. Let us now explore the process of critiquing one's own work.

10.1 The Process of Self-Review

The process of self-review, or critiquing, involves a new focus—quality communication. The creative part of the decomposition process develops a diagram that decomposes a box and its arrows. During that process, the analyst tries to describe the subject to himself or herself. Not surprisingly, the result may be a bit esoteric. Jargon and subtleties naturally creep into one's work. During a critique, the analyst steps back from the work. This distance gives a fresh perspective—one that can expand the story told by a diagram into something understandable not just by the author but by many people. The process of self-review goes as follows:

1. Criticize the new diagram.
2. Try alternate decompositions.
3. Revise the new diagram.
4. Revise all related diagrams.

Often, during critique, alternate decompositions are tried to see if the original sketch was the best way to tell the intended story. Also, critiquing checks a diagram against its parent and all other related diagrams. Necessary changes are then made to all diagrams that came under scrutiny. By the way, much of what an author does in critiquing is similar to what readers do in reviewing. So please also consider the ideas in Chapter 14 about reading diagrams and models, as well as the discussion on SADT notes. Let's now walk through a sample critique.

10.2 Criticizing the New Diagram

A critique is a question-and-answer activity, and thus there are no firm step-by-step rules to follow when critiquing a diagram. Good SADT analysts, however, ask a certain set of questions when critiquing diagrams—questions about the boxes, questions about the connections made with the parent diagram, and questions about the internal arrows—and let the answer to each question guide the rest of the critique. That is, they let the flow of questions and answers dictate how each part of the diagram is reviewed. Figure 10-1 is the result of critiquing the diagram made in Chapter 9. (Notes are indicated in light gray.) Please refer to it as we go through a question and answer session on it.

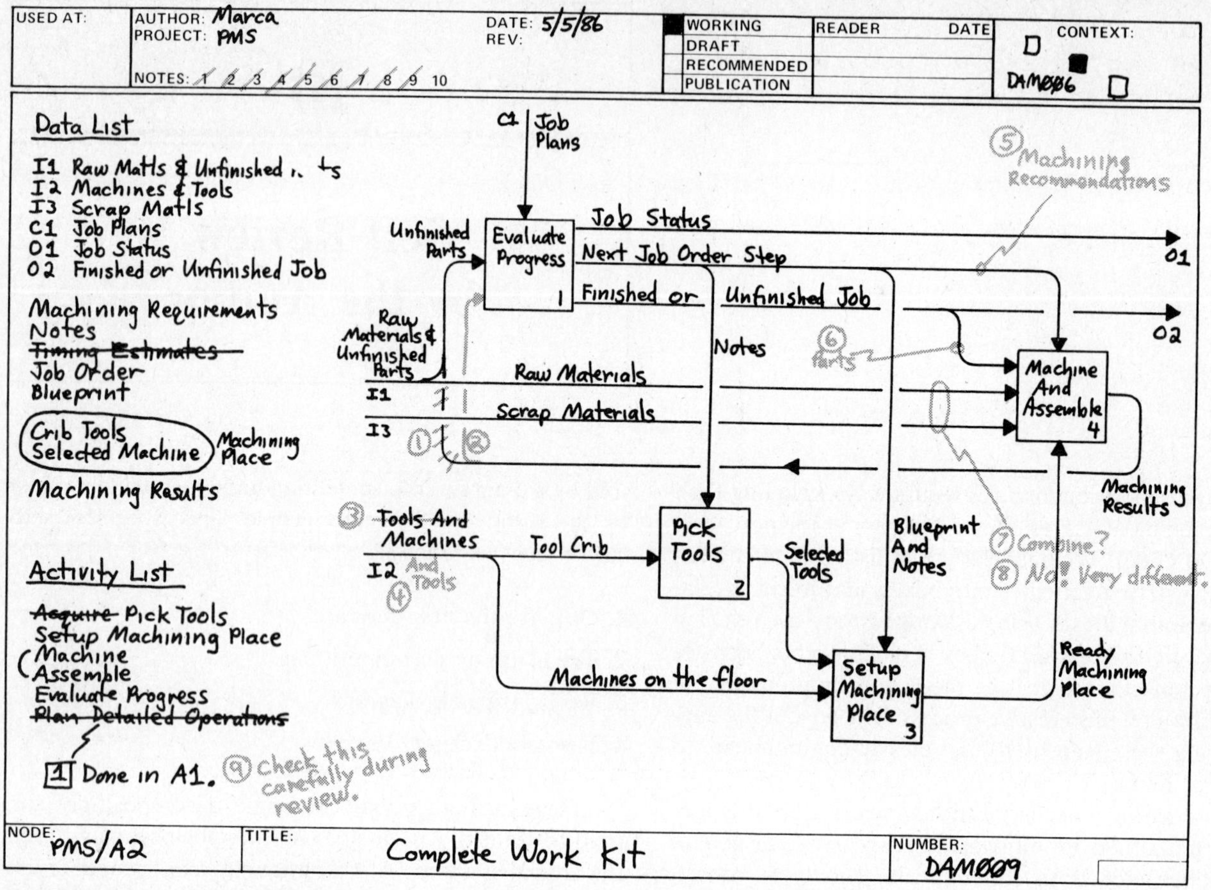

FIGURE 10-1 A Critiqued Diagram

10.2.1
Questioning the Boxes

We first try to critique the boxes of the diagram. We evaluate the functional foundation of the diagram by asking questions like:

- Do the boxes meaningfully decompose the activity?
- Does the diagram look cluttered?
- Do all boxes fit the viewpoint of the model?
- Do the boxes offer a reasonable amount of new information?
- Are all the boxes of a similar level of detail?
- Are all the boxes of similar complexity?
- Does each box detail some aspect of the box on the parent diagram?

For example, though no notes were explicitly put on the diagram of Figure 10-1 regarding its boxes, alternate names could be thought of for *evaluate progress* and *machine and assemble.* Also, the box label *pick tools* might imply too simple an activity, and thus another name like *select hand tools* might give insight into the nature of tool selection.

10.2.2
Questioning the Connections to the Parent

We next try to question the way the diagram connects to its parent. By doing so, we verify how the diagram fits into the model by asking questions like:

- Do all external arrows have ICOM's?
- Do all ICOM's connect arrows having the same meaning?
- Do external arrow labels help the diagram tell its story?
- Does this diagram misinterpret the story told by the parent diagram?

For example, notes 3 and 4 catch a spelling mistake on the *machines & tools* arrow *(I2),* and notes 7 and 8 document a running discussion this particular author had with himself regarding the nature of *scrap* and *raw materials*, an issue that frequently arises in manufacturing processes.

10.2.3 Questioning the Internal Arrows

Internal arrows are usually questioned last, unless there are glaring errors. By doing this step last, we postpone considering the details of the diagram until after the major issues have been resolved. We might ask questions like:

- Are there too many internal arrows?
- Are there boxes without controls?
- Are there boxes without outputs?
- Do constraint arrows properly reflect box dominance?
- Is the main path through the diagram correct?
- Are all important feedbacks described?
- Are all error situations covered?

For example, notes 1 and 2 show that the author reevaluated how well ***machining results*** really fit with the external input of ***raw materials & unfinished parts.*** The decision was that they are not very similar, and so should be separated. Also, notice that the label ***machining recommendations*** adds clarity to what controls the actual ***machine and assemble*** activity.

10.3 Trying Alternate Decompositions

At this point, you have a lot more information about the diagram just sketched. Use the information wisely! Try out some alternate decompositions with these new facts, and see if you can create a stronger diagram. Even if you cannot create an entirely new diagram that tells the intended story better, you can often improve a part of your original sketch. Some techniques are discussed below.

10.3.1 Splitting and Clustering Functions Differently

Sometimes, an analyst has concern over the boxes in a diagram. There are several important characteristics about the boxes in good SADT diagrams. The boxes of a particular diagram should:

- Perform clearly identifiable functions.
- Be of equal complexity.
- Be at the same level of detail.
- Connect simply with the other boxes on the diagram.
- Operate on controls, inputs, and outputs in ways that make sense.
- Work with the other boxes to accomplish the diagram's function.

Try clustering functions and data differently, or try developing a new function list if the original set of boxes makes an awkward decomposition. You can also do this to verify that the original split and cluster were correct. For example, try splitting ***machine and assemble*** into two functions instead of one. You will find that this is quite easy, but that the amount of new information presented is very small. In this case, the original clustering of machining and assembly into one box offered the best description at this level in the model.

Whenever you apply this technique, remember that the criterion for good boxes conflicts with itself. For example, making all boxes exactly equal in complexity may make box connections more complex, or simplifying the connection between two boxes may hide some important fact that should be brought out at this level of detail. Your first priority must always be to describe well the subject being decomposed. Your second priority should be to make good boxes by reaching a balance between the conflicting criteria of box connection complexity and consistent level of detail.

10.3.2 Bundling and Unbundling Arrows Differently

Sometimes, one discovers two arrows that originate and terminate at the same places in a diagram. That is, both arrows start at the same box and end at another box (see Figure 10-2). When this happens, look closely at the two arrows. You may find that they can be combined (i.e., bundled) into one arrow. If you can find a good name that collects the two individual names, bundle the two into one. If it makes more sense to show both arrows, don't bundle them. Bundling hides detail, so don't automatically bundle if hiding the detail will hurt the diagram. For example, notes 7 and 8 in Figure 10-1 document an attempt to bundle ***scrap*** and ***raw materials*** that was abandoned because they were found to be dissimilar things.

You may also discover one arrow describing two very different pieces of data. When this occurs, study the arrow to see whether separating it into two arrows will reveal detail important to the diagram. But be very care-

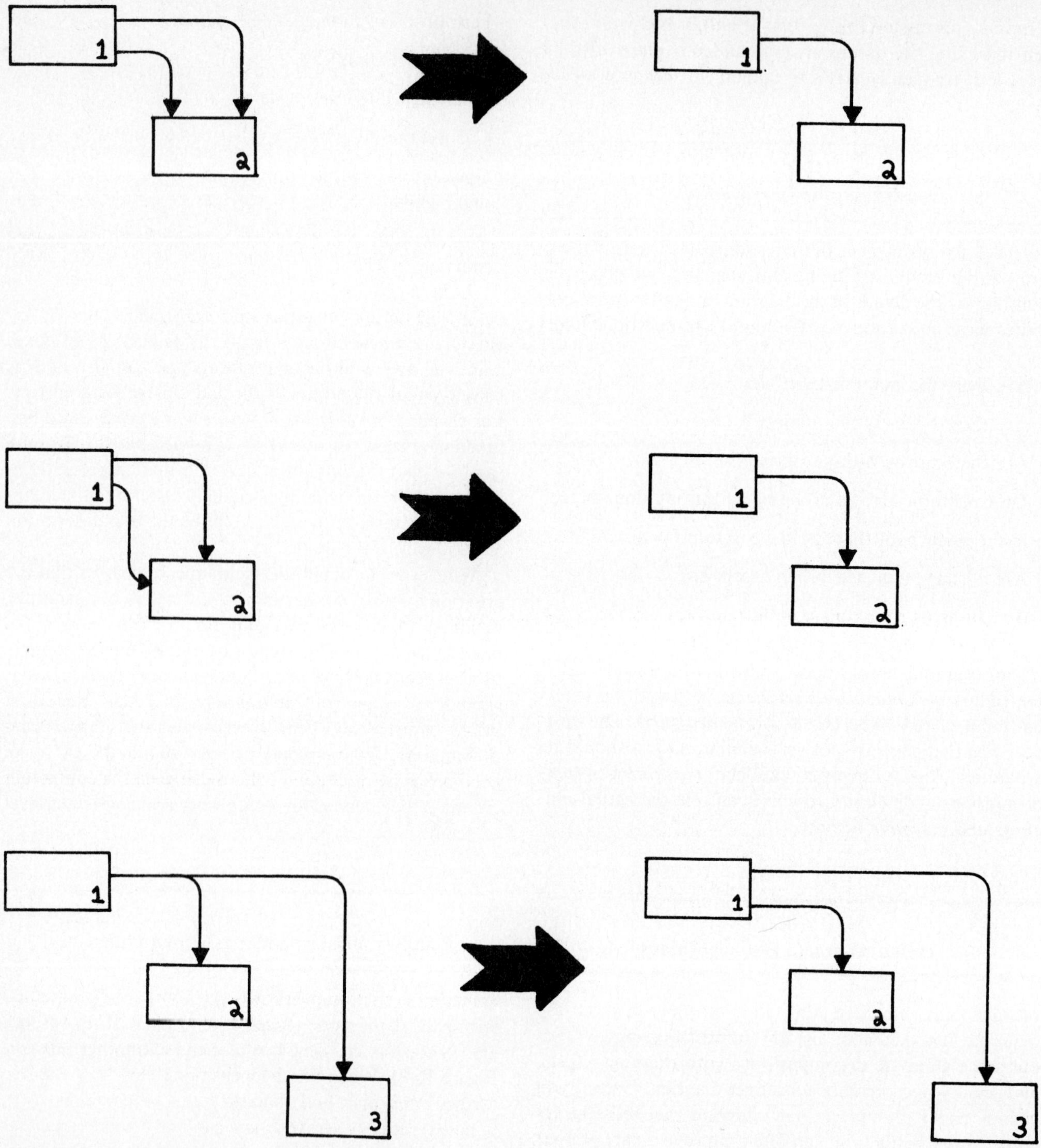

FIGURE 10-2 Examples of Bundling and Unbundling Arrows

ful to make a good tradeoff between presenting more detail and keeping the diagram simple. For example, ***job status*** and ***unfinished job*** were not bundled into one output arrow. The two are very different kinds of items (the former conceptual, the latter actual), and each affects different parts of the parent diagram (***manage job*** and ***inspect job,*** respectively). So keeping them unbundled gives a clearer picture of what results ***evaluate job*** produces.

10.3.3
Trying Test Cases

A good way to critique a diagram is to consider scenarios of its operation. That is, imagine a plausible situation and then walk through the diagram to see how it operates under those conditions. Take notes on the diagram when you walk through a scenario. This way, you can always recreate the scenario, and the information can help you decompose the boxes of the diagram. By doing so, you test both the accuracy and the clarity of the story told by the diagram.

For example, consider what the ***complete work kit*** diagram does when starting a brand new job: the ***job plans*** are evaluated and the first ***job order step*** is selected. This determines which tools are picked and how the ***machining place*** is set up. Then, the ***raw*** and ***scrap materials*** are ***machined and assembled,*** producing a ***machining result.*** This result is then evaluated, and the ***next job order step*** is determined.

10.3.4
Sketching Next-Level Decompositions

Another good way to verify the correctness of a diagram is to decompose one or two of its boxes. (When you do, save the sketches, so that future decompositions of the diagram will be easier to make.) Detailing a part of the new diagram can help determine how well the decomposition is balanced and can spot problems in poor functional distribution among the new boxes.

For example, ***complete work kit*** is a four-box diagram. Sometimes, four-box diagrams can be decomposed differently, producing a five- or six-box diagram with just a little more complexity and better information. However, try decomposing any of the boxes of ***complete work kit*** and you will discover that each is quite involved. Even though ***complete work kit*** has four boxes, each box is quite complex in its own right. It makes sense not to split and cluster these boxes to produce an alternate diagram. Keeping the four boxes and decomposing each into a diagram seems the best way to continue this part of the model.

10.4
Revising the New Diagram

An author typically revises a diagram after spending time questioning it, trying test cases, and sketching alternates. When you do this, be sure to evaluate the dominance of the boxes, choose good box names, make the arrows communicate well, and give explanations. Remember, you are now drawing the diagram to tell an understandable, as well as accurate, story to your Reader Community. Refer to Figure 10-3 as we cover specific techniques of revision in more detail.

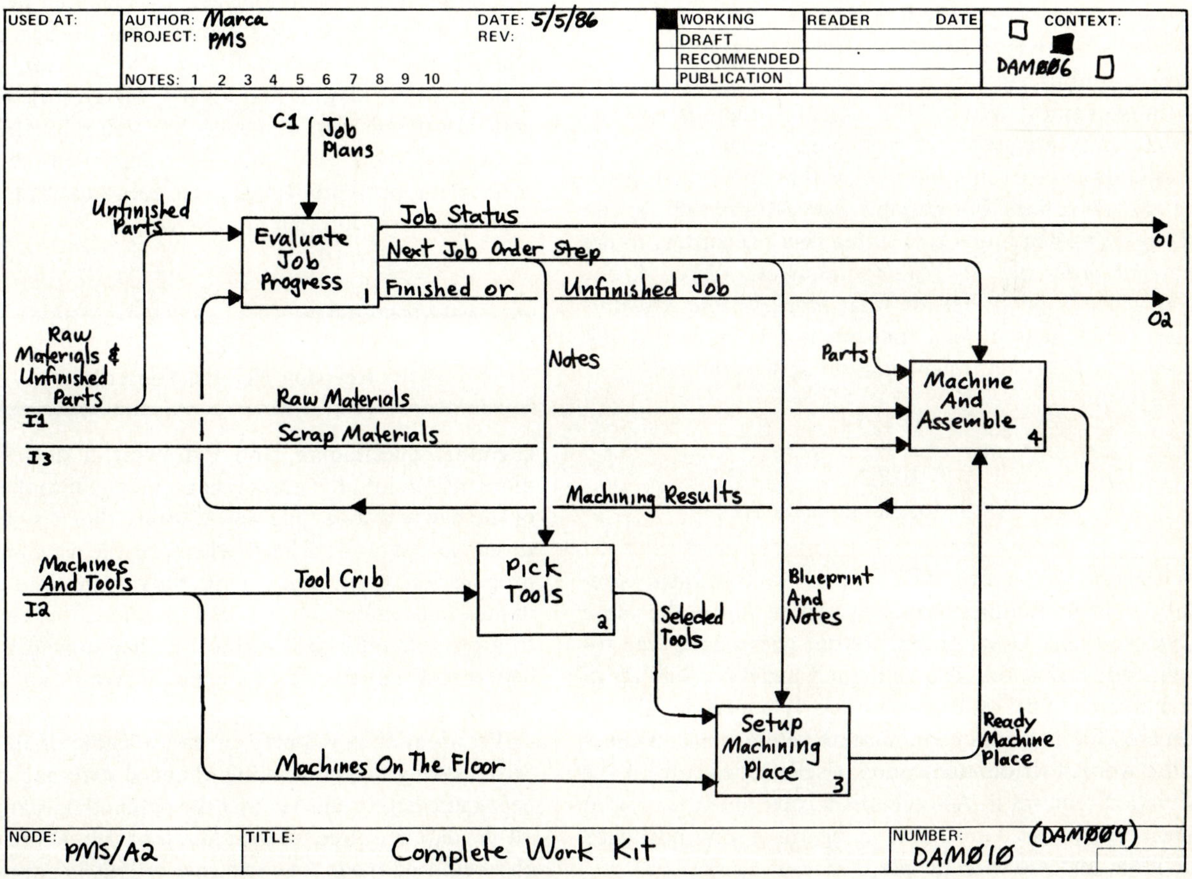

FIGURE 10-3 A Revised Diagram

10.4.1
Reevaluating Dominance

A reader of your diagram will read the box names in order to get a first impression of what the functions do. Be sure your readers will get the right first impression of the diagram by organizing boxes to indicate, as closely as possible, how they influence each other. One commonly practiced technique is to position the box having the most influence in the upper left-hand corner, the box with the least influence in the lower right-hand corner, etc. Remember too that box position can simplify or complicate arrow layout. Sometimes, you will have to make a tradeoff between the conflicting goals of box dominance and arrow layout complexity to achieve a diagram that is easy to read. For example, ***pick tools*** and ***setup machining place*** were determined to be more dominant than ***machine and assemble*** for the simple reason that they often must be done before the machines on the shop floor can be properly used. So even though they produce a mechanism to ***machine and assemble, pick tools*** and ***setup machining place*** are very important steps prior to machining and hence were numbered to reflect that importance.

10.4.2
Good Box Names

When naming boxes, one typically takes care to pick descriptive names. With SADT, however, there is no need to say everything with the box names, because the labels on the arrows surrounding a box will do much to describe what the box does. For example, ***parts, raw materials,*** and ***scrap*** are turned into a ***machining result*** according to the ***next job order step.*** By reading the arrows of box 4, one gets a pretty good feel for what ***machine and assemble*** does. Thus, a more elaborate name for the box is not needed and, in fact, might impede understanding.

10.4.3
Arrows that Communicate Well

When drawing arrows, take care to position them carefully so as to minimize crossings and maximize the space between them. Good graphic layout goes a long way toward improving the readability and understandability of a diagram. Label your arrows clearly and concisely. In general, fewer words communicate better; but don't sacrifice words for communication. Use an extra word if the situation requires it. Also, now is a good time to consider the ideas discussed in Chapter 5 regarding some advanced diagram and model concepts.

Draw arrows in the order of their significance to the message of the diagram. This way, you avoid the tendency to mechanically connect all arrows to all boxes, which gives you the opportunity to critique as you draw. If you are unsure what to draw next, follow the procedure given in Chapter 8: draw constraint arrows first, external arrows second, main path third, and feedbacks last.

Finally, we strongly recommend you avoid labeling arrows with a list of names. Such lists weaken a diagram because they contain less information than a name which summarizes the data collection. (SADT says that good functional decomposition starts first with good data decomposition.) Also, failure to name data aggregates weakens a functional decomposition. Think hard about the names of data collections, and review the data and activity lists of prior diagrams to help you generate possible names.

10.4.4
Explanations

Finally, after the diagram is drawn, explain important points about the diagram with SADT notes or supplemental material. Be careful about these explanations, however: Do not use them as a crutch for bad diagramming. Instead, use then to communicate concepts that just cannot be shown by the box and arrow notation. For example, labeling the ***I1*** arrow in Figure 10-3 as ***materials*** and making a note stating that such items are ***raw materials*** and ***unfinished parts*** is weaker than simply labeling the arrow with those words. On the other hand, attaching a picture of typical unfinished parts goes a long way toward explaining how jobs get completed.

10.5
Revising Related Diagrams

Creating, questioning, and redrawing a diagram gives more insight into the parent diagram, and into the siblings of the newly drawn diagram. Capture that insight as you revise the diagram. (This is where, by the way, bottom-up propagation of information correctly comes into the technique of decomposition.) Three situations force an author to propagate information to other diagrams: changed labels on external arrows, new external arrows, and function migration.

Propagation is required when the name of an external arrow changes. Propagating changed external arrow labels immediately ensures that the parent diagram provides all the data required by the child diagram. Propagation is also required when new external output or input arrows are drawn on the new diagram. These new arrows must, somehow, appear on the parent diagram. There are two ways to do this: draw new arrows on the parent, or bundle the arrows on the new diagram into an aggregate and alter the arrow label on the parent accordingly. Follow the rules of bundling and unbundling arrows when doing this.

Box migration is the most complicated situation. This occurs when a function (usually at a low level in the model) does not, but should, occur in the diagram you are drawing, yet it appears in another diagram in the model, or vice versa. Moving such a function, represented by a box and all its arrows, from one diagram to another is not an easy matter. Usually it requires many changes to arrow labels, causes many new arrows to be drawn, and causes several old arrows to be deleted. Sometimes, a box migration causes other boxes to migrate to different diagrams, producing a ripple effect of changes. Basically, box migration involves a lot of plain hard work and is prone to error if changes are not thoroughly verified.

10.6 Summary

Self-review is the process of critiquing one's own work. Normally, the diagram just drawn is first criticized and commented. Then, alternate decompositions are tried to see if they can better present the intended story. Even if they are inferior to the original, alternate decompositions often bring to light new facts through the bundling and rebundling of both activities and data. Last, after the diagram is checked against its parent and all other related diagrams, all required changes are made to the diagram and possibly to its parent diagram and to other related diagrams.

Suggested Readings

Cohen, G.: "A New Way to Test Writing," 22nd International Technical Communications Conference, 1975.

Elbow, P.: *Writing with Power,* Oxford University Press, Oxford, England, 1982.

Freedman, D., and G. Weinberg: *Walkthroughs, Inspections, and Technical Reviews,* Little, Brown, Boston, 1982.

——— and G. Weinberg: "Reviews, Walkthroughs, and Inspections," *IEEE Transactions on Software Engineering,* vol. 10, no. 1, January 1984.

Hale, R.: "Inspections in Application Development—Introduction and Implementation Guidelines," IBM Report TNL GN20-3814, August 1978.

IBM: "Code Reading, Structured Walkthroughs, and Inspections," IBM Report GE-19-5200, 1976.

Kohli, R.: "High Level Design Inspection Specification," IBM Report TR21.601, July 1975.

Lannon, J.: *Technical Writing,* Little, Brown, Boston, 1982

Yourdon, E.: *Structured Walkthroughs,* Yourdon Press, New York, 1977.

Chapter 11

Diagram Layout Conventions

While evaluating diagram content during the critiquing process, an author also evaluates diagram complexity brought about by poor box and arrow layout. SADT diagram layout conventions help authors draw more readable diagrams, just as structured programming constructs help people write readable code. This is why SADT recommends several kinds of layout conventions: one for just boxes, one for just arrows, and one for box and arrow combinations.

This chapter will cover each of these categories separately, delineate the rules pertaining to each category, and give an example for each rule. While the rules are elaborated and explained to some degree, we strongly recommend you study the examples carefully, for they graphically illustrate the layout rules. Also, we encourage you to scan the figures in this book for other examples of how these layout style guidelines produce easy-to-read SADT diagrams.

11.1 Box Layout Conventions

1. Lay out and number boxes diagonally from upper left to lower right. The resulting "waterfall" pattern is ideal for emphasizing dominance, minimizing arrows bends and crossovers, and simplifying feedbacks.

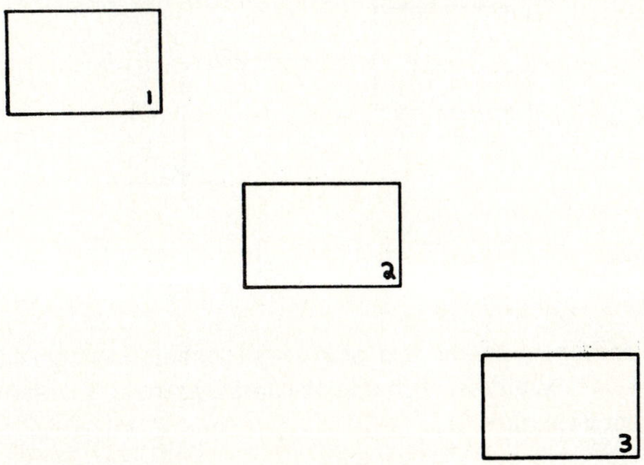

2. Place the number of each box in the lower right-hand corner. A standard location for numbers minimizes the eye movement required to locate them.

3. Write the C-Number of the diagram that decomposes a box below its lower right-hand corner. This minimizes eye movement, but it also visually associates a box number with the diagram that details it.

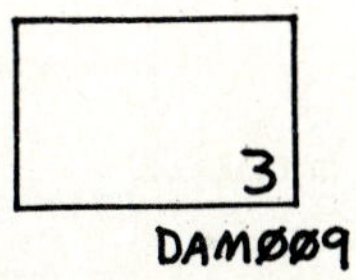

11.2 Arrow Layout Conventions

1. Draw arrows along only vertical and horizontal lines. This visually makes boxes stand out as collection points for arrows, which they are. Such layout also makes easier the task of following arrow paths.

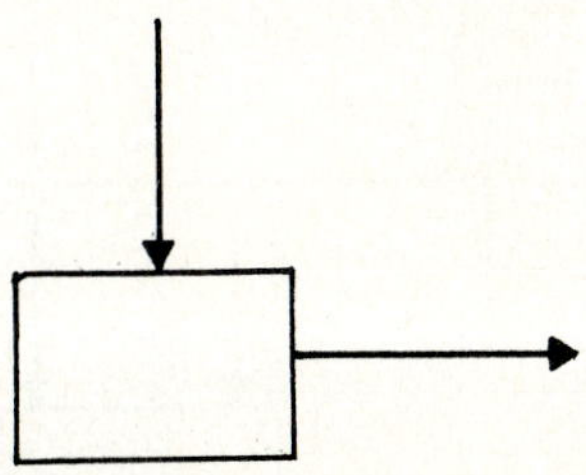

2. Boxes always have control arrows. They may or may not have input arrows. Control arrows constrain and

trigger system functions. Without them, a system cannot operate.

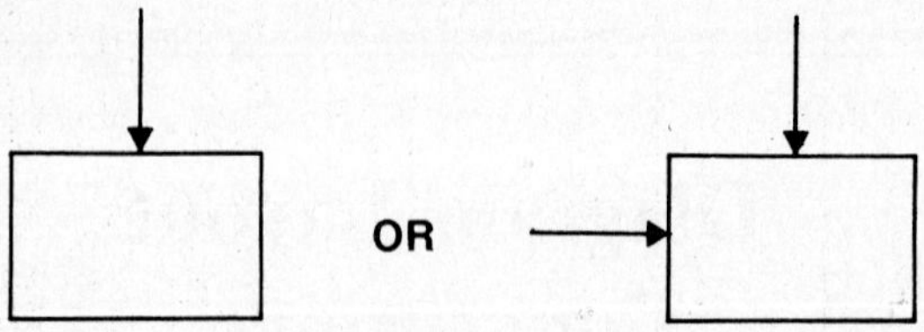

3. Draw only a control arrow when the data serves as both control and input. This reduces the complexity of the overall picture, and makes evident the control nature of the data.

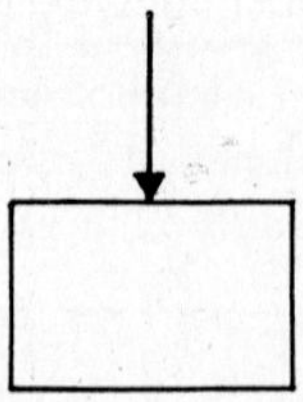

4. Maximize space between parallel arrows. This gives more room for labels, helps the eye distinguish the actual number of arrows, and makes it easier to follow arrow paths.

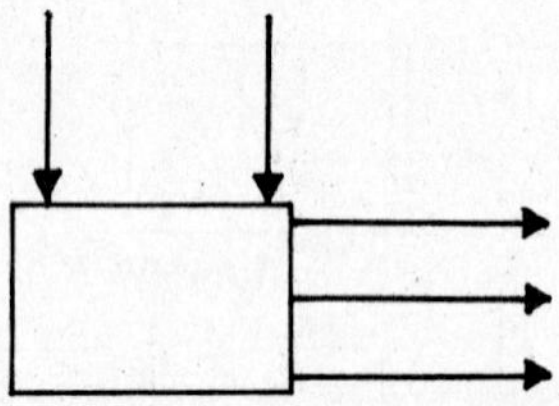

5. Maximize the distance between boxes and arrow corners, and between boxes and arrow intersections. Again, more space means easier reading and less chance to confuse two different arrows.

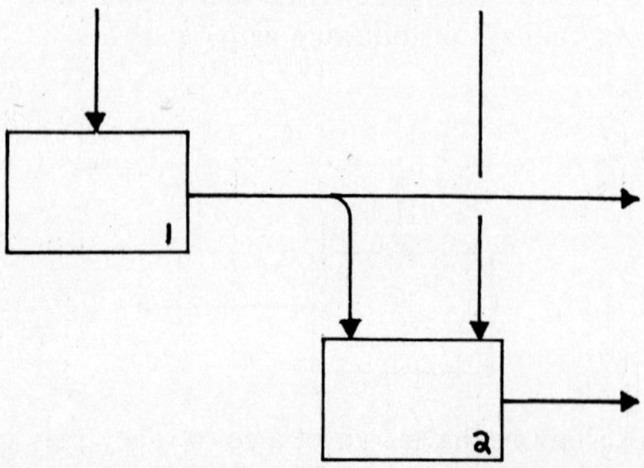

6. Join open-ended arrows when they represent the same data. This correctly shows in a graphic way the single source of similar data.

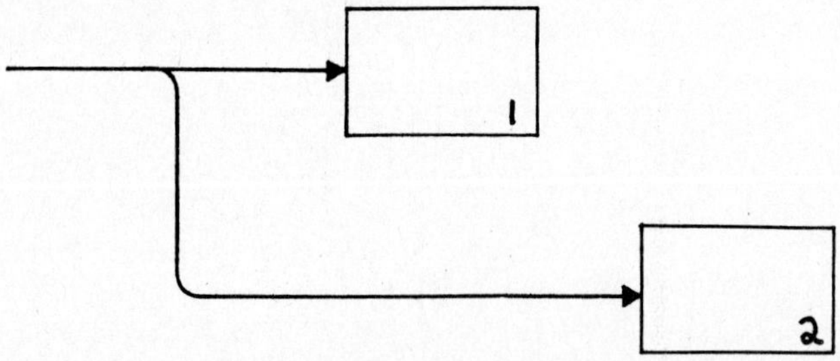

7. Draw cyclic feedbacks to the same box only for emphasis. Typically, one depicts feedback on the diagram that decomposes the box. However, buffers and reusable things sometimes require emphasis—though this is rare.

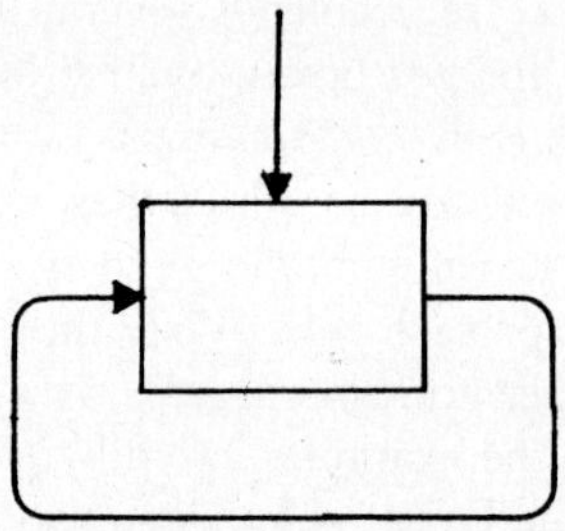

11.3
Box and Arrow Layout Conventions

1. Bundle arrows with the same source or same destination if the data is related. Doing so will yield an aggregate name that better describes the nature of the data.

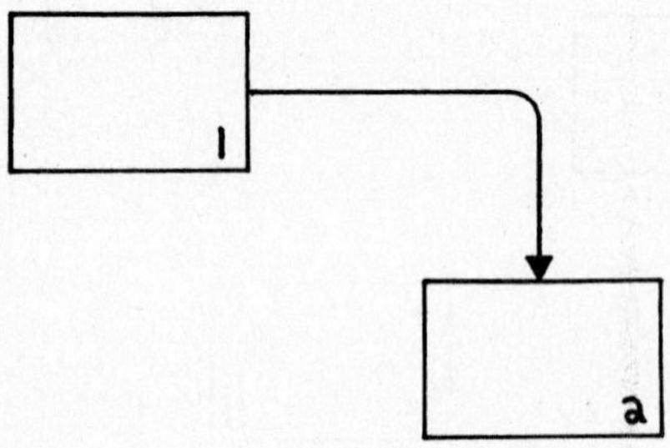

2. Minimize the number of arrows touching any one side of a box, unless, of course, the data is extremely dissimilar in nature.

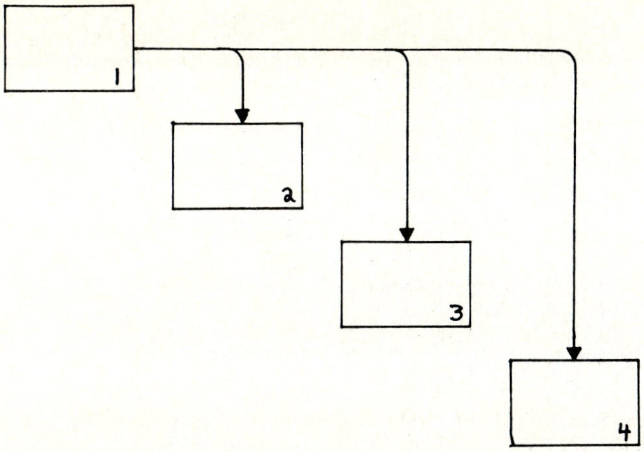

3. Control feedbacks are drawn "up and over." This correctly shows constraint feedback with the minimum of lines and crossovers, and tends to force all control arrows into the upper right-hand region of the diagram.

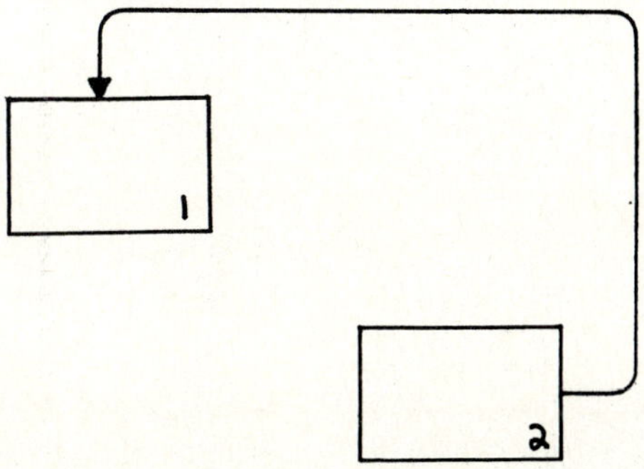

4. Input feedbacks are drawn "down and under." This correctly shows reverse dataflow with the minimum of line crossovers, and tends to force all input arrows into the lower left-hand region of the diagram.

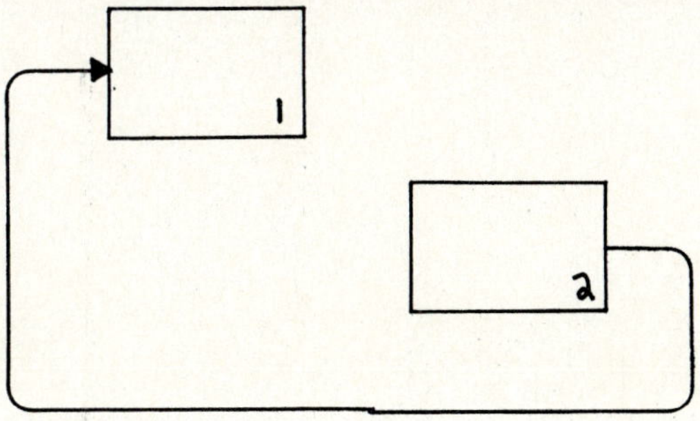

5. Connect arrows that touch several boxes at the same relative ICOM position whenever possible. This makes consistent the impact location of a particular arrow, which will simplify reading the diagram.

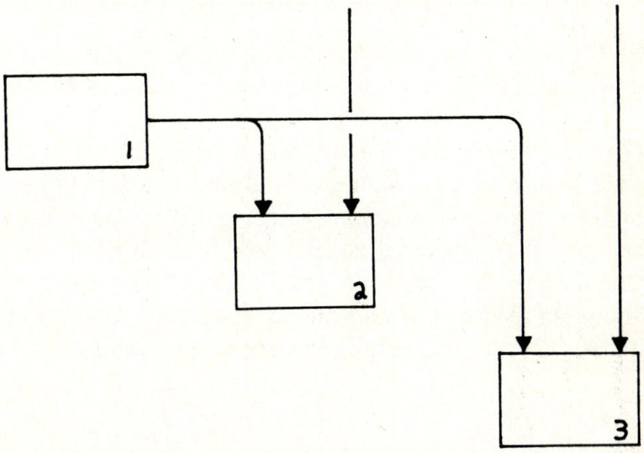

6. Avoid unnecessary arrow crossings when interconnecting many boxes. This is probably the simplest rule and most obvious guideline, yet it probably has the biggest impact on reducing diagram complexity.

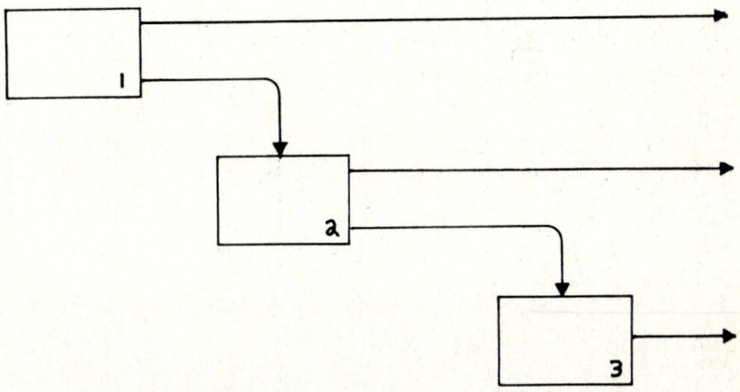

7. Minimize the curves and corners for each arrow. Again, this convention reduces diagram complexity.

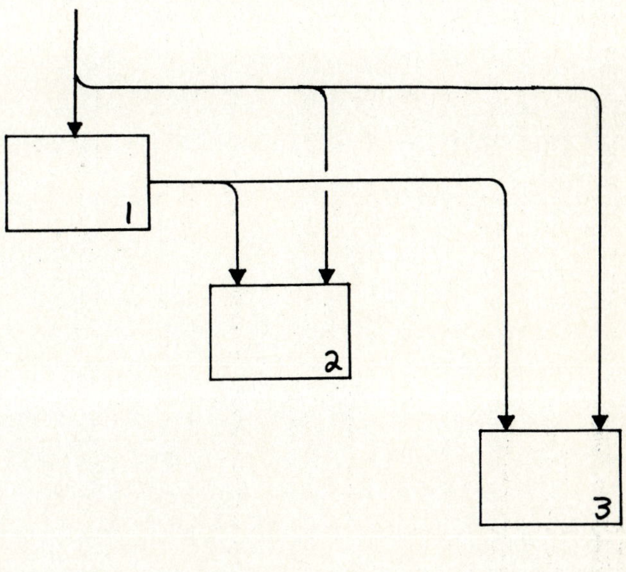

11.4 Summary

SADT diagram layout conventions strictly follow a top-down and left-to-right drawing pattern. Boxes are normally put into a staircase pattern, with arrows touching boxes at right angles. The spacing between arrows is always kept to a maximum and is always uniform. When multiple arrows of a similar nature go from one box to another, they are often summarized by a single arrow. Control feedback arrows are always drawn "up and over," and input feedback arrows are always drawn "down and under." Cyclic feedback arrows are rarely drawn.

Suggested Readings

Alexander, C.: *Notes on the Synthesis of Form,* Harvard University Press, Cambridge, Mass., 1964.

Bertin, J.: *Semiology of Graphics,* University of Wisconsin Press, Madison, 1983.

Dickover, M.: "Principles of Coupling and Cohesion for Use in the Practice of SADT," SofTech Technical Paper no. TP039, March 1976.

Minsky, M.: "Form and Content in Computer Science," *CACM,* vol. 17, no. 2, April 1970.

Ross, D.: "An Essay on Activity Diagramming," SofTech Deliverable no. 7104, 1976.

SofTech, Inc.: "IDEF0 Author's Guide to Creating Activity Diagrams," SofTech Deliverable no. 7500-11, September 1979.

Part

III

The SADT Review Process

This part of the book describes how the quality of diagrams and models is improved when an SADT author sends small packets of reviewable material called "kits" to a group of interested people for their review. The SADT Review Process emphasizes kit reading, commenting, and reacting to resolve differing opinions and to converge on a consensus. This process of asynchronous review, called the "SADT Author/Reader Cycle," has been concentrated into this part of the book to simplify its application during any systems analysis or engineering project, whether that project uses SADT or not.

The major steps of the SADT Author/Reader Cycle are described in five chapters. Chapter 12 gives an overview of the review process. Chapter 13 defines "kit" and describes how to build one. Chapter 14 discusses the skills involved in reading a kit, something that everyone on an SADT project does. Chapter 15 shows how a reader records concerns, recommendations, and issues while reviewing a kit. Chapter 16 shows how an author responds to and consolidates the comments made by a group of reviewers.

Through these chapters, a kit describing one portion of the Prototype Machine Shop model is taken entirely through one Author/Reader Cycle. This will illustrate (1) how a Kit Cover Page documents the progress of the kit through the Author/Reader Cycle, (2) how a reader reads a kit, (3) how a reader writes comments on a kit, and (4) how an author responds in writing to the comments of readers. Following this cycle will give you an accurate feel for the people, skills, and procedures required to get quick and useful review on an SADT kit.

Chapter

12

The Author/Reader Cycle

Systems analysis in a vacuum never works, because only validated models can be used to develop a system that works for its users. For example, implementing a management information system from specifications never seen by end users will result in an unusable system. Data will be wrong or missing, and the format of reports will not match the information needs of the people. In order to develop a useful system description, a successful systems analysis effort must get accurate and timely information from the people who will ultimately use the system.

The gathering of accurate and timely information starts at requirements definition time, the first step in the overall system development cycle. Also, as with any modeling process, several iterations are necessary before initial ideas are developed into concepts that can be agreed upon by the users. This is why SADT has a well-defined process for iteratively reviewing the models made during a project. As previously mentioned, this process of iterative review is called the "SADT Author/Reader Cycle."

The SADT Author/Reader Cycle was developed to facilitate the asynchronous and concurrent review of the work of many SADT analysts. That is, it was designed to maximize the amount of feedback end users give to one or more analysts in the shortest possible time and with the least amount of effort. The Author/Reader Cycle makes allowances for individuals by letting people fit the tasks of authoring and reviewing into their own work schedules. This is critical when trying to get feedback from users who are busy with their regular work, and when trying to coordinate several SADT authors during the cooperative authoring of a single model or set of interrelated models. This is why SADT advocates written review—to better coordinate work and to document the evolving ideas of everyone involved in an analysis project.

Several kinds of people participate in an SADT Author/Reader Cycle: authors develop models, readers read and make comments about the work of authors, a Technical Review Committee approves project results, and a Librarian organizes and distributes materials. These roles, and they way they interact with each other, are pictured in Figure 12-1. A detailed description of the steps in an SADT Author/Reader Cycle follows in this chapter.

12.1 Documenting a Problem

The major role of an SADT analyst is to document a proper understanding of a system by creating several SADT diagrams that fit together to form a model. Sometimes, this model may have to fit into a network of interrelated models. This process, called "authoring" in SADT, is both a knowledge-acquisition and a knowledge-representation activity. Knowledge is acquired by reading documents, interviewing experts, observing current operations, or inventing scenarios. Very often, authors interview experts to acquire vital facts about the system under study. Interview notes, existing documents, and any other kind of acquired knowledge are used as grist in developing SADT diagrams and models. The acquired knowledge is first represented in the SADT graphic language and is often further annotated with other graphics and languages, like those given in Chapter 19.

Also, the project plan for the entire modeling effort guides the author by helping to fit the individual diagrams she or he makes into the evolving model or models. These diagrams and models are sent to the Librarian, who files them into the project's model files. When appropriate, authors collect their work into small packets, or kits, which are the basic units of reviewable material. A kit is sent to the Librarian for dissemination to project participants. Before submitting a kit for distribution, the author fills out the distribution list, return date, and other required fields on the Kit Cover Page, as shown in Figure 12-2. The Author also receives a copy of a kit from the Librarian.

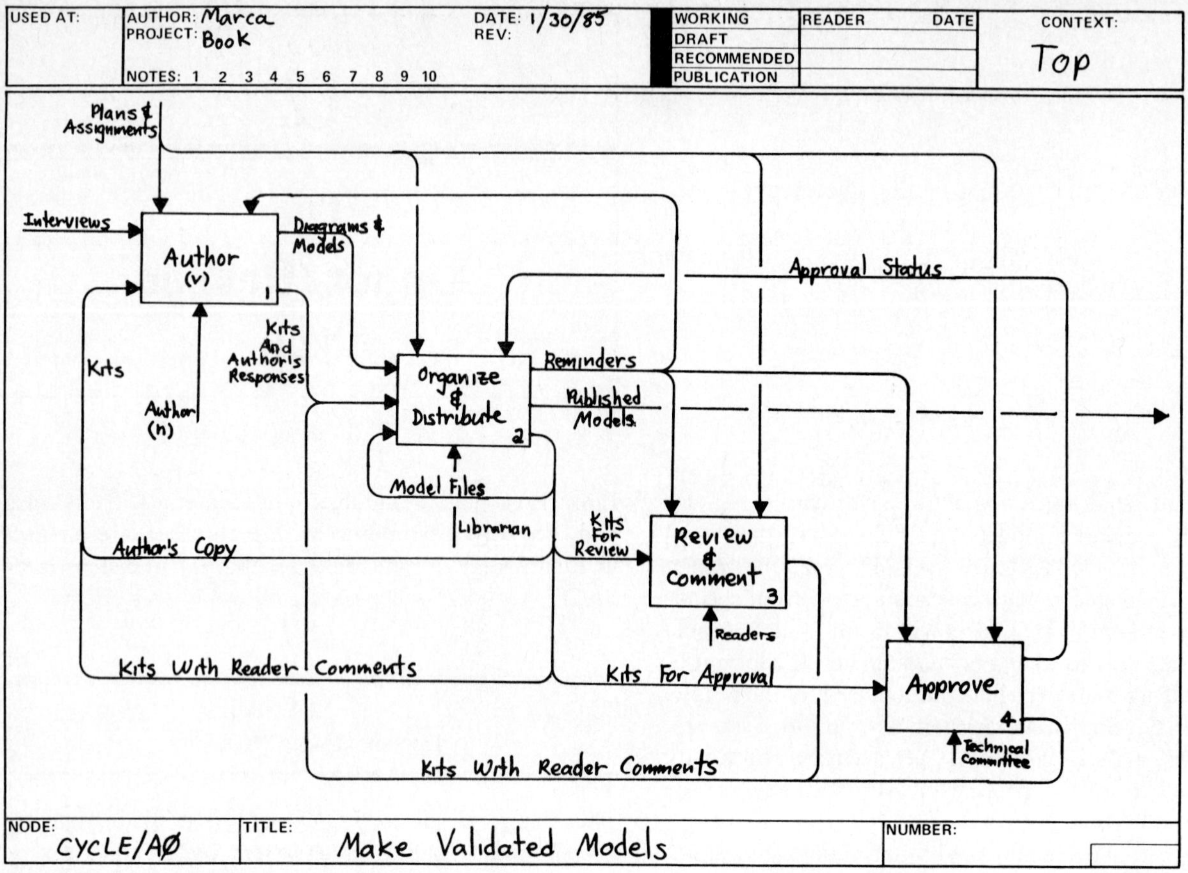

FIGURE 12-1 The SADT Process of Making Validated Models

12.2 Commenting on Work

Kits received by the Librarian are logged, duplicated, sent to the author, and distributed to the Reader Community according to the distribution list on the kit. Dates are recorded for when the kit was sent and when the readers must reply to the author. This way, the Librarian knows who received a kit, when they received it, and when they must return it. This information is required for properly controlling the document so that the feedback from readers will be timely. Readers, upon receiving a kit from the Librarian, read its contents, write down their comments, and return it to the Librarian. These commented kits are received and logged by the Librarian, and are then sent immediately to the author. The Librarian also reminds tardy readers to return overdue kits. This way, authors have a good chance of receiving timely feedback from their Reader Community.

Comments, in the form of SADT notes, are written in red (indicated in light gray in all the figures of this book) as soon as a question or issue comes to the reader's mind. These notes start a written dialog between the reader and the author by immediately capturing the concerns of the reader as they are conceived. This mode of writing down concerns as one reads is critical for ensuring that all the ideas conceived during a project are never lost. The details of a spoken dialog between two people are often forgotten, but a written dialog can always be referred to when resolving or revisiting issues. For example, notice in Figure 12-3 that the reader agreed to help the author by making a check mark on the Kit Cover Page, putting comments on the diagram, dating the comments on the Kit Cover Page, and giving a deadline for when the author should respond.

12.3 Reacting to Comments

Commented kits are reviewed and reacted to by the author. Reacted kits are given to the Librarian, who logs them in and then returns them to the readers. After this happens, each copy of that kit is in the hands of its owner, so the Librarian never processes that kit again. Tardy authors are reminded by the librarian to respond to over-

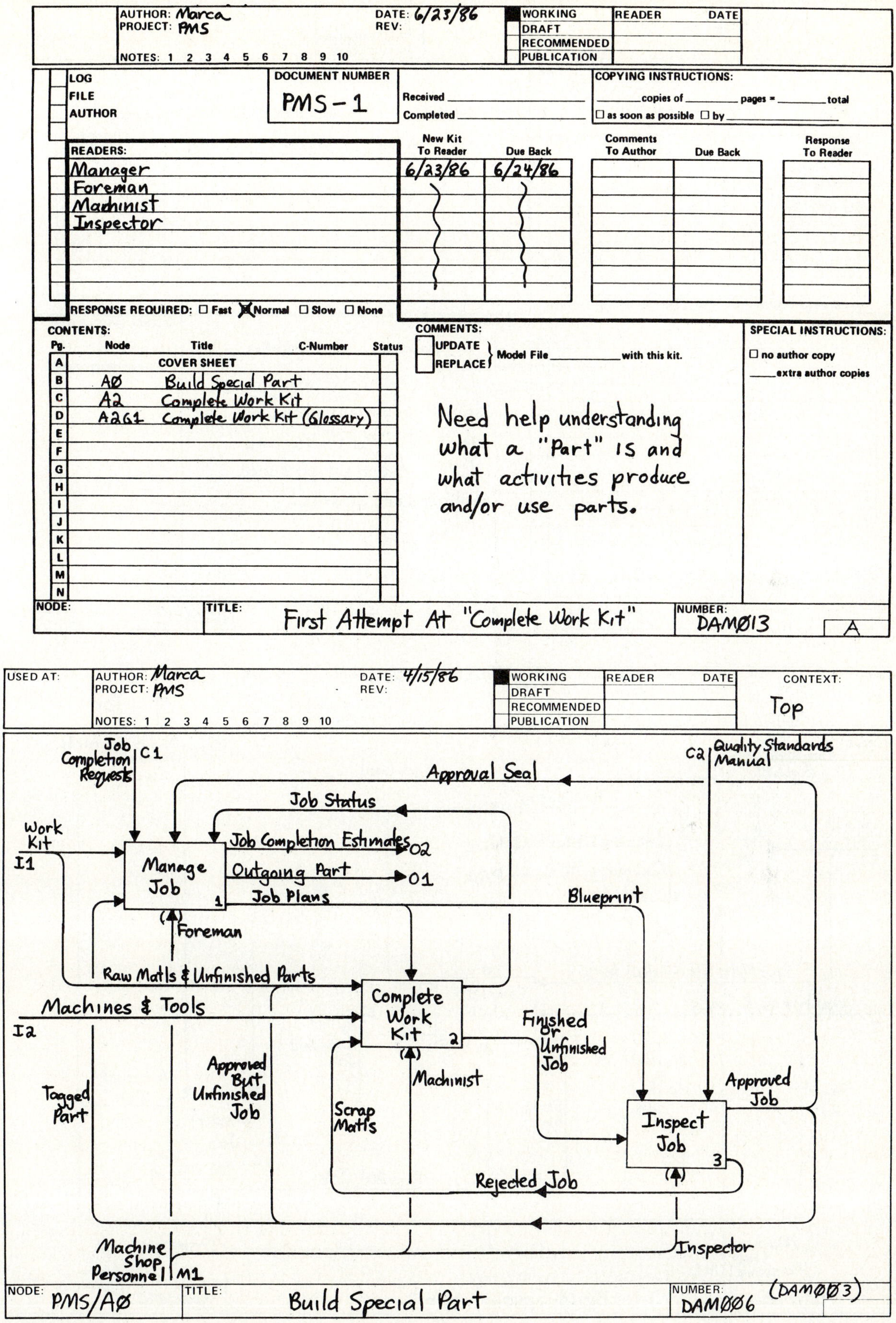

FIGURE 12-2 A Kit Cover Page and Diagram Made by an Author

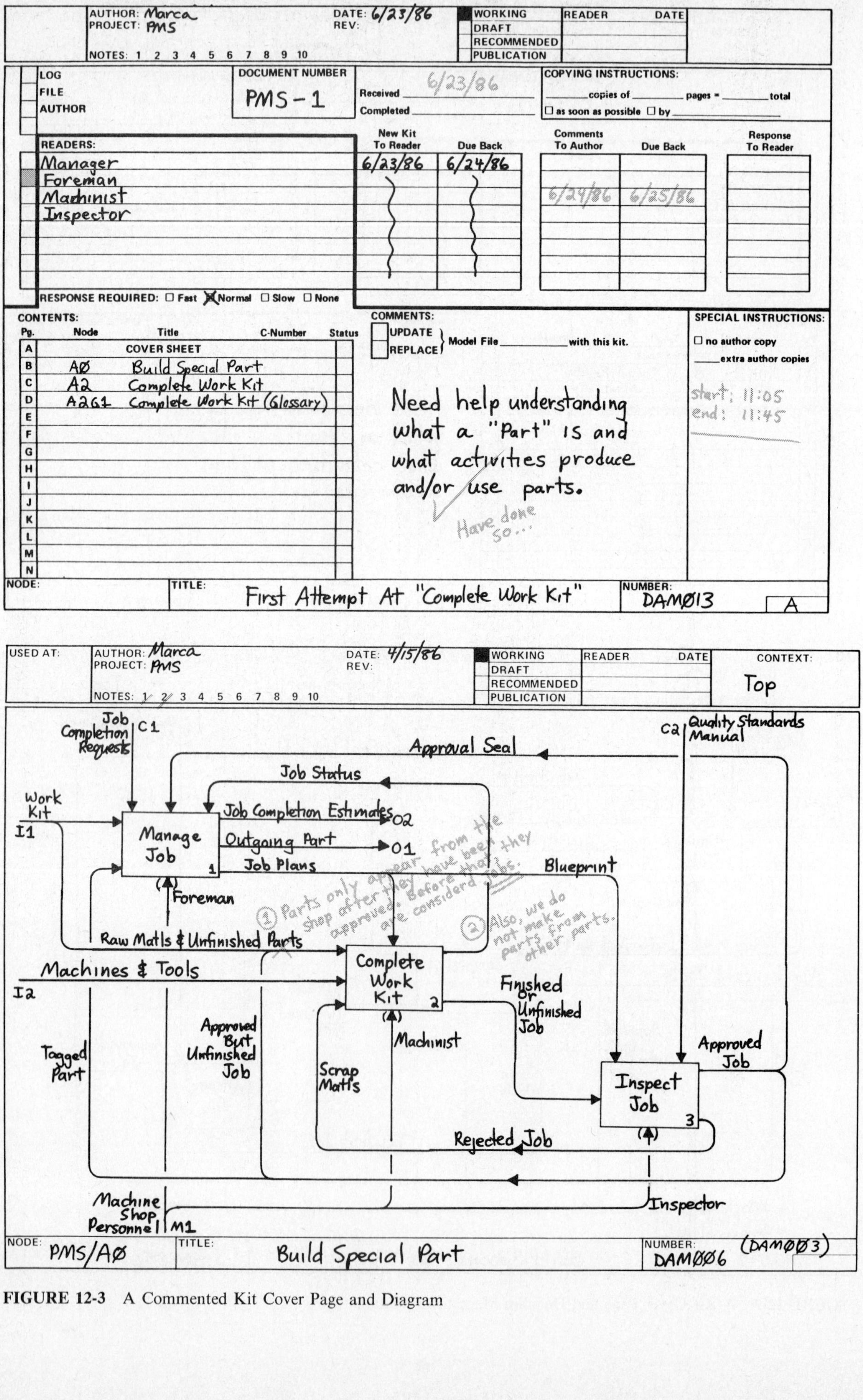

FIGURE 12-3 A Commented Kit Cover Page and Diagram

due kits. This way, timely feedback happens in both directions: from reader to author and vice versa.

When reviewing a commented kit, the author responds to each comment by writing a response in blue (indicated by dark gray in the figures of this book) next to each comment note. This way, comments are associated with and are distinguished from responses using a different color. For example, notice in Figure 12-4 that the author acknowledged the reader's check mark, responded to the reader's advice about *parts* on the diagram, and dated the response on the Kit Cover Page. Other informational notes may also be added during the reading to provide explanation about a reaction or to indicate the author's intent to revise the diagrams based on the review. In this way, the author completes his or her part of the written dialog.

During the process of responding to everyone's comments on a particular kit, an author also updates his or her own copy of the kit. Comments from all readers are consolidated onto the author's copy. This way, the author analyzes and attempts to resolve the many and possibly conflicting views of several readers. Consolidations are used to revise diagrams, thus making a better version of the evolving model. Revised diagrams are given to the Librarian, who puts them into the project's model files. Once this happens, that particular review cycle is over and the author may begin another one on the revised material (see Chapters 14 to 16 for a worked example).

12.4
Approving Models

After a few Author/Reader Cycles, the Reader Community usually begins to reach a consensus about what is described by a particular model, by converging on a common understanding of how a particular system works or will work. In essence, cooperation and agreement is achieved through an agreed-upon process, in which a group of people collectively voice and share their ideas in a way that brings all of them closer to a particular goal. In the case of an SADT analysis project, this goal is the common understanding of a system.

Authors use standard SADT approval-level key words to indicate the level of consensus reached for a particular diagram. The standard SADT approval-level key words "working," "draft," "recommended," and "publication" are defined when an SADT project starts. This gives each analysis effort a chance to tailor its own levels of formal document control by associating a particular approval key word with a metric of group consensus. For example: "working" could refer to the first attempt at a diagram, "draft" could refer to subsequent versions that require correction, "recommended" could mean the diagram is ready for final approval, and "publication" could mean the diagram is frozen.

Diagrams are appropriately marked by the author to indicate the level of reader approval, and are then sent to the Librarian who records the change in the project's model files. Once an author feels that a set of diagrams which make up a well-defined part of a model have reached a "recommended" level, he or she sends them to the Technical Review Committee for final approval. The diagrams are packaged into a kit, and the kit follows the same commenting and responding cycle that was done for earlier versions. The Technical Review Committee records its concerns with written comments, the author answers these comments with written responses, and the Librarian ensures timely review by sending reminders.

Authors revise diagrams according to the comments of the Technical Review Committee. At some point, the Technical Review Committee decides that the set of diagrams can be approved. Approval decisions are conveyed to the Librarian, who then notifies the author. At appropriate times later in the project, approved diagrams are published and distributed to people both inside and outside the project. Also, approved models are published in documents for use in later phases of the project.

12.5
Using the Author/Reader Cycle

The overview just presented summarizes the complete and formal set of procedures that make up the SADT Author/Reader Cycle. These procedures are complete in that they handle every possible situation encountered during the review of a single piece of work by many people at the same time. They are formal in that they prescribe rules for coordinating the work of many people so that they reach a consensus—that is, a common understanding. Experience has shown these procedures to be easily tailorable. Since each project has different needs regarding the level of completeness and formalism in the way it gets review, every project can satisfy its particular review needs with these procedures.

A few examples from the work experiences of the authors of this book might serve to illustrate the utility of the Author/Reader Cycle: We both worked on a government project that used approval levels to identify required approval from different military rank in order to satisfy the need to classify sensitive military information. We also worked on many projects that used the full SADT Author/Reader Cycle to distribute and review all its written material, including software, user manuals, meeting notes, and relevant technical papers. We found the members of those projects well informed, though never burdened with too much to read. Last, we typically manage our own

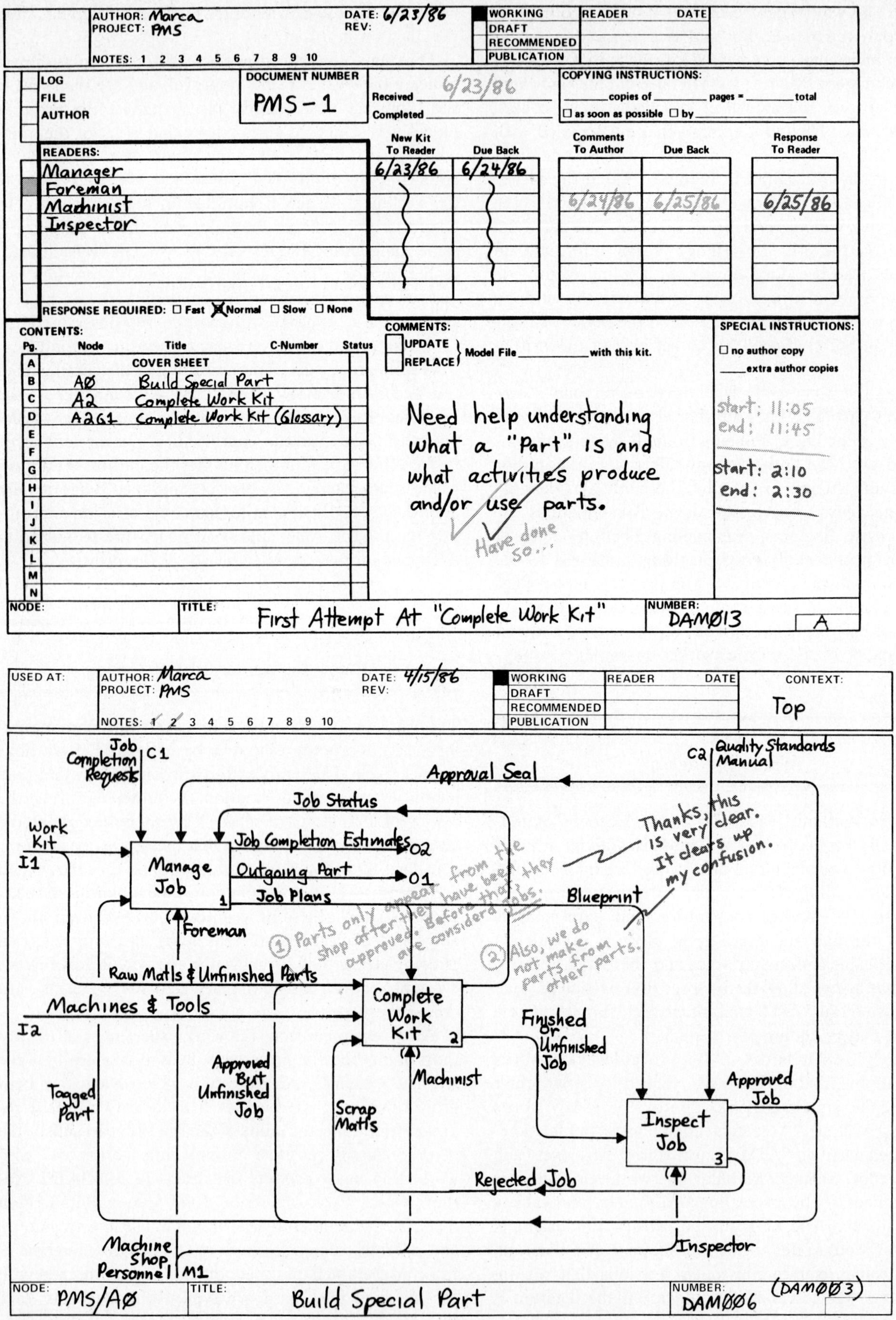

FIGURE 12-4 A Reacted Kit Cover Page and Diagram

project library when we work on one-person analysis efforts that have just a few readers.

These are just some of the ways the Author/Reader Cycle can been tailored to facilitate review leading to a consensus among a group of people about the way a system works. We would like to stress that we do not advocate the use of a full SADT Author/Reader Cycle for every systems analysis project undertaken. Instead, we recommend you always consider how much of this process best satisfies your authors, your Reader Community, the size and complexity of your system, and the scope of your analysis project.

12.6 Summary

The goal of an SADT analyst is to develop a correct description of a system. SADT contends that correctness can only come from peer review. So, the SADT methodology includes a peer-review process called the "Author/Reader Cycle": Authors create small work packets called "kits." The Librarian distributes these kits to readers, who write their concerns on the actual kits. Authors respond to these written comments by writing reactions next to each one, return reacted kits to the readers, and consolidate the various and sometimes diverse comments on their own copies of the kits. Issues not resolved by the written commenting and reacting process get resolved later by focused discussions called "Author/Reader Talks." After many of these Author/Reader Cycles, a model reaches a level of consensus at which it can be approved for use.

Suggested Readings

Curtis, B. (ed.): "Human Factors in Software Development," IEEE Catalog no. EHO 185-9, IEEE Computer Society, 1981.

Connor, M.: "Structured Analysis and Design Technique—SADT," Auerbach Portfolio no. 32-04-02, 1981.

Freedman, D., and G. Weinberg: *Walkthroughs, Inspections, and Technical Reviews,* Little, Brown, Boston, 1982.

Kemmerer, R.: "Testing Formal Specifications to Detect Design Errors," *IEEE Transactions on Software Engineering,* vol. 11, no. 1, January 1985.

Mihram, A.: "The Modeling Process," *IEEE Transactions On Systems, Man, and Cybernetics,* vol. 2, no. 5, November 1972.

Mumford, E.: *Designing Human Systems for New Technology,* Manchester Business School, Manchester, England, 1983.

Ross, D., and K. Schoman: "Structured Analysis for Requirements Definition," *IEEE Transactions on Software Engineering,* vol. 3, no. 1, January 1977.

SofTech, Inc.: "IDEF0 Forms and Procedures Guide," SofTech Deliverable no. 7500-11, September 1979.

Weinberg, G.: *The Psychology of Computer Programming,* Van Nostrand Reinhold, New York, 1971.

Yourdon, E.: *Structured Walkthroughs,* Yourdon Press, New York, 1978.

Chapter 13

Kit Preparation

Typically, a model is started and sometimes decomposed a little before being sent out for review. These steps are done to ensure that a piece of work has the best chance of being understood by people other than the author. To get feedback on the validity of the work, an SADT author initiates an Author/Reader Cycle by preparing a work packet called a "kit." This chapter will define the SADT kit and show how it is created by an author. Chapters 14 to 16 detail how a kit is processed by the other members of an SADT project.

13.1 Communicating with Kits

The term "kit" in SADT refers to a unit of reviewable work. The kit concept came from the fact that SADT projects are technically organized around an author and a group of readers. An author interviews experts, creates a set of related diagrams, and then asks the experts (and possibly other authors) for help in validating that work. Those experts who receive kits are part of the author's Reader Community—the group of people who are concerned that the work of the author be both correct and useful. Kits are therefore the basic units of information passed among people in an SADT project. In other words, kits are one of the major ways the members of an SADT project communicate.

The process of communication through kits allows for asynchonous and concurrent review of a work product: An author sends a kit to a group of readers. Readers communicate their concerns about a kit to the author via notes, written on the kit itself, called "comments." The author reads the comments and responds to each with written notes called "reactions." The kits, now containing both written comments and reactions, are returned to the readers. Thus, as shown in Figure 13-1, the interaction between an author and a Reader Community can be pictured as a cycle, with kits originating from the author, being read and returned to the author by readers, and being read and returned to the readers by the author.

13.2 The Kit Cover Page

The first page of every SADT kit is the Kit Cover Page, a special form that packages together material, describes what the kit contains, and records how it was processed by project members. To accomplish these ends, the Kit Cover Page has several regions and areas. Figure 13-2 shows an SADT Kit Cover Page with bold lines drawn on it, dividing the form into three regions. The top and bottom regions of the form are filled in by the author. The middle region is reserved for the project Librarian.

This Kit Cover Page also has large numbers drawn on it, dividing it into five areas. Authors complete area 1, the identification area, just as they do for diagrams, with: name, data, project, status, and C-Number. Authors complete area 2, information about what the kit contains, with

FIGURE 13-1 Kits Are Cycled between an Author and Readers

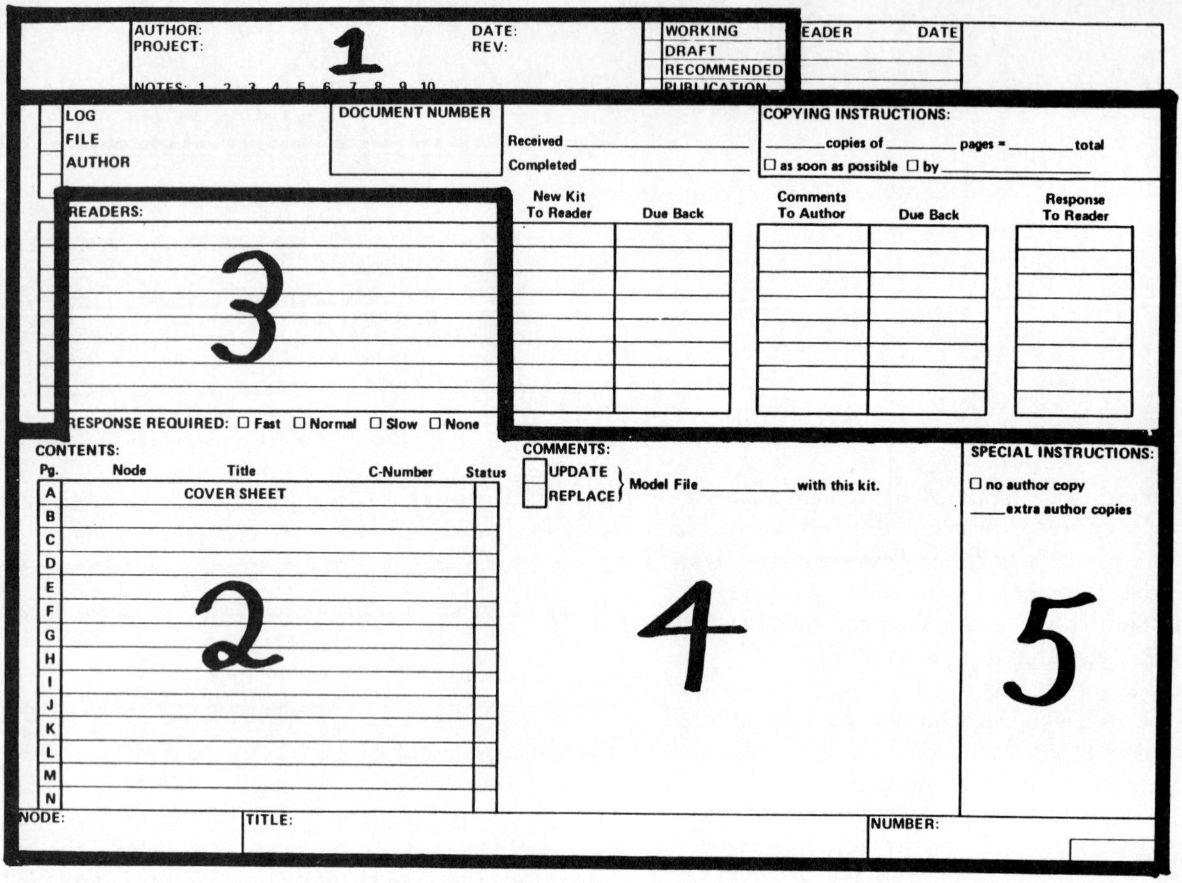

FIGURE 13-2 Regions and Areas of a Kit Cover Page

node and title for the kit, and the node, title, and C-Number for each page in the kit. Area 3 is filled in with the names of those who will get the kit. Notes about the entire kit are written in area 4, and the Author writes special instructions to the Librarian in area 5.

13.3 Kit Organization

An SADT kit consists of a Kit Cover Page, one or more diagrams, and possibly supporting material. The Kit Cover Page names and summarizes the work packet and is the medium which tracks the kit through the communication cycle between author and reader. Diagrams appear after the Kit Cover Page in node number order. Figure Pages and Glossary Pages which supplement the diagrams appear immediately after the diagrams they elaborate. (See Chapter 18 for discussion of supplements.)

For example, Figure 13-3 shows an SADT kit, which contains the results of the analysis done in Part I, where the Prototype Machine Shop model was started and partially decomposed. The Kit Cover Page tells us that this kit is concerned with the decomposition of the subject ***complete work kit,*** appearing in diagram PMS/A2. Also on the cover page is a comment, written by the author, asking for help in understanding about ***parts.*** The ***build special part*** diagram (PMS/A0) is put in front of the ***complete work kit*** diagram to provide readers with context. Glossary Page PMS/A2G1 defines the terminology introduced by the ***complete work kit*** diagram and appears after the diagram.

13.4 Kit Size

Unlike the kit in Figure 13-3, SADT kits are not always made for every new diagram of a model. The sizes of kits vary greatly and depend both on the project and on the people in the project. Also, the complexity of the system being studied, availability of experts, and experience of the SADT analysts all contribute to how much goes into a kit. However, a work packet including a context diagram, the diagram of major importance, and a Glossary Page is very common on most SADT projects. We recommend making

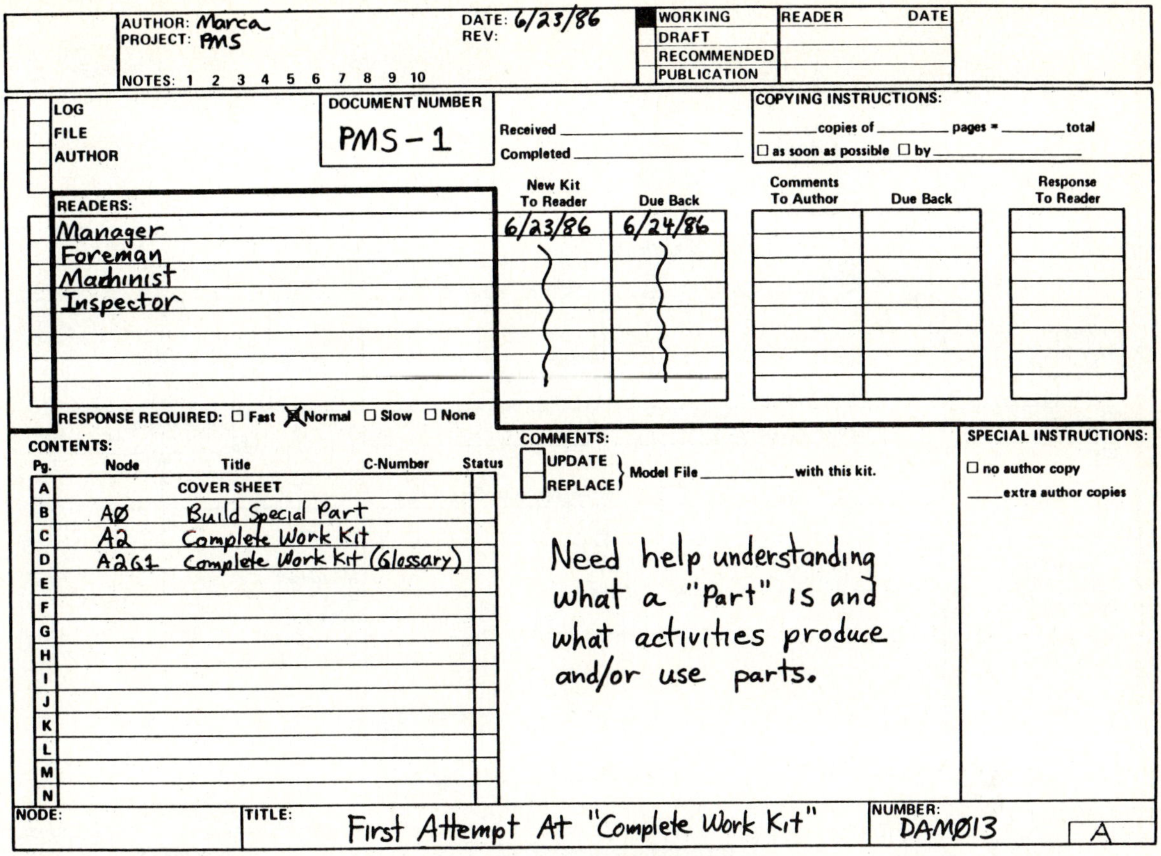

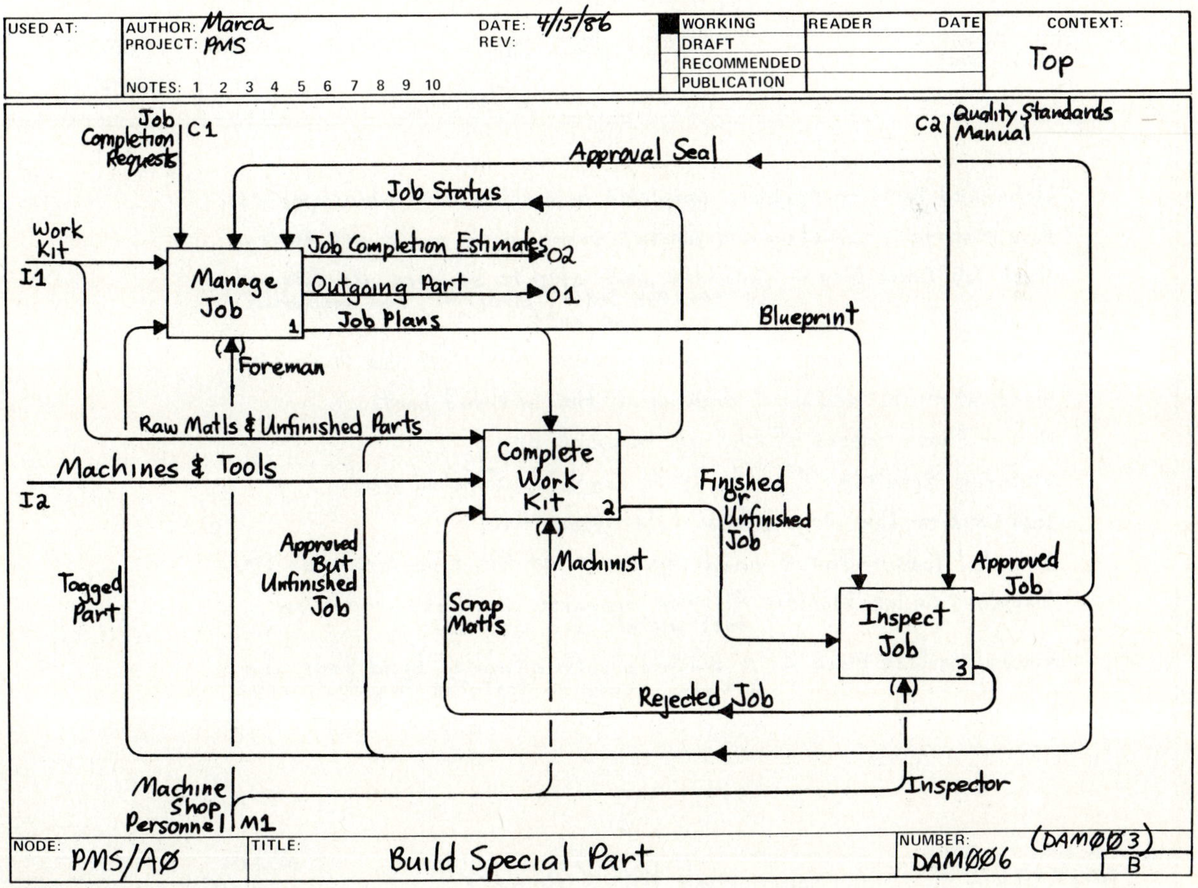

FIGURE 13-3 A Complete SADT Kit

USED AT: | AUTHOR: Marca | PROJECT: PMS | DATE: 5/5/86 | REV: | NOTES: 1 2 3 4 5 6 7 8 9 10 | WORKING / DRAFT / RECOMMENDED / PUBLICATION | READER | DATE | CONTEXT: DAM006

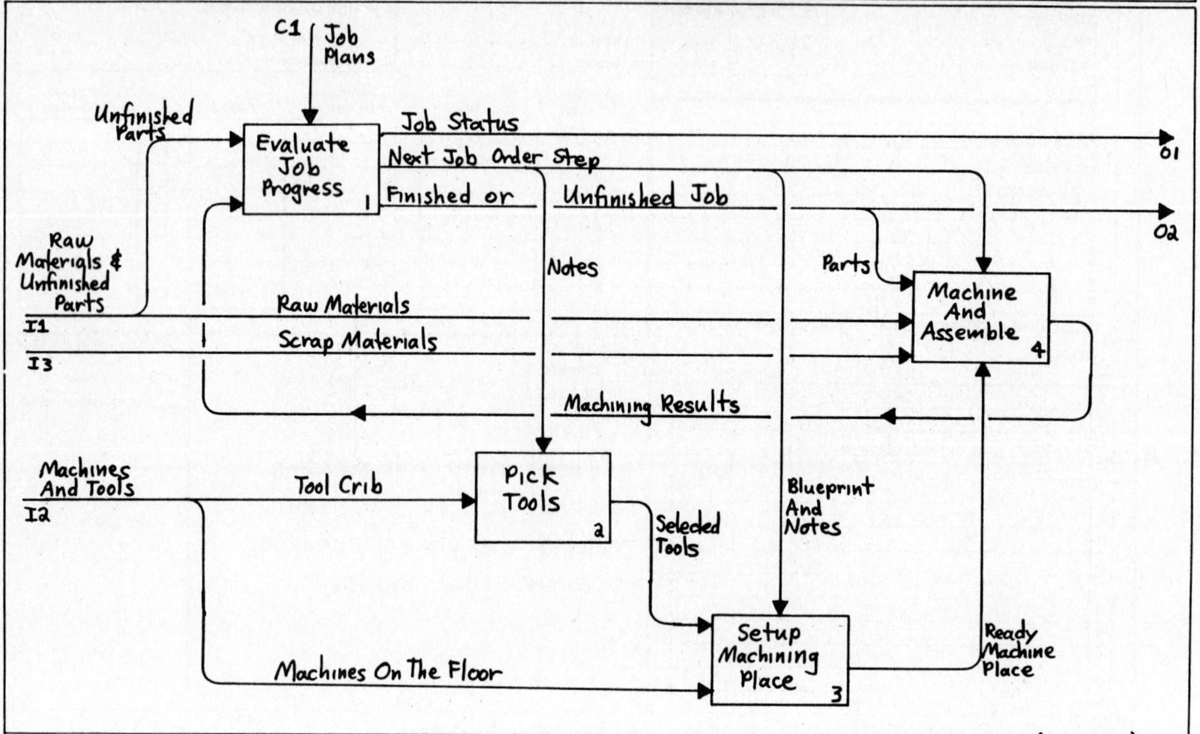

NODE: PMS/A2 | TITLE: Complete Work Kit | NUMBER: DAM010 (DAM009) | C

USED AT: | AUTHOR: Marca | PROJECT: PMS | DATE: 6/23/86 | REV: | NOTES: 1 2 3 4 5 6 7 8 9 10 | WORKING / DRAFT / RECOMMENDED / PUBLICATION | READER | DATE | CONTEXT: DAM010

Unfinished Parts — Partially completed assemblies and sub-assemblies.

Raw Materials — Metals and plastics from which parts are made.

Next Job Order Step — Only the next step to be done, as defined by the Job Order, Blueprint, and completion stage of the part.

Notes — Special things to remember about this Next Job Order Step.

Blueprint — The mechanical drawing of the intended part.

Part — Those pieces of the job to work on.

Machining Results — A partially or completely finished part.

Tool Crib — The place which holds hand tools.

Selected Tools — Those hand tools required for the Next Job Step.

Machines On The Floor - All the available machines which can tool metals and plastics.

Ready Machine Place — A machine with attached hand tools and clamps, ready to tool the Raw Materials.

NODE: PMS/A2G1 | TITLE: Complete Work Kit (Glossary) | NUMBER: DAM014 | D

kits of this size whenever you question how much information is appropriate for your readers. One diagram, with at least a decomposition of one of its boxes, and possibly a Glossary Page or a figure is a reasonable default kit size. (See Chapter 20 for a more detailed discussion about the factors affecting kit size.)

We recommend this as a default, because you should never put too much information in a kit for review. Our experience has shown that, regardless of circumstances, a kit should contain no more than one diagram and all its immediate children—roughly six diagrams. If supplemental material is also included in the kit, the number of diagrams must decrease. A kit with more than about 6 pages of information will probably swamp your readers, and you will get very little response. Thus, we think it better to err on the side of including too little information in a kit than to put unnecessary demands on your Reader Community. If large kits are a norm, you may find your Reader Community becoming less enthusiastic about giving review as time goes by. The only exception to this is the publishing of an entire model as a kit near the end of the analysis effort. Readers expect such kits to be large, but they also know that there will be very few of them.

13.5 When to Make a Kit

Kits are never created randomly during a project. In general, kits are made whenever an author feels there is enough new information for the readers to properly review. However, this is not the only time to issue a kit. A kit should also be made when an author does not have enough information to proceed or questions the validity of the work. In these cases, we recommend issuing a kit to a few very special people who can help you continue. For example, send incomplete units of work to very senior experts or senior SADT authors who have the maturity and knowledge to give you guidance on how to proceed. With their feedback, you can often complete enough of your work to issue a kit to your regular Reader Community for more feedback. Chapter 20 discusses the details of issuing a series of kits to reach a widening scope of consensus.

13.6 Summary

The SADT "kit" is a small collection of related SADT diagrams and associated material, made by an author for review by one or more readers. Kits are made whenever an author feels there is enough information for review by a Reader Community. Kit size is kept small to allow for quick review. The Kit Cover Page names the kit, summarizes its contents, and identifies who will read the kit.

Suggested Readings

Connor, M.: "Structured Analysis and Design Technique—SADT," Auerbach portfolio 32-04-02, 1981.

King, L.: "Guidelines for an SADT Chief Author," SofTech Deliverable no. 9595-2, 1980.

Ross, D., and K. Schoman: "Structured Analysis for Requirements Definition," *IEEE Transactions on Software Engineering,* vol. 3, no. 1, January 1977.

SofTech, Inc.: "IDEF0 Forms and Procedures Guide," SofTech Deliverable no. 7500-11, September 1979.

Chapter

14

Reading Diagrams and Models

The reading of the SADT diagrams in a kit follows a systematic process. The SADT process of reading is the foundation upon which an SADT reader reviews the work of an author. Reading, therefore, is one of the very basic skills for those who participate in an SADT project. With this skill, a reader can systematically understand the messages contained in the diagrams of a kit, step through each diagram to understand its details, and evaluate and validate each diagram.

14.1 The Process of Reading

The first step in reading a diagram is to acquire a general understanding of its context. Typically, the parent diagram is first quickly read and understood in order to refresh one's memory about the system at a higher level of detail from the one of current interest. Usually, a quick reading of the parent diagram takes little time and has consistently proved to be a valuable way of acquiring an introductory knowledge of what is to come. We therefore recommend that the parent be read before reading its child diagram for detail. For example, in the kit of Figure 14-1, we would skim the ***build special part*** diagram first, concentrating on the role of the ***complete work kit*** activity. This will give enough introduction to start reading the ***complete work kit*** diagram in detail.

The process of reading a diagram in detail can be broken into a sequence of four steps. The first step is to understand the details of a particular diagram. The second step is to concentrate on understanding the immediate context of that diagram. The third step is to focus on understanding how the diagram connects into its model. The fourth step involves constructively criticizing the author's message based on the acquired understanding. These four steps are performed in the order given above. Understanding the details of a particular diagram is required before one can fully understand its role in the immediate context. Understanding the immediate context of a diagram sets the stage for investigating the details of how the diagram connects to the other diagrams in the model. This complete understanding then allows one to constructively comment on the work.

14.2 Understanding the Details of a Single Diagram

Understanding of a single diagram is acquired by reading: (1) the title and node number, (2) each box, (3) the internal arrows, (4) any author notes, and (5) any associated supporting material. The most effective reading happens when each of these diagram elements are read in a step-by-step manner. Please refer to the kit in Figures 14-1 and 14-2 while we step through an example of this part of the process.

14.2.1 Read the Title and Node Number

Start reading a diagram by first scanning the form, and concentrate especially on the title and node number of the diagram. For example, because its node number is PMS/A2, the diagram ***complete work kit*** of Figure 14-2 is a second-level diagram in the Prototype Machine Shop model, and should therefore start to uncover some of the detail about the machine shop. Use the title and node number to focus your mind on the subject being discussed and its level of detail in the model.

14.2.2 Read Each Box

Now read each box separately. Concentrate on understanding a single box in its entirety—that is, what it does (its function), what it transforms (its inputs and outputs),

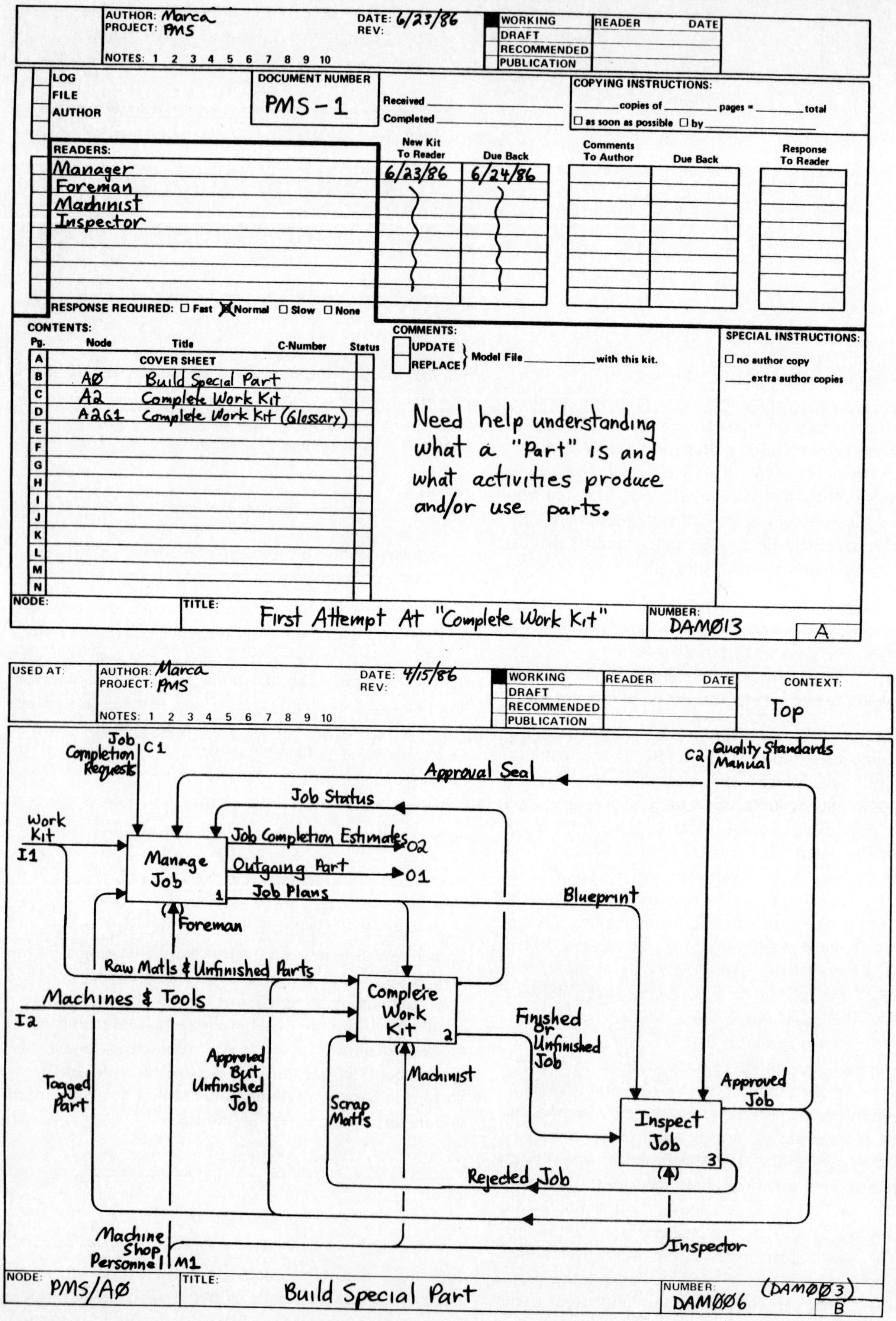

FIGURE 14-1 A Kit Cover Page and Parent Diagram

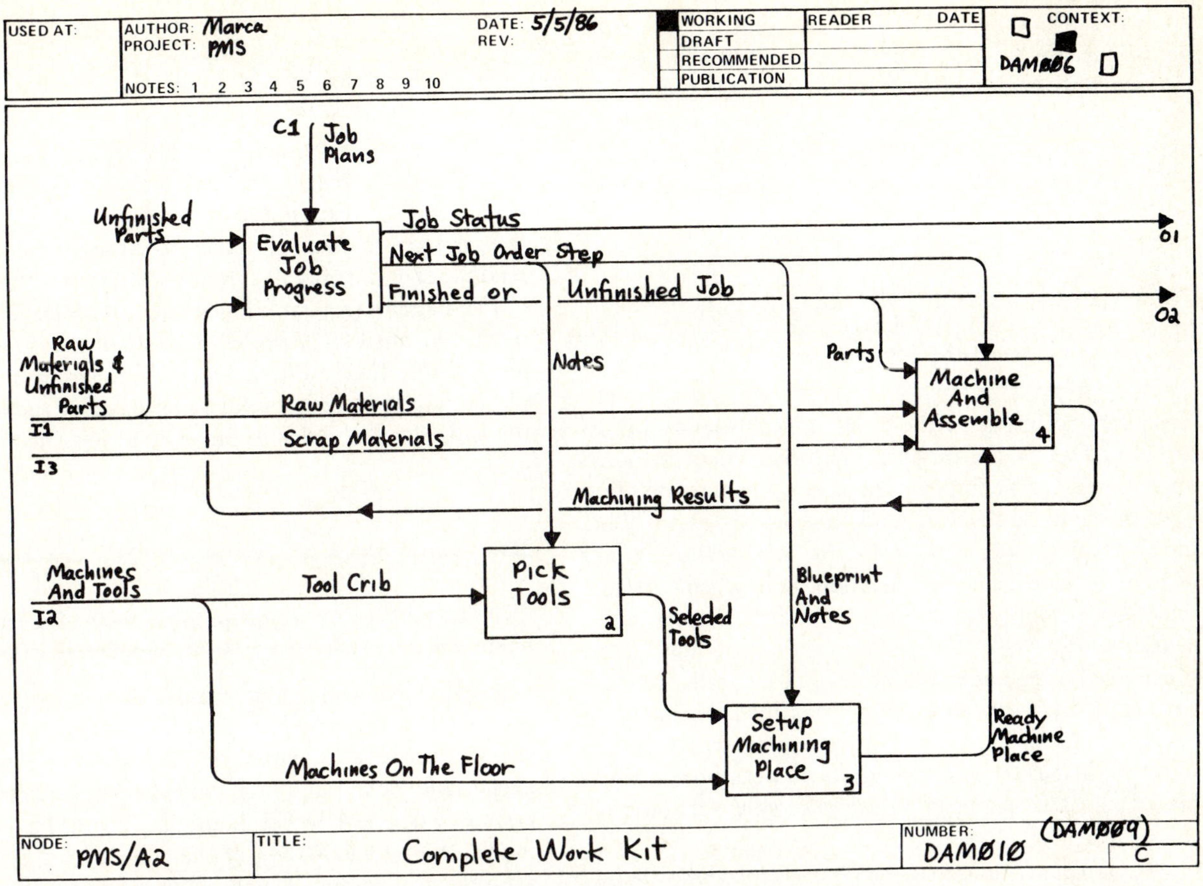

USED AT:
AUTHOR: Marca
PROJECT: PMS
DATE: 6/23/86
REV:
NOTES: 1 2 3 4 5 6 7 8 9 10
WORKING
DRAFT
RECOMMENDED
PUBLICATION
READER
DATE
CONTEXT: DAMØ1Ø

Unfinished Parts — Partially completed assemblies and sub-assemblies.

Raw Materials — Metals and plastics from which parts are made.

Next Job Order Step — Only the next step to be done, as defined by the Job Order, Blueprint, and completion stage of the part.

Notes — Special things to remember about this Next Job Order Step.

Blueprint — The mechanical drawing of the intended part.

Part — Those pieces of the job to work on.

Machining Results — A partially or completely finished part.

Tool Crib — The place which holds hand tools.

Selected Tools — Those hand tools required for the Next Job Step.

Machines On The Floor — All the available machines which can tool metals and plastics.

Ready Machine Place — A machine with attached hand tools and clamps, ready to tool the Raw Materials.

NODE: PMS/A2G1
TITLE: Complete Work Kit (Glossary)
NUMBER: DAMØ14
D

FIGURE 14-2 A Diagram with Supporting Material

and what constrains its functions (its controls). By focusing on just the box, you can understand how all arrows that touch it are related. For example, box 3 in Figure 14-2 describes the step of ***setup machining place,*** where ***selected tools*** and one of the ***machines on the floor*** are used to establish a ***ready machining place*** according to the ***blueprint and notes.***

14.2.3
Read the Internal Arrows

After concentrating on the boxes, concentrate on the internal arrows of the diagram. This will uncover details about simple and complex constraints, the main dataflow path through the diagram, and feedback situations. Simple constraint happens between functions, while complex constraint happens among functions (i.e., when data mutually constrains two or more functions). For example, ***next job order step*** is the major constraint on how all the other machining activities are done. The main dataflow path is best understood by composing the most likely, or "normal," scenario from the way the boxes transform the major inputs into the major outputs. For example, a possible "main path" scenario for ***complete work kit*** could be:

> The ***job plans*** are used to determine the ***next job order step,*** which is then used to ***pick tools, setup machining place,*** and finally ***machine and assemble*** the ***part. Machining results*** are then evaluated and this cycle repeats until the ***part*** is ready.

Recall that SADT can describe two types of feedback: dataflow and constraint. For example, the ***machining results*** arrow on diagram PMS/A2 is a dataflow feedback, while the ***approval seal*** arrow on diagram PMS/A0 is a constraint feedback. Dataflow feedbacks are the weaker feedbacks. These pipelines simply "ship" data from one function to another. Constraint feedbacks are much stronger. They signal conditions, detected in one function, that affect the way a second function must subsequently do its work. Be on the lookout for them.

14.2.4
Read Any Author Notes

Once you have studied the boxes and arrows, read any notes made by the author. These will be square notes, which usually elaborate an important point or document issues or confusion. Use the notes to gain a better understanding of what the diagram is trying to say, or use them to give specific feedback to the author. In this way, notes written by the author can focus review. After finishing the author notes, you have finished reading the diagram itself.

14.2.5
Read Associated Supporting Material

Last, read any associated supporting material. Figures, text, and glossary are often attached to a diagram to provide visual context, elaborate on a key point, or clarify terminology, respectively. Refer to this material while you read the diagram only if referring will sharpen the message of the diagram (e.g., you have a question that the supplement may answer). Otherwise, postpone reading supplements until after you understand the diagram. For example, it may have been useful to understand the layout of the shop floor in order to understand how machines are set up.

14.3
Understanding the Immediate Context of a Diagram

Once you have read the internals of the diagram, focus on their context by determining all the connections between the diagram and its parent. By doing so, you will get a deeper understanding of the diagram, because the subject boundary defines how the diagram fits with the rest of the model. An understanding of the context of a diagram is acquired by reading: (1) the boundary box and arrows appearing on the parent diagram, (2) the ICOM codes of the diagram, (3) how this diagram connects to the other boxes on the parent diagram, and (4) the supporting material of the parent diagram. These diagram elements are also read in a step-by-step fashion.

14.3.1
Read the Parent Box and Its Arrows

Start by reading just the box on the parent diagram. This will refresh your memory about the overall function of the diagram and its relationship to the rest of the model. For example, the PMS/A0 diagram in Figure 14-1 summarizes the ***complete work kit*** activity as a process that converts ***raw materials & unfinished parts*** and ***scrap*** into ***finished or unfinished jobs*** according to ***job plans.*** Also, ***machines and tools*** are used during the process and ***job status*** is sometimes reported.

14.3.2
Read ICOM Codes

Now read the external arrows on the diagram and locate their ICOM codes. Match each ICOM code to one boundary arrow on the parent diagram. Cite any mismatches or oversights. Doing this will verify that the author has not made an oversight in analysis by forgetting required data. For example, on diagram PMS/A2, ***scrap materials*** has an ICOM code of I3, which matches (counting from top to bottom) the third input arrow on box 2 of PMS/A0.

14.3.3
Read How the Diagram Connects to Its Parent

After concentrating on the appropriate box of the parent diagram, expand your reading to the other boxes. Follow each external arrow of the diagram until it touches another box on the parent diagram, and understand the reason for the connection. This will help you understand the sources and sinks of the data for the diagram under review. For example, ***scrap materials*** come from ***rejected jobs,*** and ***job status*** is reported to the ***foreman.***

14.3.4
Read the Supporting Material of the Parent

Last, read any supporting material associated with the parent diagram. Sometimes, these supplements will provide insights into the reasons why the subject boundary connects to the other boxes in the parent diagram in the way it does. Usually the Glossary Page for the parent diagram is not included in a kit. Go back to the latest kit for the parent diagram, and read its Glossary to get solid definitions for boundary arrows.

14.4
Understanding How a Diagram Fits into a Model

Once you have understood the immediate context of the diagram, focus on how its data fits into the rest of the model. Do this by tracing dataflows and constraints from the top of the model down to the diagram under study. This will give you a complete understanding of the diagram, because a top-down reading reviews the successive functional details that led to developing the diagram and reveals the way arrows were decomposed.

The process of understanding how a diagram fits into a model is repetitive. One starts at the top of the model and reads all the diagrams that link the diagram being reviewed with the context diagram for the model (i.e., the A-0 diagram). A top-down reading of this nature entails reading just the box that forms the boundary for the next diagram in the model hierarchy. Such reading is done to understand the function of the box and the relationship of its arrows. If supporting material accompanies any of these diagrams, that material should also be read. (This is the reason for keeping all prior kits and versions of diagrams. Without them, quick referral would be difficult, and one could not read the new kit in light of prior work. For example, it is often vital to know what has changed on a particular diagram.)

For example, the A-0, A0, and A2 diagrams for the Prototype Machine Shop model appear in Figures 6-1, 14-1, and 14-2, respectively. By reading these diagrams in this order we notice that: (1) ***machines and tools*** are considered to be major inputs of the shop as viewed by the manager, (2) the ***work kit*** seems to be the only source of ***job plans,*** and (3) the ***quality standards manual*** does not seem to affect machining activities. The Glossary Pages for the A-0 and A0 diagrams would explain in more detail the nature of these things, and hence might clarify why the diagrams say what they do.

14.5
Criticizing the Author's Message

At this point in the reading process, the SADT reader has acquired an understanding of the diagram, its immediate context, and how the diagram fits into its model. Good SADT readers know only what is written on paper; they assume nothing. Thus, their acquired understanding comes from only the model and its supporting material. Now comes the time when the author's message is constructively criticized according to the reader's newly acquired understanding. Criticizing means questioning the message told by a diagram. SADT readers ask three basic kinds of questions:

1. Is the syntax of the diagram correct?

2. Do I understand the author's message?

3. Do I agree with the author's message?

These questions are asked in the order given above to ensure that small issues are covered first and large issues are dealt with last. Syntax issues, though small in nature, are important to note because good communication starts with a proper use of the SADT graphic language. Issues of understanding come second because criticism is useless if a person does not have the facts straight. Issues of agreement are raised last because they are the most involved. They are often complex, requiring a good deal of thought and explanation. The remainder of this chapter delineates the specific questions to ask while criticizing. Chapter 15 will discuss how to document an issue stemming from these questions with written and numbered comments.

14.5.1
Questioning Syntax

When reading to understand the details of the diagram itself, first ask the following questions, especially if you are just beginning to read SADT diagrams or if the author is just starting to practice SADT:

- Are all boxes numbered correctly?
- Do all boxes have verbal names?
- Are there missing arrowheads?
- Do all arrows have noun labels?
- Are all labels clearly connected to one arrow?
- Are there extra labels on long arrows?
- Are there unlabeled arrows?

When reading to understand the immediate context of the diagram, ask the following questions:

- Do all external arrows have an ICOM code?
- Do the ICOM codes correctly tie external arrows with boundary arrows on the parent?
- Are the labels of external arrows compatible with the labels of boundary arrows on the parent?
- Is tunneling (parenthesized arrows) used excessively or incorrectly?

14.5.2
Questioning Your Understanding of the Message

One understands the message of a diagram by following its story line and continually checking how the boxes work, how and why boxes constrain each other, and why data is transformed in particular ways. By doing these things, you begin to verify the decomposition. When reading each box, ask yourself:

- What is the role of each box in the diagram?
- How is each box activated?
- Is the role of each arrow clear?
- How does the box transform its inputs to outputs?
- Is important error handling clear?

When reading the internal arrows to understand the main dataflow path, ask yourself:

- Is the main story line clear?
- Are alternate dataflow paths understood?
- Does the terminology support the story line?

When reading to understand the immediate context of the diagram, ask yourself:

- How do the boxes decompose the parent box?
- What are the source and target of all external arrows?
- Are the main inputs, controls, and outputs clear?

Understanding also depends on the simplicity of the story being told. Checking for conformance to proper diagramming layout conventions is also good to do, for a diagram that is overly complex hinders communication and ultimately impedes understanding. To do this when reading the details of a diagram, ask yourself the following questions:

- Are there too many or too few boxes?
- Should the boxes be reordered?
- Is part of the diagram unusually busy or inactive?
- Are there too many arrows?
- Are arrow crossings confusing?
- Are there multiple arrows with the same ICOM?
- Are labels too long or too wordy?
- Is jargon heavily used?
- Does the terminology match the viewpoint of the intended audience?

14.5.3
Agreeing with the Author's Message

Agreement comes from evaluating a diagram based on: (1) your understanding of what it says and (2) your understanding of what you believe it should say. Agreement is, therefore, a validation step. Validation is done on the decomposition, purpose and viewpoint, correctness, precision, and box activations. To evaluate the decomposition of a diagram, ask yourself:

- Is the decomposition complete?
- Is there a missing box?
- Is there a box that doesn't belong?
- Are there any surprises in the decomposition?

- Would I have come up with a radically different decomposition?

To evaluate the purpose and viewpoint of the diagram, ask yourself:

- What questions will this diagram answer?
- Does this match the purpose of the model?
- From whose viewpoint is the system being described?
- Does this viewpoint match the viewpoint of the model?

To evaluate consistency, ask yourself:

- Is the diagram too vague or detailed to answer the question set implied by the purpose of the model?
- Does the diagram answer questions outside the purpose of the model?
- Are all terms used from the same viewpoint?
- Are the facts relevant to the viewpoint of the model?

To evaluate correctness, ask yourself:

- Does the model reflect reality?
- Are boxes properly placed in order of their dominance?
- Are there extraneous or missing arrows between boxes?

To evaluate precision, ask yourself:

- Are titles and labels misleading?
- Do arrow branches contain only the data which is actually required by a box?
- Do two boxes overlap in their functionality?
- Are there unnecessary arrows touching a box?

To evaluate how the boxes are activated (i.e., how they may be sequenced during actual system operation), ask yourself:

- Do "normal" dataflow paths seem to work?
- How will wrong data affect a box?
- Are essential error paths accounted for?
- Should a function do more work than is implied by the arrows touching it?

In summary, one of the most useful questions to ask about a diagram is, "What new facts have I learned by reading the diagram?" This then leads to the ultimate question, "Was the diagram worth reading?" If a diagram is worth reading, it is definitely worth understanding and criticizing, and is probably worthy of becoming part of a model.

14.6 Summary

Reading is a basic skill of all who participate in an SADT project. The first goal of reading is to understand the details of a single diagram. To do this, the title and node number, boxes, internal arrows, author notes, and supporting material are first read. The second goal of reading is to understand the immediate context of a diagram. To do this, the boundary box and its arrows, the ICOM codes, and the parent diagram and its supporting material are read. This helps determine how the diagram fits into the whole model. The last goal of reading is to criticize the author's message. To do this, a reader questions the syntax used, the arrow and box labels, the main story line, and box activations. All these steps are done by readers in order to form opinions about the validity of a diagram. It is in this way that a group of readers evaluate the work of an author.

Suggested Readings

Freedman, D., and G. Weinberg: *Walkthroughs, Inspections, and Technical Reviews,* Little, Brown, Boston, 1982.

MacKay, D.: *Information, Mechanism and Meaning,* MIT Press, Cambridge, Mass., 1969.

Macnamara, J.: *Names for Things,* MIT Press, Cambridge, Mass., 1982.

O'Rourke, J.: "Writing for the Reader," Digital Equipment Corporation, 1976.

SofTech, Inc.: "IDEF0 Forms and Procedures Guide," SofTech Deliverable no. 7500-11, September 1979.

Chapter

15

Commenting Constructively

When reading an SADT diagram, one also documents issues as they arise. SADT has the following systematic process, called "commenting," for recording all issues: (1) record work times, (2) check the contents of the diagram's frame, (3) use simple agreement and disagreement markings whenever appropriate, (4) use notes to record substantive and constructive comments, (5) use the SADT Reference Language whenever possible, and (6) reread the kit before returning it to the author. We will now present the mechanics of the SADT commenting process and conclude with a discussion of how to make effective and constructive comments.

A small but important point should be discussed before we begin. SADT views comments as entities separate from the actual diagram. One can think of comments as being made on a clear piece of plastic placed over the diagram. In other words, comments are overlays put on the diagram by the reader; comments are never considered part of the original diagram. SADT requires the use of red for all comment markings so they can be distinguished from the actual diagram. (Light gray is used to denote the color red in all figures of this book.) Figures 15-1 and 15-2 show a completely commented kit. Notice that the comment markings are overlaid on top of the original graphics and text.

15.1 Recording Work Times

SADT recommends that each reader record the time taken to read and comment on a kit. This allows for monitoring and improving one's skills, judging the complexity of a kit, and estimating the time required to complete work outstanding. To do this, simply label and record in red the time when you start and finish a kit in the Special Instructions area of the Kit Cover Page. If you are just beginning to use SADT on the job, we particularly recommend that you record your work times. If you do, you will notice a steady drop in the time it takes you, as a new SADT practitioner, to process a kit. After a while, the time will remain constant at around 20 minutes, depending on the complexity of the kit.

15.2 Checking Frame Contents

Checking frame contents is a quick task. First check the author's name, project, date, revision, and C-Number to distinguish this diagram from all others. For example, the PMS/A2 diagram in Figure 15-2, made for the Prototype Machine Shop (PMS) project on 5/5/86, has a C-Number of DAM010. Second, check the context box, title, and node number to see how the diagram fits into its model. For example, the ***complete work kit*** diagram decomposes the second box of its parent and is at the A2 level in the Prototype Machine Shop model. Third, check the status area to determine the level of approval this diagram has received. For example, the ***complete work kit*** diagram is just a working diagram and thus has received little or no review. Last, place your initials and the date of your review in the Reader Area. If you do these things first, you will find that they soon become automatic.

15.3 Simple Agreement and Disagreement Markings

Simple agreement is recorded with a red check mark, simple disagreement with a red cross mark. Normally, these markings are used quickly to comment on a particular part of a diagram or author note. For example, the control arrow for box 2 on diagram PMS/A2 received a check mark because the reader agreed that only ***notes*** are

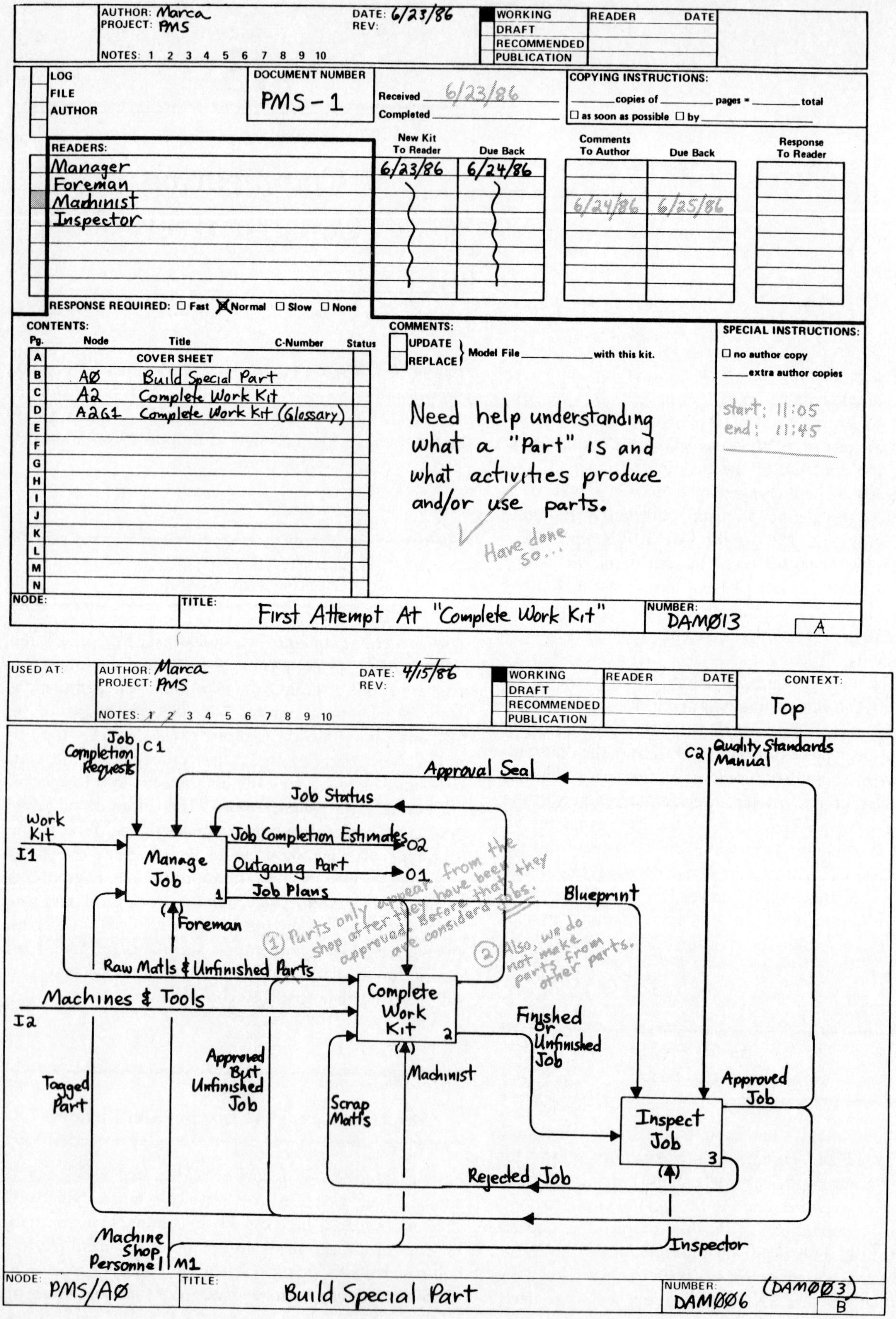

FIGURE 15-1 Commented Kit Cover Page and Parent Diagram

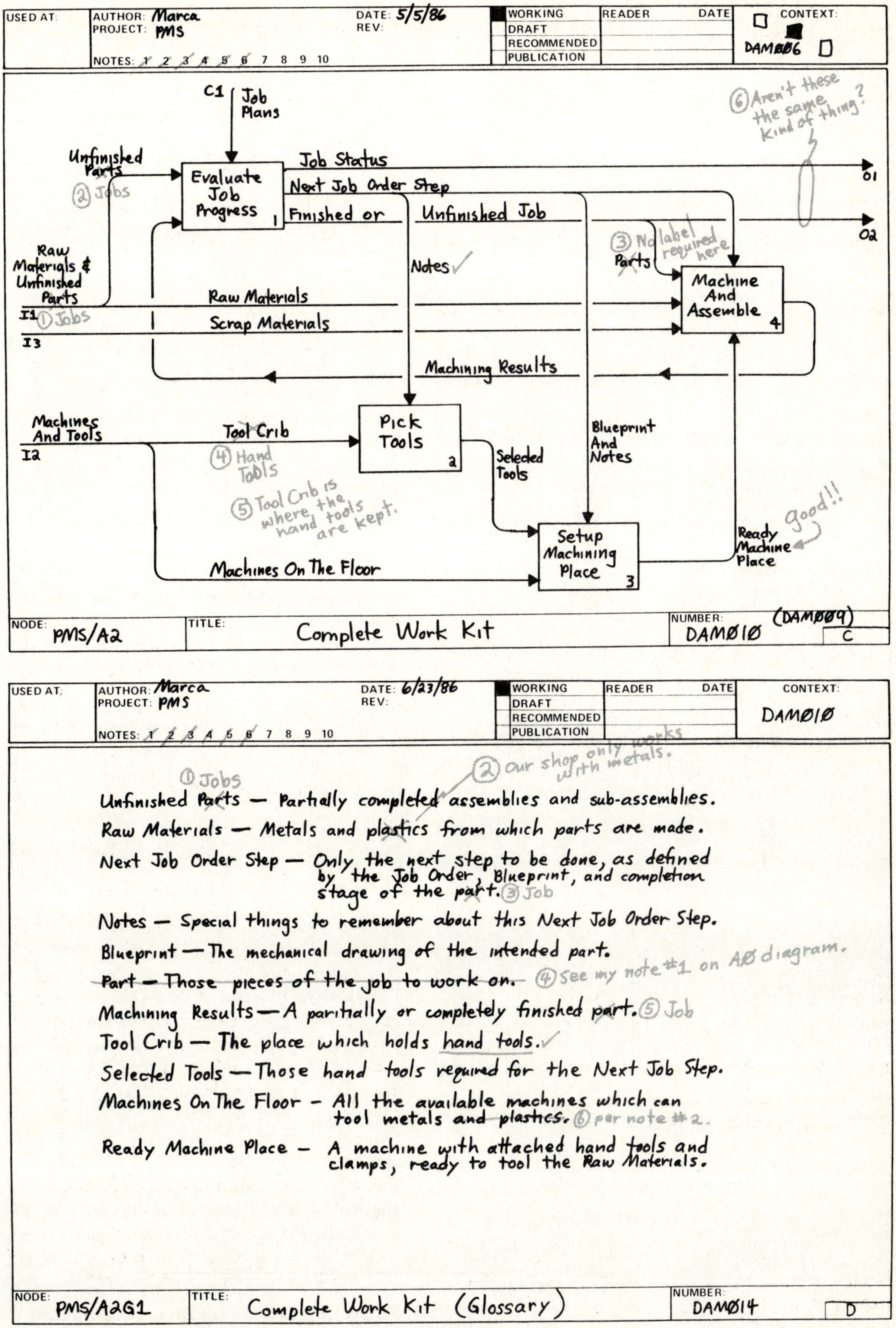

FIGURE 15-2 Commented Diagram and Supporting Material

used to determine what tools to pick. Also the label for the input to box 2 was crossed out, indicating the reader did not agree with the name given to that arrow. At the very least, every diagram and page of supporting material should receive a check or cross mark. These marks tell the author that the diagram or page was read and that the reader either generally agrees or disagrees with the message being presented. We recommend, however, putting more substantive comments on every diagram or page in a kit.

15.4 Making SADT Notes

Substantive comments are recorded using SADT round notes. To make a note: (1) make a tick mark over the next available note number, (2) write that number down and circle it, (3) write down your issue next to the circled number, and (4) if necessary, connect the note to the appropriate part of the diagram with a squiggle. By using note numbers, you uniquely identify each note made. By circling note numbers, you distinguish your notes from notes the author may have also written on the same page. Connecting each note to a diagram with a squiggle pinpoints where the note applies. For example, the ***complete work kit*** diagram in Figure 15-2 received six reader notes. Notice that the squiggle of note 6 encircles two arrows, indicating the note applies to both arrows. Notice too that notes 4 and 5 pertain to the same issue. One is a correction and the other is an explanation.

SADT distinguishes between round notes and square notes. Round notes are overlays put on the diagram by readers and authors during the Author/Reader Cycle. Therefore they appear in both red and blue. Square notes, on the other hand, are put on the original diagram by the author, usually in order to make a strong point about the diagram. Square notes are therefore a permanent part of a diagram, are always drawn in black, and do not appear often in a model. For example, refer to the models in Part VI to see how square notes are used to convey key points about a model.

15.5 The SADT Reference Language

Too much text clutters a diagram and its meaning. The same is true for making notes. The SADT methodology therefore has a Reference Language to simplify note writing. The Reference Language helps one identify a specific box, arrowhead, or note on a specific diagram. A reference code is built from a diagram node number or kit page letter and an ICOM code or note number. ICOM codes are used to refer to specific arrows for a particular box. Table 15-1 gives some examples.

The Reference Language is also helpful when writing text about a diagram. For example, in section 14.2.3, Read the Internal Arrows, the description of the main dataflow path through ***complete work kit*** given in English could have been written using the SADT Reference Language as: "C1 to 1C1 to 1O2 to 2C1 and 3C1 and 4C1, repeating when 4O1 to 1I2." Because the notation is compact, it is common to see English narrative followed by a parenthesized Reference Language expression which details the navigation path taken by the story through the diagram. We recommend using the Reference Language after you and your Reader Community get comfortable with the rest of the Author/Reader Cycle. In fact, you will probably find yourself wanting to use the reference language once you are in the middle of your first SADT project.

15.6 Rereading the Kit

After a kit has been commented, it is wise to go back and reread it. Rereading is not busywork. As with any kind of writing, SADT first drafts are rarely good enough to publish. So by going over your comments, you can mark important notes with a check mark, mark irrelevant notes with a cross, clarify weak notes, add new notes, and summarize your review on the Kit Cover Page. There are few reviews that could not be strengthened by rereading. Our experience has shown that the quality of review greatly increases if one invests a couple of minutes in rereading the entire kit.

TABLE 15-1 Examples of the SADT Reference Language

Code	Format	Meaning
3	Box	Box 3.
5	Note	Note 5.
O1	Arrowhead	The boundary arrow O1.
3I2	Box + arrowhead	Second input arrow of box 3.
.	.	See.
E.2	Page + box	On page E, see box 2.
A.1	Page + note	On page A, see note 1.
A42.3	Diagram + box	On diagram A42, see box 3.
B.4O1	Page + arrowhead	On page B, see first output arrow of box 4.

15.7 Making Constructive Criticism

No one likes to receive comments about one's work that criticize without suggesting ways to improve what was done. SADT strongly advocates a positive and constructive commenting style. Constructive comments are brief, clear, positive, and specific. For example, notes 4 and 5 in Figure 15-2 not only correct a mistake made by the author but also explain why the original arrow label was incorrect.

It is especially important to be helpful when an author makes a large oversight, has misunderstood basic terminology, or has drawn a diagram outside the purpose or viewpoint of the model. In such cases, we recommend you make a note on the Kit Cover Page describing the problem as you see it and request time with the author to discuss the matter. During the subsequent Author/Reader Talk, use the kit as the agenda for the meeting and talk through your concerns in note number order. In this way, the kit will help focus discussion and keep the meeting short.

Good SADT readers are not satisfied if they see nothing wrong. They must also see everything right because important points could have been overlooked, parts of a diagram could be irrelevant, or a diagram could be drawn from the wrong viewpoint or for the wrong purpose. That is why one should point out the good as well as the bad. Authors need to know they are on the right track, even if they have missed some minor points. For example, in Figure 15-2 the annotation on the ***ready machine place*** arrow praises the author for insight and for the very descriptive name.

Good SADT readers also make the most of their comments. One of the best ways to make comments is to use the SADT graphic language. By doing so, a reader can quickly: correct syntax problems, suggest alternate labels, reroute arrows to different boxes, and offer alternate decompositions. We recommend using graphics whenever possible, because fewer words often lead to more communication. Figure 16-3 gives an example of a graphic note.

15.8 Summary

The documenting of issues as they arise when reading a diagram is called "commenting." Commenting is a skill, and constructive comments—those that criticize with the goal of improving a diagram—are the only kind readers should make. Comments are written directly on the pages of a kit in red. Simple agreement and disagreement marking conventions, the SADT Reference Language, and the SADT graphic language make comments and references to diagram parts easy to write. Numbered notes are used to keep track of each comment and the sequence in which they were written.

Suggested Readings

Elbow, P.: *Writing with Power,* Oxford University Press, Oxford, England, 1981.

Freedman, D., and G. Weinberg: *Walkthroughs, Inspections, and Technical Reviews,* Little, Brown, Boston, 1982.

Macnamara, J.: *Names for Things,* MIT Press, Cambridge, Mass., 1982.

O'Rourke, J.: "Writing for the Reader," Digital Equipment Corporation, 1976.

SofTech, Inc.: "IDEF0 Forms and Procedures Guide," SofTech Deliverable no. 7500-11, September 1979.

Weinberg, G., and D. Freedman; "Reviews, Walkthroughs, and Inspections," *IEEE Transactions on Software Engineering,* vol. 10, no. 1, January 1984.

Chapter

16

Reacting to and Consolidating Comments

The process of responding to the comments of readers is called "reacting." Reacting is done by an author after receiving a commented kit from the Librarian. SADT advocates following a systematic process of reacting to ensure that authors give consistent and quality feedback to their Reader Community. To react to a commented kit, the author: (1) reads and responds to every comment made by the reader, (2) may decide to hold an Author/Reader Talk to discuss major issues, (3) consolidates all comments on the master copy of the kit, and then (4) redraws the diagram according to his or her set of consolidated notes.

A small but important point should be discussed before we begin. SADT views reactions as entities separate from both the diagram and the reader comments. That is, they never become a permanent part of the diagram. Reactions are considered overlays put on the commented diagram by an author. (One can think of reactions as being made on a clear piece of plastic placed over the commented diagram.) SADT requires the use of blue for all reaction markings so they can be distinguished from both the actual diagram and the reader comments. (In this book, dark gray is used to denote the color blue in all figures.) Figures 16-1 and 16-2 give an example of a reacted kit.

16.1 Reading and Responding to Comments

First read quickly through the kit, noting the locations and quantity of comments. This way you can get a feel for those parts of the kit about which there is general agreement and those parts of the kit that sparked issues. Also, check the reader times to see how difficult it was for the readers to comment on the kit. Record your work times on the kit cover page too. After reading all the comments, go back and respond to each one separately. To do this, read the comment thoroughly and understand what the reader is trying to say. Then evaluate the recommendation. If you agree, respond with a check mark. If you disagree, respond with a cross mark. For example, notice how on diagram PMS/A2 of Figure 16-2 the author has indicated agreement with notes 1 through 5 and disagreement with note 6.

After marking the comment, add explanation with text if needed. Answer the reader's question, clarify the understanding, or explain why you rejected the comment. For example, the author's explanation of why note 6 in Figure 16-1 was not accepted explains a distinction between the physical *parts* and information about *parts.* For major reactions or new ideas caused by reading the comment, make a new round note in blue near the comment and attach it to the comment with a squiggle. There is nothing wrong with making new notes on an already commented kit.

When reacting or making additional notes, remember the guidelines for constructive commenting. They apply here as well, because authors must be sensitive to their readers when reacting. Especially important is remembering that all your readers will project their own viewpoints onto your work. For example, the *manager* might have had no trouble with the *complete work kit* diagram, but the *machinist* who reviewed the kit certainly did. Also, the *machinist* had some difficulty understanding that *parts* and their *status* are conceptually two different things. The *manager* would probably have less trouble with this distinction. Thus, authors must be sensitive to the viewpoint of each reader and react to comments accordingly, providing explanations and rationale where appropriate.

16.2 The Author/Reader Talk

Sometimes, major disagreements arise, major misunderstandings happen, or important comments cannot be understood. In all cases, record these problems on the Kit Cover Page with a note. We recommend that you always

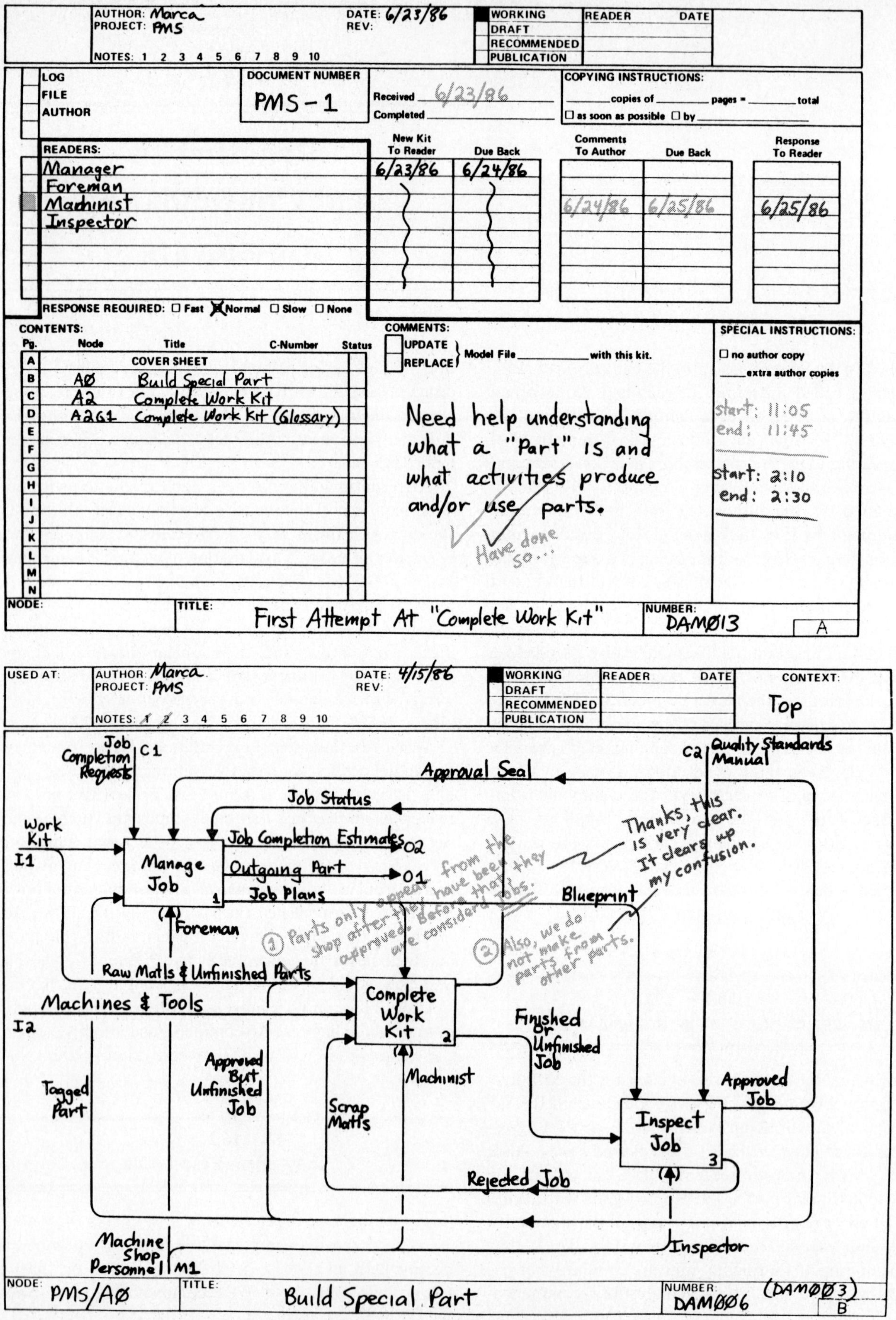

FIGURE 16-1 Reacted Kit Cover Page and Parent Diagram

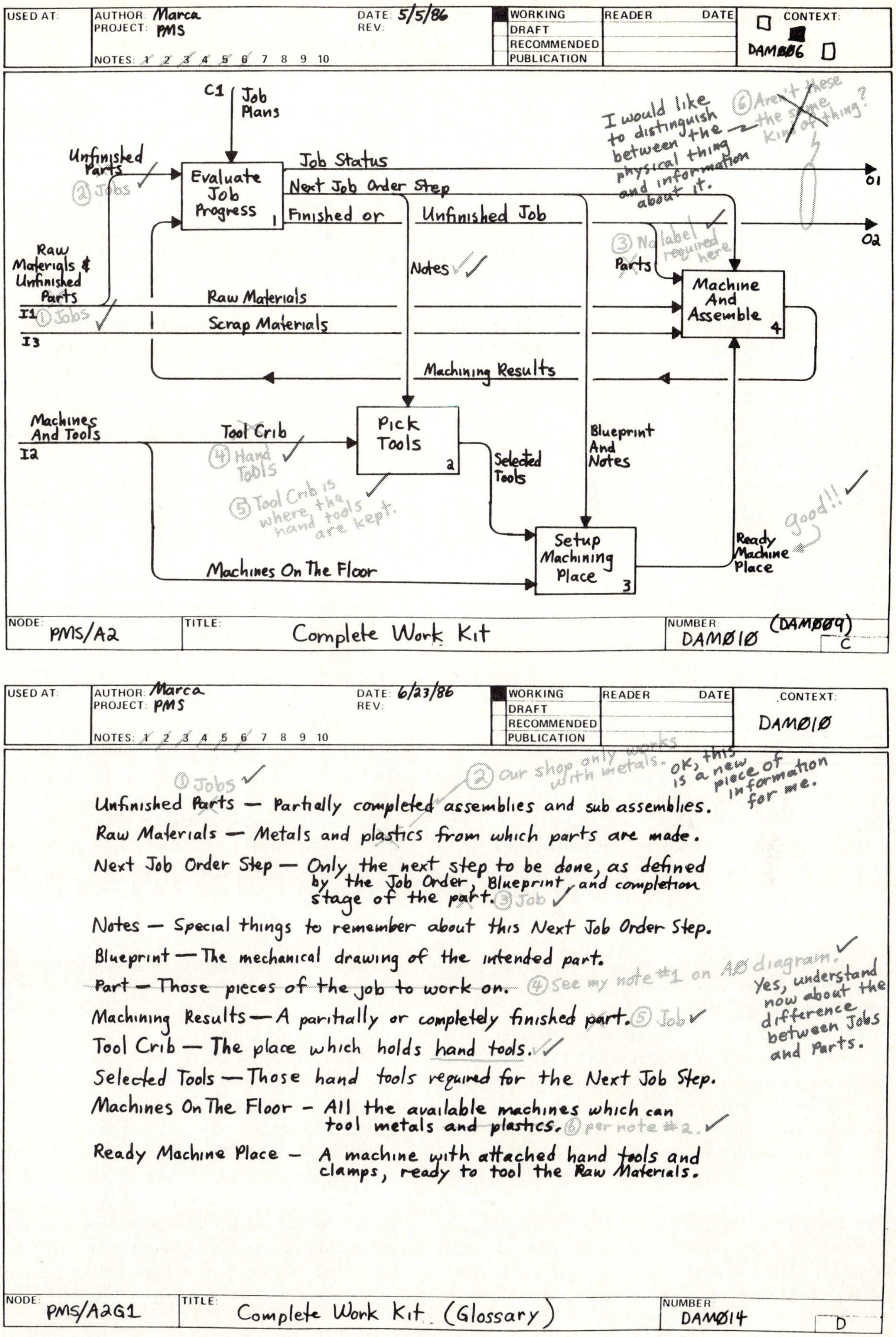

FIGURE 16-2 Reacted Diagram and Supporting Material

add the capitalized words *LET'S TALK* at the end of such a note. This tells the reader that you want to hold an Author/Reader Talk to more thoroughly resolve any issues. By doing this, you encourage the reader to give more feedback to your work. But before you get together and talk, prepare an agenda, including just those topics requiring an Author/Reader Talk. Hold the meeting and address all the topics one by one. Decide on actions that will resolve each issue, and record these on your kit with round notes.

When you meet, stick to topics and keep the meeting to 1 hour. You may find that the meeting takes longer than this because the issues require much deliberation and new topics may arise. Record new topics and quickly get back to the original discussion. If long deliberations are required, we do not recommend you let the meeting continue too much beyond an hour. Our experience shows that everyone's productivity starts to drop after an hour of intense deliberation on hard issues. Instead, set a date and make a new agenda for another meeting. We have found this style of concentrated interaction very productive, because: (1) the SADT Author/Reader Cycle quickly resolves small issues so they need no further discussion, and (2) following the agenda keeps the meetings focused and short.

16.3 Consolidating Reader Comments

Every note you agree to causes a change in the original diagram. Capture these changes by making round notes on your own copy of the kit. Use the complete power of the SADT graphic and reference languages so your annotations can be made quickly and so you can redraw the original diagram accurately. For example, Figure 16-3 shows the annotations made to the ***complete work kit*** diagram after reader comments were processed. Notice that accepting changes in terminology altered arrow labels, and that this resulted in the author's deciding to make a stronger connection between ***machining results*** and ***unfinished jobs.***

Often during the process of reacting to several commented kits, two readers will both give an author valid but conflicting comments. When this happens, the author chooses the one that best pertains to the purpose and viewpoint of the model. An author who has trouble doing this should hold an Author/Reader Talk with both readers to resolve the issue. The final resolution, usually a compromise in the case of conflicting comments, is written onto the author's copy of the kit. Even when there is no conflict, the author may not do exactly what the reader suggested. This can happen when a comment identifies a large problem to solve, suggests a big impact on the model, or can be resolved in a way very different from the original suggestion of the reader.

Keep in mind that annotating your own copy of the kit is not busywork. Most authors find that during this time they add qualitative changes to their diagrams beyond what was recommended by the readers. This happens because authors have the opportunity to interrelate and synthesize the ideas of all readers during consolidation. This often produces still more good ideas, which are realized by more and better qualitative changes to the original diagram. For example, in Figure 16-3, after making stronger connections between ***machining results*** and ***unfinished jobs,*** the author thought harder about the feedback between ***machine and assemble*** and ***evaluate job progress.*** This caused the author to add a new arrow, called ***plan alterations,*** which is indicated by note 7.

16.4 Redrawing Diagrams

The process of annotating your own copy of the kit is done for every accepted comment for every person commenting on a particular kit. After you have completed annotating your copy of the kit, you are ready to redraw the diagram and any related diagrams that were affected. For major changes, we recommend redrawing the diagram from scratch. For example, Figure 16-4 shows the results of a redraw done according to the annotations made on the ***complete work kit*** diagram of Figure 16-3. Notice that note 7 caused the author to draw a control feedback loop from box 4 to box 1, and notice that the author gave it a different label when redrawing the diagram.

When you redraw a diagram from scratch, be sure to indicate the replaced diagram by putting its C-Number in parentheses next to the C-Number of the new diagram, like that shown in Figure 16-4. Notice that the C-Number of original diagram (DAM010) is parenthesized and is placed adjacent to the C-Number of the redrawn diagram (DAM015). It is in this way that SADT links old versions of a particular diagram together through a set of backward pointers, so one can always trace the evolution of ideas during an analysis project.

When redrawing a diagram, be careful to avoid introducing copying errors and omissions. These are common things to do during a complete redraw. When required changes are simple, obtain the original diagram from the Librarian, add the date of the revision in the frame, make the necessary changes, and issue the diagram in a new kit. The Librarian will make the necessary changes to the model files to keep all diagram versions in order. Also, we strongly recommend waiting until you have reacted to all your readers before you actually redraw the original diagram. This way, you avoid wasting your time, your

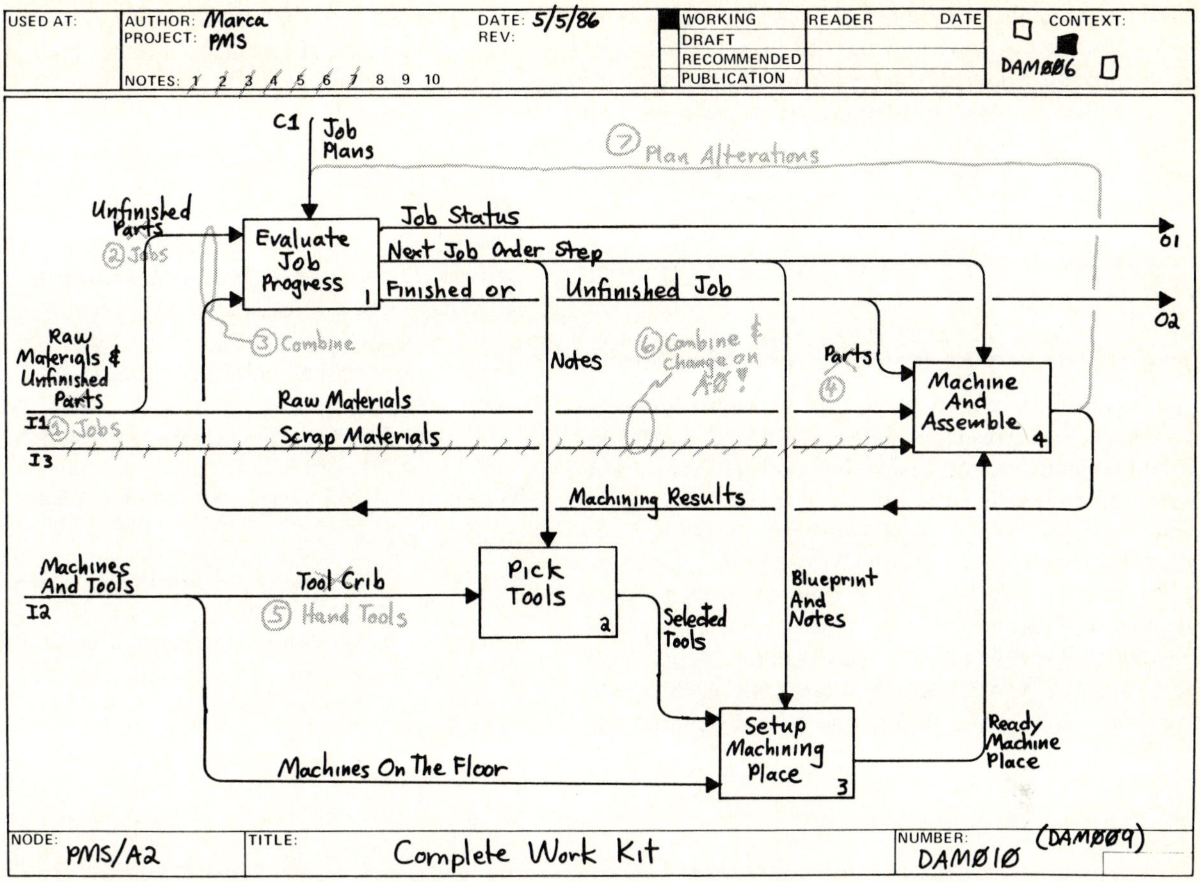

FIGURE 16-3 A Diagram with Consolations

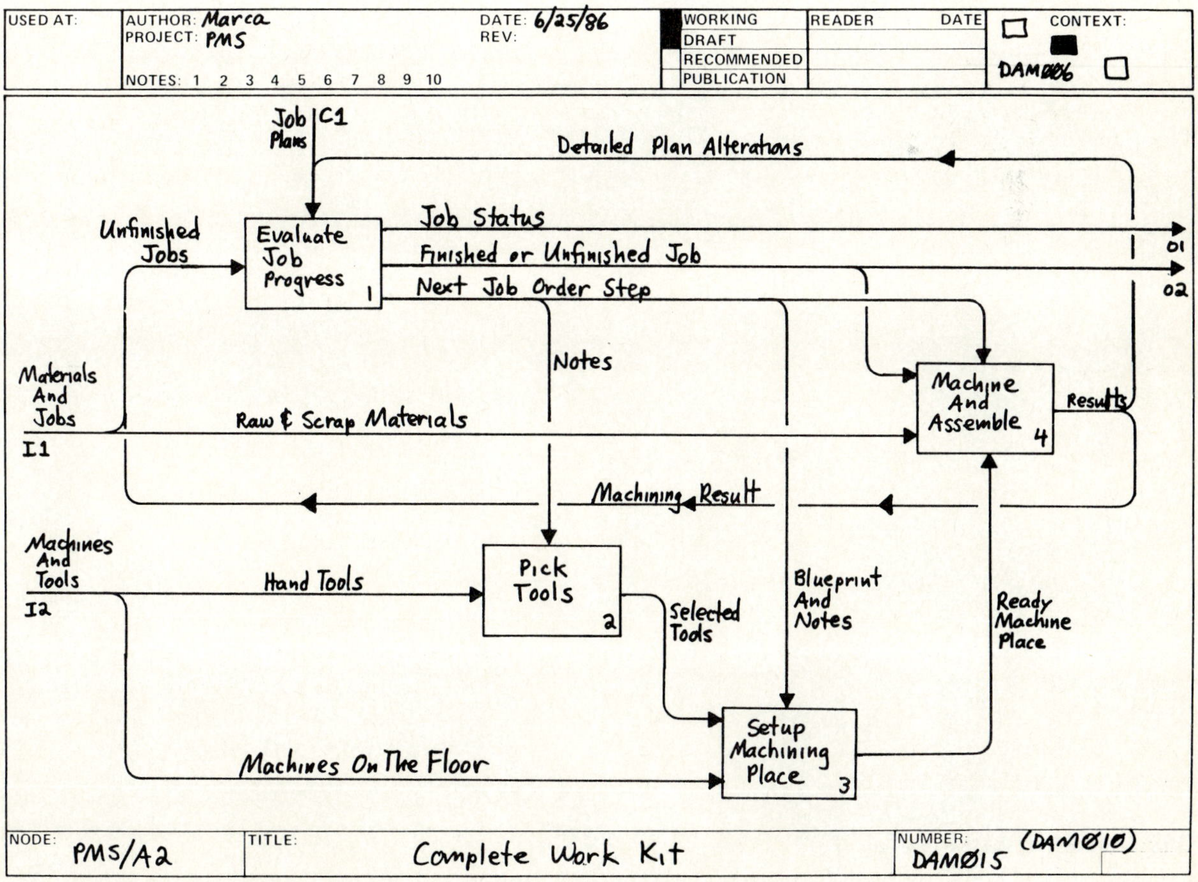

FIGURE 16-4 A Redrawn Diagram

reader's time, and the Librarian's time by issuing a diagram with trivial changes. Wait until you have considered and reacted to all comments, and have resolved all issues and conflicting recommendations, before redrawing the original diagram. After the redraw is finished, issue the redrawn diagram in a new kit.

16.5 Summary

Authors react to comments by writing their response in blue next to each reader comment. They first quickly scan all the comments in the kit and then read and react to each comment in detail. Valid issues are recorded on the author's copy of the kit. Major issues are reacted to by indicating that a talk is required. When all commented kits have been reacted to by the author, the author's kit is worked on again to resolve conflicting reader viewpoints. To do this may require one or more Author/Reader Talks for major issues. After all issues are resolved, the author redraws the diagrams, corrects associated material, and usually issues a new kit. Again, just as during commenting, the reactions of an SADT author should be critical but always constructive.

Suggested Readings

Elbow, P.: *Writing with Power,* Oxford University Press, 1981.

Lannon, J.: *Technical Writing,* Little, Brown, Boston, 1982.

Macnamara, J.: *Names for Things,* MIT Press, Cambridge, Mass., 1982.

Minsky, M.: "Form and Content in Computer Science," *CACM,* vol. 17, no. 2, April 1970.

O'Rourke, J.: "Writing for the Reader," Digital Equipment Corporation, 1976.

SofTech, Inc.: "IDEF0 Forms and Procedures Guide," SofTech Deliverable no. 7500-11, September 1979.

Weinberg, G., and D. Freedman: "Reviews, Walkthroughs, and Inspections," *IEEE Transactions on Software Engineering,* vol. 10, no. 1, January 1984.

Weinberg, G.: *Understanding the Professional Programmer,* Little, Brown, Boston, 1982.

Weinberg, G.: *Rethinking Systems Analysis and Design,* Little, Brown, Boston, 1982.

Weizenbaum, J.: *Computer Power and Human Reason,* W. H. Freeman, 1976.

Part

IV

Finishing and Managing a Modeling Effort

This part of the book discusses the techniques used to determine when to stop a model, how to produce supplementary materials, and how to manage an SADT modeling effort. These topics complete our coverage of making SADT activity models. To illustrate some of the techniques, we have presented some of our experiences with coordinating teams of SADT analysts using quantitative measures like model size, kit size, and review cycles per kit. Also discussed are a couple of the currently available SADT support tools that hint at the future of more productive model building. We collected all this information on SADT project management into one place in the hopes of making it more easily accessible to the managers of system engineering projects.

Chapter 17 presents the set of heuristics senior SADT authors use to determine if a model is complete and should be stopped. Chapter 18 shows how a model can be supplemented with definitions of technical terms, written narrative, and figures. Chapter 19 covers the ways SADT diagrams can be annotated to make explicit the details regarding size and execution of the system being described. Chapter 20 summarizes how an SADT project is typically run. Chapter 21 summarizes the capabilities of two currently available automated SADT tools, and gives examples of their output.

Throughout these chapters we discuss the metrics of good SADT model building—that is, the criteria for managing the process. We do not claim these metrics to be exhaustive. Rather, they are facts about SADT modeling currently in the public domain or are based on our personal experiences and observations. They are presented here both to give closure to the middle of this book which deals with the practical concepts behind building one SADT activity model and to set the stage for a very detailed look at how an SADT activity model is actually developed: the subject of Part V.

Chapter

17

Stopping a Model

One of the questions most asked during SADT projects is when a particular model should be stopped. This question is not always easy to answer, though some heuristics exist for determining reasonable completeness. This chapter presents the rules used by senior SADT authors when deciding whether or not a model should be stopped. (Those rules pertaining to interconnected SADT models are not presented here, because that topic is outside the scope of this book.) We caution the reader at this point to keep in mind that the rules presented here are guidelines for when to stop a model. Sometimes, even senior SADT authors apply these rules and find out later in the project that they made the wrong decision. Only through continued practice will you begin to develop the knowledge needed to make good decisions about stopping a model.

17.1 Growth Patterns of SADT Models

Before discussing the criteria used to stop modeling, let us look at how an SADT model grows. The mathematics of growth state that hierarchic models like those developed with SADT expand geometrically. For example, Table 17-1 shows the size of a complete four-level SADT model, in which each diagram contains four boxes and each box is decomposed by a similar diagram. Such a model has a total of 1365 boxes, and a four-level model having six boxes per diagram has a total of 9331 boxes.

TABLE 17–1 Hierarchic Models Grow Geometrically

Level in Model	Total Boxes in Model	
	Four Boxes per Diagram	Six Boxes per Diagram
Top	1	1
0	5	7
1	21	43
2	85	259
3	341	1555
4	1365	9331

Though the mathematics are correct, SADT models of such size are never built, for several reasons. First, no SADT model is deep in its entirety. Typically, a model is developed in layers and most of it is shallow, generally stopping somewhere around three levels. Experience has shown that it is very common to develop several second- and third-level diagrams only to discover there is enough detail on the first level to satisfy the purpose of the model.

It is also typical, however, for a single portion of an SADT model to be decomposed down to five or six levels of detail. When this happens, usually one box on the A0 diagram is decomposed that far. Functions that require such detailing are often critical ones, and revealing their details usually unlocks key secrets about the way a system works. But though critical functions may require much detail, one usually encounters very few such functions when building a model. Models that have a critical function exhibit an umbrellalike growth pattern by having a shallow and even top and a long stem where the detailing occurs. Thus, a second reason why SADT models do not grow geometrically is that, while it not uncommon for models to have even five or six levels, few are decomposed completely to such a level of detail.

A third reason stems from the fact that large analysis projects are usually broken into several smaller modeling efforts, in which each model emphasizes one particular aspect of the whole problem. A network of several small models is made instead of one gigantic model. For example, one analysis project the authors of this book worked on involved describing the defensive weapon system for a submarine. Instead of making one large model of the boat defending itself, we used a separate model to describe each kind of weapon (e.g., a torpedo), countermeasure (e.g., a decoy), delivery device (e.g., a launching tube), and operator console. This way, a set of small, manageable, and understandable models was developed in lieu of a huge and unmanageable model that would be difficult to read and understand. Do not, however, be misled by these last remarks. Extremely large projects, like the one described in Chapter 29, can produce a high-quality model comprising thousands of boxes, but this is a rare occurrence. Fortunately, most systems do not require models of that size for adequate description.

17.2 When to Stop Decomposition

We recommend stopping a model when there is sufficient detail in the model to satisfy its purpose. In other words, you should stop a model when you feel that going further will not satisfy or will conflict with the information needs of the project. While this general rule makes intuitive sense, it is hard to follow without evaluating the model against a specific set of metrics. The first 10 years of using SADT to develop models during requirements definition in a variety of application domains uncovered some specific model-stopping criteria. This experience showed that for a single model which is being developed independently of any other model, the decomposition of one of its boxes should stop when:

1. The box contains sufficient detail.
2. A change in abstraction level is required to detail the box.
3. A change in viewpoint is required to detail the box.
4. The box is very similar to another box in the same model.
5. The box is very similar to a box in another model.
6. The box is a trivial function.

These rules emphasizes the practical aspects of employing SADT to describe a real-world system for a particular purpose (e.g., understand a telephonic switching station in order to define its software requirements). Since we assume the majority of our readers will use SADT for such purposes, we will now discuss each of the above rules in some detail. We will use the diagram in Figure 17-1 to illustrate these rules and their fine points, since it is a diagram at the "bottom" of the Prototype Machine Shop model.

17.3 Sufficient Detail

The most common situation encountered near the end of a modeling effort is a box that describes the system at the desired level of detail. Testing for sufficient detail is usually quite easy. Simply ask yourself if the box answers all or some of the questions that make up the purpose of the model. If the box does help answer one or more of those questions, then further decomposition may not be required.

For example, the purpose of the Prototype Machine Shop model was to describe the machine shop well enough so a set of job descriptions for machine shop personnel could be written. For this model, a box contains sufficient detail when it describes an aspect of ***building special parts*** to a level where a written procedure about that box can be understood and executed. The ***layout hand tools*** box is a good example, because its function and associated things are simple and detailed enough to write the following description:

> To ***layout hand tools***, first review your ***machine operation notes*** to determine what detailed steps will be done. Lay out the ***hand tools*** around the machine in the same order as the steps. Take into account the ***machine orientation*** so the tools are close at hand during each step.

Since this description seems to fit well with the purpose of the Prototype Machine Shop model, we would choose not to decompose the ***layout hand tools*** box further. The text is an excellent substitute for further decomposition.

If this was our first attempt at stopping the Prototype Machine Shop model, we would use the ***layout hand tools*** box, its arrows, and their labels as a pattern of sufficient detail. Then we would evaluate other boxes on this or future diagrams to see if they fit this pattern of simplicity. If they did, they would immediately be considered candidates for a paragraph of text instead of decomposition. For example, the ***pick machine*** and ***adjust machine*** boxes seem to have the same level of simplicity as the ***layout hand tools*** box. In this way, an SADT author builds a set of heuristics about stopping a particular model when the low-level boxes begin to satisfy the purpose of the model.

17.4 Change in Level of Abstraction

Boxes get decomposed when they are not detailed enough to meet the purpose of the model. But sometimes, one decomposes a box only to find that the diagram begins to talk about *how* the box operates instead of *what* the box does. In this case, we say that a change in level of abstraction has occurred. Thus one can think of jumping levels of abstraction as changing the nature of the story being told by the model (i.e., changing the way the system is being described). In SADT, changing a level of abstraction

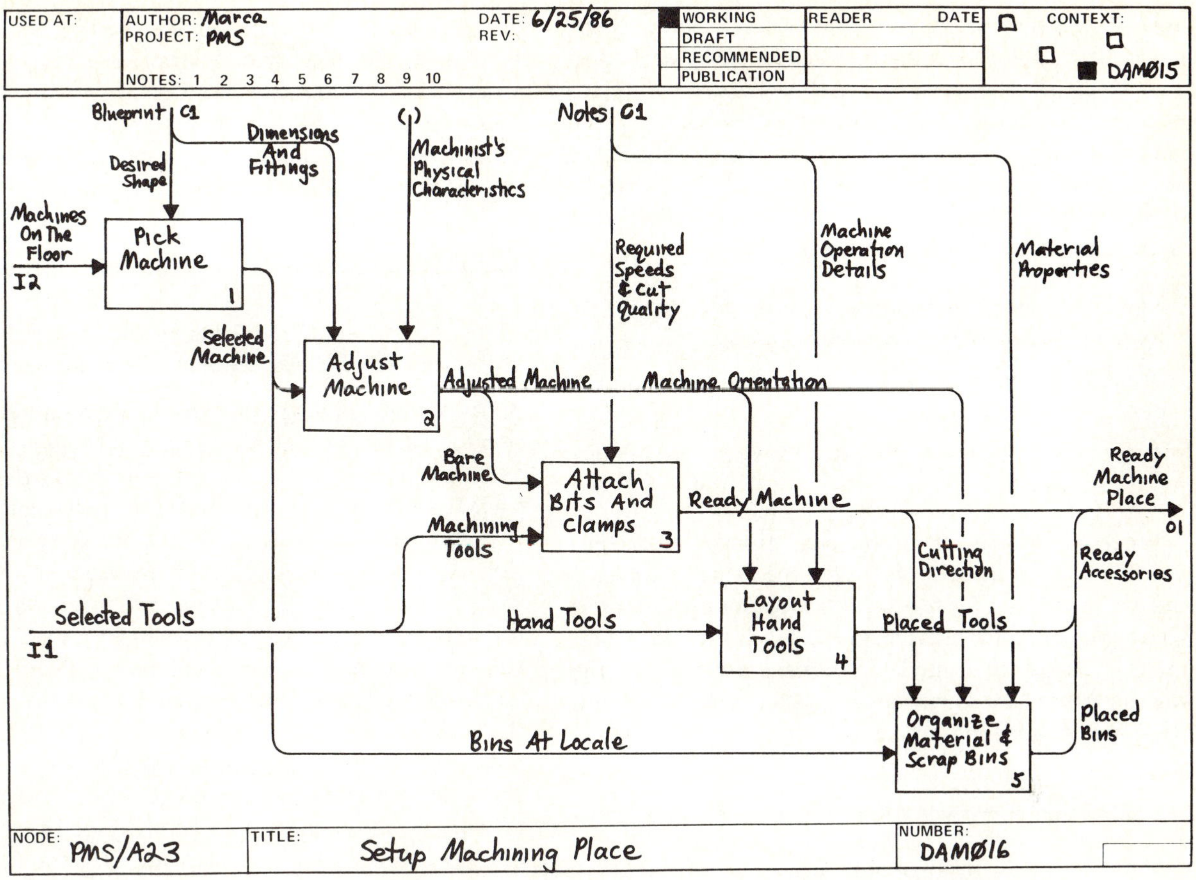

FIGURE 17-1 Boxes Needing No Further Decomposition

often means going outside the purpose of the model and thus signals a stop to decomposition.

For example, upon trying to decompose the ***adjust machine*** box, activities like ***rotate, slide, secure,*** and ***roll*** come to mind. These activities are saying *how* to adjust the machine. Thus they lie outside the model, because they are in fact part of a job description. Getting an understanding of the machine shop so job descriptions could be written was the purpose of the model. In the case of the Prototype Machine Shop model, if a decomposition is actually a job description, then the box should not be decomposed. Instead, a job description paragraph substitutes for the decomposition:

> Observe the current orientation and height of the machine you selected. Now raise or lower it so you can work comfortably. Then change its orientation so you can easily and safely load, tool, and remove the part. You may have to slide, roll, or tip the machine to do this. Study the machine housing before you attempt to adjust the entire machine.

Changes in level of abstraction usually occur after a model is two or three levels deep, and are often easy to spot when you look for them. Subtle changes can still, however, go undetected by authors. Usually, these mistakes are caught when the diagram is sent out for review. The fresh perspective of the Reader Community usually spots the change from "what" to "how." If the change is not caught during the Author/Reader Cycle, decomposing that diagram one more level usually makes apparent the change in abstraction.

17.5 Change in Viewpoint

Changes in viewpoint happen in ways similar to changes in abstraction level. A change in viewpoint happens when the viewpoint of the model cannot be used to decompose a particular box. That is, the box can only be decomposed when one looks at it from a different perspective. This is most often signaled by a noticeable change in terminology.

For example, turn to diagram FF/A23 in Appendix C and consider the function ***travel to store,*** a task within the general area of buying food in order to feed a family. Trying to decompose this box leads to activities like ***catch bus*** and ***ride taxi,*** and to things like ***crosswalk*** and ***bus stop.***

Suddenly, a new set of terminology has crept into the model. In order to decompose ***travel to store,*** we had to view a person as someone being moved about by a complex transportation system and not as the feeder of the family. Since our viewpoint shifted, we would choose not to decompose this activity. This example is a good one, for it points out several commonly occurring situations that signal a possible change in viewpoint:

- You find it hard to decompose a "well-understood" box from the viewpoint of the model.
- Brand-new terminology must be used to decompose a box.
- A system not under study is described by the decomposition.

When any of these things happen, criticize the decomposition to see if it really belongs in the model. If you are unsure, send it out for review with explicit instructions to read the diagram for viewpoint shift. Since a Reader Community usually sees a model from many perspectives, the readers are the best source of information to help you with this kind of modeling problem.

17.6 Similar Functions

Sometimes one encounters a box that is extremely similar to another box in the model. Two boxes are similar if they perform roughly the same function and have almost the same number and types of inputs, controls, and outputs. If the other box is already decomposed, it is wise to defer decomposition and examine the two boxes carefully. If only small changes are required to match the first box with the second, making those changes will save decomposition effort and improve the modularity of the model (i.e., similar functions are detailed in a consistent manner).

For example, consider the ***pick machine*** box in Figure 17-1 and the ***pick tools*** box in Figure 16-3. They both are concerned with picking from a set of given devices, resulting in a single selection. ***Pick tools*** does this according to ***notes,*** while ***pick machine*** does this according to the ***blueprint.*** A more careful study of the two activities might uncover similarity between how ***notes*** and ***blueprints*** are used. If this happens, these two activities could both be further decomposed by a ***pick device*** diagram.

Sometimes this examination and retrofit produces a more general-purpose function. (In fact, this is exactly what happens when subroutines are created to improve the modularity of a software program.) General-purpose functions are often difficult to detect, and the retrofits required to incorporate them into a model do take time. Therefore, we encourage the development of models with general-purpose functions, but we add the caution that it is difficult and expensive to attain generality on a very large scale. We do, however, foresee that the automated SADT tools of the future will help analysts in this and other vital areas which would greatly improve the quality of models and reduce the time required to develop them.

17.7 Trivial Functions

A trivial function is one that absolutely needs no explanation to understand. The advisability of forgoing decomposition in this case seems obvious to us, because the role of SADT is to make a difficult problem understandable, not to pedantically elaborate details that are truly understood. For example, consider the ***pick machine*** box in Figure 17-1. It may very well be looked upon as a trivial function, because machinists know very well which machine is best suited to tool metal in a particular way. Thus, this "trivial" box should not be decomposed, for that could insult the machinists. It is wise to avoid decomposing trivial boxes, especially in very politically sensitive organizations, where the decomposition of particular boxes may actually do more harm than good. Trivial functions are best described with a small amount of text.

It is worth noting that "trivial" does not imply "useless." Trivial functions do very important work in providing details to more complex functions, and sometimes they even tie together major subsystems. Trivial functions should not, therefore, be passed over during analysis. Instead, their existence should be documented, and they should be detailed as any other function is. We do, however, caution the reader not to spend inordinate amounts of time analyzing the trivial functions of a system. Concentrating too much on details may result in a model that lacks good abstraction, making it hard to understand and use.

17.8 Making Good Decisions about Stopping a Model

The chances of making bad decisions about when to stop a model can be lessened if you evaluate each box you wish to detail according to the above rules. Since most SADT models usually comprise from 10 to 30 diagrams, the best time to start evaluating boxes is when the model begins to reach these proportions. Once a section of the model requires no further decomposition, always review your model-stopping criteria before starting to decompose a box that has not yet been detailed.

Keep in mind that the criteria given in this chapter are

guidelines, hence they cannot be mechanically applied. In fact, you may even find occasions when two rules conflict with each other in a particular modeling situation. Resolving such conflicts requires good judgment and discretion—the products of experience. If such a situation happens and you are really unsure of how to proceed, send a special kit out for review. Use the Author/Reader Cycle to enlist the efforts of a senior SADT analyst, who will be most likely to give you the extra information you need to make a sound decision. Also, we advise keeping copious notes on the issue, how it was resolved, and the reasons why. These notes will be useful later if it is discovered that the model should have been stopped earlier. Such notes also become a guide for identifying and selecting subjects for future modeling efforts.

17.9 Summary

SADT models are hierarchic, and so they grow geometrically. Though many SADT models have been decomposed down to five or six levels, models are often made up of tens of diagrams and rarely go beyond the 100-diagram limit. This is because the decomposition of a model or particular portions of it is always stopped when the model is detailed enough to suit its purpose. This happens when the model has enough detail to answer all the questions summarized by the model's purpose. This can also happen when a change in abstraction level or viewpoint occurs. Also, decomposition of a box can be stopped when it is noticed that the box functions very similarly to another part of the model that is already decomposed. Thus, sufficient detail, change in level of abstraction, change in viewpoint, and similar functionality are the major criteria SADT analysts use to determine when to stop decomposition.

Suggested Readings

Alexander, C.: *Notes on the Synthesis of Form,* Harvard University Press, Cambridge, Mass., 1964.

Connor, M.: "Structured Analysis and Design Technique—SADT," Auerbach Portfolio no. 32-04-02, 1981.

Doczi, G.: *The Power of Limits,* Shambhala Press, 1981.

Mihram, G.: "The Modeling Process," *IEEE Transactions on Software Engineering,* vol. 2, no. 5, November 1972.

Nishio, T., and K. Nogi: "A Stepwise Composition Technique for User Requirements Definition," International Workshop on Models and Languages for Software Specification and Design, University of Laval, Laval, Canada, March 1984.

Ross, D.: "Principles of Structuring," SofTech Technical Paper no. TP082, November 1978.

Ross, D., and K. Schoman: "Structured Analysis for Requirements Definition," *IEEE Transactions on Software Engineering,* vol. 3, no. 1, January 1977.

SofTech, Inc.: "IDEF0 Author's Guide to Creating Activity Diagrams," SofTech Deliverable no. 7500-13, September 1979.

SofTech, Inc.: "The DWS/CS Emergency Preset Structured Specification," Technical Paper no. 1083-1, August 1981.

Chapter

18

Supplementing Diagrams and Models

One value of an SADT model is the way it organizes and presents information: The top diagram of the model summarizes the entire system being studied, first-level diagrams present the major subsystems and their interconnections, and diagrams at the lowest levels present the detailed functions that actually make the system work. Once an SADT model is finished, its diagrams organize all the important abstractions and details of a system in an orderly way. Good analysts take advantage of this organized collection of information by supplementing it in a variety of ways. Supplements, elaborations which are not placed on the diagrams themselves, enrich the informational content of the model. Since supplemental information is not formally part of the model, SADT recommends putting such facts on their own pages and linking them to the diagrams of the model.

18.1 Supplementing Diagrams

Good SADT analysts keep their supplemental information focused by letting a particular aspect or part of a single diagram bound their extra descriptions. This way, they maximize their writing or drawing efforts, because they need provide only small amounts of detail. (The model provides the context needed to relate the supplementary information to the system.) For example, notice in Figure 18-3 how well a small amount of very focused text communicates how to set up a machining place.

The types of supplemental information for SADT diagrams are text, figures, and glossary. Text is narrative usually written about part of a diagram. Figures are pictures drawn to clarify particular points. Glossary is a collection of definitions about the things and activities appearing on a diagram. To show how SADT analysts supplement a diagram, we will develop text, glossary, and a figure for the ***setup machining place*** diagram of the Prototype Machine Shop model given in Figure 18-1. Though our sample diagram is at the bottom of the Prototype Machine Shop model, supplements are typically developed for all the diagrams in a model.

Supplementary information is written or drawn on the standard SADT form. Since supplements are concerned with elaborating particular diagrams in a model, the SADT node numbering scheme is used to identify and connect a supplement page to the diagram being elaborated. To do this, the node number of the diagram is appended with a single letter followed by an integer. The letter stands for the type of supplement (T for text, F for figure, and G for glossary), and the integer sequences this page with other supplemental pages for that diagram. For example, notice how the node numbers for the supplement pages of Figures 18-2 to 18-4 tie themselves to the ***setup machining place*** diagram.

18.2 Defining Terminology with Glossary

Glossary is used to collect and define the new details uncovered by a diagram that decomposes a box, especially if the diagram is the first decomposition of its parent box. For SADT activity diagrams, these new details are either new activities, new things represented by arrows that connect only the boxes on the diagram, or decompositions of external arrows. Figure 18-2 is a Glossary Page that defines some of the new things uncovered by the ***setup machining place*** diagram. For instance, ***selected machine*** goes only between the ***pick machine*** and ***adjust machine*** boxes on the diagram, and it has not appeared before in the Prototype Machine Shop model, so it is considered a new detail and hence requires definition.

Typically, a Glossary Page is made every time a diagram details a box which has never been decomposed. One or two pages usually suffice, because only the new terms are defined. (External arrows need no definition, because they are defined on the Glossary Pages of the parent dia-

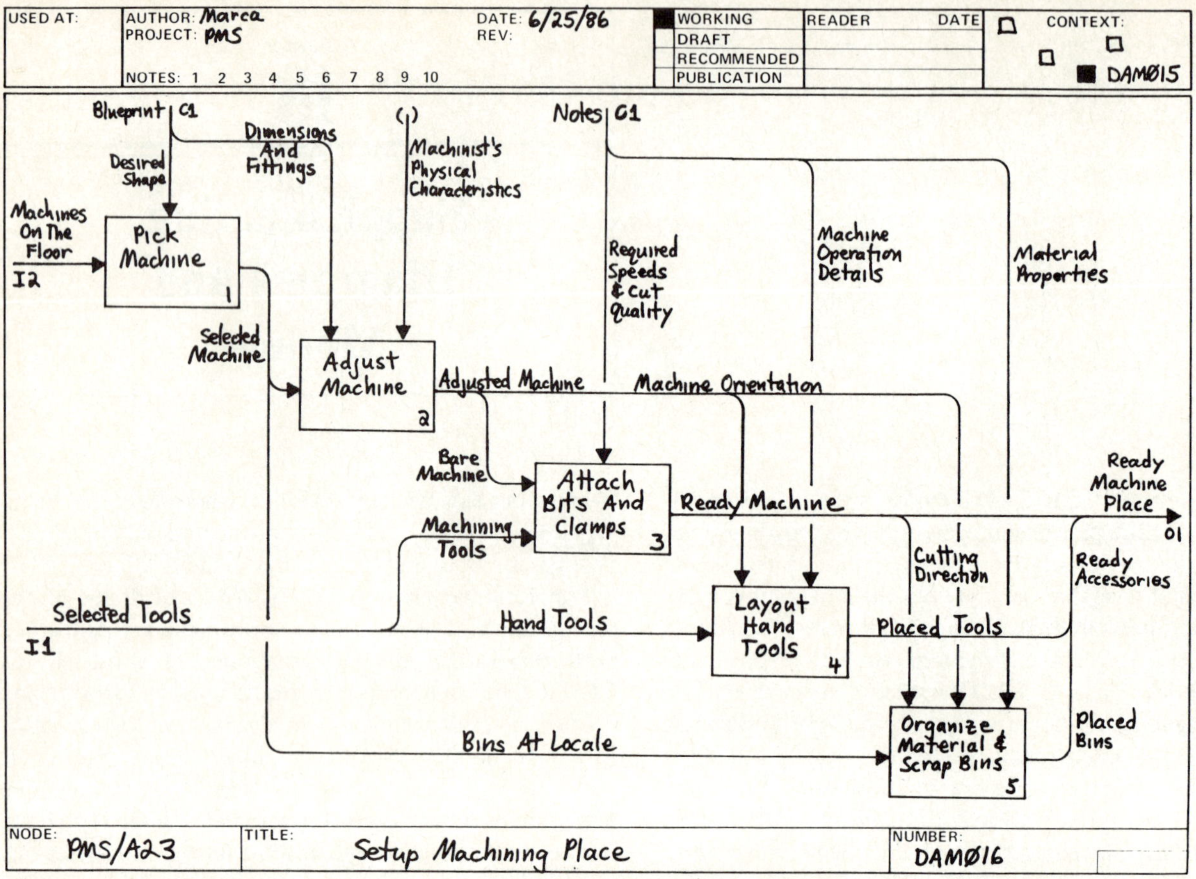

FIGURE 18-1 A Diagram Ready for Supplements

USED AT: | AUTHOR: Marca | PROJECT: PMS | DATE: 6/29/86 | REV: | NOTES: 1 2 3 4 5 6 7 8 9 10 | WORKING | DRAFT | RECOMMENDED | PUBLICATION | READER | DATE | CONTEXT: DAMØ16

Desired Shape — the overall appearance of the part.

Dimensions And Fittings — the size of the part and the way its pieces fit together.

Machinist's Physical Characteristics — anything about the machinist himself that influences how to orient and adjust a machine on the floor.

Require Speeds & Cut Quality — the critical machine performance requirements that affect what kind of bits and clamps to use.

Machine Operation Details — the exact set of tooling steps to be done.

Material Properties — the characteristics of the material being machined that affect how the part can be tooled.

Selected Machine — the machine appropriate for this next job step.

Adjusted Machine — a machine that can be used by the machinist.

Bare Machine — a usable machine without tooling attachments.

Machining Tools — machine attachments that will shape the part.

Hand Tools — tools held by the machinist during a tooling operation.

Bins at Locale — scrap and material containers.

Machine Orientation — the position of the machine.

Ready Machine — a machine ready to tool a part.

NODE: PMS/A23G1 | TITLE: Setup Machining Place (Glossary) | NUMBER: DAMØ17

FIGURE 18-2 A Glossary of New Terminology

gram.) For example, the external arrow labeled ***blueprint and notes*** was first uncovered and defined by the PMS/A2 diagram and its Glossary Page. So, this contextual information is considered known to the ***setup machining place*** diagram, and thus needs no definition on the PMS/A23G1 Glossary Page.

SADT Glossary Pages define the words, phrases, synonyms, and abbreviations technically used within a particular project. These definitions expand upon the terms appearing in the diagram and thus permit briefer box and arrow labels, which can make diagrams easier to read and understand. Typically, simple lists are made like the one shown in Figure 18-2. We know this Glossary Page defines terms for the ***setup machining place*** diagram because of its node number. But also, we know which version of the diagram these terms apply by the C-Number placed in the Context Area of the SADT form. In this way, the author ties the Glossary Page to a specific version of the ***setup machining place*** diagram, so that it will always be associated with the correct version of the diagram it supplements.

It is typical to create a Glossary Page that captures technical terms as one builds a model, especially when creating brand-new diagrams. Be sure to capture those words and phrases that have special meaning beyond the normal conversational meaning. If you do, you will not loose the details most important about a system—what is meant by its terminology. This method of making Glossary Pages is an excellent way to progressively build a collection of interrelated terminology, commonly called a Data Dictionary, which contains both definitions and structural descriptions of the things and activities of the system being studied. As this set of terminology grows, special kits containing the current Data Dictionary are periodically issued.

Their importance during diagram creation makes Glossary Pages the first supplements to accompany a diagram. SADT recommends that a Glossary Page come with every diagram that details a new section of a model. Also, because new terminology is being defined, the diagram and its Glossary Page are usually enough new information for one person to effectively review. This is why SADT authors usually issue a kit comprising only the new diagram, its parent diagram, and associated Glossary Pages. For example, notice in Part III how a new diagram, its parent, and its Glossary Page make a small but meaningful kit.

18.3 Providing Narrative with Text

Text supplements are usually written to provide a general narrative for a diagram, to help match the orientation of the reader with that of the author, and to describe how the lowest activities in a model transform their inputs to outputs according to their controls. Text supplements help walk readers through diagrams in ways that reduce the chance for misinterpretation and that elaborate system details. For example, notice how the text in Figure 18-3 describes the detailed steps to perform when setting up a machining place. This text is at a sufficient level of detail to be considered becoming part of the job description manual for the machine shop.

Writing good text takes time, even for a single box. So, one usually produces text when a diagram is stable, although text is sometimes used to elaborate a critical point during diagram development. We strongly recommend whenever possible using the graphics and the Author/Reader Cycle of SADT to flush out unclear thinking and produce a diagram that accurately describes reality. Then you can summarize or detail the diagram with well-organized and factual text without wasted effort. For example, if we wrote text right after making the first sketch of the ***complete work kit*** diagram of Figure 9-2, we would have to rewrite it several times because, as shown in Part III, the diagram underwent many and varied changes once it was reviewed.

Good text is also brief. A few carefully chosen words can really illuminate a diagram, and too many can cloud its meaning. Normally, one page of text is all that is needed to elaborate a diagram. More than one page lessens quality by adding unnecessary detail. For example, the text of Figure 18-3 focuses on how a machinist sets up a machine and the place around it. Notice that the text does not discuss extraneous topics like safety standards nor does it discuss exceptional situations like the absence of scrap bins. In such circumstances, it is wise to make separate pages of text for important but secondary topics and for exceptional situations and how to handle them.

Text can also be used to explain why a particular decomposition or a particular viewpoint was chosen, tell how some difficult operation occurs, or describe the assumptions the author made when drawing the diagram. But under no circumstances is text used to describe what can be learned by reading the diagram, detail a box that is already decomposed, identify the components of decomposed arrows, describe box activations, or define the names of boxes and arrows. (These names appear on a Glossary Page.) Follow these guidelines, and your text supplements will always be small, crisp, and informative; and most important, they will get read.

18.4 Clarifying Points with Figures

Though written language is used to define terminology and add narrative, one should always use words sparingly.

USED AT:	AUTHOR: Marca PROJECT: PMS NOTES: 1 2 3 4 5 6 7 8 9 10	DATE: 6/29/86 REV:	■ WORKING DRAFT RECOMMENDED PUBLICATION	READER DATE	CONTEXT: DAMØ16

A231 Pick Machine

Review the overall desired shape of the part, and determine the machine best suited for this kind of tooling operation.

A232 Adjust Machine

Observe the current orientation and height of the machine, raising or lowering it so you can work comfortably. Then change its orientation so loading, tooling and removing the part is easy and safe.

A233 Attach Bits And Clamps

According to the kind of tooling required, attach the neccessary bits and clamps to the machine. Be particularly aware of required machine speeds and tooling quality.

A234 Layout Hand Tools

Review your machine operation notes to determine what detailed steps must be done. Lay out the hand tools around the machine in the same order as the steps. Take into account the machine orientation so the tools are always close at hand.

A235 Organize Material & Scrap Bins

Place bins to easily load raw materials and catch scrap. Take into account material properties and the direction of cuts to determine where scrap will be thrown.

NODE: PMS/A23T1	TITLE: How To Setup A Machining Place	NUMBER: DAMØ18

FIGURE 18-3 Text Descriptions that Detail System Activities

USED AT:	AUTHOR: MARCA PROJECT: PMS NOTES: 1 2 3 4 5 6 7 8 9 10	DATE: 6/29/86 REV:	■ WORKING DRAFT RECOMMENDED PUBLICATION	READER DATE	CONTEXT: DAMØ16

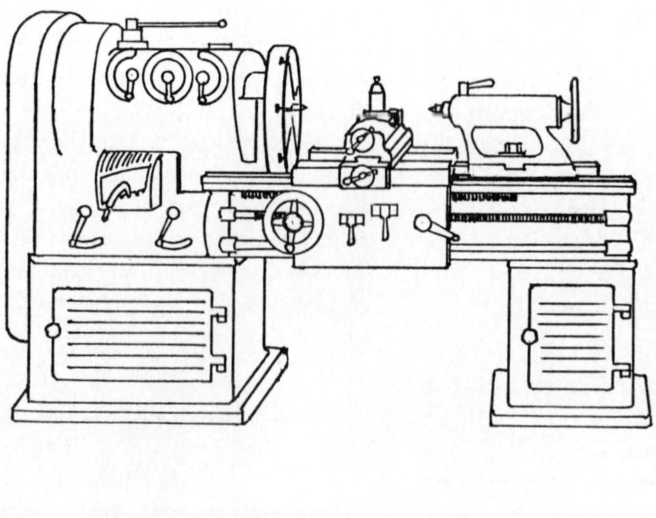

NODE: PMS/A23F1	TITLE: A Ready Machine	NUMBER: DAMØ19

FIGURE 18-4 A Figure of What a Particular Arrow Represents

We advocate using pictures whenever possible to communicate subtle points and to explain what would be difficult to say with words. In SADT, any kind of picture that is not a formal diagram is considered a figure supplement. Usually, SADT figures are either pictures or highlighted diagrams tied to a specific diagram in a model. Ties are made through the node number of the figure by appending the letter F and a sequence number to the node number of the diagram being elaborated, like the one in Figure 18-4.

A figure supplement is any visualization drawn to complement a model. Figures are often used to visualize what a particular part of the system looks like (e.g., a machine ready to use), how two system parts fit together (e.g., how raw materials fit into clamps before tooling), how to properly use a system part (e.g., the correct way to lathe aluminum) a particular event during system operation (e.g., machine startup), and the passage of time (e.g., a calendar of upcoming jobs). For example, Figure 18-4 gives a picture of what a typical ***ready machine place*** looks like.

Good pictures are indispensable but may take time to draw. However, there is always an abundant supply of pictures already available for an author to use—the diagrams in the SADT model itself. It is very common for SADT authors to highlight a few diagrams in the model to visually stress important points. A highlighted diagram can elaborate detail by giving a cross-sectional view of arrow composition, describing box activation patterns, describing system operation in various environments, showing the main dataflow path, detailing the role of feedbacks, describing pathelogical and error-handling cases, or simply documenting an important change to a diagram. For example, Figure 18-5 visually highlights the transformations the initial hardware (i.e., ***machines on the floor*** and ***selected tools***) undergoes to become a ***ready machine place.*** Compare this to the textual narrative of setting up the machining place in Figure 18-3.

Thick lines are most often used to highlight a diagram. However, we use color as often as we can because color superimposes another important dimension of information on a diagram—abstraction by type. For example, try coloring the arrows of the ***setup machining place*** diagram in the following way: color plans red, hardware blue, and other information green. Notice how the use of color summarizes the diagram in a novel way—by categorizing the things represented by arrows. Highlighting an entire model in this way helps describe the impact certain data has on system activities (e.g., which steps require the ***blueprint***) and also points out those activities that operate on a particular kind of thing (e.g., which steps handle ***raw materials***). While color is a limited medium, it is very effective when the number of categories is around 10.

Note that while SADT figures are typically drawings and highlighted diagrams, any language can be used to make a figure. For example, we have used the languages of decision tables and finite-state machines to elaborate SADT models of telephone switching systems and program evaluation and review technique (PERT) charts to convert an SADT planning model into a format understood by the managers of a very large government project. We therefore recommend you use any and all the visual media at your disposal to elaborate your SADT models.

18.5 Supplementing Models

The hierarchic collections of SADT diagrams called "models" introduce their subject and detail it in a very regular, controlled, and understandable manner. They do so because the diagrams of a model are always organized in what is called "node number" order. That is, node A-0 comes first, node A0 comes second, node A1 comes third, node A11 comes fourth, and so on. The ordering of SADT diagrams in this manner is patterned after the "tree" orderings common in mathematics and computer science. A complete SADT model is typically read in two ways. The first way is for overview, in which a person reads all the high-level diagrams before proceeding to the next level of detail. This is termed a "breadth-first" reading. The second way is reading for detail, in which a person reads directly down a particular branch of the tree to the lowest-level diagrams. This is called a "depth-first" reading.

Model supplements are designed to help people navigate through the tree-ordered collection of diagrams while they read the model. The two supplements that do this are called a Diagram Index and a Node Index. Both are indented outlines which define the contents of a model. The Diagram Index outlines all the diagrams in the model by listing their node numbers and names on separate lines. The Node Index outlines all the boxes in a model by listing their node numbers and names on separate lines. Thus, a Node Index is simply a more detailed outline than a Diagram Index. For examples, please refer to the Diagram Indexes we have provided for the models in Part VI. When you do, notice how the outlines provide both an interesting written system summary and a way to rapidly pinpoint discussion about particular system functionality.

18.6 Summary

SADT diagrams sometimes need supplemental information to describe detailed information critical for understanding a system. SADT allows for three kinds of diagram supplements: glossary, text, and figures. Glossary

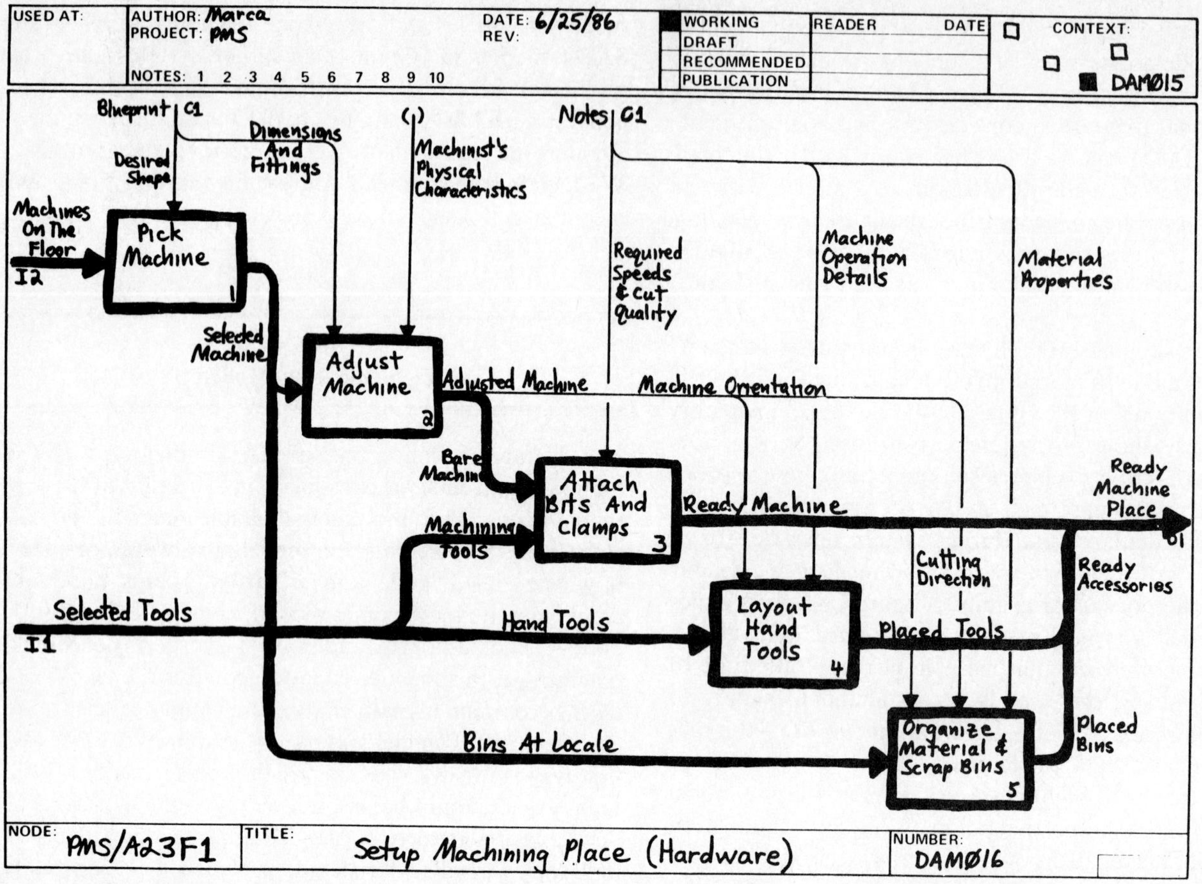

FIGURE 18-5 A Figure Highlighting the Main Path

Pages contain the definitions and often the structural descriptions of system data. Glossary Pages are used to build what is called the "Data Dictionary" for the model. Text pages provide the narrative for a diagram, which is often used to detail the low-level boxes in a model. Figure pages contain illustrations that explain an important aspect of the system (e.g., a finished part, a storage area, a map). Though very helpful at certain times, we strongly recommend that model supplements never be used in place of diagrams to decompose a model. They should be used only to enhance diagrams to meet the purpose of the model. Diagram Indexes and Node Indexes are outlines which provide both a written system summary and a way to rapidly pinpoint discussion about particular system functionality.

Suggested Readings

Bertin, J.: *Semiology of Graphics,* University of Wisconsin Press, Madison, 1983.

DeMarco, T.: *Structured Analysis and System Specification,* Yourdon Press, New York, 1978.

Hurley, R.: *Decision Tables in Software Engineering,* Van Nostrand Reinhold, New York, 1983.

Lannon, J.: *Technical Writing,* Little, Brown, Boston, 1979.

Marca, D.: *Applying Software Engineering Principles,* Little, Brown, Boston, 1984.

Martin, J., and C. McClure: *Diagramming Techniques for Analysts and Programmers,* Prentice-Hall, Englewood Cliffs, N.J., 1985.

Weinberg, G.: *Rethinking Systems Analysis and Design,* Little, Brown, Boston, 1982.

Chapter 19

Annotating Diagrams and Models

The SADT box and arrow language describes all the activities and things of a system, and thus helps to describe the functionality of a system. Sometimes, however, just the graphics of an SADT activity model is an insufficient description of how a system actually operates. Notice that we said "insufficient," not "inaccurate." If properly developed, a validated SADT activity model accurately describes system functionality from a static perspective, but this may not be enough to describe the other aspects of system operation and behavior required to get a complete understanding of how a system is expected to work. Two such common aspects of a system are its properties and the activation rules for its functions. Properties are named numeric and textual descriptions about the characteristics of the functions and data of a system. Activations define what is required in order for the functions to properly execute and what results happen after execution. With properties and activations, SADT allows us to elaborate a diagram with notations that describe its dynamics. In doing so, these annotations bring the model to life.

19.1 Property Labels

Properties are nonfunctional characteristics. They are named values, describing some important aspect of the things and activities of a system. For example, properties of things often include rates, volumes, and access frequency. Properties of activities often include place, time, and execution frequency. Properties are often static over time. For example, it is common for databases to have fixed sizes with expected and steady growth rates over a period of time. To explain further, the shop floor can be considered a database, for it is a repository with a finite amount of room that is filled to a particular level of capacity by the machines and their work areas.

Properties are very important, because they are system requirements that must be fulfilled, or the system will not operate properly. For example, the time a coal liquefication reactor spends processing a particular batch of coal is dependent upon thc weight and quality of the coal. For this system, properties of the coal and liquification times are critical requirements. The same holds true for inspection within the machine shop. The ***quality standards manual*** and the ***inspection points*** on the ***blueprint*** are properties that control the quality of the ***parts*** made by the prototype machine shop. Without these constraints, the shop could not produce parts that would fit into the new aircraft prototypes being built.

It is not surprising that the understanding of a system may not be complete until the properties of both activities and things are defined. SADT requires properties to be defined using "property labels." A property label is a square note, attached to either a box or an arrow by a squiggle, that describes a single property. Not every property of a particular box or arrow is put on a diagram—only the ones that make the story of the diagram clearer. For example, Figure 19-1 shows how property labels are used to describe the typical frequency with which tasks are done during the completion of a work kit. These property labels are attached to the boxes of the diagram. Also annotated on the diagram are property labels for use when particular things are given by the machinist to others in the shop.

Figure 19-1 shows how just a few property labels can make the meaning of a diagram stand out. For completeness, however, a Glossary must be kept on every box and arrow regarding its properties. Our experience has shown that putting Glossary information of this kind into a Data Dictionary is a reliable way to record these facts. This way, descriptions of properties are readily available and easy to find.

19.2 Activation Rules

Recall that SADT activity diagrams are constraint diagrams, because they emphasize the constraints among the

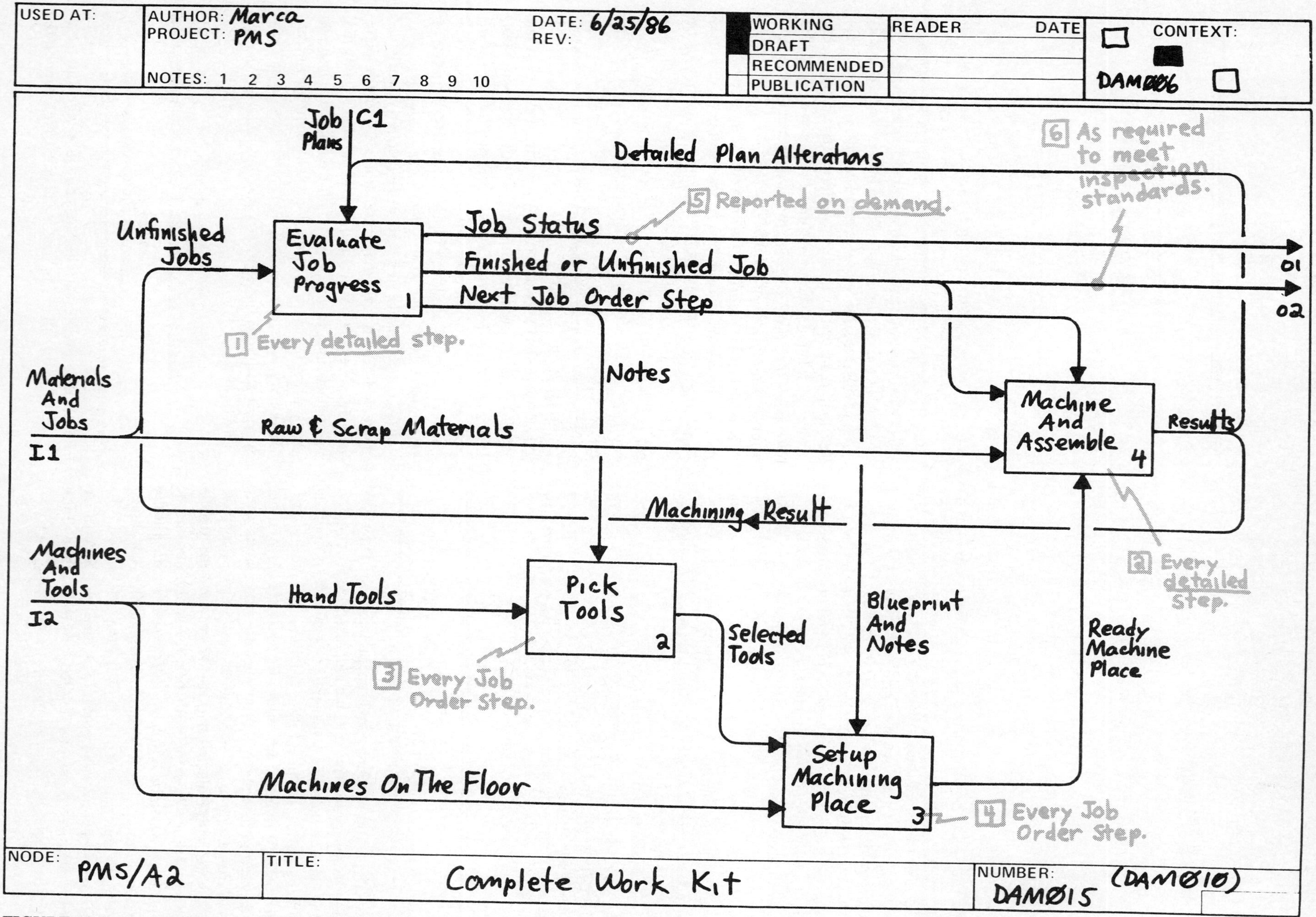

FIGURE 19-1 Adding Property Labels

functions of a system. A completed activity diagram gives a snapshot of all the control arrows that constrain a function at any given moment during the operation of the system. Thus, in general, not all the constaints on a function affect it at the same time. More often than not, only some of the constraints on a particular function are "active" at any point in time. SADT defines an "activation" as the way a function operates when it is "triggered" by the arrival of some of its controls and inputs, to generate some of its outputs. Thus, for any particular activation, not all the possible controls and inputs are used and not all the possible outputs are produced. The number of possible activations of a function depends solely upon the number of possible interactions of its data, not on the number of arrows coming into the box. More about this later.

An SADT activation rule is a notation (similar to abstract grammars) that describes a single combination of control, input, and output for a specific activity of a particular model. Activation rules are made up of a box number, a unique activation identifier, preconditions, and postconditions. Each activation rule for a model is stated on its own line with the following syntax:

> [*Model/*] *box* * *activation* : *preconditions* – – > *postconditions*

Box and ***activation*** uniquely name the rule. ***Box*** identifies a particular box in a model. ***Activation*** identifies one activation rule for that particular box. Thus, the combination of ***box*** and ***activation*** uniquely identifies one activation rule in a model. Activation rules from more than one model can be discussed at the same time. When this happens, ***model,*** the name of the model to which the rule applies is put in front of the rule. ***Preconditions*** and ***postconditions*** describe what is required for, and what results from, the activation. To summarize, each activation rule is interpreted to mean:

> For function ***box,*** activation rule ***activation*** is defined to be: when ***preconditions*** are true, function ***box*** is performed and makes ***postconditions*** true.

Both ***preconditions*** and ***postconditions*** are logical expressions of ICOM codes, where each ICOM code identifies a single control, input, or output arrow for that particular box. The logic operators of AND, OR, and NOT, along with parentheses, provide the means for expressing arbitrarily complex logical expressions. For example, Figure 19-2 shows a box on the PMS/A2 diagram which has only one Activation Rule. This rule states that ***setup machine place*** must have ***selected tools, machines on the floor,*** and ***blueprint and notes*** in order for the machinist to create a ***ready machine place.*** This is an example of an Activation Rule that mandates involvement from every input, control, and output arrow in order to activate a particular box.

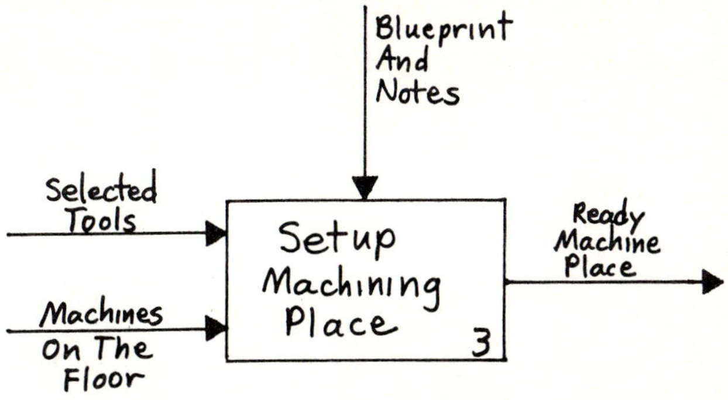

23*1: I1 and I2 and C1 ⟶ O1

FIGURE 19-2 An Activity with Only One Activation Rule

Often, situations arise in which the absence of one or more arrows is required for the proper activation of a box. Those arrows that must not participate in a particular activation are marked with a horizontal bar (standing for "NOT") over their ICOM code when they are in a ***precondition.*** This means that the things represented by the arrow must not be present in order for the activation to happen. For example, activation 3 in Figure 19-3 states that ***job plans*** must not be present during job evaluations that precede inspections.

Situations also arise in which only some of the arrows are used during the activation to produce outputs. When an input or control arrow does not participate in an activation, it is simply omitted from the ***precondition.*** Similarly, when only some of the outputs of a box are produced during an activation, the ICOM codes for those outputs not generated are omitted from the ***postcondition.*** For example, the omission of O2 and O3 from activation 4 means that only ***job status*** is produced during that activation.

Sometimes activities use up their inputs and controls during an activation. When this happens, these input and control arrows are listed in the ***postcondition*** with a NOT bar placed over them. This means that the things represented by those arrows are consumed during that activation, and are unavailable to other boxes or to future activations.

19.3 Generating Activation Rules

Sometimes, it is hard for analysts to define all the ways a function is activated during system operation. When this happens, we recommend you employ the technique of building truth tables—a technique originally developed from propositional logic theory and frequently used in

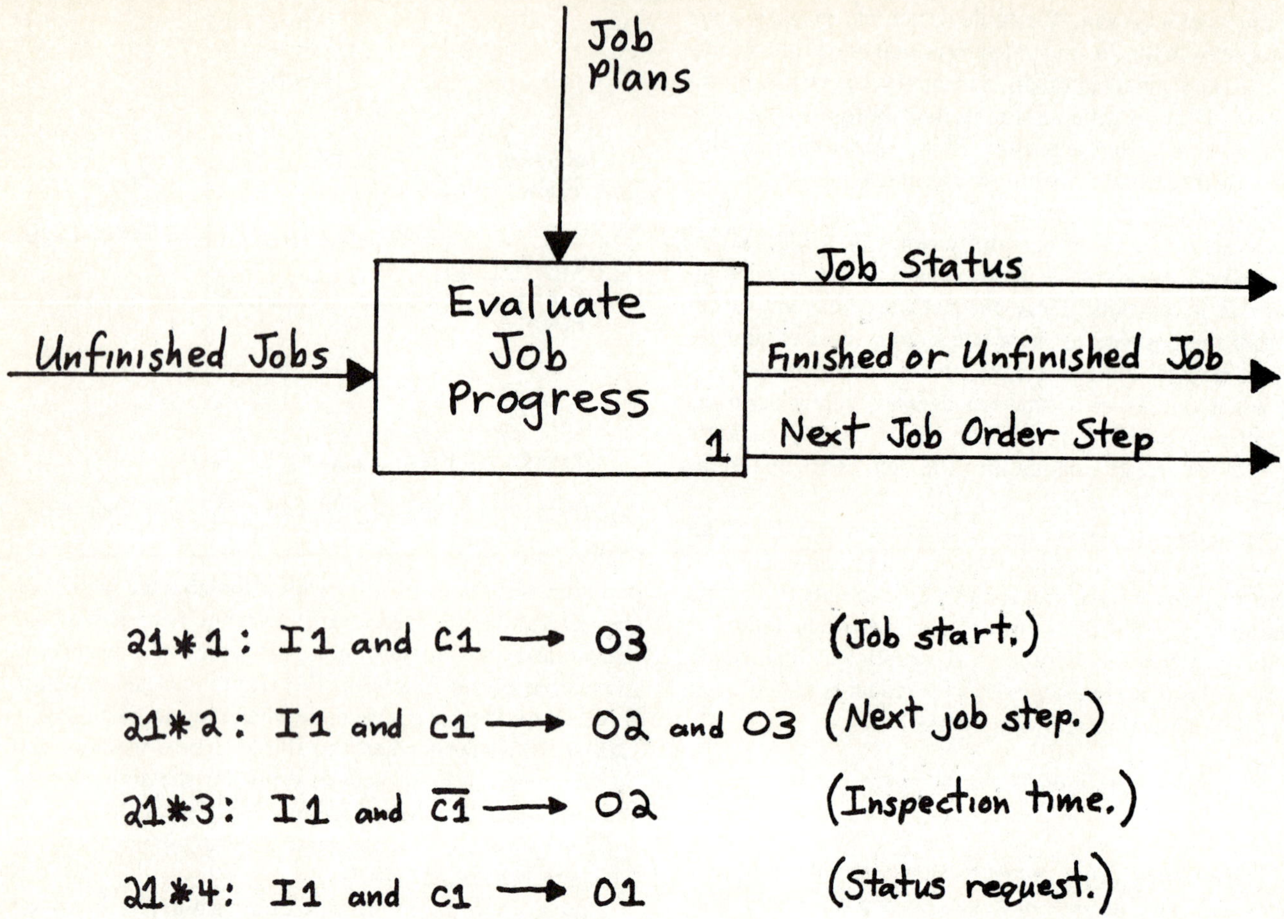

FIGURE 19-3 An Activity with Several Activation Rules

general systems theory. To summarize, the activation of every box is described with a truth table containing the Cartesian product of all possible combinations of the presence (indicated by "true" or T) and mandatory absence (indicated by "false" or F) of input, control, and output arrows. Each column in the table then becomes a potential activation rule. (By the way, sometimes one does not care if a particular arrow participates in an activation. Letting D stand for these cases seems reasonable, but remember that the tables will become much larger to completely handle the Cartesian product.)

Usually, not all entries in the table become activation rules, but each real activation is defined somewhere in that table. For example, the Cartesian product table for the ***setup machining place*** box, where arrows either participate or are absent, is given in Table 19-1. For the prototype machine shop, only case 1, in which all input and control arrows participate in the activation, actually occurs. Consider using this technique whenever you are unsure how a particular activity works in the system you are modeling.

19.4 Summary

SADT property labels describe particular system aspects like timing and volume. They are put on a diagram as annotations, usually as a stepping-stone to developing a specification from the model. SADT Activation Rules codify the set of possible activations for the boxes of a model. Sometimes, these rules are hard to define for a particular box. When this happens, we suggest employing the techniques of propositional logic to develop a complete set of possible activations.

Suggested Readings

Dickover, M., and C. McGowan: "Software Design Using SADT," SofTech Technical Paper TP061, August 1977.

Mendelson, E.: *Introduction to Mathematical Logic,* Van Nostrand Reinhold, New York, 1964.

TABLE 19–1 All Possible Activations of the Setup Machining Place Box

	Activation Possibilities															
Activation	**1**	**2**	**3**	**4**	**5**	**6**	**7**	**8**	**9**	**10**	**11**	**12**	**13**	**14**	**15**	**16**
Selected tools	T	F	T	F	T	F	T	F	T	F	T	F	T	F	T	F
Machines on the floor	T	T	F	F	T	T	F	F	T	T	F	F	T	T	F	F
Blueprint and notes	T	T	T	T	F	F	F	F	T	T	T	T	F	F	F	F
Ready machine place	T	T	T	T	T	T	T	T	F	F	F	F	F	F	F	F

Martin, J., and C. McClure: *Diagramming Techniques for Analysts and Programmers,* Prentice-Hall, Englewood Cliffs, N.J., 1985.

Parnas, D.: "Software Requirements for the A-7E Aircraft," Tecnical Report no. 3876, Naval Research Laboratory, November 1978.

Ross, D.: "An Essay on Activity Diagramming," SofTech Deliverable no. 7104, November 1976.

———: "Structured Analysis (SA): A Language for Communicating Ideas," *IEEE Transactions on Software Engineering,* vol. 3, no. 1, January 1977.

Schoman, K.: "SADT and PERT," SofTech Deliverable no. CLIN#0-02AG, November 1977.

SofTech, Inc.: "The DWS/CS Emergency Preset Structured Specification," SofTech Deliverable no. 1083-1, August 1981.

Savitch, W.: *Abstract Machines and Grammars,* Little, Brown, Boston, 1982.

Weinberg, G.: *An Introduction to General System Theory,* John Wiley, New York, 1975.

———: *Rethinking Systems Analysis and Design,* Little, Brown, Boston, 1982.

Chapter

20

Managing an SADT Project

An SADT project involves a lot more than just making a set of diagrams. Our experiences have shown that a successful SADT project follows a process which coordinates the work of experts, authors, readers, and approvers to develop work products on time and on budget. Successful projects also get a good start, make usable work products, develop only the models they need, and support their models with the right kinds of supplements. Our experience with running SADT projects is presented in this chapter, and we will concentrate on how we typically manage SADT projects by coordinating individuals and teams.

20.1 Starting a Project

Successful SADT projects always get off to a good start, which typically involves establishing a project charter, initially choosing what models to make, organizing the analysis team, selecting the technical review committee, and scheduling milestones. Such a kickoff results in a project plan that coordinates the efforts of many people and makes the project tractable.

The charter for a project is made during a project initiation phase. (In some organizations, the project charter is referred to as the "project concept document.") The charter identifies the mission, scope, constraints, deliverables, and expected benefits of the project. A charter may also include a sketch of the project plan, which helps document the assumptions made by the organizers and technical leaders of the project. Typically, small teams are assembled for this initiation phase. However, if the project is very small, it can be initiated by a single author.

Milestones—what models to make (i.e., deliverables) and when to make them (i.e., schedule)—are selected after the charter is issued. Milestones help to identify the subject or subjects of importance, to clarify the objectives of the project, and to form a detailed project plan. Before selecting milestones, we recommend you evaluate what is currently known about the problem, what new information must be gathered, and what information resources are available to the project. You may even have to develop a few small SADT models to get all these facts. For example, we often make a project model in SADT to identify which models will be made (e.g., current operations, future operations, conversion plan, system architecture). A plan formulated based on this information has a better chance of producing accurate and usable models.

Firm milestones pave the way for picking the right team to get the job done. Determine which and how many experts, authors, readers, and support staff (i.e., Librarian and clerical and publication staff) are required to generate and publish the deliverables. (You can get a feel for level of effort by examining the project summaries in Part VI.) Also, determine who will sit on the Technical Review Committee. These people are responsible for resolving conflicts, setting project policy, approving work products, and understanding and maintaining the charter for the project. Make sure this board has several high-level people so they will have the ability to enforce their decisions. When through, your project may very well have a structure similar to that given in Figure 20-1.

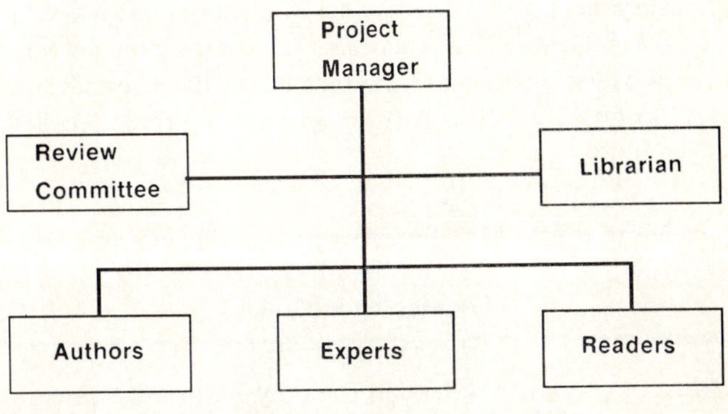

FIGURE 20-1 Typical SADT Project Structure

20.2
Making and Reviewing Work Products

One secret to making correct and useful SADT models is to always produce kits that convey the right amount of information to the right audience. A kit made up of a Context Diagram, the diagram being reviewed, and a Glossary Page is very common in SADT projects. However, not all SADT kits are so small. We recommend making a kit whenever you feel you have a reviewable unit of information for your readers. A reviewable unit of information varies depending on both the project and the skills and expertise of its members. Here are some situations in which kit size can really affect productivity:

- Projects involved with new technology often require extensive review of every new idea. Because so much new information is being discovered, analysts make better progress if their work packets are kept small.
- Projects that study well-known systems or systems with little complexity can move more quickly when larger kits are made.
- Projects comprising many novice SADT analysts make better progress when they keep their kits small. Small kits provide opportunity for correcting mistakes and changing work assignments.
- Projects comprising just a few senior SADT analysts often progress better and faster if their kits contain several diagrams—often a diagram and all its children.

Kit size is one project concern. The number of kits distributed on a particular subject is equally important. We recommend that not one but a series of kits be distributed for a reviewable unit of information. To do this, first issue a kit to a very senior expert and reach consensus with just this person. Then issue a second kit on the same topic to a few more experts, and reach consensus with that small group. Then distribute a revised kit to your whole Reader Community to achieve global consensus on the validity of your work. This process of achieving increasing levels of consensus, from one individual to a small group to everyone, is the way we typically acquire knowledge in SADT. Figure 20-2 summarizes the process, showing levels of wider consensus. Note that it usually takes several kits to finally reach consensus among the whole Reader Community.

20.3
Developing a Model

Making good SADT models does not mean just drawing correct diagrams. It also involves using group techniques

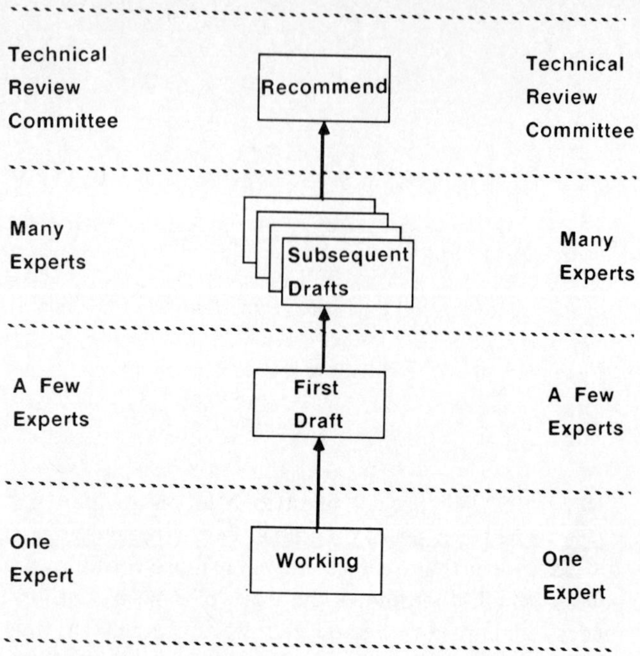

FIGURE 20-2 Levels of Consensus

to arrive at stable A-0 and A0 diagrams, assigning authors to decompose particular boxes on the A0 diagram, setting primary goals for decomposition, and then grinding out the model. We usually start a modeling effort by holding a technical "kickoff" meeting. We bring the experts and project team together for a group work session. The chief author leads the session by starting a flow of information from experts to authors, encouraging interchange among experts and authors, and guiding discussions until a rough A0 and A-0 with purpose and viewpoint have been drawn. (See the workshops in Chapter 22 for a detailed example of how to run such a session.)

These initial diagrams, with associated Glossary, become the first kit, which then goes through several Author/Reader Cycles until stable A0 and A-0 diagrams emerge. After that, the analysis team develops decomposition goals for each of the boxes on the A0 diagram and assigns an author to each box. Each author then details one part of the model and resolves interfaces on the A0 diagram with other authors. In this way, a team of authors can quickly build a model without interfering with each other (See the workshops in Chapters 23 and 24 for a detailed example of how this can be done.)

During this cooperative authoring process, each author details an assigned part of the model and reviews it with the experts and authors. Mostly an individual effort, detailing requires an author to gather information (mostly by reading documents and interviewing experts), to bound each new subject for decomposition, to recognize what perspective to choose for decomposition, to produce a diagram, and to review that diagram with experts. (See the workshops in Chapters 25 and 26 for details on how this is done.) Such work requires disciplined habits and an awareness of potential problem areas. Like all SADT authors, we have developed some personal disciplines over the years to avoid pitfalls. Some of the most important ones from our experiences are deciding which task is most appropriate to do next, developing a sense for personal productivity, and monitoring critical limits.

Critical to the success of any analysis effort is the responsibility each author has in deciding what is technically the next best task to perform. Authors do many things during an analysis effort. They think, read background material, interview experts, pick boxes for decomposition, draw diagrams, redraw diagrams, make kits, read kits, react to comments, talk to reviewers, write supplements, edit supplements, resolve interface issues, proofread material for publication, and brainstorm with other authors. We make it a point to think very hard about which of these activities is the most appropriate and most important one to do next.

Important to the long-term technical development of SADT authors is their personal perception of their productivity and the quality of their work. Extremely valuable is developing a feel for your own productivity by recording the time you spend doing particular modeling tasks in the margins of the SADT form. By reviewing these times, you can determine which activities you do well, which ones give you trouble, and which ones you are doing less than optimally. You can use Table 20-1 as a set of goals to achieve.

Important to the success of a project from a schedule perspective is the rate at which quality work is accomplished. Be on the lookout for kits and models that exceed the critical limits defined in Table 20-2. Work products that exceed these limits may be hard to use or to review, or may not be in concert with the goals of the project. We always reevaluate both the contents of such work products and how they fit into our project plans. Sometimes large models or kits are required, but often they indicate potential problem areas for the project.

TABLE 20–1 Limits to Making SADT Diagrams

Task	Brand New	New Version
Draw	15–60 minutes	10–30 minutes
Criticize	10–30 minutes	10–30 minutes
Comment	10–40 minutes	5–20 minutes
Diagrams made per day	2–5	5–10

TABLE 20–2 Critical SADT Product Sizes

Work Product	Typical Numbers
Kit	2–6 diagrams
Kit	3–6 review cycles
Model	10–30 diagrams

20.4 Strategies for Supplementing a Model

Describing a system with SADT produces a model that identifies and describes its functions and things. But it usually takes more than an SADT model to develop a specification. Usually, a model is enhanced with supplements in order to make, or get a start on, a complete system specification. For example, Chapter 27 shows how the Feed Family model can be supplemented with narrative to develop a readable and executable specification of how to feed a family. We will now identify what we think are useful supplements to a model. We refrain from discussing format, however, because specification standards differ widely in the industry.

A system specification often contains descriptions of functions seen by users, design guidance and constraints, acceptance criteria, and reference material. We get the most from our SADT models when we document this kind of information using strategies like the following:

- The functions of the system seen by the user, like all system functions, are defined by the SADT model. Highlight these functions on the model and tie them to discussions about human factors issues.
- Design constraints identified during analysis (e.g., budget, schedule, memory limits, processing limits, predefined hardware, software, and personnel resources) should become annotations placed on the functions they affect.
- Design guidance (e.g., which functions are separable, which functions form a kernel, which functions can be postponed to later software builds) should be described with narrative tied to annotations on the model (e.g., highlighted kernel functions).
- Acceptance criteria (e.g., the independent evaluators, the tests themselves, the location and duration of testing) should be organized around the structure of the SADT model when no better organization can be found. Use mechanisms to document these facts.
- Good reference material (e.g., background documents, dictionary of glossary terms, descriptions of special notations) are vital to the correct understanding and usage of a specification. So, incorporate this information into appropriate sections of the specification. Also, put the bulky material (e.g., an aircraft blueprint) and common material (e.g., the Data Dictionary) in one or more appendixes.

20.5 Summary

SADT projects involve the coordination of teams of authors, readers, and the Technical Review Committee—a body of people who technically approve models for use. These groups are coordinated through the judicious creation and use of kits, through special group working sessions, and through individual efforts. SADT projects also involve careful management, just as any other kind of engineering project does. Forming the right team, developing a project charter, and meeting deadlines all make for a successful SADT project. Thus, one of the most important responsibilities of the SADT project manager is to monitor the kits and project plans to ensure a smooth-running effort.

Suggested Readings

Brackett, J.: "Structuring the Analysis and Design Process," *GUIDE Productivity Symposium,* 1976.

Burack, E., and F. Torda: "The Manager's Guide to Change," Lifetime Learning Publications, 1979.

Demarco, T.: "Breaking the Language Barrier," *Computerworld,* August 1978.

———: *Controlling Software Projects,* Yourdon Press, New York, 1982.

Parkin, A.: *Data Processing Management,* Little Brown, Boston, 1980.

Ross, D., and K. Schoman: "Structured Analysis for Requirements Definition," *IEEE Transactions on Software Engineering,* vol. 3, no. 1, January 1977.

Schoman, K.: "SADT and PERT," SofTech Deliverable no. CLIN#0-02AG, November 1977.

Weinberg, G.: *Understanding the Professional Programmer,* Little, Brown, Boston, 1982.

Yourdon, E.: *How to Manage Structured Programming,* Yourdon Press, New York, 1976.

Chapter 21

Automated Support

The early 1970s saw a strictly paper implementation of the SADT methodology. This implementation, practiced by SADT analysts on Diagram Forms and Kit Cover Pages, is what we have presented throughout this book. The unique and efficient way SADT codifies the connection of decompositions using ICOM's and the way SADT organizes peer review with the Author/Reader Cycle greatly simplifies its paper implementation. In our opinion, these advantages put SADT far above all other structured methods that have paper implementations.

The late 1970s saw computers with the necessary power and bandwidth to do graphics at reasonable speeds. With that capability came the possibility of automating structured methods that heavily relied on graphics, as SADT does. Though such technology was just emerging, the U.S. Air Force funded the successful development of the first automated SADT tool (and, by the way, the first automated structured analysis tool emphasizing graphics), called AUTOIDEF0.

The early 1980s saw the personal computer, with graphics capabilities that was small enough to fit on a desktop. This led to personal workstations for several graphic structured analysis methods. Several early attempts at implementing SADT on mini- and microcomputers were also made during this time in the United States, Europe, and Scandinavia. One of the products to come out of these efforts was an SADT workstation from France called SPECIF_X. This chapter describes both the AUTOIDEF0 and the SPECIF_X tools. We will summarize their capabilities, describe their user interface, give examples of their output, and discuss the hardware they require.

21.1 AUTOIDEF0

The U.S. Air Force Program for Integrated Computer-Aided Manufacturing, commonly called ICAM, adopted a subset of the full SADT methodology they call IDEF0. (In fact, the description of SADT in this book is almost exactly the IDEF0 subset of SADT.) One goal of the ICAM program was to standardize the description of aerospace manufacturing across government contractors. Adopting the IDEF0 language was a big step toward this standardization.

AUTOIDEF0 was intended to make easy the process of developing and reviewing IDEF0 diagrams and models among the geographically dispersed aerospace contractors. Because IDEF0 models were often reviewed by contractors spread around the United States, the Air Force required AUTOIDEF0 to run on interactive devices and a telecommunication network that were commonly used or could be easily acquired at the time. The initial system configuration had vector graphics displays and plotters connected to a large mainframe, which could be connected into a commonly used telecommunication network.

The AUTOIDEF0 tool that ran under this configuration presented its users with a menu-driven, command-oriented graphics manipulation system. Users would first select graphic operations from a hierarchic menu (e.g., ***add box, delete icom***), and would then select a particular graphic object with a horsehair cursor. As shown in Figure 21-1, the hierarchic menu made it easy to manage a library of diagrams, start new models, and draw diagrams. The storage tube device and plotters produced SADT diagrams with a reasonable graphic resolution.

Much to the credit of AUTOIDEF0, SADT project management and the Author/Reader Cycle were also supported. For example, a project manager would specify kit distribution lists when a new project was initially set up. This supported the distribution of kits to many different contractors in widely diverse locations, allowing people to annotate diagrams with comment and reaction overlays. The distribution of kits happened automatically after the kit was created, commented, or reacted.

Thus, AUTOIDEF0 was not just an automated diagramming tool. It also supported model development. By selecting particular commands, an analyst could build a file of SADT diagrams that represented a model by initial-

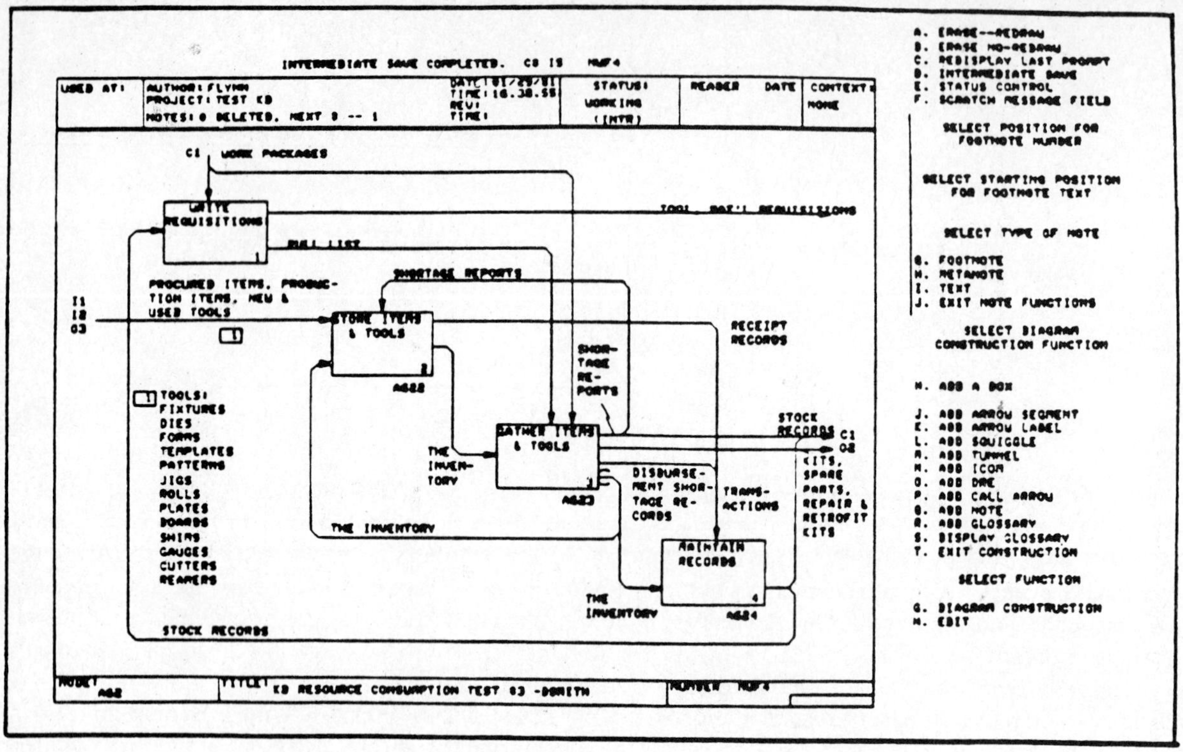

FIGURE 21-1 An AUTOIDEF0 Terminal Session

izing it with an A-0 diagram, and by adding the other diagrams in node number order. In this way, many models could be created, stored, worked on, published, and archived at the same time. This capability also allowed one to link together several related models. For example, the model for an aircraft wing could be linked with the model of the fuselage at the exact points where the two must connect.

Performance and usability factors of the hardware and software technology at that time caused the Air Force to requisition the development of a second version of AUTOIDEF0. The new version will utilize newer bit-mapped devices and workstations that network together. With these new and more effective technologies, the second version of AUTOIDEF0 should be a much better fit between price and performance, while preserving prior functionality.

21.2 SPECIF_X

Though initially developed in the United States, the SADT methodology underwent its first major "field test" at a large telecommunications company in Europe during the early 1970s. The project specified a large telephonic supervisory system. Since then, SADT has become an extremely popular requirements specification methodology throughout Europe. Most recently, the use of SADT to specify requirements is supported by the ESPRIT program—the 10-year software initiative effort of the European Economic Community. Not surprisingly, several SADT automation efforts sprang up in greater Europe during the early 1980s, resulting in tools like SPECIF_X. The goal of SPECIF_X is to support the widely used SADT technique so that better-quality specifications can be produced more quickly and accurately than with paper and pencil.

SPECIF_X was initially developed by the French company Institut de Genie Logiciel (IGL) to run on current bit-mapped workstation technology, with attached plotting devices or laser printers. A less ergonomic version of the tool, which has the same functionality, can operate on computers having certain terminals with extended graphics capabilities and a line printer. The graphic workstation version of SPECIF_X presents an iconic menu, from which the user selects the basic building blocks of SADT (e.g., box, arrow, label) and combines them to build a diagram. The user interface is highly interactive, allowing one to automatically or manually route arrows around boxes and to move boxes while dragging attached arrows. As Figure 21-2 shows, the resolution of resulting diagrams is quite good, especially when output on laser printers.

In addition to its graphical diagram editing, SPECIF_X also supports diagram and model management. Diagram variants are tracked using C-Numbers, and dia-

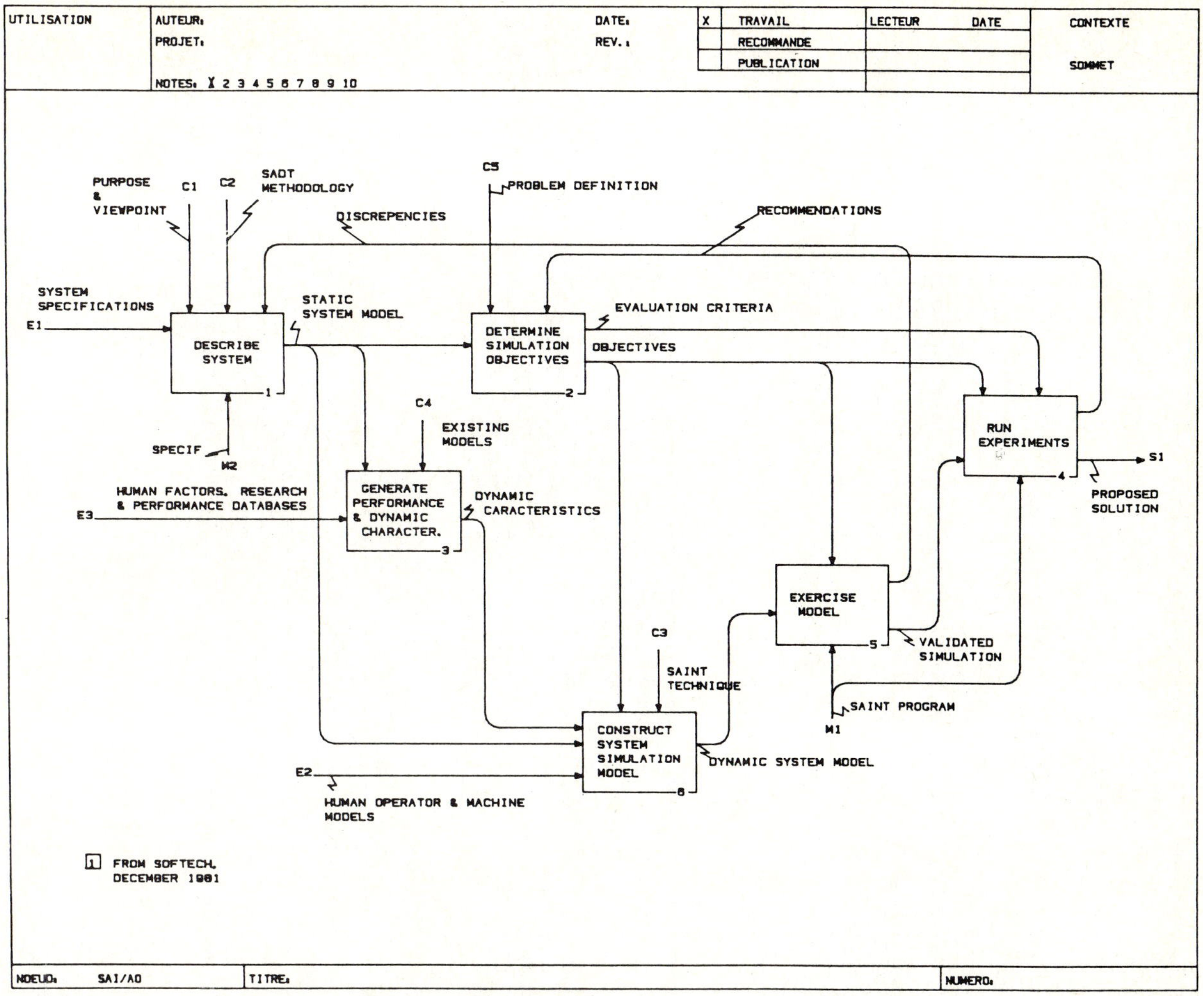

FIGURE 21-2 An SADT Diagram Drawn with SPECIF_X

grams are collected into models using node numbers. Consistency checks of the model are made, some while drawing the diagram and others when "inserting" a diagram into a model. Assistance is also given when modifying a diagram. For example, SPECIF_X can evaluate how deleting a particular arrow will affect the entire model. One can also create Glossary, tie terms to their associated diagrams, and put terms into a Data Dictionary. The entire Data Dictionary can be managed and printed separately from the diagrams, allowing development of a dictionary of technical terms for each model. Last, SPECIF_X also supports the Author/Reader Cycle by allowing diagrams to be collected into kits which can be read by others using the tool.

21.3
A Checklist for Automated Support of SADT

Developing an automated systems analysis tool like AUTOIDEF0 or SPECIF_X is no easy undertaking. For SADT, such a task is not just handled by creating a picture editor. Even though the semantics of SADT graphics are quite complicated, a large number of other aspects of SADT must go into an automated tool for it to properly and fully support the entire methodology. For this reason, we feel that good automated support of the SADT methodology must be built upon a central database of the concepts behind SADT. (We use the term "database" here to mean repository, whether that place is in main or secondary memory, central or distributed.) Building such a database of SADT concepts requires a good summary of the methodology. Table 21-1 presents a checklist of the SADT concepts covered in this book along with how they are represented in the methodology. The checklist summarizes SADT by organizing it into the major categories of product (i.e., what is ultimately created), language (i.e., how ideas are expressed), and process (i.e., how products are made). Subcategories have been included to help identify some strongly related groupings in a category.

You should be able to use this list in several ways. First you can use it as a checklist to evaluate existing SADT tools by determining what portions of the methodology

TABLE 21–1 SADT Methodology Checklist for Activity Diagramming

Category	Concept	Representation in SADT
Product:		
Models	Subject	Bounded subject
	Question set	Purpose
	Perspective	Viewpoint
	Hierarchy	Node numbers, box numbers
Diagrams	Decomposition	Diagram, data, and activity lists
	Versions, variants	C-Numbers
	Complexity	Three-to-six boxes rule
	Clarity	Diagram layout rules
Supplements	Explanation	Text page
	Emphasis	Figure page
	Terminology	Glossary page, Data Dictionary
	Summary	Node index
Language:		
Functions	Function	Box
	Input	Left side of box
	Output	Right side of box
	Constraint	Top side of box
	Mechanism	Bottom side of box
	Names	Labels
Data	Data	Arrow
	Constraint	Arrow
	Flow	Arrowhead
	Pipeline	Arrow branch and join
	Names	Labels
Interfaces	Bounded subject	Box and its arrows
	Interface tie	ICOM codes
	Suppression	Tunneled arrow tip or tail
Annotations	Comments	Round notes
	Reactions	Round notes
	Decisions	Round notes
	Properties	Property labels
	Sequencing	Activation rules
	Notes	Square notes
Process	Explain	Experts
	Interview	Authors
	Describe	Authors
	Criticize	Reviewers; Author/Reader Cycle
	Distribute	Kits; Librarian
	Approve	Technical Review Committee
	Publish	Librarian; model publication format

they do and do not support. For example, many automated analysis tools these days concentrate on diagram editing (i.e., language) and do little to help produce final products or to manage the process of making models (e.g., the Author/Reader Cycle). By using this checklist you can determine how well a tool implements a particular aspect of SADT. For example, a tool that does not implement comments and reactions as diagram overlays puts extra burdens on its users to manage diagram evolution. As another example, a tool that does not let you define a Glossary entry at the time you name a box or label an arrow burdens its users by requiring them to always scan diagrams for undefined terminology.

Also, you can use the list as a checklist to estimate the amount of effort it would take your organization to build automated support for the SADT. First, determine the conceptual schema for the underlying database. Second, determine an architectural design that correctly separates but efficiently integrates diagram editing, the process of making a model, keeping a Data Dictionary of terminology, and producing a publication-quality model with its supplements. Third, determine an overall user interface that properly visualizes and integrates each of these functions. After doing these things, we feel you will have enough knowledge to accurately estimate the level of effort for developing an SADT tool. We think you will find such a project to be large and expensive.

21.4 Summary

Current hardware technology has the bandwidth to build automated support for SADT. Two SADT tools available today are AUTOIDEF0, initially built for the Integrated Computer-Aided Manufacturing effort of the United States Air Force, and SPECIF_X, currently being used in Europe as part of ESPRIT software initiative. From our personal perspective, we see automated support of this kind evolving from simple drawing tools to software which runs on a knowledge base of more general modeling concepts. Such advanced tools would be capable of understanding the semantics of an interrelated network of SADT diagrams and multiple models, and would be capable of integrating this network of facts and rules with other software engineering and artificial intelligence technologies.

Suggested Readings

Bernus, P., and J. Hatvany: *Computer Aids to the Design of Integrated Manufacturing Systems,* Computers in Industry, North Holland, 1979.

Boehm, B., R. McClean, and D. Urfrig: "Some Experience with Automated Aids to the Design of Large-Scale Reliable Software," *IEEE Transactions on Software Engineering,* vol. 1, no. 1, March 1975.

ESPRIT project: "ESPRIT '84: Status Report of Ongoing Work," North Holland, 1985.

Foley, J., V. Wallace, and P. Chan: "The Human Factors of Computer Graphics Interaction Techniques," *Computer Graphics and Applications,* November 1984.

Hopgood, F., and R. Witty: "PERQ and Advanced Raster Graphics Workstations," *Computer Graphics and Applications,* April 1984.

IGL: "SPECIF-Simulation Integration Project," Technical Publication no. T/0053/DT, Institut De Genie Logiciel, October 1985.

Machover, C.: "Engineering Workstations," *Computer Graphics and Applications,* August 1984.

Nelson, D., and P. Leach: "The Architecture of the Apollo Domain," *Computer Graphics and Applications,* April 1984.

Pederson, J, and J. Buckle: "Kongsberg's Road to an Industrial Software Methodology," *IEEE Transactions on Software Engineering,* vol. 4, no. 4, July 1978.

Ross, D.: "Applications and Extensions of SADT," *IEEE Computer,* April 1985.

Reed, J.: "Graphics in Workstations—Emerging Systems and Solutions," *Computer Graphics and Applications,* August 1984.

Smith, D.: "AUTOIDEF0: A New Tool for Functional Modeling," SofTech Technical Paper no. 125, September 1981.

Schoman, K.: "SADT and Simulation," SofTech Technical Paper no. TP072, May 1979.

Teichroew, D., and E. Hershey: "PSL/PSA: A Computer-Aided Technique for Structured Documentation and Analysis of Information Processing Systems," *IEEE Transactions on Software Engineering,* vol. 3, no. 1, January 1977.

Part

V

Workshops— Developing an Activity Model and a Specification

This part of the book collects a series of 25 workshops that guides you through the major steps in developing an SADT activity model and writing a specification based on that model. (The problem statement for these workshops appears on the next page.) The workshops are designed for either an individual learning alone or a classroom environment. Each workshop is broken into four sections: Goal, Activities, Notes, and Sample. The Goal section summarizes what you will accomplish in the workshop. The Activities section gives a list of steps to follow to reach the goal. The Notes section highlights some important points about how SADT should be used during the workshop. The Sample section illustrates representative results for the workshop taken from an actual class.

If you are learning by yourself, read the first page of each chapter before reading and following the steps and explanations for workshops it contains. For each workshop, read the Goal, Activities, and Notes sections first, then try to do each step in the Activities section. After you finish each workshop, check your result with that given in the Sample section. Revise your work only if you did not do what the workshop asked. Do not revise your results if they are the natural artifact of your analysis, and may not agree totally with the samples. Continue with your analysis by doing the other workshops, and keep a notebook of all your results. When you finish all the workshops, you will have a very good record of your first SADT modeling effort.

If these workshops are used in a classroom setting, the instructor must read the Notes for Instructors Running the Workshops, immediately following the problem statement. These notes contain essential directions on how best to give instruction in the workshops. The instructor should have the students read only the Goal and Activities sections before starting a workshop. For a workshop that requires the instructor to lead the class, the points given in the Notes section should be discussed while doing the workshop. Otherwise, the instructor should summarize those points at the end of the workshop. Also, the instructor should use the Sample section as a guideline to what the class should produce. We say "guideline," because we do not want instructors to inhibit the analytical ability of a class by forcing it to produce an exact duplicate of the sample model.

In fact, instructors who have used this problem to teach SADT in several countries have found out that radically different models of feeding a family are produced, depending on the cultural patterns of a country. Also, please do not think there is just one "right" answer for even a single country. Cultural differences exist even between different regions of a country (see Appendix C). However these workshops are used, and in whatever country, developing a model of feeding a family has proved to be a reliable benchmark and challenging domain for learning how to use SADT for the first time.

The Feed Family Problem

Systems analysis work is as varied as its problem domains. Analysts are often faced with difficult technical problems, even though the system being studied may have appeared simple on the surface. Some hold the view that only current technological problems are worthy of testing the skill and techniques of an analyst. Our experience with teaching SADT to many commercial firms around the world and to many domestic military organizations has shown otherwise. We have found that the best analysts can apply their skills to a wide variety of domains. To these analysts, how well they can apply their method to solve a particular problem, not the problem itself nor the domain, is the real challenge.

We have consistently challenged every group of systems analysts we have taught with the Feed Family problem given below. The decision to choose this particular problem was based on several factors. First, the Feed Family problem is not as easy as it may appear to be. It has more than adequate complexity to test out any systems analysis methodology. Second, the problem is easy to describe and understand. Third, everyone is an expert on this problem to some degree. Thus, one can concentrate on learning to use SADT rather than struggling with understanding the details of a particular technical domain or problem. Since SADT is also taught to managers and to a wide variety of disciplines in various universities, we feel strongly about Feed Family being an extremely good general workshop problem. Here it is:

> Write a description of how to feed a family of two adults, two young children, and two teenagers, so that this description can be understood and followed by the teenage members of the family without assistance.

The problem asks for a narrative description—a specification—of how to feed a family. The problem is a realistic one: There is an existing (and critical!) process, a system not unlike a manufacturing process, which must be learned by members of the organization. Some kind of document is required to teach the process. Also, like other kinds of system documentation, the specification must also double as a reference document. Such problems are encountered every day in the working world.

The series of workshops in this part of the book is designed to guide you through the steps of developing this specification, using SADT to guide your analytical thinking and shape your specification. The workshops describe how to attack the problem analytically, document and describe the process, convey complex interactions among process activities, and use the resulting SADT model to write the specification. After this step-by-step exposition of modeling, you will have a very good feel for what a systems analyst can do with the SADT graphic language and methodology.

Notes for Instructors Running the Workshops

The following notes have been collected here for instructors who wish to teach SADT to a class of no more than 20 people in about 5 consecutive days. Each paragraph of notes applies to one entire chapter in Part V. The workshops were designed to be done one right after the other, with natural breaks occurring at the end of each chapter, and were designed to stress class, group, and individual participation, in that order. Please read the notes with these concepts in mind.

Chapter 22
Workshops 1 to 7

Lead the entire class through all seven workshops. Go around the room and solicit contributions from individuals in the class whenever possible. Always write notes, make lists, and sketch diagrams in front of everyone so the class sees its ideas evolving. Always be the one who resolves conflicts over diagram content and interpretation of terminology, because you are the chief author. Take time to draw the A-0 and A0 diagrams of Workshop 7 very neatly, because they will become the de facto drawing standard throughout the rest of the course. Use the simplest graphics and shortest labels when drawing diagrams. Avoid squiggles.

Chapter 23
Workshops 8 to 10

These workshops are to be done by groups of people. Split the class into three to six groups, one for each box on the A0 diagram. Try to balance the technical composition of each group. Strive to have at least one senior person and one junior person in each group. Then assign one box on the A0 diagram to each group. Let the groups work out problems of logistics and dynamics (e.g., who will scribe and who will lead), but encourage everyone to participate. Remind each group that their goal is to produce a diagram, then a critique, and last a kit. Also, remind the groups not to share information with other groups during these workshops. The time for sharing results will come later.

Chapter 24
Workshops 11 to 14

These workshops are to be done by the three to six groups that drew the first-level decompositions. Remind the groups that this is the time to share information. Get dialogs going among the groups. You may have to talk the class through the A0 diagram at some time to keep the discussions focused. Point out that the groups must work together to jointly improve the A0 diagram. Make sure that one group is not giving undue work to another group. Be the final judge in resolving interface issues between groups.

Chapter 25
Workshops 15 to 17

These workshops are to be done by the individuals in the three to six groups that developed first-level decompositions. Let each group decide how it assigns work to its members, but make sure two people are not working on the same box. Remind all the class members that their goal is to produce, on their own, both a single diagram and a kit. Also, permit the class to share information with each other during these workshops.

Chapter 26
Workshops 18 to 21

These workshops are to be done by individuals in the three to six groups that drew the first-level decompositions. Tell the groups that the last round of interface resolution was at the class level (i.e., the whole class resolved issues), and that this round of interface resolution will be at the group level (i.e., interfaces are resolved inside each group). Keep dialogs going inside the groups. Walk through the A0 diagram if needed. Point out that the groups must work together to jointly improve their group decompositions. Make sure that individuals are not being given too much work to do on their own diagrams by others in the group. Be the final judge in resolving interface issues between individuals.

Chapter 27
Workshops 22 to 25

Lead the class through Workshop 22. Groups take themselves through Workshop 23. Individuals do Workshop 24 on their own. Text for Workshop 25 is written by individuals from their own Axx diagrams. If there is time, groups can compose text for their Ax diagrams, and the instructor can write text for the A-0 and A0 diagrams as well as the cover page to the specification. Remind the class that the goal is to produce a single specification document which reflects the collective effort of the class throughout the course. Copy and distribute this document to everyone sometime after the course has finished. This will help each person remember what was learned, especially the dynamics of working in groups and as individuals.

Chapter

22

Starting a Model

Recall that an SADT model is started by bounding the system, determining purpose and viewpoint, and developing high-level diagrams. This chapter of seven workshops is designed to guide you through the steps SADT analysts most often do to start making an activity model: Workshop 1 bounds the context of the problem, Workshop 2 defines the purpose and viewpoint of the model, Workshop 3 produces an A0 diagram, Workshop 4 produces an A-0 diagram, Workshop 5 critiques the A-0 diagram, Workshop 6 critiques the A0 diagram, and Workshop 7 revises both diagrams.

Ideally, you should go through all seven workshops without a break. This will give you an accurate feel for the amount of work required to start a model. At a minimum, we recommend you do Workshops 1 to 4, take a break, and then do workshops 5 to 7. Allocate about ½ hour to read, understand, and do each workshop. Don't be concerned if you take more than ½ hour. Your goal is to learn the SADT methodology, not to race against the clock. As you become more experienced with SADT, you will become more productive because, as with any language, the more you use the SADT language, the more skilled you become with it.

Workshop

1

Bounding the Subject

Goal Develop a bounded context for the Feed Family problem.

Activities

1. Before starting, review the modeling concepts of SADT. See how they apply to bounding the subject of a modeling effort.
2. Start listing the names of any major things you think are part of the system. Free-associate. Do not worry about correctness for the present.
3. Now critique the initial list. Cross out the names of things that do not belong to the system. If appropriate, group names into larger collections (i.e., aggregates), connecting related names with lines or encircling them. Add more names as your ideas evolve.
4. Stop when your flow of ideas slows down.
5. Now do the same for system activities. Use the Data List to generate activity names, and then critique the list. Cross out those names that belong outside the system. Group similar functions into aggregates, connecting their names with lines or encircling them. Change the Data List as you evolve what the system does.
6. Stop when you have named enough things and activities to draw a diagram.

Notes

1. This is a process in which an initial consensus about the boundary around "the system" is achieved.
2. The things of the system will start to become clear only after an initial list is made, some things are thrown out, and other things are added.
3. Sometimes things that were originally thrown away are brought back into the bounded context.
4. The Data List will change once the listing of activities starts. You may find yourself "bouncing" between the lists when ideas start to flow.

Sample Notice that the attributes of "teenager" were listed while the data and activity lists were being developed. This is an example of how to document assumptions during analysis.

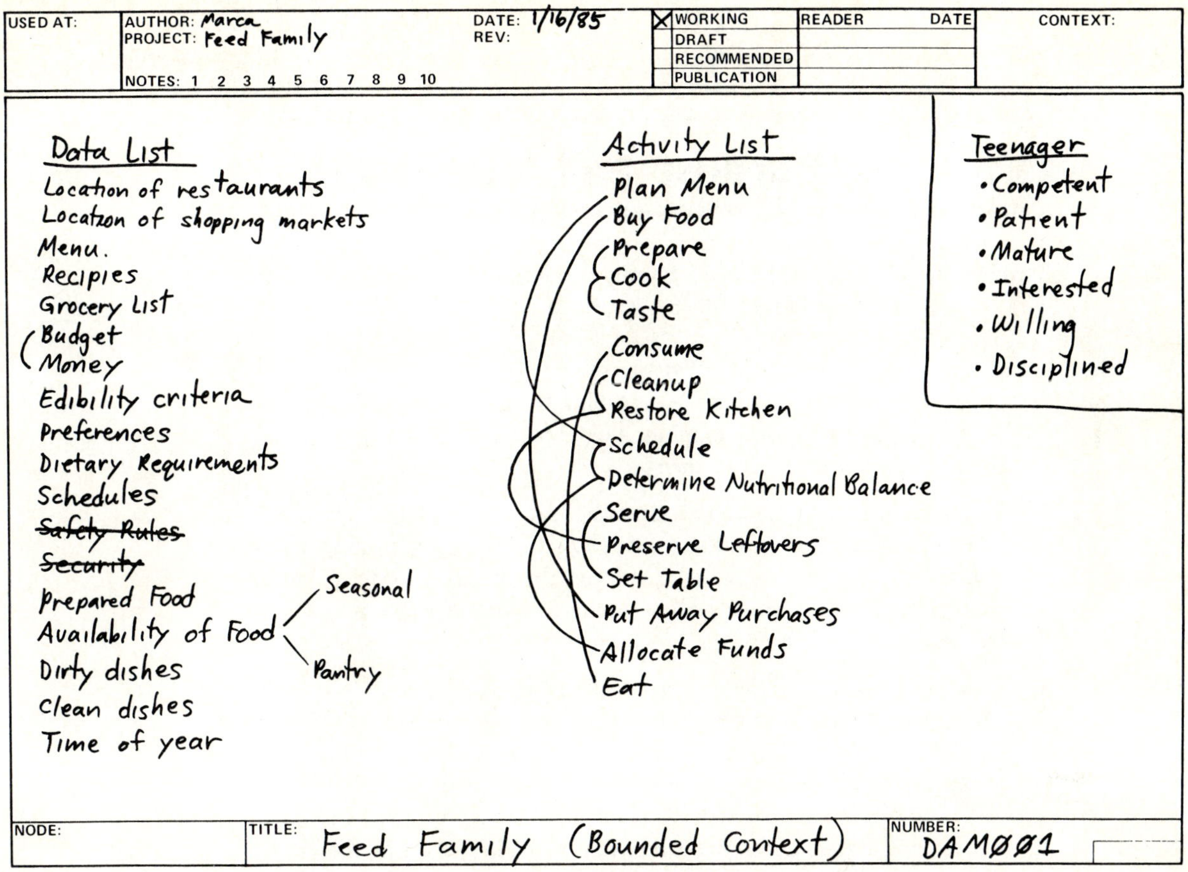

USED AT: | AUTHOR: Marca | DATE: 1/16/85 | X WORKING | READER | DATE | CONTEXT:

PROJECT: Feed Family | REV: | DRAFT | RECOMMENDED | PUBLICATION

NOTES: 1 2 3 4 5 6 7 8 9 10

Data List

Location of restaurants
Locatzon of shopping markets
Menu.
Recipies
Grocery List
Budget
Money
Edibility criteria
Preferences
Dietary Requirements
Schedules
~~Safety Rules~~
~~Security~~
Prepared Food
Availability of Food — Seasonal, Pantry
Dirty dishes
Clean dishes
Time of year

Activity List

Plan Menu
Buy Food
Prepare
Cook
Taste
Consume
Cleanup
Restore Kitchen
Schedule
Determine Nutritional Balance
Serve
Preserve Leftovers
Set Table
Put Away Purchases
Allocate Funds
Eat

Teenager

- Competent
- Patient
- Mature
- Interested
- Willing
- Disciplined

NODE: | TITLE: Feed Family (Bounded Context) | NUMBER: DAMØØ1

Workshop

2

Determining Purpose and Viewpoint

Goal Develop the purpose of the Feed Family model, and decide from which viewpoint feeding the family will be described.

Activities

1. Develop a set of questions that the model will answer. Sharpen this question set by determining who is asking the questions. Write at least 5 to 10 questions before stopping. Then specify the level of precision required to answer each of them.
2. Now use this question set to determine how the model will be used. If you cannot state how the model will be used, try writing more questions or try to imagine who will use the model. In one sentence, summarize how the model will be used. This becomes the purpose of the model.
3. Now decide who is telling the story of feeding the family. Generate a list of candidates so you are confident that your choice is the best viewpoint. Choose the one from your list that will answer the most questions. This becomes the viewpoint of the model.

Notes

1. Developing a good question set helps give a feel for the way the model will be used. The more questions, the better idea you will have of what story the model must tell.
2. The viewpoint can be one person (e.g., the father) or more than one person (e.g., the parents). Remember, each viewpoint will emphasize different aspects of the problem. The best viewpoint is the one that will answer all your questions in the simplest way.

Sample

1. Notice that many detailed questions about the process of feeding the family were written. This shaped the purpose of the model to be one of understanding the entire range of activities required to feed the family.
2. Notice that part of the purpose is that the model must be good enough to describe how the teenagers should feed the family, even when the parents cannot be consulted.
3. The sample also shows some candidate viewpoints (e.g., neighbors). The parents were chosen to be the viewpoint, because they are the authorities in the household.

USED AT:	AUTHOR: Marca PROJECT: Feed Family NOTES: ~~1~~ 2 3 4 5 6 7 8 9 10	DATE: 1/16/85 REV:	☒ WORKING ☐ DRAFT ☐ RECOMMENDED ☐ PUBLICATION	READER DATE	CONTEXT:

Questions:

- How do I know what food to buy?
- What do I do first?
- How do I prepare the kitchen so I can cook?
- What is the usual way to set the table?
- What do I do with leftovers?

Purpose: Understand the functions involved in feeding a family, while the parents are away.

Perspectives:

Young children
Teenagers
Adults
Neighbors
Grocers

Viewpoint: Parents of the family.

[1] Authority required!!

NODE:	TITLE: Feed Family (Purpose & Viewpoint)	NUMBER: DAM002

Workshop

3

Drawing the Top Diagram

Goal Describe feeding the family as a single diagram, comprising from three to six important activities, interrelated and constrained by information (e.g., menu) and tangible things (e.g., dishes).

Activities

1. Aggregate the functions in the activity list of your bounded context into from three to six activities, and rank them according to their relative dominance. Draw and label boxes, one for each activity, according to your dominance ranking.
2. Draw and label constraints among just the boxes, using the Data List of the bounded context. To do this, analyze the function of each box and ask yourself if any of its outputs influence other activities.
3. Now draw and label the constraints from "outside" the system, using the Data List of the bounded context. Imagine what kinds of things influence the feeding of the family but come from someone other than the teenagers or somewhere other than the home.
4. Last, draw the major flow of data, weaving your way through the boxes. Use the Data List of the bounded context, and imagine yourself telling the teenagers a story about how people get fed.

Notes

1. Watch out for feedbacks between boxes. Feedbacks often happen in process-oriented systems (like this one) in which some functions frequently repeat. Feedback also occurs in systems (again like this one) that efficiently utilize their by-products or waste.

Sample

1. Planning is centralized in the first box. It could have been distributed inside all the other boxes, but the feeling was that planning should be emphasized immediately to the teenagers.
2. Notice that ***budget*** (an abstract thing) constrains ***plan,*** while ***money*** (a real thing) is given to and consumed by the activity that maintains the food supply.
3. Notice that ***dishes*** and ***leftovers*** feed back from ***clean up*** to the activities that require them.
4. Mechanism arrows were purposely omitted, because they will not help achieve the model's purpose.

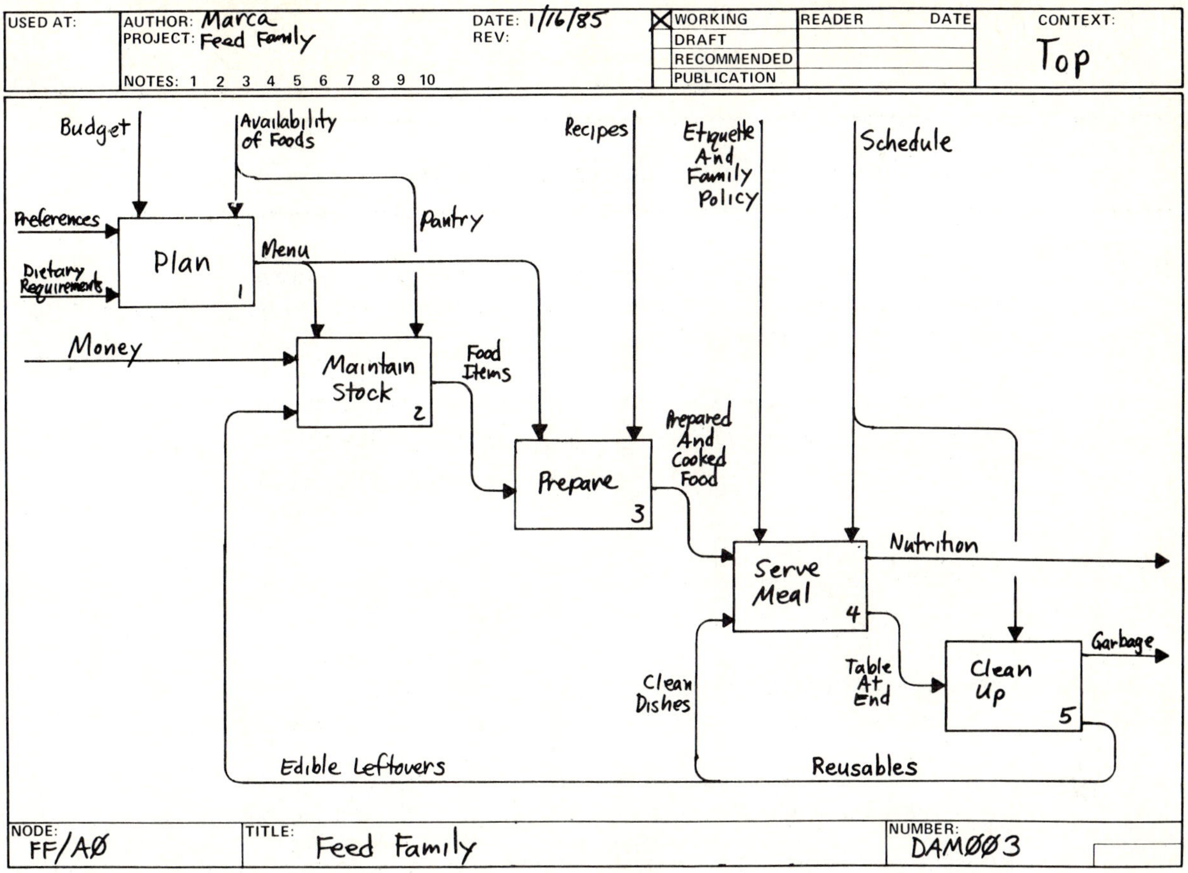
USED AT:
AUTHOR: Marca
PROJECT: Feed Family
DATE: 1/16/85
REV:
NOTES: 1 2 3 4 5 6 7 8 9 10
WORKING
DRAFT
RECOMMENDED
PUBLICATION
READER
DATE
CONTEXT:
Top
Budget
Availability of Foods
Recipes
Etiquette And Family Policy
Schedule
Preferences
Dietary Requirements
Plan
1
Menu
Pantry
Money
Maintain Stock
2
Food Items
Prepare
3
Prepared And Cooked Food
Serve Meal
4
Nutrition
Clean Up
5
Garbage
Table At End
Clean Dishes
Edible Leftovers
Reusables
NODE:
FF/AØ
TITLE:
Feed Family
NUMBER:
DAMØØ3

Workshop

4

Summarizing the Top Diagram

Goal Draw a single box, with its inputs, controls, and outputs, which summarizes the entire A0 diagram just drawn.

Activities

1. Draw a single big box in the center of the page, and label it with the title of the A0 diagram. This summarizes the functions of the system.
2. Now draw and label input, control, and output arrows—one for each external arrow on the A0 diagram. This makes a consistent tie between the two pictures.
3. Last, rewrite and label the purpose and viewpoint of the model underneath the big box. This will provide both a reason and a perspective for the model as soon as people start reading it.

Notes This single box with its arrows summarizes the "interface" to the Feed Family system.

Sample

1. Notice how this picture emphasizes what the system does, its external data, the purpose of the model, and the viewpoint of the model all on 1 page. This is why an A-0 diagram is used to introduce an SADT activity model.
2. Notice too that the C-Number of the A0 diagram is put below the big box on the right-hand side. This indicates the exact version of the A0 diagram that currently details the box.

USED AT:	AUTHOR: Marca PROJECT: Feed Family NOTES: 1 2 3 4 5 6 7 8 9 10	DATE: 1/16/85 REV:	☒ WORKING DRAFT RECOMMENDED PUBLICATION	READER DATE	CONTEXT: None

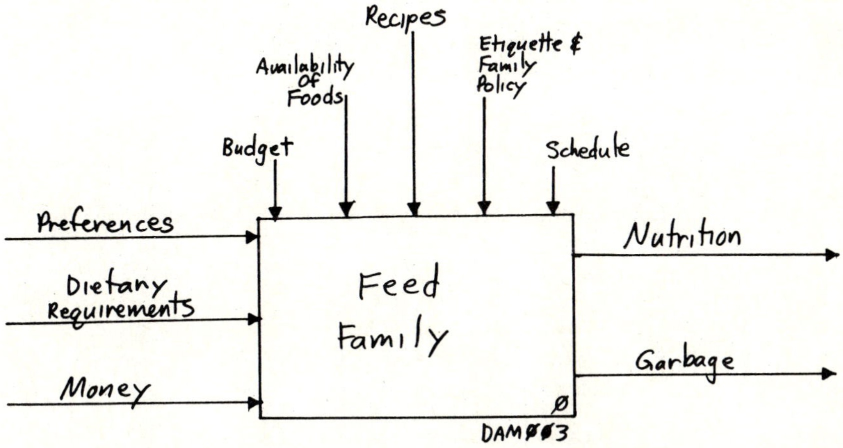

Purpose: Understand the functions involved in feeding a family, while the parents are away.

Viewpoint: The parents of the family.

NODE: FF/A-Ø	TITLE: Feed Family	NUMBER: DAMØØ4

Workshop

5

Critiquing the Summary Diagram

Goal Document all problems with the A-0 diagram.

Activities

1. Read the diagram out loud, using a storyboard template like "The function of (box name) transforms (input arrow names) into (output arrow names) according to (control arrow names)."
2. Evaluate the story being told to see if it makes sense. Locate problems in the story, document where these problems are, and revise it accordingly (e.g., rewrite labels, cluster arrows).
3. Also, critique the purpose and viewpoint after critiquing the picture. Document problems, and revise the purpose and viewpoint accordingly.

Notes Talking through the story, especially out loud, will make more apparent the shortcomings of the diagram.

Sample

1. The diagram reads: "*Feed family* transforms *preferences, dietary requirements,* and *money* into *nutrition* and *garbage,* according to *budget, availability of foods, recipes, etiquette & family policy,* and *schedule.*" This does not make complete sense, and it can also be told more simply: "*Feed family* transforms *unfed people* and *money* into *fed people* and *garbage* according to. . . ."
2. Notice that the constraints of *recipes, etiquette & family policy,* and *schedule* were clustered into the aggregate *family policy,* and that the purpose of the model was made more specific.
3. Notice that the use of note numbers chronologically identifies each problem of and revision to the diagram.

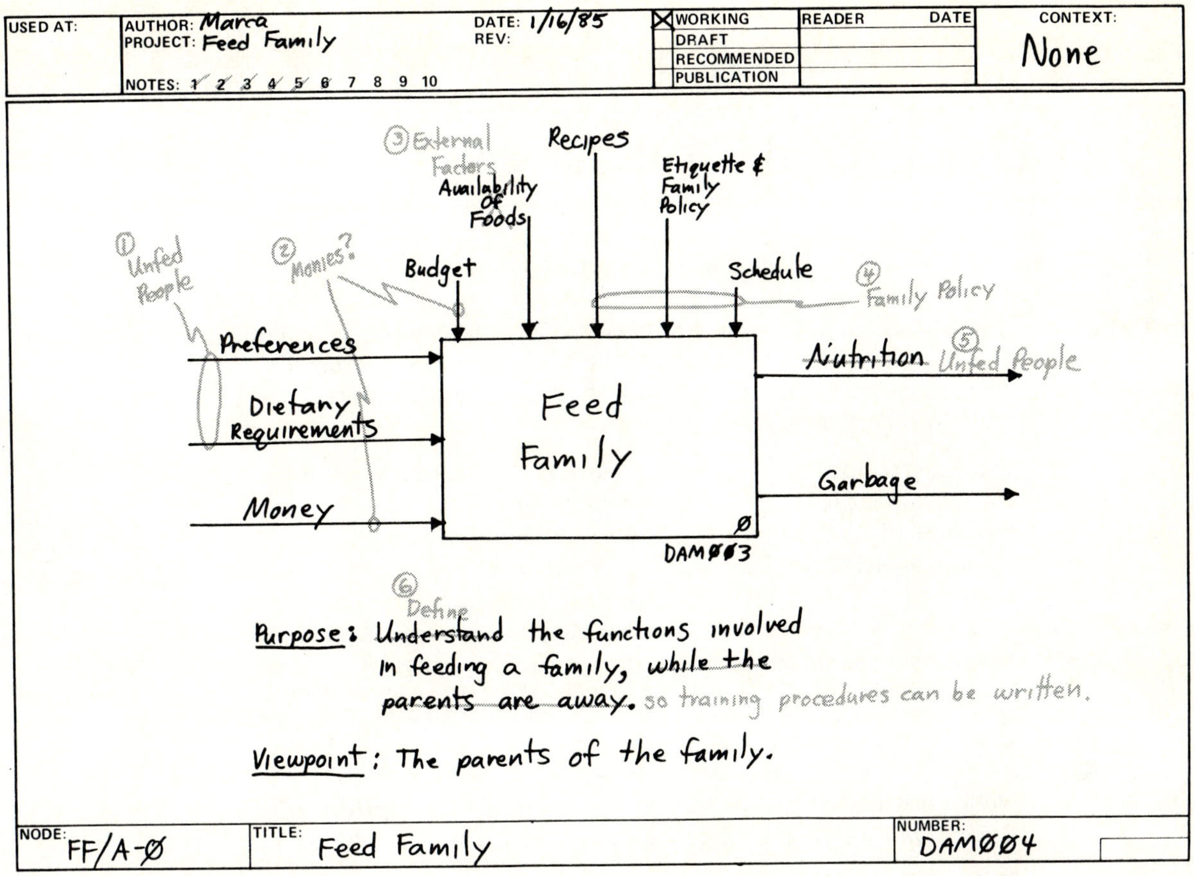
USED AT:
AUTHOR: Marca
PROJECT: Feed Family
DATE: 1/16/85
REV:
NOTES: 1 2 3 4 5 6 7 8 9 10
WORKING
DRAFT
RECOMMENDED
PUBLICATION
READER
DATE
CONTEXT:
None
③ External Factors
Recipes
Availability Of Foods
Etiquette & Family Policy
① Unfed People
② Monies?
Budget
Schedule
④ Family Policy
Preferences
Dietary Requirements
Money
Feed Family
Ø
DAMØØ3
Nutrition
⑤ Unfed People
Garbage
⑥ Define
Purpose: Understand the functions involved in feeding a family, while the parents are away. so training procedures can be written.
Viewpoint: The parents of the family.
NODE: FF/A-Ø
TITLE: Feed Family
NUMBER: DAMØØ4

Workshop

6

Critiquing the Top Diagram

Goal Document all problems with the A0 diagram.

Activities

1. Consistently copy the problems raised on the A-0 summary diagram onto this diagram. For example, if the label of a control arrow was changed on the A-0 diagram, change the corresponding external control arrow on this diagram.
2. Evaluate the story this diagram tells after all A-0 problems have been consistently copied. See if the corrected story makes sense. Locate problems with the new story, document these problems, and revise accordingly (e.g., rewrite labels, cluster arrows).

Notes

1. You will make fewer errors if you first copy onto the A0 diagram all the changes you made to the A-0 diagram.
2. By then talking through the revised story, you will make apparent the shortcomings of the revised A0 diagram.

Sample

1. Notice that the external arrows of ***recipes, etiquette & family policy,*** and ***schedule*** are clustered into the new external arrow ***family policy.***
2. Notice that the change of clustering ***preferences*** and ***dietary requirements*** into ***unfed people*** caused a new external arrow to be input to both ***plan*** and ***serve.***
3. Notice the use of note numbers to chronologically identify each problem and revision.

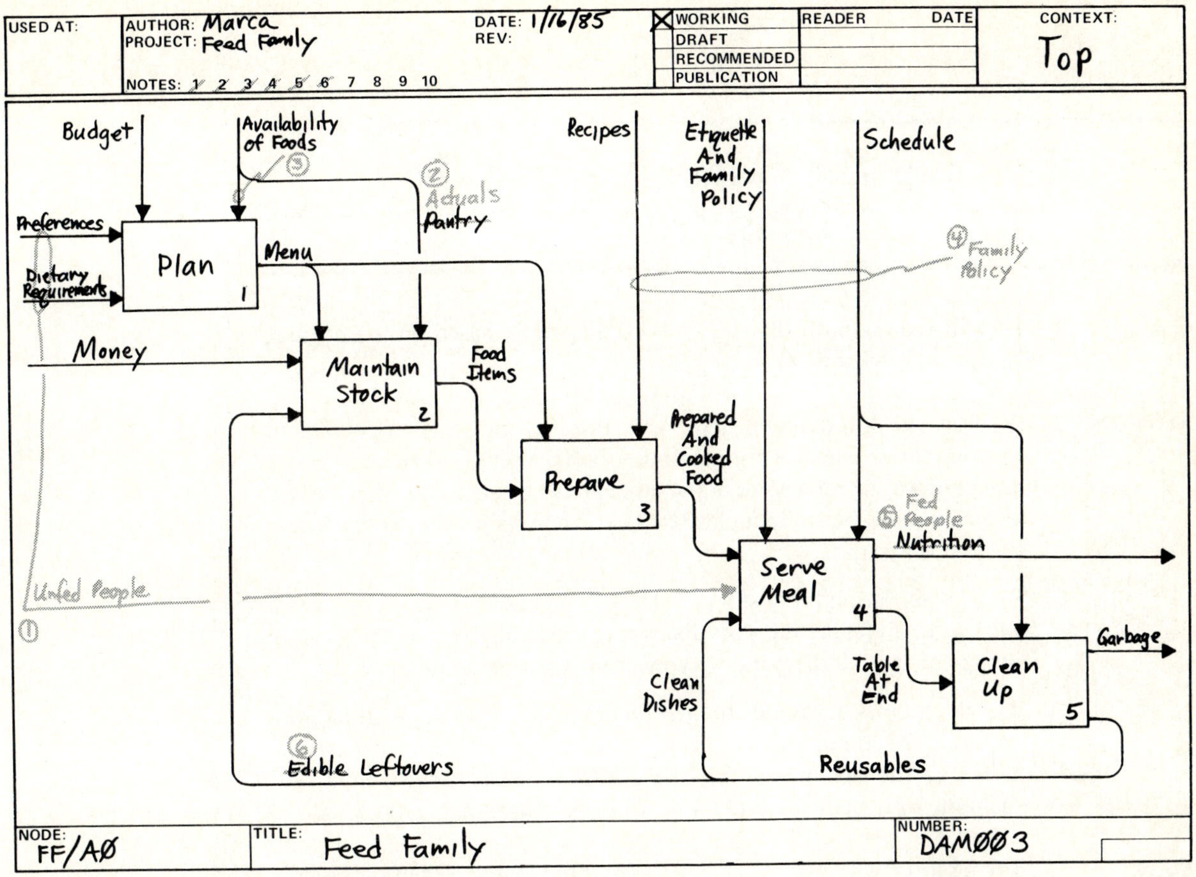
USED AT:
AUTHOR: Marca
PROJECT: Feed Family
DATE: 1/16/85
REV:
NOTES: 1 2 3 4 5 6 7 8 9 10
WORKING
DRAFT
RECOMMENDED
PUBLICATION
READER
DATE
CONTEXT:
Top
Budget
Availability of Foods
Recipes
Etiquette And Family Policy
Schedule
Preferences
Dietary Requirements
Plan
1
Menu
Actuals
Pantry
Family Policy
Money
Maintain Stock
2
Food Items
Prepare
3
Prepared And Cooked Food
Fed People
Nutrition
Serve Meal
4
Unfed People
Clean Dishes
Table At End
Clean Up
5
Garbage
Edible Leftovers
Reusables
NODE:
FF/AØ
TITLE:
Feed Family
NUMBER:
DAMØØ3

Workshop

7

Revising the Summary and Top Diagrams

Goal Revise and redraw both the A-0 and A0 diagrams, based on the critique you made in Workshop 6.

Activities

1. Redraw the A-0 diagram first. Walk through the story in your mind as you redraw, and see if it continues to make sense. Rewrite or revise the purpose and viewpoint of the model if necessary. Then put the A-0 diagram aside, but keep it close by so you can refer to it while revising the A0 diagram.

2. Now, walk through the story of the A0 diagram in your mind as you redraw it. Refer to the A-0 diagram periodically, to verify that the details of the A0 diagram are consistent with its context.

3. Tie all external arrows of the A0 diagram to its parent A-0 diagram using ICOM codes. This will keep you from forgetting external arrows. When you do this, match the labels of the external arrows on the A0 diagram with the arrow labels on the A-0 diagram.

Notes

1. Saying more with less is the key to good modeling. When redrawing, try to preserve accuracy while using fewer words and simpler graphics.

2. External arrows are vital to the decomposition, because they link the more general story of the A-0 diagram with the more detailed story of the A0 diagram. Make sure they are consistent in name and number between the A-0 and A0 diagrams.

Sample

1. Notice how fewer and better-labeled arrows sharpen the overall story. This illustrates why "less is more"—something analysts know from experience. That is, the simpler the message, the better it can be understood.

2. Notice how the two control arrows of ***external factors*** and ***family policy*** split the five activities into two groups. The first arrow influences the preparation functions: ***plan*** and ***maintain stock.*** The second arrow influences the functions concerned with an actual feeding: ***prepare, serve meal,*** and ***clean up.*** This is an example of how control arrows influence decomposition. That is, controls explain why a particular decomposition was chosen.

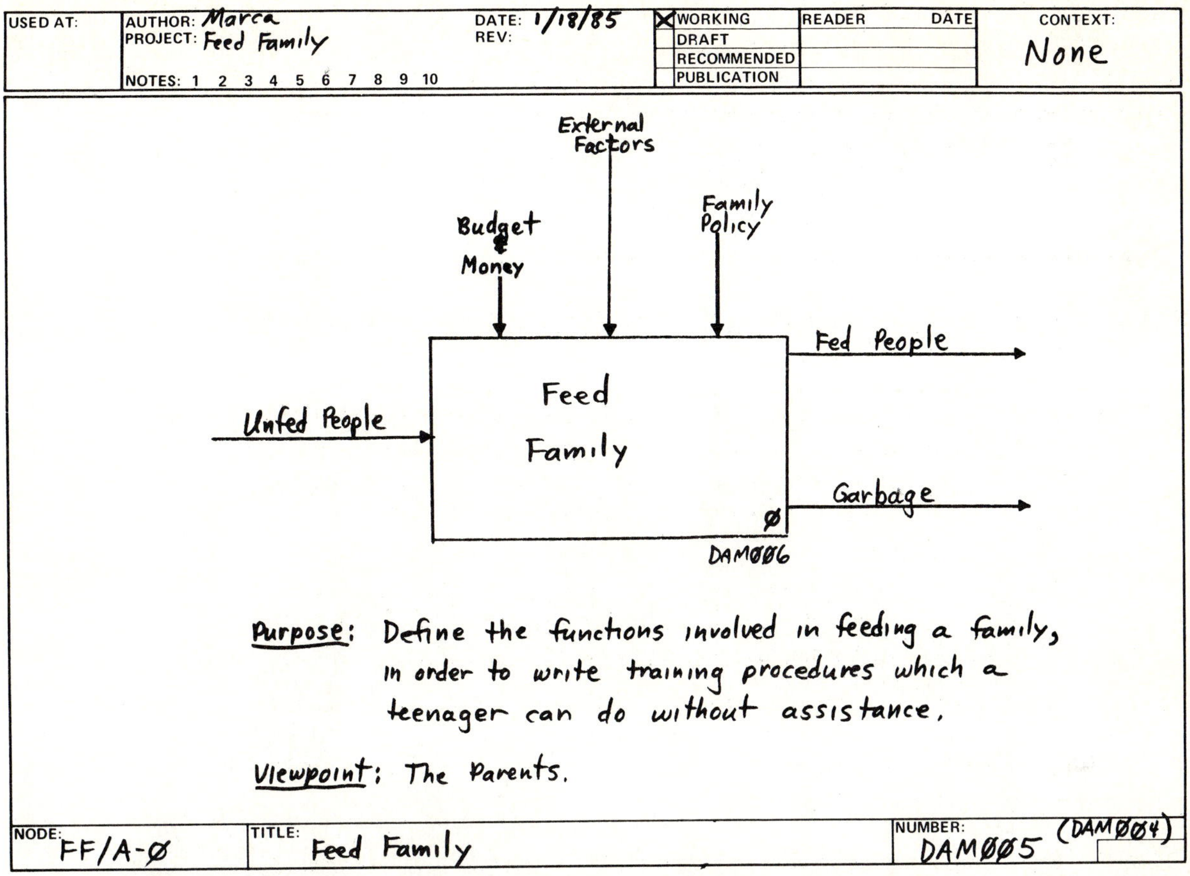
USED AT:
AUTHOR: Marca
PROJECT: Feed Family
DATE: 1/18/85
REV:
NOTES: 1 2 3 4 5 6 7 8 9 10
WORKING
DRAFT
RECOMMENDED
PUBLICATION
READER
DATE
CONTEXT:
None
External Factors
Budget & Money
Family Policy
Unfed People
Feed Family
Ø
Fed People
Garbage
DAMØØ6
Purpose: Define the functions involved in feeding a family, in order to write training procedures which a teenager can do without assistance.
Viewpoint: The Parents.
NODE: FF/A-Ø
TITLE: Feed Family
NUMBER: DAMØØ5 (DAMØØ4)

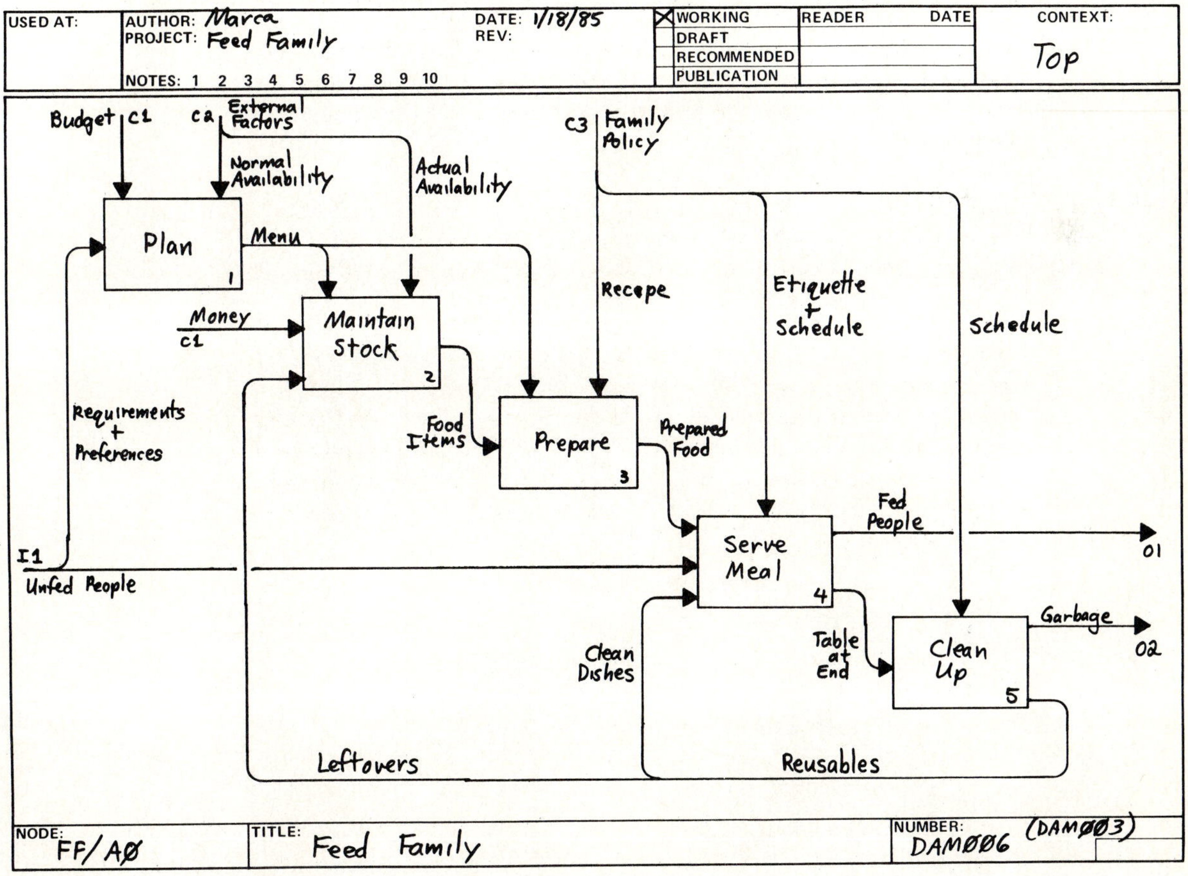
USED AT:
AUTHOR: Marca
PROJECT: Feed Family
DATE: 1/18/85
REV:
NOTES: 1 2 3 4 5 6 7 8 9 10
WORKING
DRAFT
RECOMMENDED
PUBLICATION
READER
DATE
CONTEXT:
Top
Budget C1
C2 External Factors
Normal Availability
Actual Availability
C3 Family Policy
Plan
1
Menu
Money
C1
Maintain Stock
2
Recepe
Etiquette + Schedule
Schedule
Requirements + Preferences
Food Items
Prepare
3
Prepared Food
Fed People
O1
I1
Unfed People
Serve Meal
4
Clean Dishes
Table at End
Clean Up
5
Garbage
O2
Leftovers
Reusables
NODE: FF/AØ
TITLE: Feed Family
NUMBER: DAMØØ6 (DAMØØ3)

Chapter
23

Developing First-Level Decompositions

Recall that an SADT analyst sketches a decomposition, critiques it, redraws it, and then packages it into a kit. This chapter consists of three workshops which will guide you through the steps involved in developing first-level decompositions—that is, the steps of decomposing a box on the A0 diagram. In Workshop 8, a diagram is developed which decomposes one box on the A0 diagram. Workshop 9 guides you through self-critique and revision of the diagram. Workshop 10 walks you through the steps of packaging this work into a kit for review by others.

Go through all three workshops without a break. This will give you an accurate feel for the amount of work required to decompose an already bounded subject. (After the workshops, reflect on the amount of effort it took to start the model versus the amount of effort it took to decompose one box.) Allocate about ½ hour for each workshop, but don't be concerned if you take more time. Finish all three workshops, even if you have difficulty keeping to the exact description of the system as told by the A0 diagram.

Workshop

8

Group Diagramming

Goal Choose and decompose a single box on the A0 diagram.

Activities

1. Pick a box on the A0 diagram. This box becomes the complete context for this entire workshop. Do not stray outside this boundary.
2. Mentally review the steps of drawing a diagram: list things first, list activities second, cluster activities into from three to six boxes, draw boxes from the most to least dominant, draw external arrows, then draw control arrows, and last draw input and output arrows.
3. Read the A0 diagram again, focusing on the way your box fits with the rest of its boxes. Use the boundary arrows of the box you picked to start your data list.
4. Now go through all the other steps of developing a diagram. Try to put your data and activity lists on the left side of the form, and try to draw the diagram on the right side of the form. (Left-right placement is only a suggestion.)
5. As you draw, make notes to yourself and define terminology.
6. After you finish, go back and check ICOM's. Make sure you did not forget to use any boundary data.

Notes

1. Don't stray outside the bounded context defined by the box you picked on the A0 diagram. You may find it hard to make the diagram, because first decompositions often point out difficulties with the parent diagram. Record difficulties by making notes on the A0 diagram.
2. Do not worry about the correctness of this diagram for now. First-level decompositions are rarely drawn correctly the first time.

Sample

1. Notice that first the external input, control, and output arrows were listed, and then subsequent data was listed that detailed the components of those arrows. This is a reliable way to make sure you do not forget the context in which you are working.
2. Notice the control feedback from ***buy*** to ***make shopping list.*** This describes the situation in which the shopping list gets immediately changed when you see the grocery store lacks the items you want.

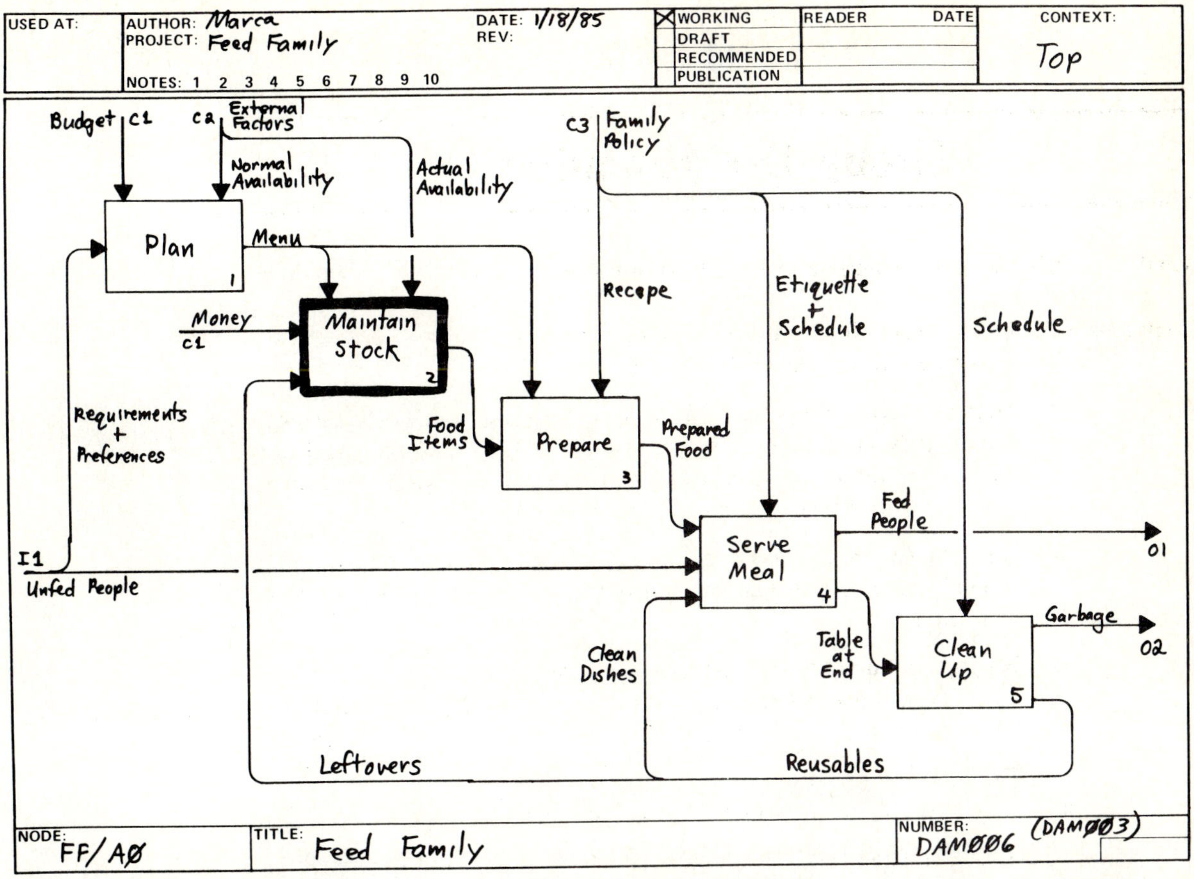
USED AT:
AUTHOR: Marca
PROJECT: Feed Family
DATE: 1/18/85
REV:
NOTES: 1 2 3 4 5 6 7 8 9 10
WORKING
DRAFT
RECOMMENDED
PUBLICATION
READER
DATE
CONTEXT:
Top
Budget C1
C2 External Factors
Normal Availability
Actual Availability
C3 Family Policy
Plan 1
Menu
Money C1
Maintain Stock 2
Recepe
Etiquette + Schedule
Schedule
Requirements + Preferences
Food Items
Prepare 3
Prepared Food
Fed People
O1
I1
Unfed People
Serve Meal 4
Table at End
Clean Up 5
Garbage
O2
Clean Dishes
Leftovers
Reusables
NODE: FF/AØ
TITLE: Feed Family
NUMBER: DAMØØ6 (DAMØØ3)

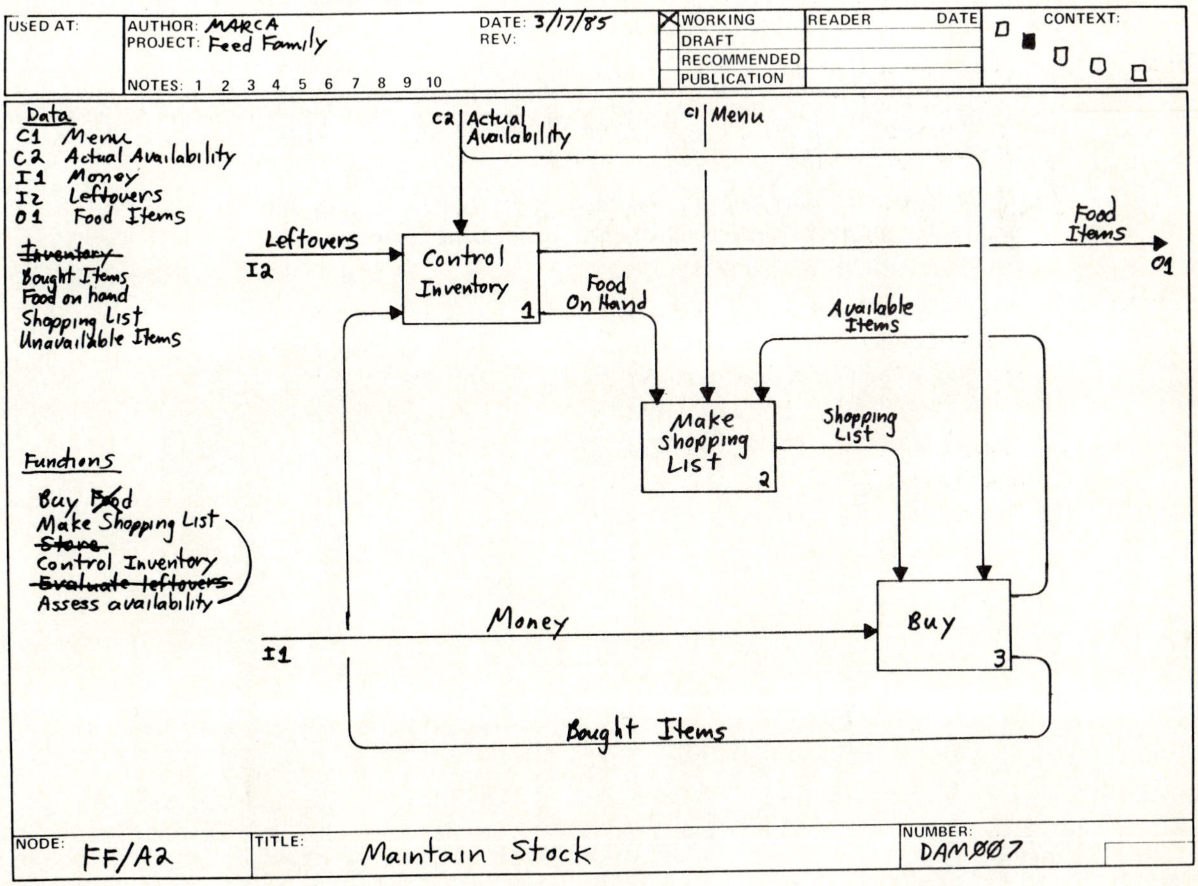
USED AT:
AUTHOR: MARCA
PROJECT: Feed Family
DATE: 3/17/85
REV:
NOTES: 1 2 3 4 5 6 7 8 9 10
WORKING
DRAFT
RECOMMENDED
PUBLICATION
READER
DATE
CONTEXT:
Data
C1 Menu
C2 Actual Availability
I1 Money
I2 Leftovers
O1 Food Items
Inventory
Bought Items
Food on hand
Shopping List
Unavailable Items
Functions
Buy Food
Make Shopping List
Store
Control Inventory
Evaluate leftovers
Assess availability
C2 Actual Availability
C1 Menu
Leftovers
I2
Control Inventory 1
Food On Hand
Food Items
O1
Available Items
Make Shopping List 2
Shopping List
Money
I1
Buy 3
Bought Items
NODE: FF/A2
TITLE: Maintain Stock
NUMBER: DAMØØ7

Workshop

9

Critiquing First-Level Decompositions

Goal Criticize the diagram you drew in Workshop 8 for the way it details its parent box on the A0 diagram.

Activities

1. Walk through the diagram you drew in Workshop 8, and try to tell the story of how it accomplishes part of feeding the family. Start the story at a logical beginning, with the arrival of one or more things from a box on the A0 diagram. Refer to the A0 diagram continuously, and take notes whenever you find the story incorrect or weak.
2. Evaluate the way you split external arrows and the way you clustered activities into boxes. See if the data can be decomposed differently or if activities can be clustered into a different set of boxes.
3. Remember to note intended changes on both your diagram and the A0 diagram. Change or refine your terminology as you criticize.
4. Draw a brand-new diagram from the notes you took, and redraw the A0 diagram if necessary. Remember to check the ICOM connections between this diagram and the A0 diagram.

Notes Try alternate decompositions, or try changing box and arrow names, to see if you can make the story sharper. Sometimes, different decompositions (or just using synonyms) can lead to new insights.

Sample

1. Notice how box and arrow names were changed to better describe ***maintain stock.*** For example, the ***stock the pantry*** label is a better description using less technical terminology.
2. Notice that ***make shopping list*** was made more general. Doing this means that the ***plan shopping*** activity now handles the complexities of selecting possibly several stores.
3. Note 4 reminds the author to check with the author of ***plan*** on the meaning of the term ***actual availability.*** This happens frequently when a model is built by several authors.

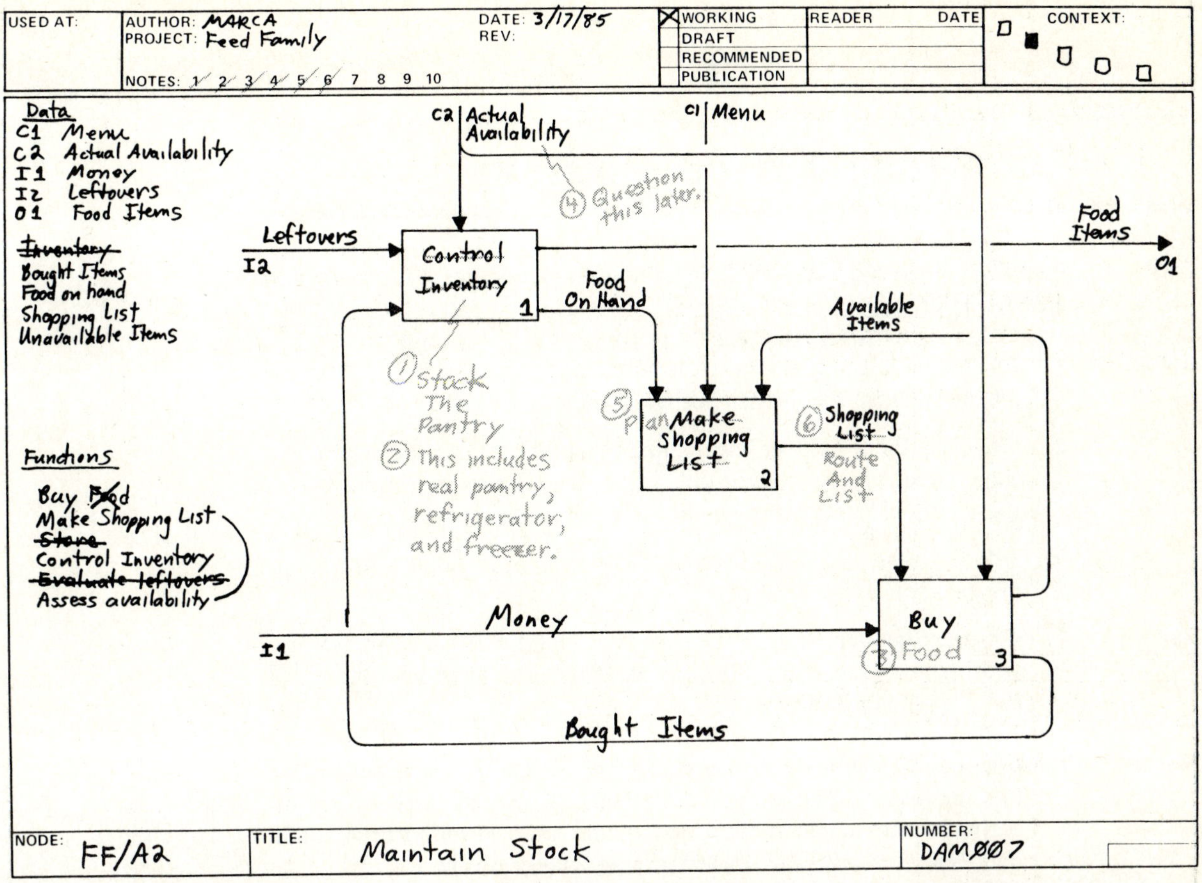
USED AT:
AUTHOR: MARCA
PROJECT: Feed Family
DATE: 3/17/85
REV:
NOTES: 1 2 3 4 5 6 7 8 9 10
WORKING
DRAFT
RECOMMENDED
PUBLICATION
READER
DATE
CONTEXT:
Data
C1 Menu
C2 Actual Availability
I1 Money
I2 Leftovers
O1 Food Items
Inventory
Bought Items
Food on hand
Shopping List
Unavailable Items
Functions
Buy Food
Make Shopping List
Store
Control Inventory
Evaluate leftovers
Assess availability
C2 Actual Availability
C1 Menu
4 Question this later.
Leftovers
I2
Control Inventory 1
Food Items
O1
Food On Hand
Available Items
1 Stock The Pantry
2 This includes real pantry, refrigerator, and freezer.
5 Plan
Make Shopping List 2
6 Shopping List
Route And List
Money
I1
Buy Food 3
Bought Items
NODE: FF/A2
TITLE: Maintain Stock
NUMBER: DAM007

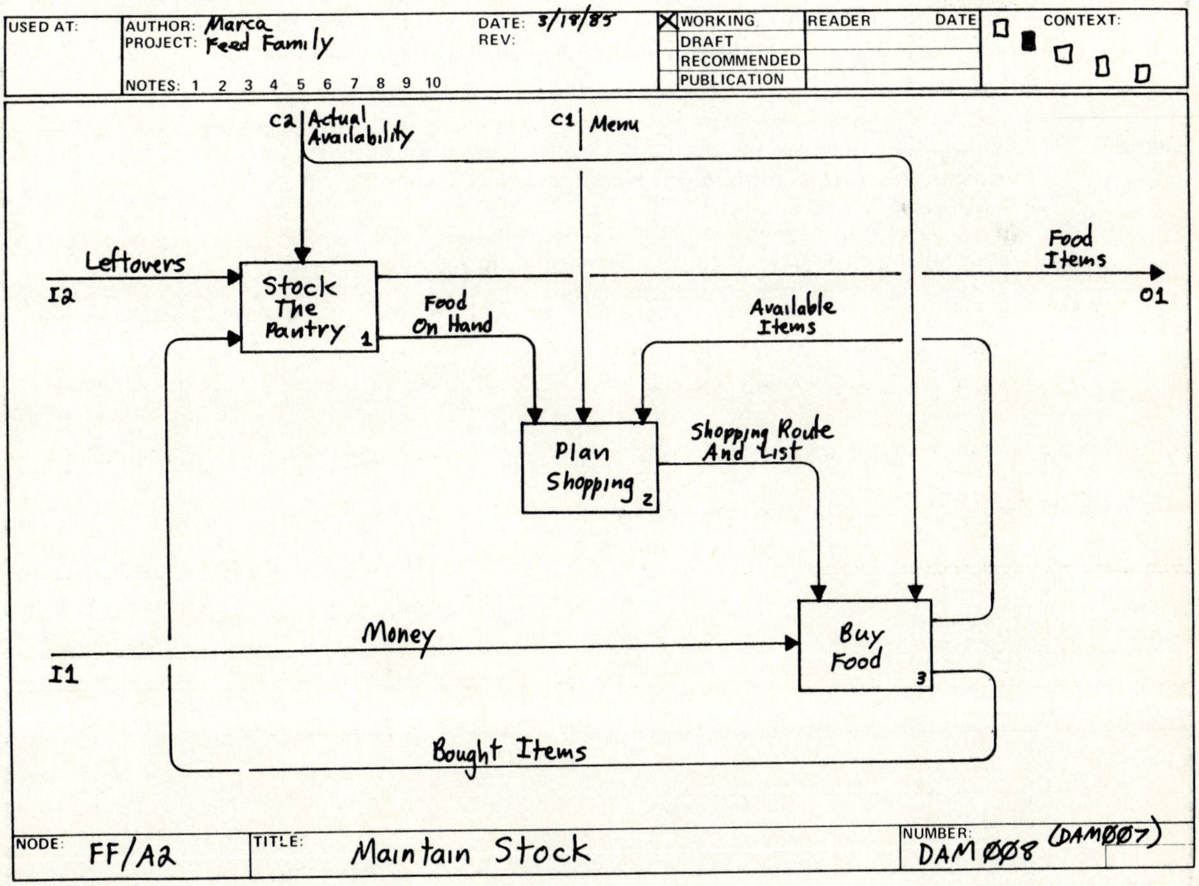
USED AT:
AUTHOR: Marca
PROJECT: Feed Family
DATE: 3/18/85
REV:
NOTES: 1 2 3 4 5 6 7 8 9 10
WORKING
DRAFT
RECOMMENDED
PUBLICATION
READER
DATE
CONTEXT:
C2 Actual Availability
C1 Menu
Leftovers
I2
Stock The Pantry 1
Food Items
O1
Food On Hand
Available Items
Plan Shopping 2
Shopping Route And List
Money
I1
Buy Food 3
Bought Items
NODE: FF/A2
TITLE: Maintain Stock
NUMBER: DAM008 (DAM007)

Workshop

10

Group Kit Preparation

Goal Package your revised first-level diagram and any associated Glossary into an SADT kit.

Activities

1. Lay out both your diagram and its Glossary Page, and check to see that all information is consistent between them.
2. Complete the Kit Cover Page, filling in identification information (author, project, and date), the title of the kit, what is in the kit, whom the kit should go to, and when the kit must be returned.
3. Add any other special comments or notes to your readers in the Comments Area of the Kit Cover Page.
4. Clip the pages together—Kit Cover Page first, the A0 diagram second, your diagram third, and the Glossary Page last. Then send the kit to the project Librarian.

Notes

1. Use the Comments Area to communicate special points about the kit to your readers. These notes can often help you get good review by asking readers to watch out for particular areas of concern. (A single note can go a long way toward getting quality review.)
2. The wiggly line below the dates on the Kit Cover Page means that the return date applies to all listed readers. We will consistently use this notation on all Kit Cover Pages throughout the workshops.

Sample

1. Notice that a note was put on the Kit Cover Page, asking readers to define the term ***actual availability***. Notice how the Glossary Page flags this concern.
2. Notice that both ***pantry*** and ***shopping route and list*** were defined in the Glossary by giving their constituent parts.

	AUTHOR: Marca PROJECT: Feed Family NOTES: ~~1~~ 2 3 4 5 6 7 8 9 10	DATE: 3/21/85 REV:	☒ WORKING ☐ DRAFT ☐ RECOMMENDED ☐ PUBLICATION	READER	DATE

☐ LOG
☐ FILE
☐ AUTHOR

DOCUMENT NUMBER: Feed – 1

Received ________
Completed ________

COPYING INSTRUCTIONS: ______copies of ______ pages = ______total
☐ as soon as possible ☐ by ________

READERS:	New Kit To Reader	Due Back	Comments To Author	Due Back	Response To Reader
Group 1	3/21/85	3/22/85			
Group 3	}	}			
Group 4	}	}			
Group 5	}	}			

RESPONSE REQUIRED: ☒ Fast ☐ Normal ☐ Slow ☐ None

CONTENTS:

Pg.	Node	Title	C-Number	Status
A		COVER SHEET		
B	FF/A2	Maintain Stock	DAMØØ8	
C	FF/A2G1	Glossary	DAMØØ9	
D				
E				
F				
G				
H				
I				
J				
K				
L				
M				
N				

COMMENTS:
☐ UPDATE / ☐ REPLACE } Model File ________ with this kit.

[1] Please help with defining what "Actual Availability" means.

SPECIAL INSTRUCTIONS:
☐ no author copy
____extra author copies

NODE: | TITLE: First Decomposition of "Maintain Stock" | NUMBER: DAMØ1Ø

USED AT:	AUTHOR: Marca PROJECT: Feed Family NOTES: 1 2 3 4 5 6 7 8 9 10	DATE: 1/18/85 REV:	☒ WORKING ☐ DRAFT ☐ RECOMMENDED ☐ PUBLICATION	READER	DATE	CONTEXT: Top

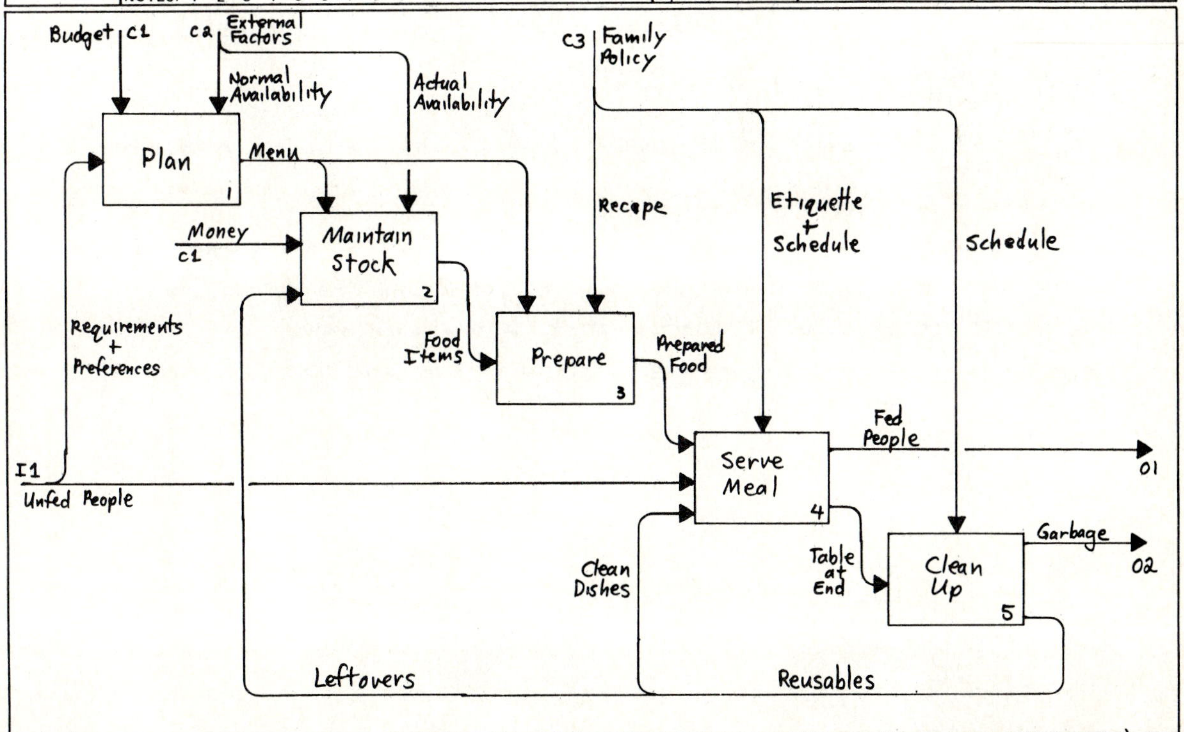

NODE: FF/AØ | TITLE: Feed Family | NUMBER: DAMØØ6 (DAMØØ3)

USED AT:	AUTHOR: Marca PROJECT: Feed Family NOTES: 1 2 3 4 5 6 7 8 9 10	DATE: 3/18/85 REV:	☒ WORKING DRAFT RECOMMENDED PUBLICATION	READER	DATE	CONTEXT:

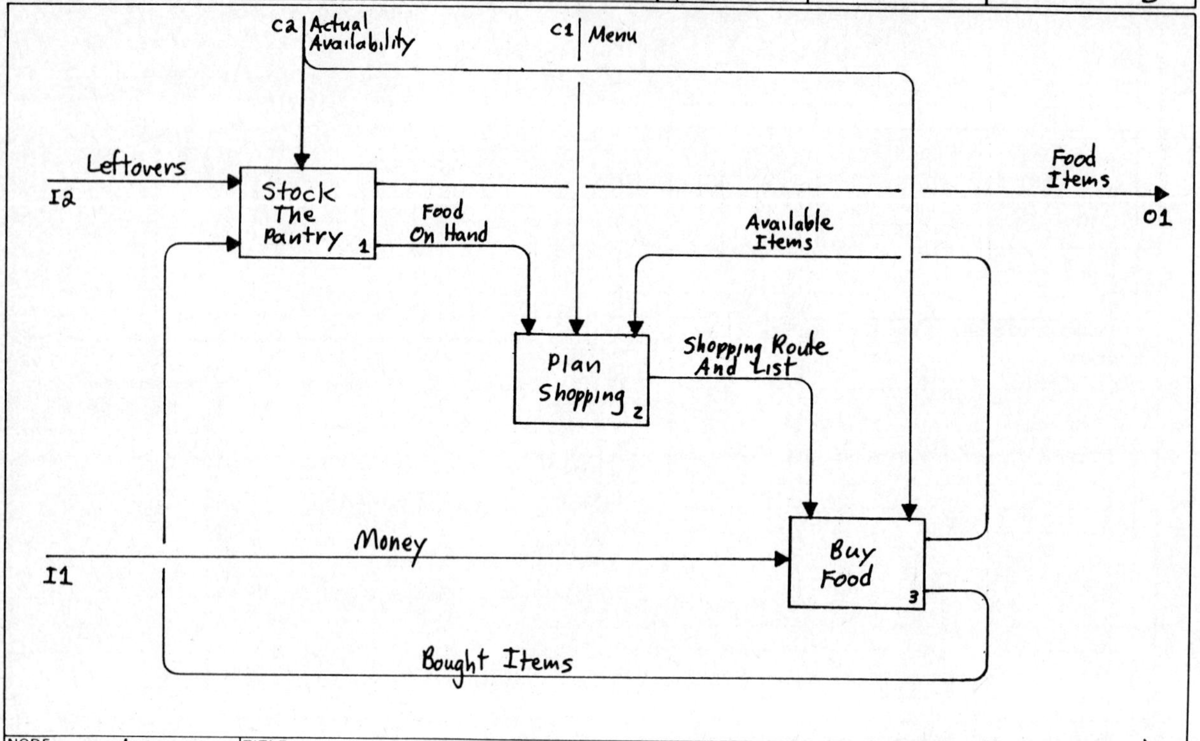

NODE: FF/A2	TITLE: Maintain Stock	NUMBER: DAM008 (DAM007)

USED AT:	AUTHOR: Marca PROJECT: Feed Family NOTES: 1 2 3 4 5 6 7 8 9 10	DATE: 3/20/85 REV:	☒ WORKING DRAFT RECOMMENDED PUBLICATION	READER	DATE	CONTEXT: DAM008

Actual Availability – ?? needs clarification ??

Pantry – The places where all the food is kept. Includes the larder, refrigerator, and freezer.

Food on Hand – the foods in the pantry.

Available Items – Those desired items that are and are not on the grocer's shelves.

Shopping Route – A map of the best way to get to all the desired stores.

Shopping List – A written list of what to buy in each shop.

Bought Items – Food in shopping bags that have been paid for.

NODE: FF/A2G1	TITLE: Maintain Stock (Glossary)	NUMBER: DAM009

Chapter

24

Resolving Top-Level Interfaces

Recall that the first Author/Reader Cycle annotates kits with comments and reactions, and often impacts the interfaces among boxes on the A0 level diagram. This chapter comprises four workshops, designed to guide you through the steps involved in resolving the interfaces among first-level decompositions. In Workshop 11, first-level decompositions are reviewed. This is when interface issues get raised. Workshop 12 lets you react to the comments of others. Comments that are accepted become verified interface issues that require resolution. In Workshop 13, the A0 diagram is revised, thus documenting the resolutions regarding interface issues. Workshop 14 then steps you through redrawing a first-level decomposition, showing how the new interfaces are reflected in the revised diagram.

Go through at least the first two workshops without a break. This will give you the opportunity to properly think about, raise, and verify interface issues. Then go through the last two workshops without a break, so that when you resolve an interface issue you can properly reflect required changes on first-level decompositions. Allocate about ½ hour for each workshop, but don't be concerned if you take more time.

Workshop

11

Group Commenting

Goal Comment on the kits of other groups.

Activities

1. Read the Kit Cover Page. Notice any special message the author may have written in the Comments Area.
2. Read the Ax diagram in the kit, comparing its message with that of its context box on the A0 diagram. Be sure that the messages are consistent. Document problems with reader notes in red.
3. Read the Ax diagram, noting problems with the details of its message or its terminology. (An incorrect story may still be consistent with the parent box.) Check the Glossary frequently, and document problems with reader notes in red.
4. If the Ax diagram you drew has a direct connection with this Ax diagram, place both diagrams side by side, then read the pair. Study the ways they should and do connect, and document problems with reader notes in red.
5. Send the commented kit back to the author.

Notes

1. Remember, be critical but make your comments constructive.
2. Write comments on every page. (No comments means you either haven't read the page or you have nothing to say.) At least make a check mark to indicate you have read the page and agree with the author.

Sample

1. Notice on the Kit Cover Page: (1) when the kit was received by the Librarian, (2) when the kit was read, and (3) when the reader expects the kit back from the author. Also, notice the check mark after the author's note, meaning the reader will try to help.
2. Note 4 on the Ax diagram is a statement about something done very well by the author. (Remember, comments can strongly support an author's work as well as document problems.)

AUTHOR: Marca	DATE: 3/21/85	☒ WORKING	READER	DATE
PROJECT: Feed Family	REV:	DRAFT		
		RECOMMENDED		
NOTES: ~~1~~ 2 3 4 5 6 7 8 9 10		PUBLICATION		

LOG
FILE
AUTHOR

DOCUMENT NUMBER: Feed – 1

Received 3/21/85
Completed

COPYING INSTRUCTIONS:
____ copies of ____ pages = ____ total
☐ as soon as possible ☐ by ____

READERS:	New Kit To Reader	Due Back	Comments To Author	Due Back	Response To Reader
Group 1	3/21/85	3/22/85	3/22/85	3/23/85	
Group 3	}	}	}	}	
Group 4	}	}	}	}	
Group 5	}	}	}	}	

RESPONSE REQUIRED: ☒ Fast ☐ Normal ☐ Slow ☐ None

CONTENTS:

Pg.	Node	Title	C-Number	Status
A		COVER SHEET		
B	FF/A2	Maintain Stock	DAM008	
C	FF/A2G1	Glossary	DAM009	
D				
E				
F				
G				
H				
I				
J				
K				
L				
M				
N				

COMMENTS:
☐ UPDATE / ☐ REPLACE } Model File ____ with this kit.

1 Please help with defining what "Actual Availability" means. ✓

SPECIAL INSTRUCTIONS:
☐ no author copy
____ extra author copies

NODE:	TITLE: First Decomposition of "Maintain Stock"	NUMBER: DAM010

USED AT:	AUTHOR: Marca	DATE: 1/18/85	☒ WORKING	READER	DATE	CONTEXT:
	PROJECT: Feed Family	REV:	DRAFT			Top
			RECOMMENDED			
	NOTES: 1 2 3 4 5 6 7 8 9 10		PUBLICATION			

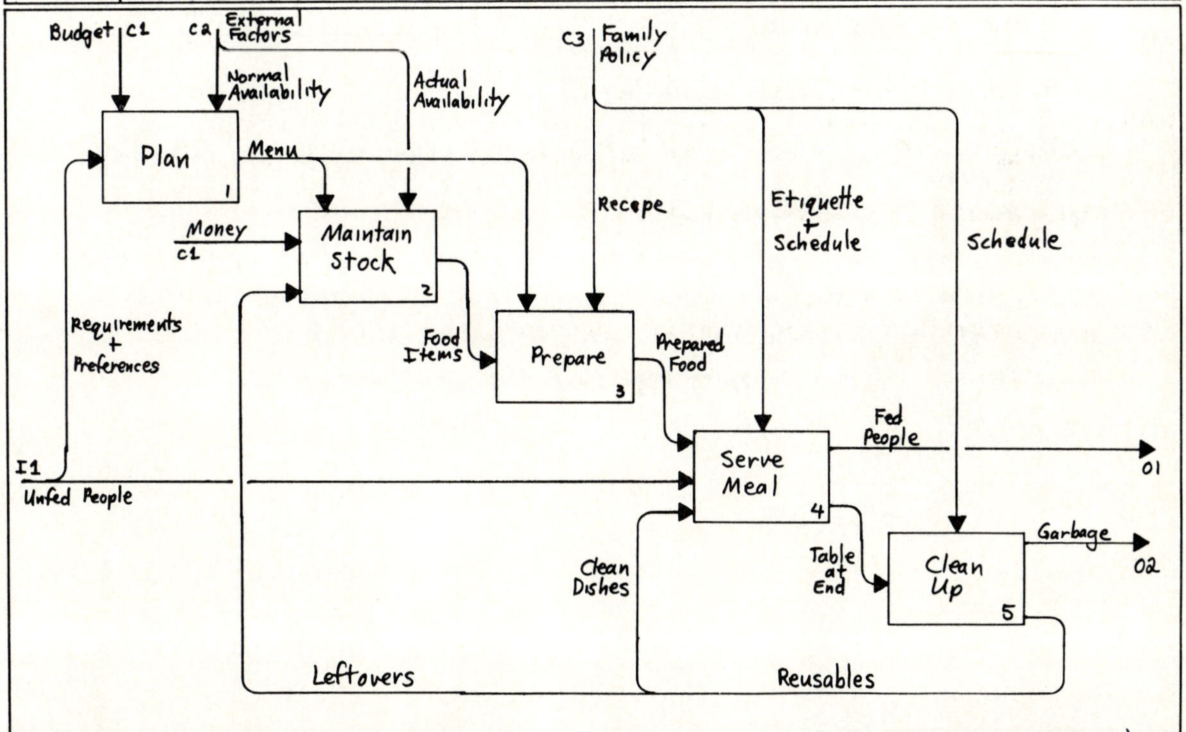

NODE: FF/A0	TITLE: Feed Family	NUMBER: DAM006 (DAM003)

USED AT: | AUTHOR: Marca | PROJECT: Feed Family | DATE: 3/18/85 | REV: | ☒ WORKING | DRAFT | RECOMMENDED | PUBLICATION | READER | DATE | CONTEXT:

NOTES: ~~1 2 3 4~~ 5 6 7 8 9 10

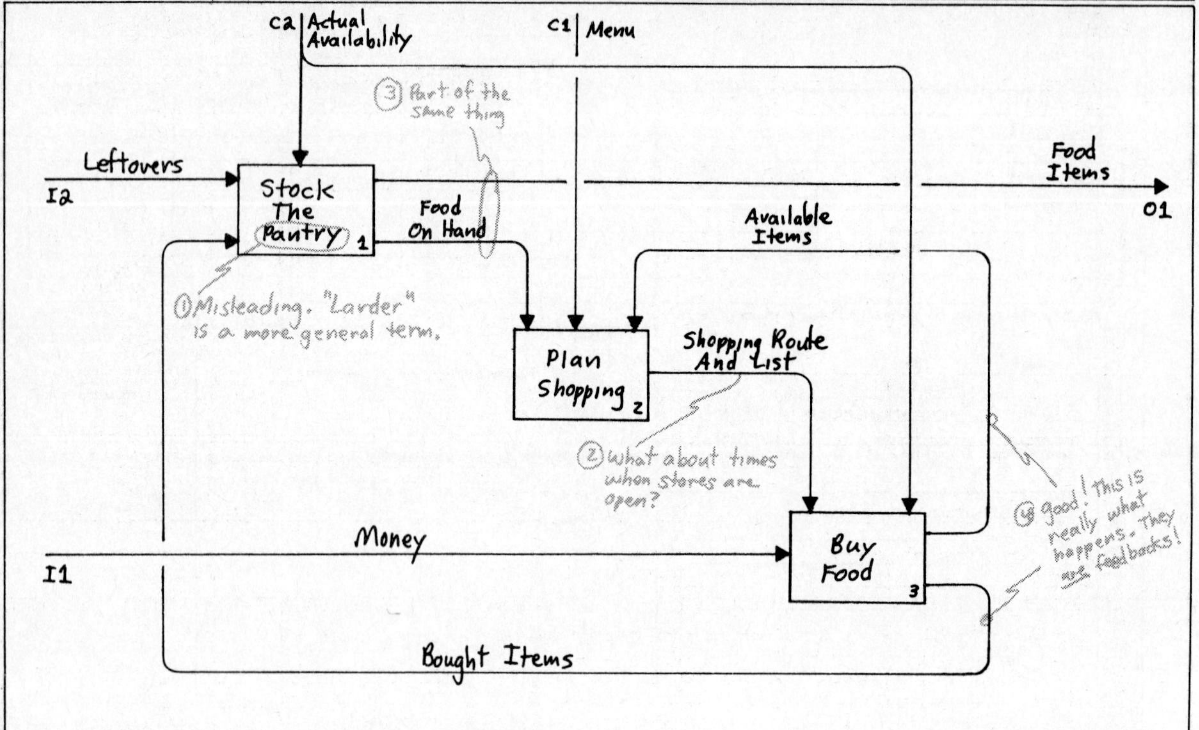

NODE: FF/A2 | TITLE: Maintain Stock | NUMBER: DAMØØ8 (DAMØØ7)

USED AT: | AUTHOR: Marca | PROJECT: Feed Family | DATE: 3/20/85 | REV: | ☒ WORKING | DRAFT | RECOMMENDED | PUBLICATION | READER | DATE | CONTEXT: DAMØØ8

NOTES: ~~1 2 3~~ 4 5 6 7 8 9 10

<u>Actual Availability</u> – ~~?? needs clarification ??~~ ① availability of food on hand

Pantry – The places where all the food is kept. Includes the larder, refrigerator, and freezer.

Food on Hand – the foods in the pantry.

Available Items – Those desired items that are and are not on the grocer's shelves.

Shopping Route – A map of the best way to get to all the desired stores.

Shopping List – A written list of what to buy in each shop.

Bought Items – Food in shopping bags that have been paid for.

② Shopping Schedule – Times when stores are open.

③ This will require some feedback to "plan"?

NODE: FF/A2G1 | TITLE: Maintain Stock (Glossary) | NUMBER: DAMØ09

Workshop

12

Group Reacting

Goal React to the kits commented on by the other groups.

Activities

1. For each kit, read the Kit Cover Page, noting any special comments by the reader in the Comments Area.
2. Read the comments on the Ax diagram in note number order. Respond to each as soon as you read it. Determine if the comment is valid and why. Check valid comments, cross out invalid ones, and always explain your reactions. Write your responses in blue.
3. Read and react to the comments on the Glossary Page. Do not forget this, because (a) definitions not in agreement often lead to misunderstandings, and (b) readers may redefine terminology, which can lead to useful insights.
4. Read again all the comments that dealt with diagram interfaces. Compare your interface assumptions with those of the reader and reevaluate their validity. Document all valid problems with round notes in blue on your copy of the kit.
5. Send the reacted kit back to the reader.
6. Consolidate the comments of all readers onto your copy of the kit. Make notes to yourself to explain why you resolved a set of conflicting comments in a particular way. Because you are the author, make all notes in blue.

Notes

1. Remember to react critically but constructively.
2. Write a reaction for every comment. (No reaction means you either have not read the comment or have nothing to say.) Make a check or a cross to indicate your agreement or disagreement.

Sample

1. Notice on the Kit Cover Page when the author finished the kit.
2. Note 5 on the Ax diagram documents the author's initial concern about note 1 and eventual agreement, after some deliberation and study.
3. Note 2 was incorrect, and the author noted that fact with a cross and explained why. This illustrates use of written reactions to explain an undocumented idea of the author.

AUTHOR: Marca | DATE: 3/21/85 | ☒ WORKING | READER | DATE
PROJECT: Feed Family | REV: | DRAFT
NOTES: ~~1~~ 2 3 4 5 6 7 8 9 10 | RECOMMENDED | PUBLICATION

LOG	DOCUMENT NUMBER: Feed - 1
FILE	Received 3/21/85
AUTHOR	Completed 3/23/85

COPYING INSTRUCTIONS: ______ copies of ______ pages = ______ total
☐ as soon as possible ☐ by ______

READERS:	New Kit To Reader	Due Back	Comments To Author	Due Back	Response To Reader
Group 1	3/21/85	3/22/85	3/22/85	3/23/85	3/23/85
Group 3					
Group 4					
Group 5					

RESPONSE REQUIRED: ☒ Fast ☐ Normal ☐ Slow ☐ None

CONTENTS:

Pg.	Node	Title	C-Number	Status
A		COVER SHEET		
B	FF/A2	Maintain Stock	DAM008	
C	FF/A261	Glossary	DAM009	
D				
E				
F				
G				
H				
I				
J				
K				
L				
M				
N				

COMMENTS:
☐ UPDATE / ☐ REPLACE } Model File ______ with this kit.

1 Please help with defining what "Actual Availability" means. ✓ ✓

SPECIAL INSTRUCTIONS:
☐ no author copy
______ extra author copies

NODE: | TITLE: First Decomposition of "Maintain Stock" | NUMBER: DAM010

USED AT: | AUTHOR: Marca | DATE: 1/18/85 | ☒ WORKING | READER | DATE | CONTEXT: Top
PROJECT: Feed Family | REV: | DRAFT | RECOMMENDED | PUBLICATION
NOTES: 1 2 3 4 5 6 7 8 9 10

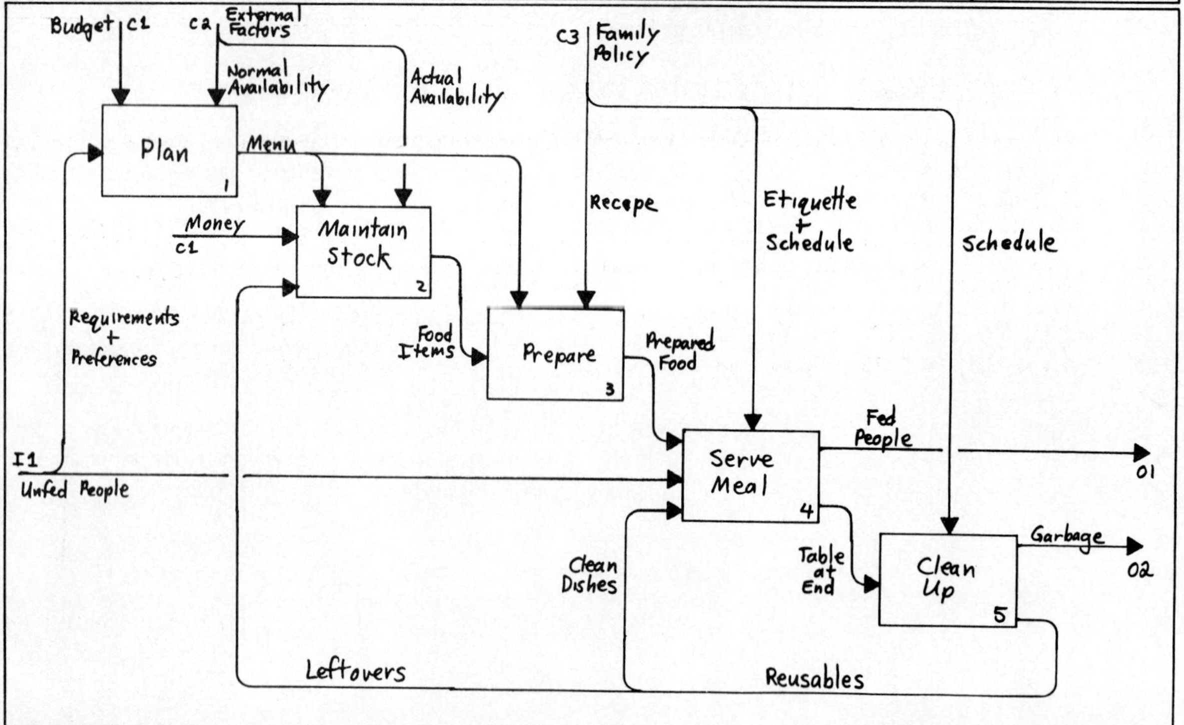

NODE: FF/A0 | TITLE: Feed Family | NUMBER: DAM006 (DAM003)

USED AT: | AUTHOR: Marca | DATE: 3/18/85 | ☒ WORKING | READER | DATE | CONTEXT:
PROJECT: Feed Family | REV: | DRAFT | RECOMMENDED | PUBLICATION
NOTES: ~~1 2 3 4 5~~ 6 7 8 9 10

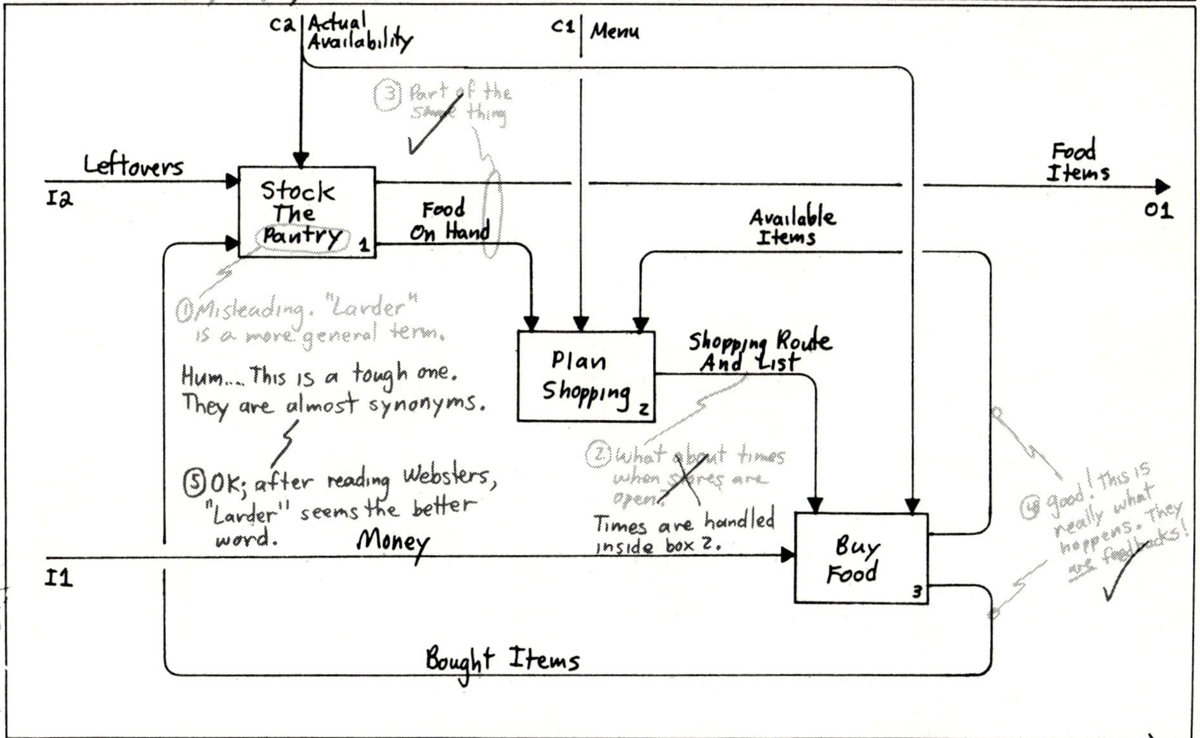

NODE: FF/A2 | TITLE: Maintain Stock | NUMBER: DAM ØØ8 (DAMØØ7)

USED AT: | AUTHOR: Marca | DATE: 3/20/85 | ☒ WORKING | READER | DATE | CONTEXT: DAMØØ8
PROJECT: Feed Family | REV: | DRAFT | RECOMMENDED | PUBLICATION
NOTES: ~~1 2 3 4~~ 5 6 7 8 9 10

Actual Availability — ~~??needs clarification??~~ ① availability of food on hand ✓ Thanks!

Pantry — The places where all the food is kept. Includes the larder, refrigerator, and freezer.

Food on Hand — The foods in the pantry.

Available Items — Those desired items that are and are not on the grocer's shelves.

Shopping Route — A map of the best way to get to all the desired stores.

Shopping List — A written list of what to buy in each shop.

Bought Items — Food in shopping bags that have been paid for.

② ~~Shopping Schedule — Times when stores are open.~~
But I'll use this when I decompose box 2.

③ This will require some feedback to "plan"?
Hum.... Maybe so!!

④ Now this idea made me think: This is more like the amount of inventory we want to keep on hand. That IS something that should come from "plan".

NODE: FF/A2G1 | TITLE: Maintain Stock (Glossary) | NUMBER: DAMØØ9

Workshop

13

Revising the Top Diagram

Goal Resolve all interface issues on the A0 diagram.

Activities

1. Consistently copy onto the A0 diagram all the issues raised during the review of first-level decompositions. For example, if box 1 incorrectly constrained box 2, put red tick marks on the arrow going from box 1 to box 2.
2. Evaluate the story now told by the marked-up A0 diagram. See if any other issues come to mind, and document them accordingly.
3. Resolve all the interface issues, one at a time. Go slowly. Identify and investigate several alternative resolutions, and backtrack your problem-solving steps when an alternative does not fit with the current set of resolutions.

Notes

1. Consistently copying all the issues will help you clearly identify the scope of the problem with the A0 diagram.
2. Talking through the story, especially out loud, will make the issues apparent and may give you clues on how best to resolve them.

Sample

1. One interface issue was raised in the last workshop. Information about the availability of food is not an ***external factor*** but comes from ***maintain stock.*** Several changes are required to resolve this issue: (a) the arrows ***actual availability*** and ***normal availability*** get crossed out, (b) ***plan*** generates ***weekly meal plan,*** (c) ***menu*** must constrain ***prepare,*** and (d) ***maintain stock*** must constrain ***plan*** with ***food on hand.***
2. Notice that the approval level for the A0 diagram has been increased from "working" to "draft." This often happens when a large number of interface issues have been resolved, because a diagram then becomes less likely to change.

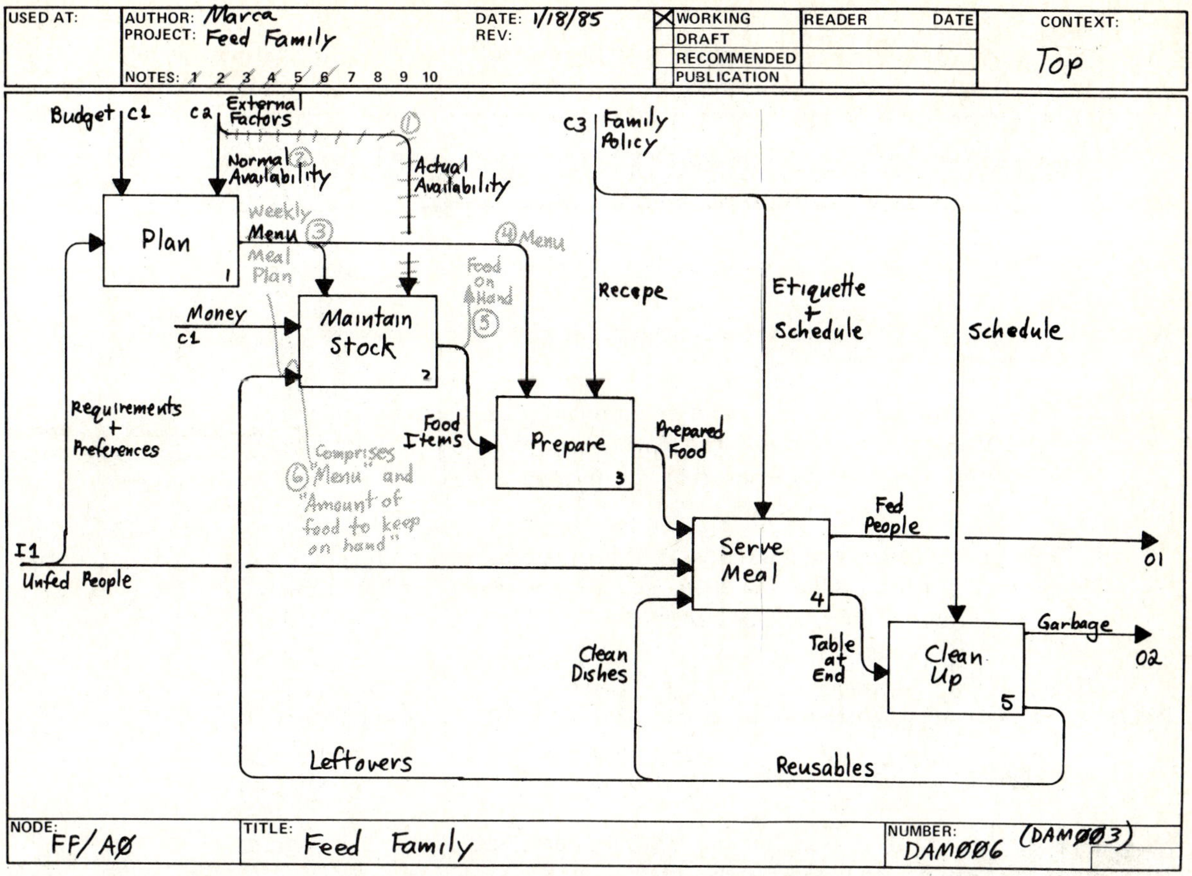
USED AT:
AUTHOR: Marca
PROJECT: Feed Family
DATE: 1/18/85
REV:
NOTES: 1 2 3 4 5 6 7 8 9 10
WORKING
DRAFT
RECOMMENDED
PUBLICATION
READER
DATE
CONTEXT:
Top
Budget
C1
C2
External Factors
Normal Availability
Actual Availability
C3
Family Policy
Plan
1
Weekly Menu Meal Plan
Menu
Food on Hand
Maintain Stock
2
Money
C1
Recepe
Etiquette + Schedule
Schedule
Requirements + Preferences
Food Items
Prepare
3
Prepared Food
Comprises "Menu" and "Amount of food to keep on hand"
I1
Unfed People
Serve Meal
4
Fed People
O1
Clean Dishes
Table at End
Clean Up
5
Garbage
O2
Leftovers
Reusables
NODE:
FF/AØ
TITLE:
Feed Family
NUMBER:
DAMØØ6
(DAMØØ3)

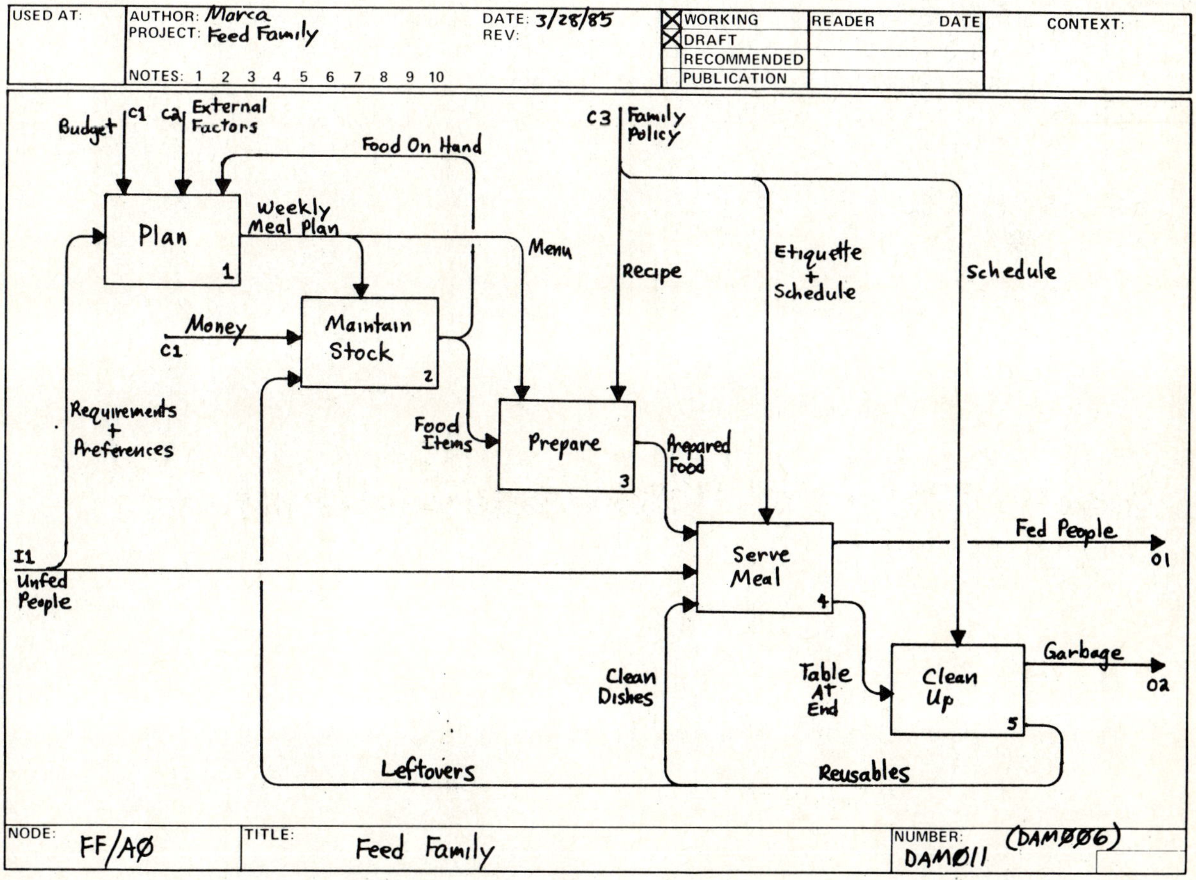
USED AT:
AUTHOR: Marca
PROJECT: Feed Family
DATE: 3/28/85
REV:
NOTES: 1 2 3 4 5 6 7 8 9 10
WORKING
DRAFT
RECOMMENDED
PUBLICATION
READER
DATE
CONTEXT:
Budget
C1
C2
External Factors
Food On Hand
C3
Family Policy
Plan
1
Weekly Meal Plan
Menu
Recipe
Etiquette + Schedule
Schedule
Maintain Stock
2
Money
C1
Requirements + Preferences
Food Items
Prepare
3
Prepared Food
I1
Unfed People
Serve Meal
4
Fed People
O1
Clean Dishes
Table At End
Clean Up
5
Garbage
O2
Leftovers
Reusables
NODE:
FF/AØ
TITLE:
Feed Family
NUMBER:
DAMØ11
(DAMØØ6)

Workshop

14

Revising First-Level Decompositions

Goal Revise and redraw both the first-level decomposition and its Glossary.

Activities

1. Keep the newly revised A0 diagram nearby while you work.
2. Entirely redraw your first-level decomposition. Walk through the new story in your mind as you draw, seeing if it continues to make sense. Redefine Glossary terms as you go. Refer to the A0 diagram often to verify that the details on your Ax diagram coincide with the parent box.
3. Tie all external arrows of your Ax diagram to the A0 diagram using ICOM codes. While doing this, match the labels of the external arrows on your diagram with the arrow labels touching the context box on the A0 diagram.

Notes Redrawing will help you simplify the diagram. When redrawing, try to be as accurate as you were in previous diagrams, but use fewer words and simpler graphics.

Sample Notice how the message of and Glossary definitions for the A2 diagram have become stronger after the interface issues on the A0 diagram were resolved. This illustrates how simplifying a diagram sharpens its details and why good analysts invest their time resolving high-level interfaces before investing much effort in lower-level decompositions.

USED AT:	AUTHOR: Marca PROJECT: Feed Family NOTES: 1 2 3 4 5 6 7 8 9 10	DATE: 3/28/85 REV:	☒ WORKING ☐ DRAFT ☐ RECOMMENDED ☐ PUBLICATION	READER DATE	CONTEXT:

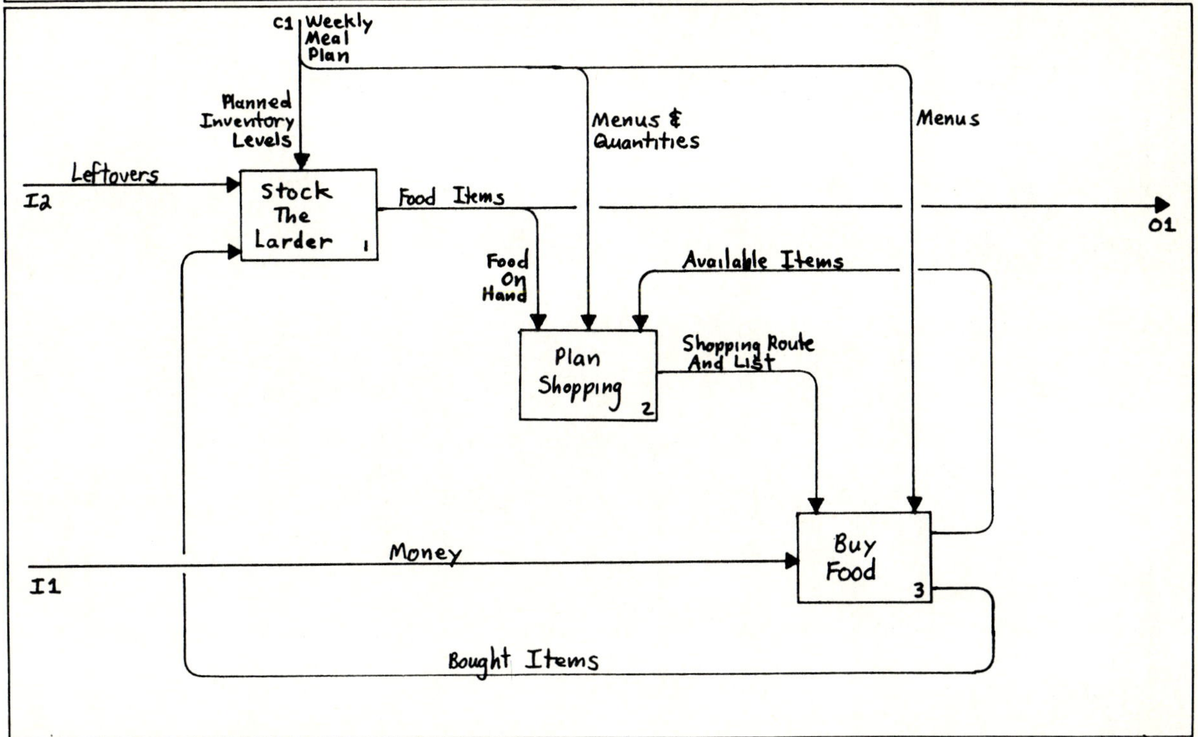

NODE: FF/A2	TITLE: Maintain Stock	NUMBER: DAMØ12 (DAMØØ8)

USED AT:	AUTHOR: Marca PROJECT: Feed Family NOTES: 1 2 3 4 5 6 7 8 9 10	DATE: 3/28/85 REV:	☒ WORKING ☐ DRAFT ☐ RECOMMENDED ☐ PUBLICATION	READER DATE	CONTEXT: DAMØ12

Larder – All the places where food is kept (e.g. pantry, refrigerator, freezer).

Food Items – Contains food on hand, because this is known when the larder is stocked.

Food on Hand – What's currently in the larder.

Available Items – Those desired items that are and are not on the grocer's shelves.

Shopping Route – A written timetable and map describing the best way to get to desired shops and when those shops are open.

Shopping List – A written list of what and how much to buy in each shop.

Bought Items – Food in shopping bags that have been paid for.

Menus – A detailed description of each planned meal.

NODE: FF/A2G1	TITLE: Maintain Stock (Glossary)	NUMBER: DAMØ13 (DAMØØ9)

Chapter

25

Developing Second-Level Decompositions

Recall that during the decomposition of first-level diagrams, you may begin to uncover details about the system under study. This chapter comprises three workshops, designed to guide you through the steps involved in developing second-level decompositions. By the end of these workshops, you will have made a diagram that details one box on the first-level decomposition. This diagram should contain sufficient detail to meet the purpose of the model. That is, the diagram should answer some of the model's question set. In Workshop 15, the diagram is developed. Workshop 16 guides you through critique and revision of the diagram. Workshop 17 walks you through the steps of packaging your work into a kit for review by others.

Go through all three workshops without a break. This will give you an accurate feel for the amount of work required to detail a box. (After you finish these three workshops, reflect on the amount of effort it took to start the model versus the amount of effort it took to make a first-level decomposition versus the amount of effort it took to make a second-level decomposition.) Allocate about ½ hour for each workshop, but don't be concerned if you take more time. Finish all three workshops, even if you have difficulty keeping to the description given in the first-level decomposition.

Workshop

15

Individual Diagramming

Goal Decompose a single box on the Ax diagram.

Activities

1. Pick a box on the Ax diagram. This becomes the complete context for this workshop. Do not stray outside this boundary.
2. Mentally review the steps of drawing a diagram: list data, list activities, cluster activities into from three to six boxes, draw boxes from most to least dominant, draw external arrows, draw primary constraint arrows, and so on.
3. Read the Ax diagram, and focus on how your context box fits with its other boxes. Use the input, output, and control arrows of your context box to start the data list.
4. Now follow the steps of developing a diagram. Put the Data List and the Activity List on the left side of the form, and draw the diagram in the remaining area.
5. As you draw, make notes to yourself and define the terminology used. After you finish, go back and check all ICOM's. Make sure you did not forget to use any data defined by the arrows touching the boundary box.

Notes

1. Do not stray outside the bounded context defined by the box you picked on the Ax diagram. You may find it hard to make your diagram because second-level decompositions usually uncover a lot of detail, and this detail often points out difficulties with the parent diagram. Record these difficulties with appropriate notes on the Ax diagram.
2. Do not worry about the correctness of this diagram. Second-level decompositions are rarely drawn correctly the first time. Strive now to uncover details. Accuracy will come when the diagram is reviewed.

Sample

1. Notice that the external input, control, and output arrow names were listed first, and subsequent data names detail their composition. This is a reliable way to help remember the context in which you are working.
2. There are many arrows with multiple branches in this diagram. This often happens when a lot of detail is exposed.
3. This decomposition was based on the attributes of food: bought, packed, stored, withdrawn, good, spoiled, etc. Decomposing based on the properties of important data is an acceptable strategy.

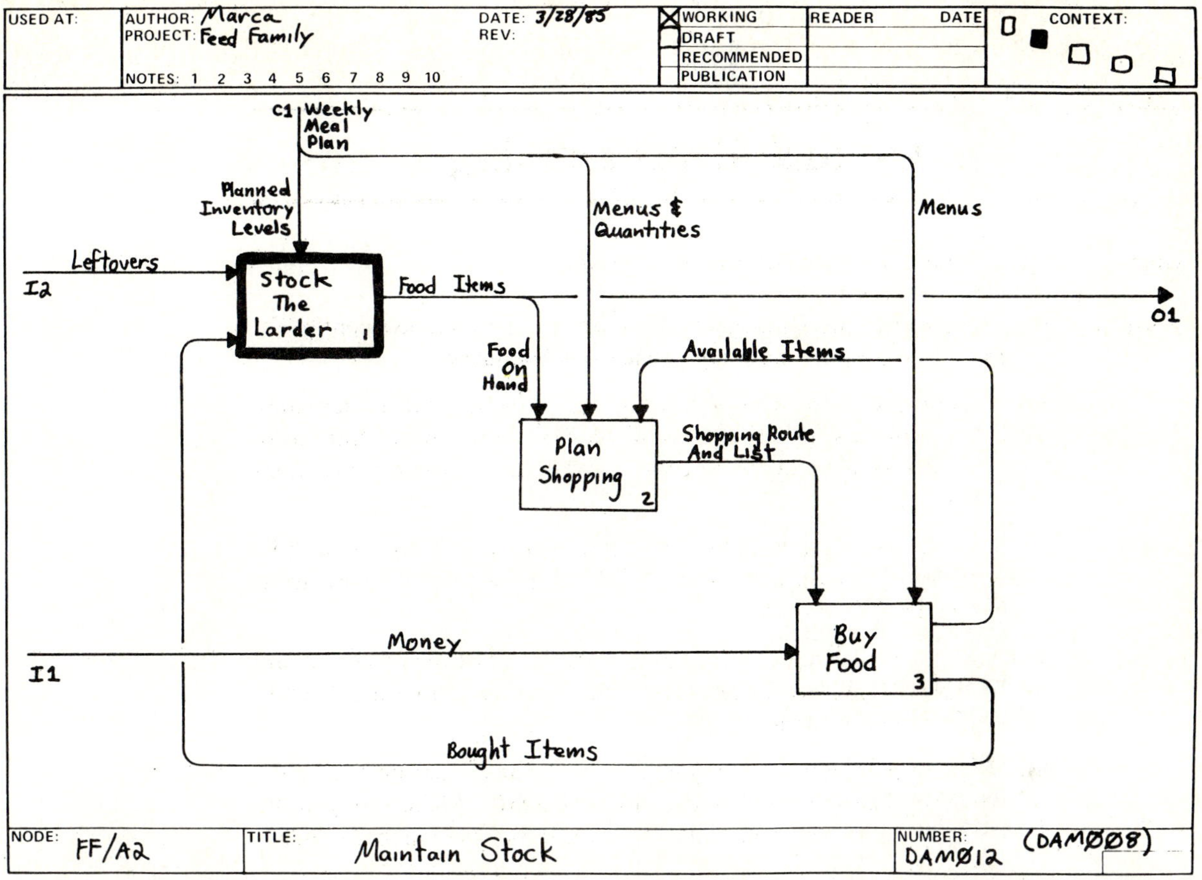
USED AT:
AUTHOR: Marca
PROJECT: Feed Family
DATE: 3/28/85
REV:
NOTES: 1 2 3 4 5 6 7 8 9 10
WORKING
DRAFT
RECOMMENDED
PUBLICATION
READER
DATE
CONTEXT:
C1 Weekly Meal Plan
Planned Inventory Levels
Menus & Quantities
Menus
Leftovers
I2
Stock The Larder 1
Food Items
O1
Food On Hand
Available Items
Plan Shopping 2
Shopping Route And List
Money
I1
Buy Food 3
Bought Items
NODE: FF/A2
TITLE: Maintain Stock
NUMBER: DAMØ12 (DAMØØ8)

ST 252

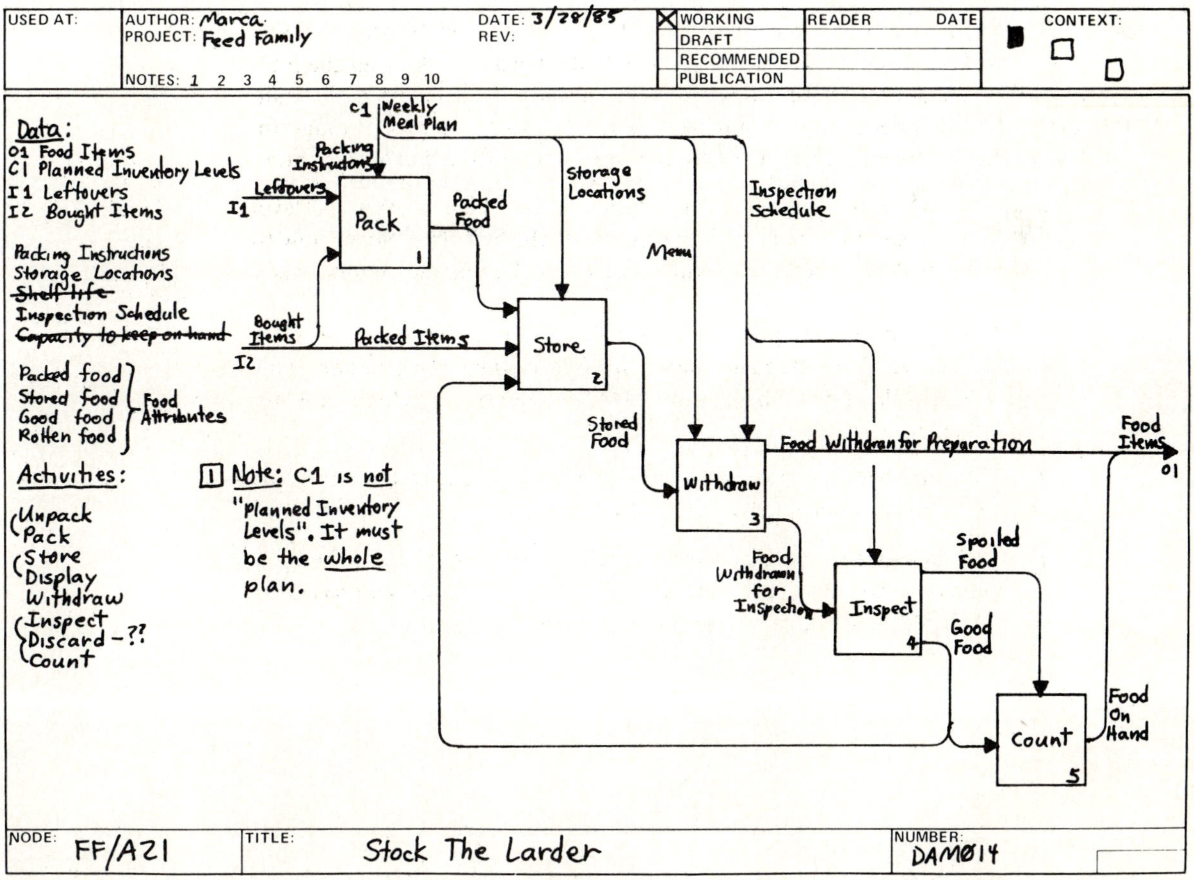
USED AT:
AUTHOR: Marca
PROJECT: Feed Family
DATE: 3/28/85
REV:
NOTES: 1 2 3 4 5 6 7 8 9 10
WORKING
DRAFT
RECOMMENDED
PUBLICATION
READER
DATE
CONTEXT:
Data:
O1 Food Items
C1 Planned Inventory Levels
I1 Leftovers
I2 Bought Items
Packing Instructions
Storage Locations
Shelf life
Inspection Schedule
Capacity to keep on hand
Packed food
Stored food
Good food
Rotten food
Food Attributes
Activities:
Unpack
Pack
Store
Display
Withdraw
Inspect
Discard -??
Count
1 Note: C1 is not "Planned Inventory Levels". It must be the whole plan.
C1 Weekly Meal Plan
Packing Instructions
Leftovers
I1
Pack 1
Packed Food
Storage Locations
Inspection Schedule
Menu
Bought Items
I2
Packed Items
Store 2
Stored Food
Withdraw 3
Food Withdrawn for Preparation
Food Items
O1
Food Withdrawn for Inspection
Inspect 4
Spoiled Food
Good Food
Count 5
Food On Hand
NODE: FF/A21
TITLE: Stock The Larder
NUMBER: DAMØ14

Workshop

16

Critiquing Second-Level Decompositions

Goal Criticize Axx diagrams for the way they detail boxes on the Ax diagram.

Activities

1. Walk through the diagram you just drew and try to tell the story of how your diagram accomplishes part of feeding the family. Start the story at a logical beginning with the arrival of an important thing (e.g., ***money***) or piece of information (e.g., ***menu***) coming from a box on the Ax diagram. Refer to the Ax diagram continuously to see if the story fits into the context. Note where you find the story incorrect or weak.
2. Evaluate the way you decomposed external arrows and evaluate how activities were clustered into boxes. See if the arrows can be decomposed in a different way, or if activities can be clustered into a better alternate set of boxes.
3. Remember to note changes on both your diagram and the Ax diagram, and to change or refine your terminology as you criticize.
4. Draw a brand-new diagram from the notes you just took, and redraw the Ax diagram if necessary. Remember to check ICOM connections with the Ax diagram.

Notes Try alternate decompositions or try changing box and arrow names to sharpen the story. Sometimes, different decompositions using synonyms can lead to useful insights.

Sample

1. Notice that notes 2 to 6 indicate places where the author split data needlessly. Too much splitting leads to unnecessary complexity, which causes misunderstandings and complicates future decompositions.
2. Note 7 reminds the author to include discarding spoiled food when ***count*** is decomposed.

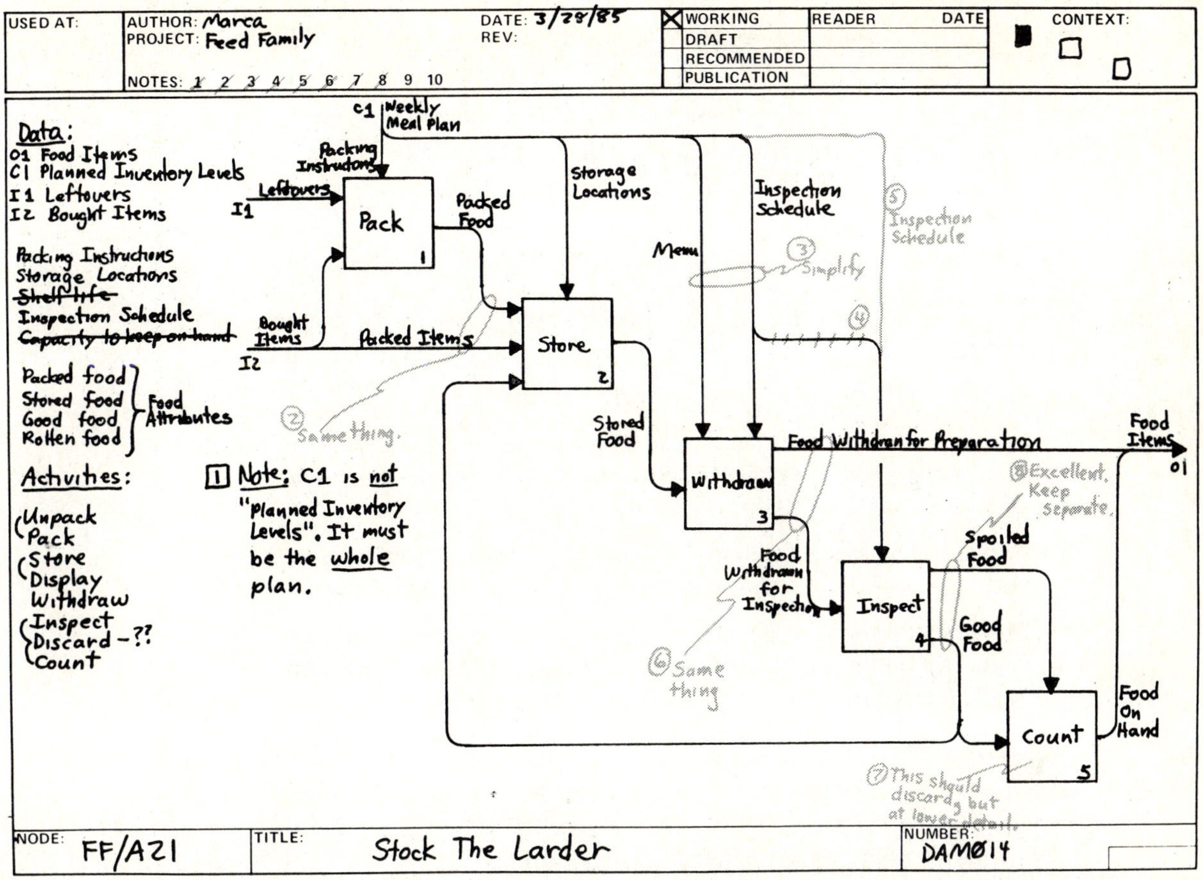
USED AT:
AUTHOR: Marca
PROJECT: Feed Family
DATE: 3/28/85
REV:
NOTES: 1 2 3 4 5 6 7 8 9 10
WORKING
DRAFT
RECOMMENDED
PUBLICATION
READER
DATE
CONTEXT:
Data:
O1 Food Items
C1 Planned Inventory Levels
I1 Leftovers
I2 Bought Items
Packing Instructions
Storage Locations
Shelf life
Inspection Schedule
Capacity to keep on hand
Packed food
Stored food
Good food
Rotten food
Food Attributes
Activities:
Unpack
Pack
Store
Display
Withdraw
Inspect
Discard -??
Count
1 Note: C1 is not "Planned Inventory Levels". It must be the whole plan.
C1 Weekly Meal Plan
Packing Instructions
Leftovers
I1
Pack
Packed Food
Storage Locations
Inspection Schedule
Menu
Bought Items
I2
Packed Items
Store
Stored Food
Withdraw
Food Withdrawn for Preparation
Food Items
O1
Food Withdrawn for Inspection
Inspect
Spoiled Food
Good Food
Count
Food On Hand
5 Inspection Schedule
3 Simplify
4
2 Same thing.
8 Excellent. Keep separate.
6 Same thing
7 This should discard, but at lower detail.
NODE: FF/A21
TITLE: Stock The Larder
NUMBER: DAMØ14

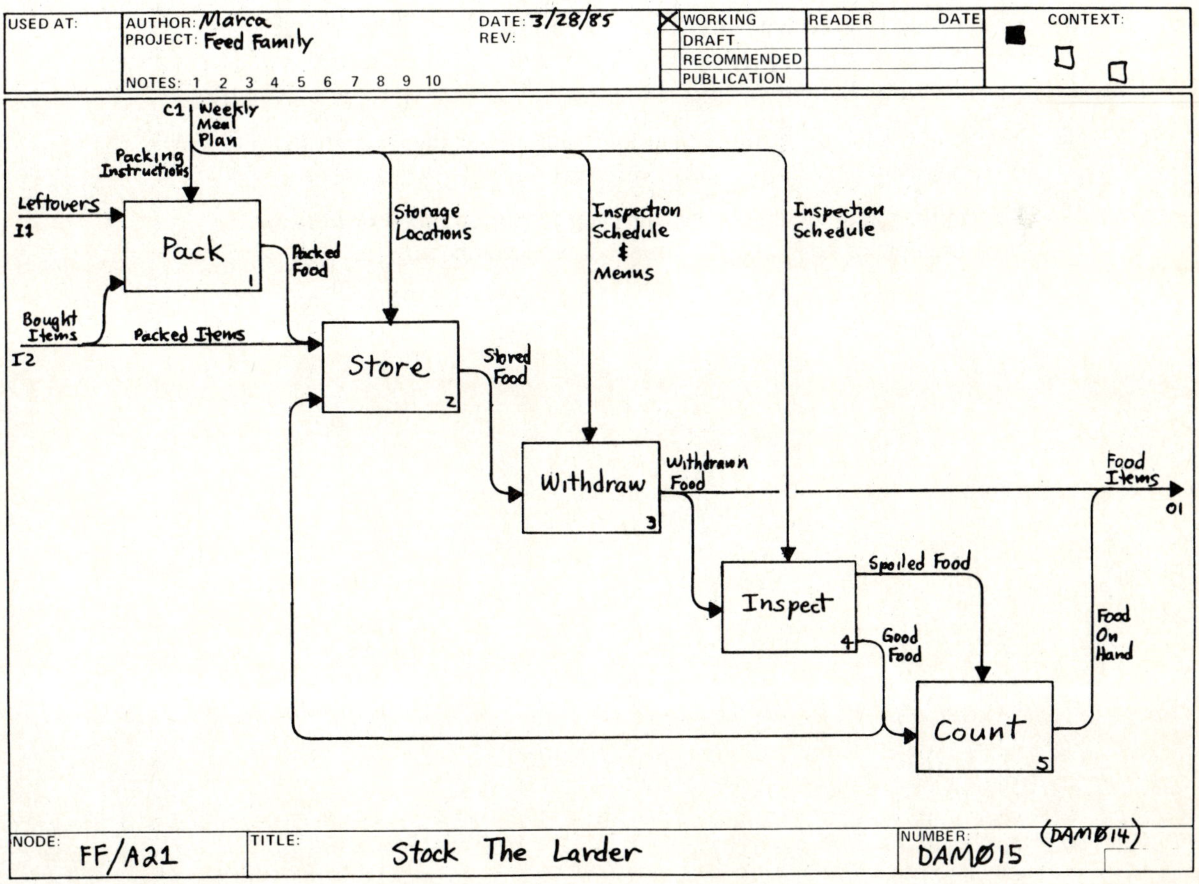
USED AT:
AUTHOR: Marca
PROJECT: Feed Family
DATE: 3/28/85
REV:
NOTES: 1 2 3 4 5 6 7 8 9 10
WORKING
DRAFT
RECOMMENDED
PUBLICATION
READER
DATE
CONTEXT:
C1 Weekly Meal Plan
Packing Instructions
Leftovers
I1
Pack
Packed Food
Bought Items
I2
Packed Items
Storage Locations
Store
Stored Food
Inspection Schedule & Menus
Inspection Schedule
Withdraw
Withdrawn Food
Food Items
O1
Inspect
Spoiled Food
Good Food
Count
Food On Hand
NODE: FF/A21
TITLE: Stock The Larder
NUMBER: DAMØ15
(DAMØ14)

Workshop

17

Individual Kit Preparation

Goal Package your revised second-level decomposition diagram and its Glossary into an SADT kit.

Activities

1. Spread out the diagram and its Glossary, and check to see that all information is consistent between the two pages.
2. Complete the Kit Cover Page, filling in identification information (author, project, and date), the title of the kit, what is in the kit, who the kit should go to, and when the kit must be returned.
3. Add any other special comments or notes to your readers in the Comments Area of the Kit Cover Page.
4. Clip the pages together: Kit Cover Page first, Context Diagram second, your diagram third, and the Glossary Page last. Then send the kit to the project Librarian.

Notes Use the Comments Area to communicate special points about the kit to your readers. These notes can often help you get good review by telling the readers to watch out for particular areas of concern or points of interest. Such notes go a long way to obtaining useful reviews.

Sample

1. Read the note on the Kit Cover Page concerning an interface issue. It explains how detailing uncovered more connectivity among the major activities of feeding the family.
2. Notice how simplifying the graphics has sharpened the story of the A21 diagram.
3. The splitting and joining of data for ***bought items*** and ***food items*** is a good example of how SADT arrows can be decomposed.
4. Notice that the Glossary definitions are becoming detailed enough to write instructions to the teenagers of the family.

AUTHOR: Marca	DATE: 3/28/85	☒ WORKING	READER	DATE
PROJECT: Feed Family	REV:	☐ DRAFT		
		☐ RECOMMENDED		
NOTES: ~~1~~ 2 3 4 5 6 7 8 9 10		☐ PUBLICATION		

LOG
FILE
AUTHOR

DOCUMENT NUMBER: Feed - 2

Received ________
Completed ________

COPYING INSTRUCTIONS:
______copies of ______ pages = ______ total
☐ as soon as possible ☐ by ______

READERS:	New Kit To Reader	Due Back	Comments To Author	Due Back	Response To Reader
Clem	3/28/85	3/29/85			
Peter					

RESPONSE REQUIRED: ☒ Fast ☐ Normal ☐ Slow ☐ None

CONTENTS:

Pg.	Node	Title	C-Number	Status
A		COVER SHEET		
B	FF/A21	Stock The Larder	DAMØ15	
C	FF/A2161	Glossary	DAMØ16	
D				
E				
F				
G				
H				
I				
J				
K				
L				
M				
N				

COMMENTS:
☐ UPDATE
☐ REPLACE } Model File ________ with this kit.

[1] Parent diagram must pass in *whole* plan. There are more information requirements than we expected.

SPECIAL INSTRUCTIONS:
☐ no author copy
____extra author copies

NODE: | TITLE: Stocking the Larder Details | NUMBER: DAMØ17

USED AT:	AUTHOR: Marca	DATE: 3/28/85	☒ WORKING	READER	DATE	CONTEXT:
	PROJECT: Feed Family	REV:	☐ DRAFT			
			☐ RECOMMENDED			
	NOTES: 1 2 3 4 5 6 7 8 9 10		☐ PUBLICATION			

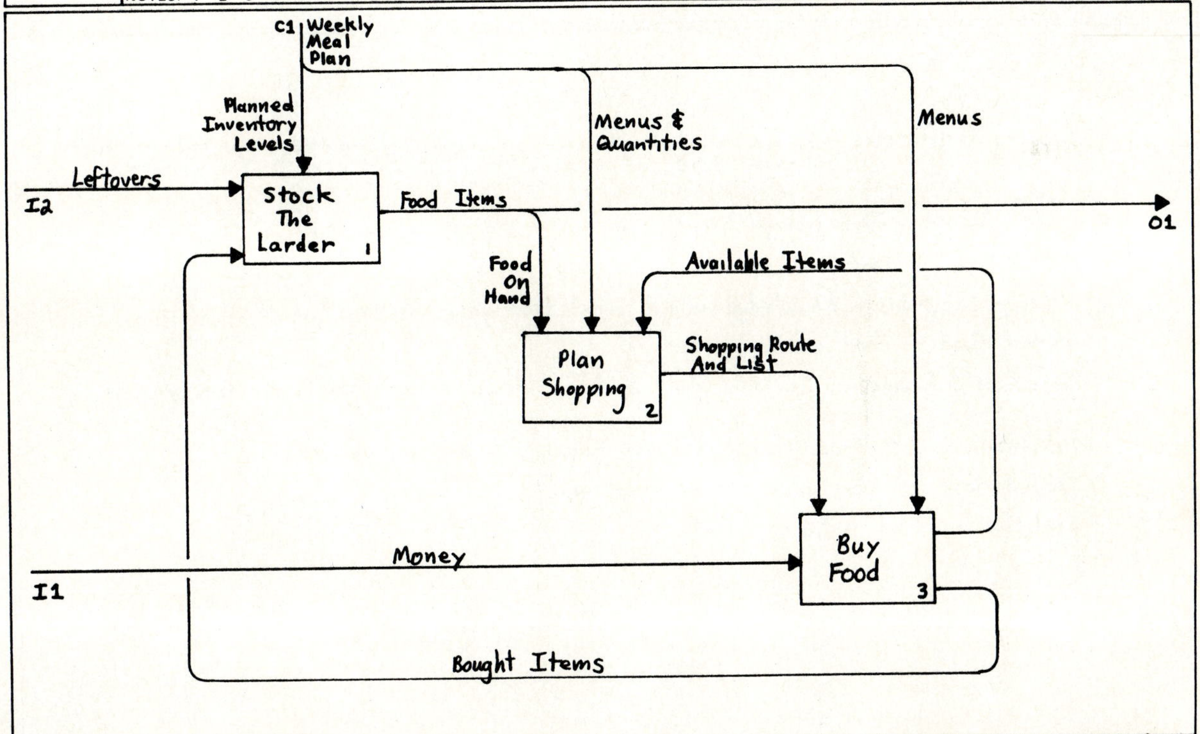

NODE: FF/A2 | TITLE: Maintain Stock | NUMBER: DAMØ12 (DAMØØ8)

USED AT:	AUTHOR: Marca PROJECT: Feed Family NOTES: 1 2 3 4 5 6 7 8 9 10	DATE: 3/28/85 REV:	☒ WORKING DRAFT RECOMMENDED PUBLICATION	READER	DATE	CONTEXT:

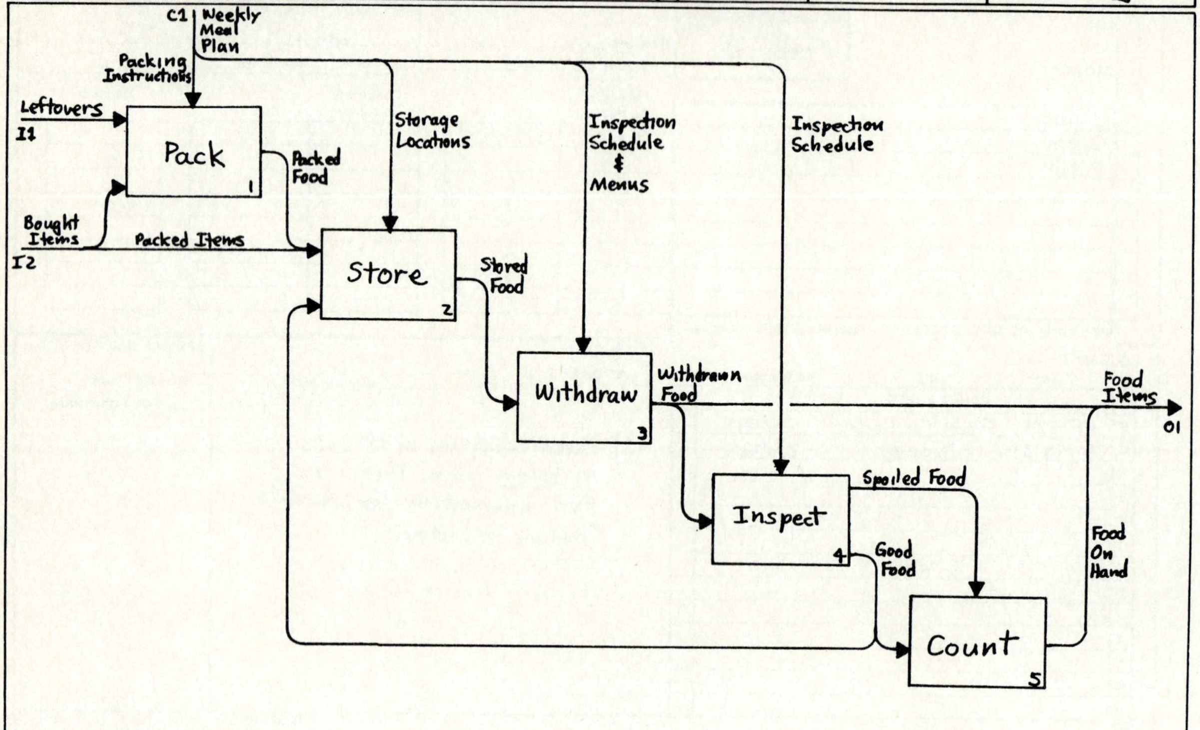

NODE: FF/A21	TITLE: Stock The Larder	NUMBER: DAMØ15 (DAMØ14)

USED AT:	AUTHOR: Marca PROJECT: Feed Family NOTES: 1 2 3 4 5 6 7 8 9 10	DATE: 3/28/85 REV:	☒ WORKING DRAFT RECOMMENDED PUBLICATION	READER	DATE	CONTEXT: DAMØ15

Packing Instructions – A written list of materials and procedures required to take bought food and package it so it will keep in the larder.

Packed Items – Those bought items suitable for storage immediately.

Packed Food – Food now suitable for storage.

Storage Locations – Knowledge about what food goes where.

Stored food – Food in the larder.

Inspection Schedule – Timetable that says how often to check the larder for spoilage.

Withdrawn food – Food that has been taken out of the larder.

Spoiled Food – Food gone bad in some way.

Good food – Food suitable to eat.

NODE: FF/A21G1	TITLE: Stock The Larder (Glossary)	NUMBER: DAMØ16

Chapter

26

Resolving First-Level Interfaces

Recall that an Author/Reader Cycle done on low-level diagrams of a model may require resolving interface issues on higher-level diagrams in the model. The four workshops of this chapter are designed to guide you through the steps involved in resolving the interfaces among second-level decompositions. In Workshop 18, first-level decompositions are reviewed. This is when interface issues get raised. Workshop 19 lets you react to the comments of others. Comments that are accepted become verified interface issues that require resolution. In Workshop 20, first-level decompositions are revised, thus resolving the interface issues. Workshop 21 then steps you through redrawing a second-level decomposition, showing how the new interfaces are reflected in the revised diagram.

Go through the first two workshops without a break. This will give you the opportunity to properly think about, raise, and verify interface issues. Then go through the last two workshops without a break, so you can properly resolve interface issues and reflect required changes in the second-level decompositions. Allocate about ½ hour for each workshop, but don't be concerned if you take more time.

Workshop

18

Individual Commenting

Goal Comment on the kits containing Axx diagrams.

Activities

1. Read the Kit Cover Page and any special messages the author may have written in the Comments Area.
2. Read the Axx diagram in the kit, comparing its message with that given by its Context Area on the Ax diagram. Do this also to ensure consistency between the Axx diagram and its bounded context. Document problems with reader notes in red.
3. Read the Axx diagram to see if there are problems with the details of its message or its terminology. (While a diagram may be consistent with its parent box, its story may be incorrect or may contain too much or inappropriate detail.) Document problems with reader notes in red.
4. If the Axx diagram you drew has a direct interface with this diagram, read it again. Study the details of your interface assumptions and compare them with what the author of this diagram assumed. Document problems with reader notes in red.
5. Send the commented kit back to the author.

Notes

1. Remember, comment critically but constructively.
2. Write comments on every page. (No comments means you either have not read the page or have nothing to say.) Make at least a check mark to indicate you have read the page and agree with the author.

Sample

1. On the Kit Cover Page, notice: (a) when the kit was received by the Librarian, (b) when the kit was read, and (c) when the reader expects the kit back. Also, notice the check mark after the author's note, indicating that the reader agrees with the issue raised.
2. Note 3 on the Axx diagram expands the interface issue by suggesting a missing component to the ***weekly meal plan.*** This term was also added to the Glossary by the reader.

AUTHOR: Marca | DATE: 3/28/85 | ☒ WORKING | READER | DATE
PROJECT: Feed Family | REV: | ☐ DRAFT
NOTES: ~~1~~ 2 3 4 5 6 7 8 9 10 | ☐ RECOMMENDED | ☐ PUBLICATION

LOG
FILE
AUTHOR

DOCUMENT NUMBER: Feed - 2

Received: 3/28/85
Completed:

COPYING INSTRUCTIONS: ____ copies of ____ pages = ____ total
☐ as soon as possible ☐ by ____

READERS:	New Kit To Reader	Due Back	Comments To Author	Due Back	Response To Reader
Clem	3/28/85	3/29/85	3/29/85	3/30/85	
Peter					

RESPONSE REQUIRED: ☒ Fast ☐ Normal ☐ Slow ☐ None

CONTENTS:

Pg.	Node	Title	C-Number	Status
A		COVER SHEET		
B	FF/A21	Stock The Larder	DAMØ15	
C	FF/A2161	Glossary	DAMØ16	
D				
E				
F				
G				
H				
I				
J				
K				
L				
M				
N				

COMMENTS:
☐ UPDATE / ☐ REPLACE } Model File ____ with this kit.

[1] Parent diagram must pass in whole plan. There are more information requirements than we expected. ✓

SPECIAL INSTRUCTIONS:
☐ no author copy
____ extra author copies

NODE: | TITLE: Stocking the Larder Details | NUMBER: DAMØ17

USED AT: | AUTHOR: Marca | DATE: 3/28/85 | ☒ WORKING | READER | DATE | CONTEXT:
PROJECT: Feed Family | REV: | ☐ DRAFT
NOTES: 1 2 3 4 5 6 7 8 9 10 | ☐ RECOMMENDED | ☐ PUBLICATION

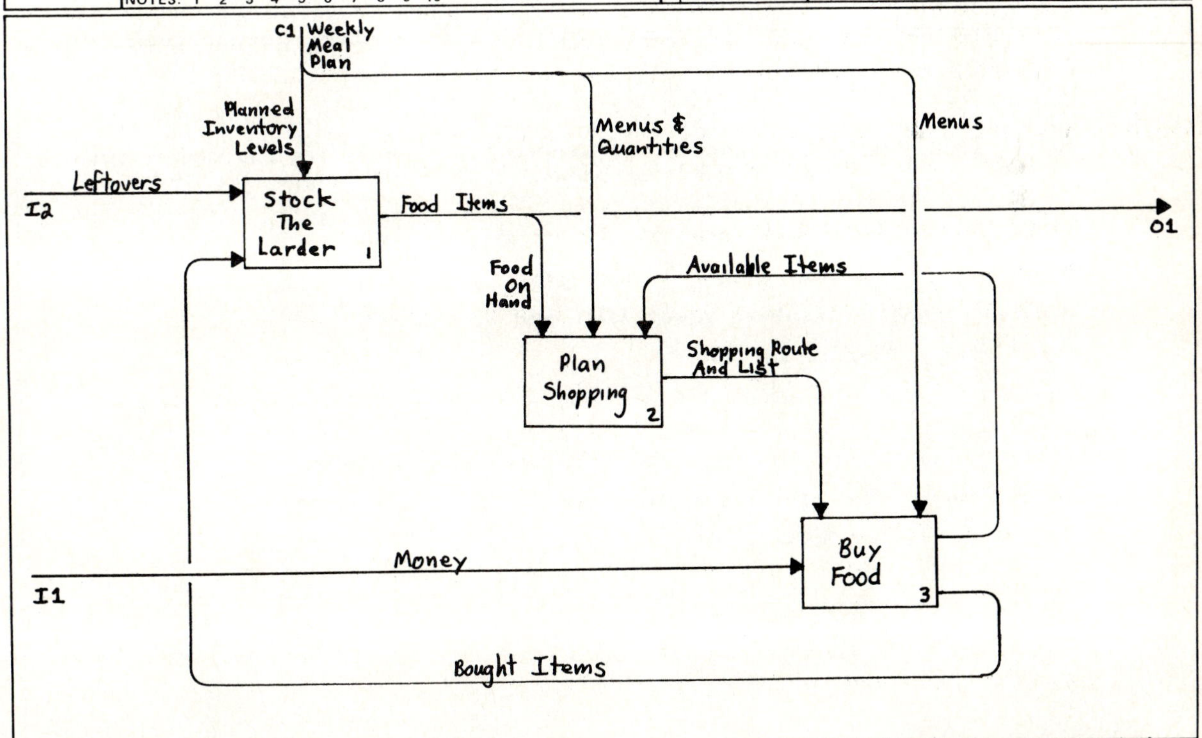

NODE: FF/A2 | TITLE: Maintain Stock | NUMBER: DAMØ12 (DAMØØ8)

USED AT: | AUTHOR: Marca | PROJECT: Feed Family | DATE: 3/28/85 | REV: | NOTES: 1 2 3 4 5 6 7 8 9 10

☒ WORKING | DRAFT | RECOMMENDED | PUBLICATION | READER | DATE | CONTEXT:

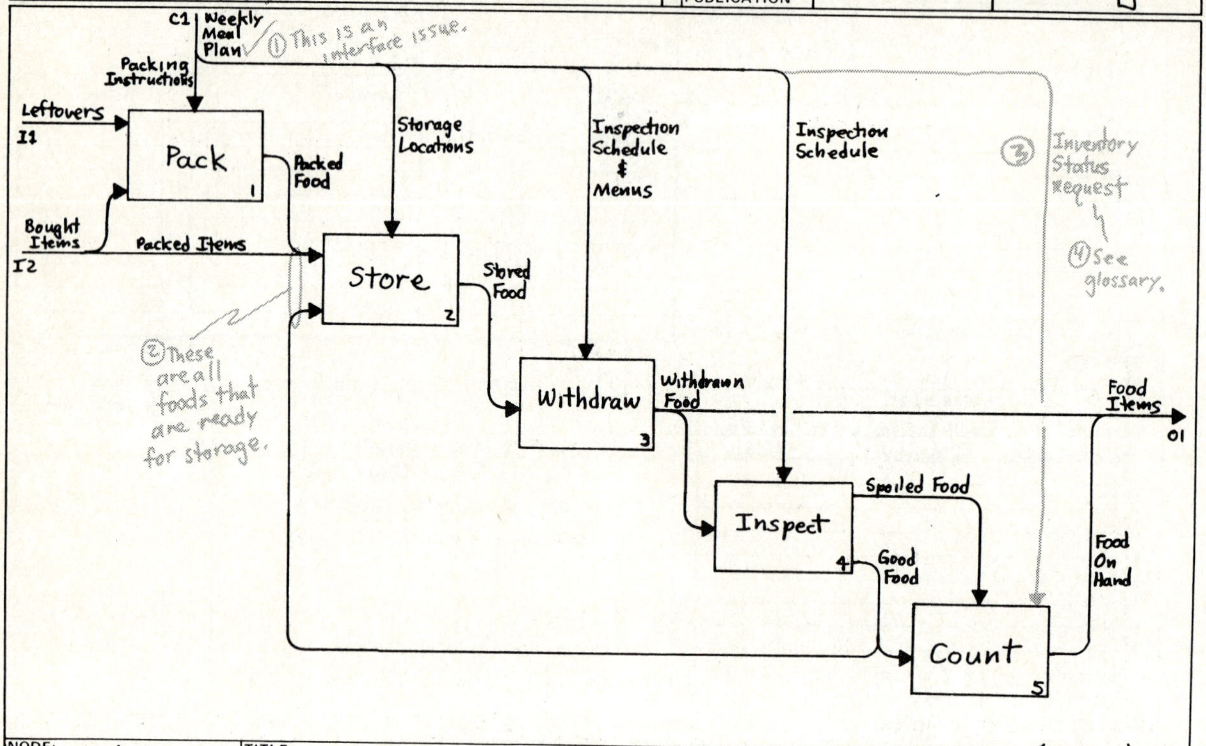

NODE: FF/A21 | TITLE: Stock The Larder | NUMBER: DAMØ15 (DAMØ14)

USED AT: | AUTHOR: Marca | PROJECT: Feed Family | DATE: 3/28/85 | REV: | NOTES: 1 2 3 4 5 6 7 8 9 10

☒ WORKING | DRAFT | RECOMMENDED | PUBLICATION | READER | DATE | CONTEXT: DAMØ15

Packing Instructions – A written list of materials and procedures required to take bought food and package it so it will keep in the larder.

Packed Items – Those bought items suitable for storage immediately.

Packed Food – Food now suitable for storage.

Storage Locations – Knowledge about what food goes where.

Stored food – Food in the larder.

Inspection Schedule – Timetable that says how often to check the larder for spoilage.

Withdrawn food – Food that has been taken out of the larder.

Spoiled Food – Food gone bad in some way.

Good food – Food suitable to eat.

① Inventory Status Request – A request to know the current food on hand.

NODE: FF/A21G1 | TITLE: Stock The Larder (Glossary) | NUMBER: DAMØ16

Workshop

19

Individual Reacting

Goal React to the commented kits containing Axx diagrams.

Activities

1. For each kit, read the Kit Cover Page, and notice any special comments about your notes in the Comments Area.
2. Read the comments on Axx diagram in the order they were written, and respond to each as you read. Determine if the comment is valid and why. If it is valid, check it. Otherwise, cross it out and give reasons for your disagreement. Write down all these responses in blue.
3. Read and react to the comments on the Glossary Page. Do not forget to do this, because (a) definitions not in agreement can often lead to misunderstandings, and (b) readers may redefine terminology, which can lead to useful insights.
4. Read again all comments that pertain to interface issues. Study and critique your interface assumptions in the process. Document all problems with reader notes in blue on your own copy of the kit.
5. Send the commented kit back to the reader.
6. Consolidate the comments of all readers of this kit onto your copy. Make extra notes to yourself to explain why you resolved conflicting reader comments a particular way. Make all notes in blue.

Notes

1. Remember, react critically but constructively.
2. Explain every reaction to all comments. (No reaction means you either have not read the comment or have nothing to say.) Make at least a check or a cross mark to indicate agreement or disagreement.

Sample Notice how the author agreed to and further refined the interface issue raised by the reader. The reacted Glossary definition for ***inventory status request*** says more clearly what is needed by ***plan***.

AUTHOR: Marca
PROJECT: Feed Family
NOTES: ~~1~~ 2 3 4 5 6 7 8 9 10

DATE: 3/28/85
REV:

☒ WORKING
☐ DRAFT
☐ RECOMMENDED
☐ PUBLICATION

READER | DATE

☐ LOG
☐ FILE
☐ AUTHOR

DOCUMENT NUMBER: Feed - 2

Received 3/28/85
Completed 3/30/85

COPYING INSTRUCTIONS:
_____ copies of _____ pages = _____ total
☐ as soon as possible ☐ by _____

READERS:	New Kit To Reader	Due Back	Comments To Author	Due Back	Response To Reader
Clem	3/28/85	3/29/85	3/29/85	3/30/85	3/30/85
Peter					

RESPONSE REQUIRED: ☒ Fast ☐ Normal ☐ Slow ☐ None

CONTENTS:

Pg.	Node	Title	C-Number	Status
A		COVER SHEET		
B	FF/A21	Stock The Larder	DAMØ15	
C	FF/A2161	Glossary	DAMØ16	
D				
E				
F				
G				
H				
I				
J				
K				
L				
M				
N				

COMMENTS:
☐ UPDATE
☐ REPLACE } Model File _____ with this kit.

[1] Parent diagram must pass in whole plan. There are more information requirements than we expected.

SPECIAL INSTRUCTIONS:
☐ no author copy
_____ extra author copies

NODE: | TITLE: Stocking the Larder Details | NUMBER: DAMØ17

USED AT:
AUTHOR: Marca
PROJECT: Feed Family
NOTES: 1 2 3 4 5 6 7 8 9 10

DATE: 3/28/85
REV:

☒ WORKING
☐ DRAFT
☐ RECOMMENDED
☐ PUBLICATION

READER | DATE

CONTEXT:

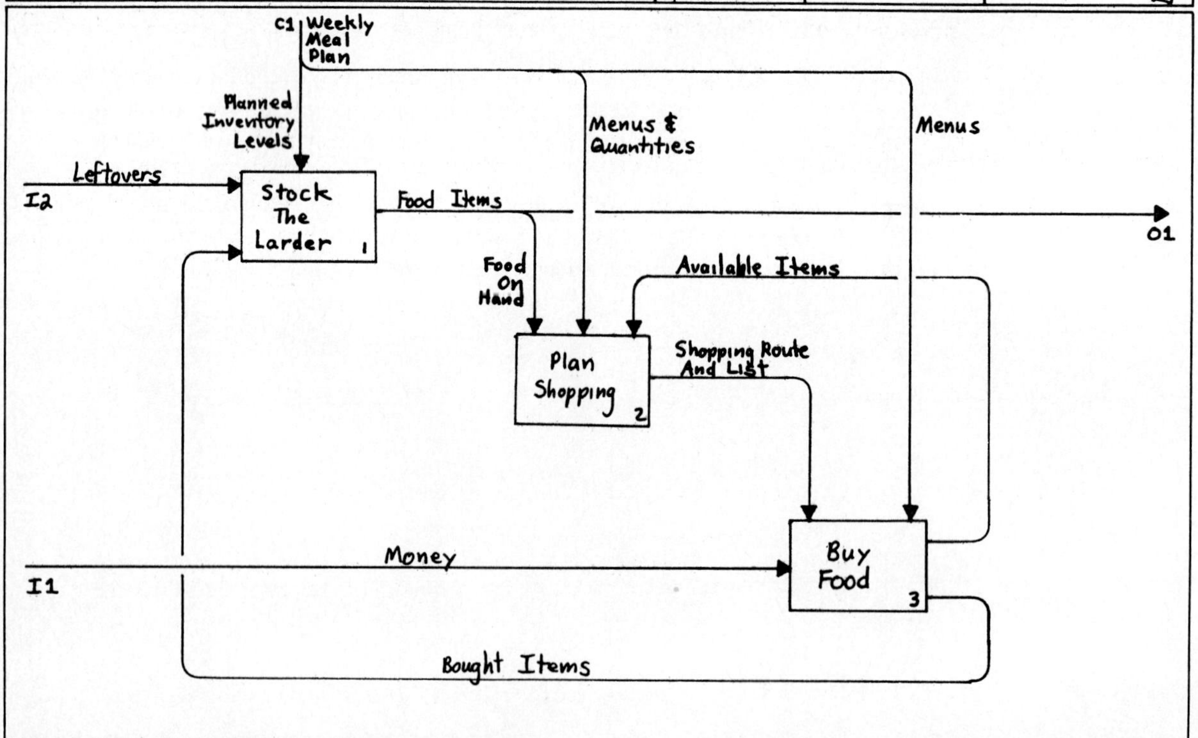

NODE: FF/A2 | TITLE: Maintain Stock | NUMBER: DAMØ12 (DAMØØ8)

USED AT:	AUTHOR: Marca PROJECT: Feed Family NOTES: ~~1~~ ~~2~~ ~~3~~ ~~4~~ 5 6 7 8 9 10	DATE: 3/28/85 REV:	☒ WORKING DRAFT RECOMMENDED PUBLICATION	READER	DATE	CONTEXT:

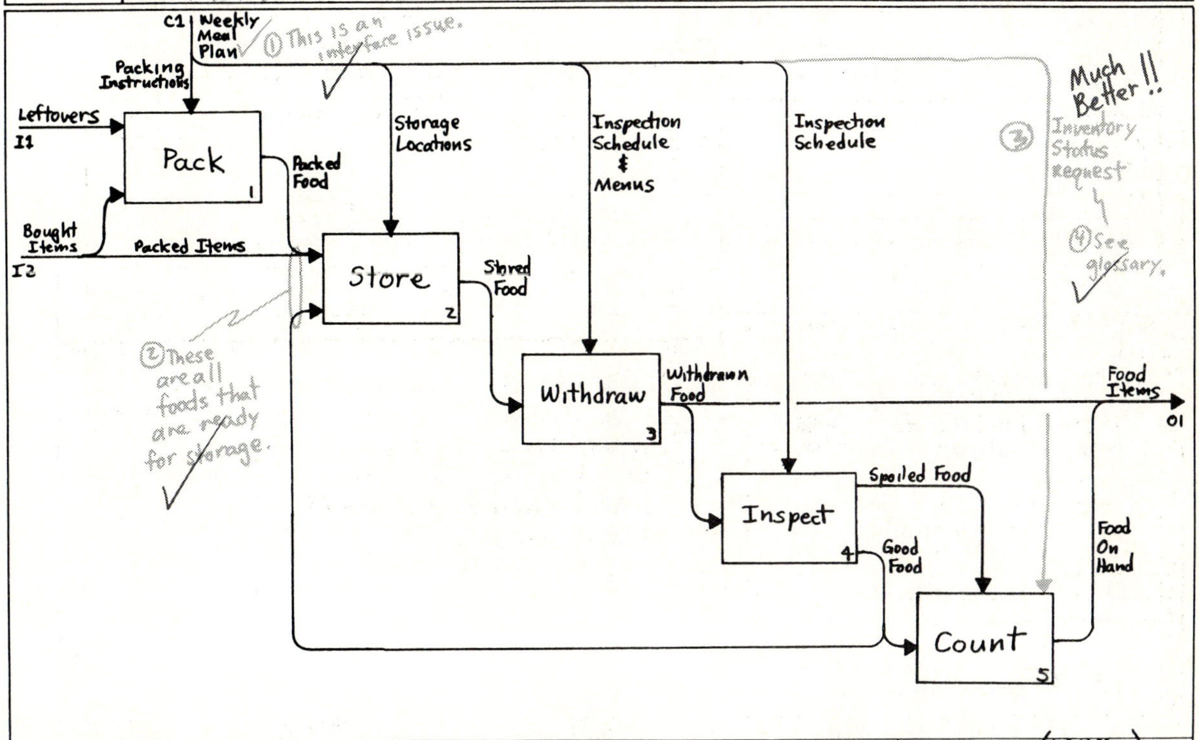

NODE: FF/A21	TITLE: Stock The Larder	NUMBER: DAMØ15 (DAMØ14)

USED AT:	AUTHOR: Marca PROJECT: Feed Family NOTES: ~~1~~ 2 3 4 5 6 7 8 9 10	DATE: 3/28/85 REV:	☒ WORKING DRAFT RECOMMENDED PUBLICATION	READER	DATE	CONTEXT: DAMØ15

Packing Instructions – A written list of materials and procedures required to take bought food and package it so it will keep in the larder.

Packed Items – Those bought items suitable for storage immediately.

Packed Food – Food now suitable for storage.

Storage Locations – Knowledge about what food goes where.

Stored food – Food in the larder.

Inspection Schedule – Timetable that says how often to check the larder for spoilage.

Withdrawn food – Food that has been taken out of the larder.

Spoiled Food – Food gone bad in some way.

Good food – Food suitable to eat.

① Inventory Status Request – A request to know ~~the current food on hand.~~ what and how much food is on hand.

NODE: FF/A21G1	TITLE: Stock The Larder (Glossary)	NUMBER: DAMØ16

Workshop

20

Revising First-Level Decompositions

Goal Resolve all interface issues on the Ax diagram.

Activities

1. Consistently copy the issues raised while reviewing second-level decompositions onto the Ax diagram. For example, if box 1 should constrain box 2, draw and label such a constraint arrow on the Ax diagram.
2. Evaluate the story the Ax diagram tells after all interface issues have been copied. See if any other issues now come to mind, and document them accordingly.
3. Resolve all interface issues, one at a time. Go slowly, because this process involves identifying and trying out several alternate ideas, as well as backtracking problem-solving steps when an alternative does not fit with the present set of resolutions.

Notes

1. Consistently copying all the issues first will help you identify and clarify all the problems with the Ax diagram.
2. Talking through the story, especially out loud, will make issues more apparent and may give you clues to how best to resolve them.

Sample

1. An interface issue raised in Workshop 19, regarding information about the status of the larder, impacts the A2 diagram in two ways: (a) the external arrow ***planned inventory levels*** gets crossed out, and (b) ***plan shopping,*** being a planning function, must request the status of the larder.
2. Notice the change in approval level for the A2 diagram (i.e., from "working" to "draft"). Once interface issues have been resolved for a diagram, it becomes less likely to change substantially.

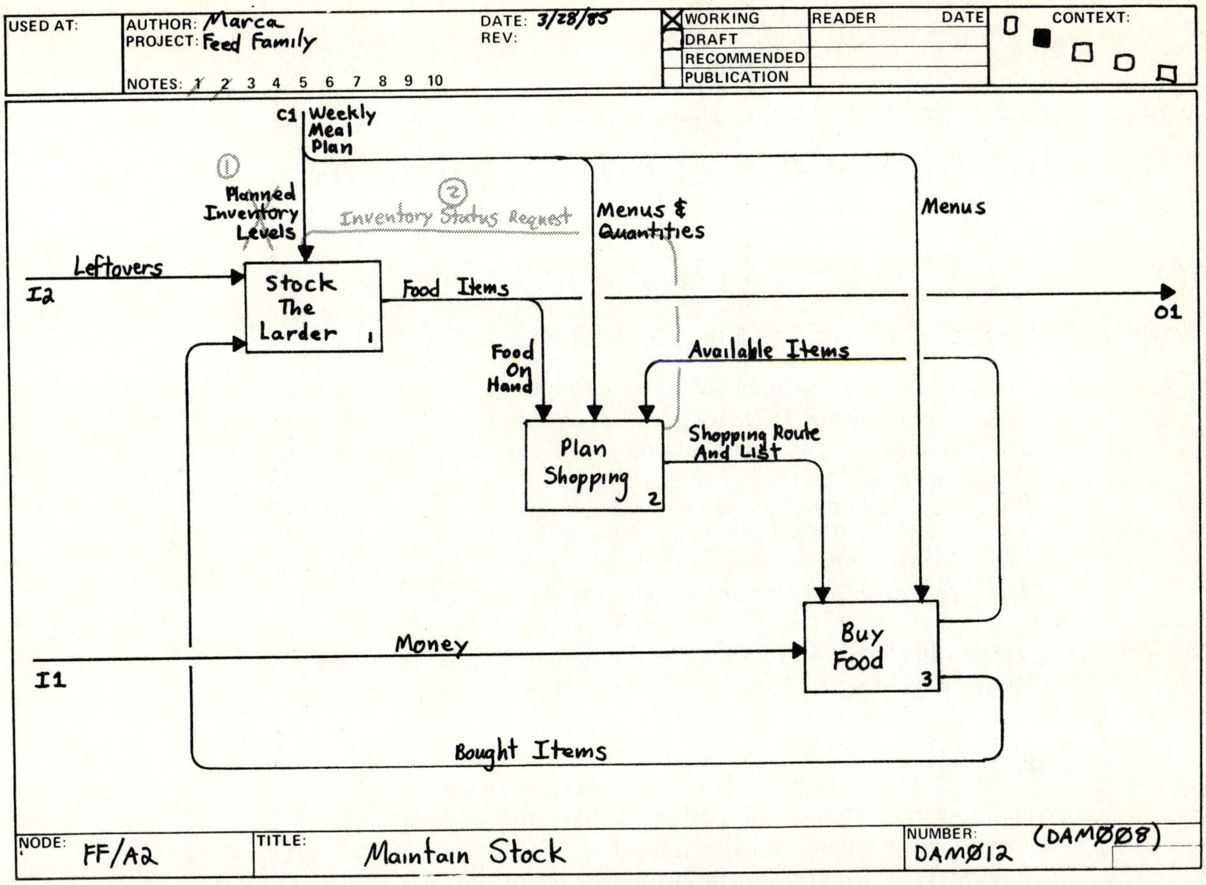
USED AT:
AUTHOR: Marca
PROJECT: Feed Family
DATE: 3/28/85
REV:
NOTES: 1 2 3 4 5 6 7 8 9 10
WORKING
DRAFT
RECOMMENDED
PUBLICATION
READER
DATE
CONTEXT:
C1 Weekly Meal Plan
① Planned Inventory Levels
② Inventory Status Request
Menus & Quantities
Menus
Leftovers
I2
Stock The Larder 1
Food Items
O1
Food On Hand
Available Items
Plan Shopping 2
Shopping Route And List
Money
I1
Buy Food 3
Bought Items
NODE: FF/A2
TITLE: Maintain Stock
NUMBER: DAMØ12
(DAMØØ8)

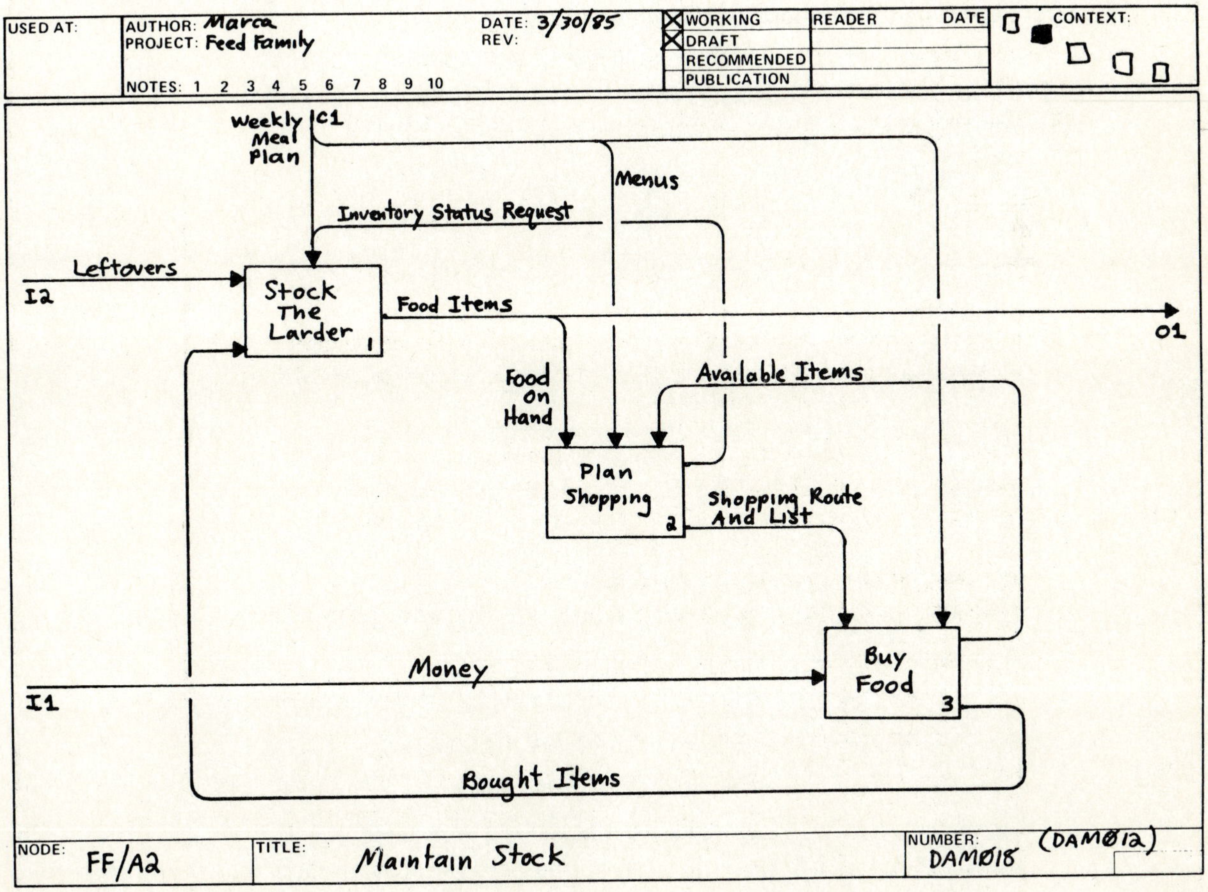
USED AT:
AUTHOR: Marca
PROJECT: Feed Family
DATE: 3/30/85
REV:
NOTES: 1 2 3 4 5 6 7 8 9 10
WORKING
DRAFT
RECOMMENDED
PUBLICATION
READER
DATE
CONTEXT:
Weekly Meal Plan
C1
Menus
Inventory Status Request
Leftovers
I2
Stock The Larder 1
Food Items
O1
Food On Hand
Available Items
Plan Shopping 2
Shopping Route And List
Money
I1
Buy Food 3
Bought Items
NODE: FF/A2
TITLE: Maintain Stock
NUMBER: DAMØ18
(DAMØ12)

Workshop

21

Revising Second-Level Decompositions

Goal Revise and redraw both the Axx diagram and its Glossary.

Activities

1. Keep the newly revised Ax diagram nearby while you work.
2. Redraw your Axx diagram entirely. Critique its story line as you redraw, and redefine Glossary terms accordingly. Refer to the Ax diagram often to verify that details on your diagram coincide with the context described by the parent box.
3. Tie all external arrows of your diagram to the Ax diagram using ICOM codes. While doing this, match the labels of the external arrows on your diagram with the arrow labels of the Ax diagram.

Notes Redrawing will help you better describe the system with less graphics and text. When redrawing, try to be as accurate as you were in previous diagrams, but use fewer words and simpler graphics.

Sample Notice that the A21 diagram and its Glossary definitions have become stronger after the interface issue on the A2 diagram was resolved. This illustrates how simplifying a diagram helps to simplify all diagrams that detail it. This is why good analysts take time to resolve high-level interfaces before investing effort in low-level decompositions.

USED AT:	AUTHOR: Marca PROJECT: Feed Family NOTES: 1 2 3 4 5 6 7 8 9 10	DATE: 3/30/85 REV:	☒ WORKING ☒ DRAFT RECOMMENDED PUBLICATION	READER	DATE	CONTEXT:

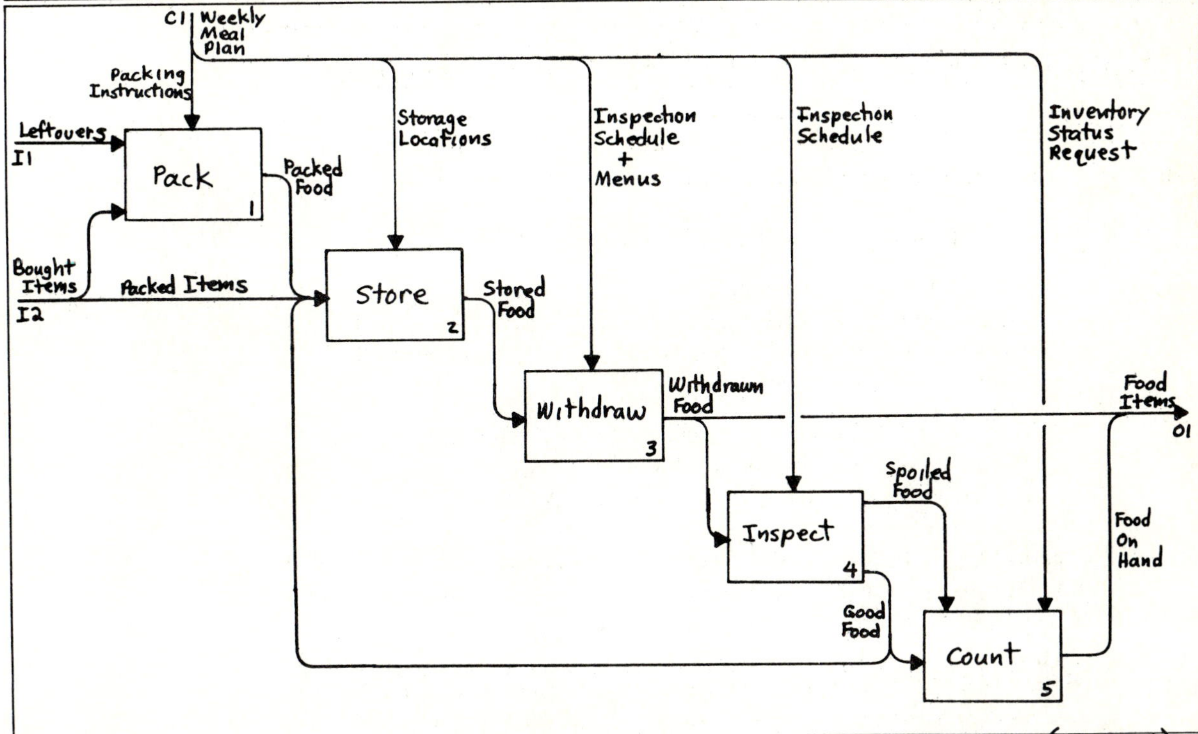

NODE: FF/A21	TITLE: Stock The Larder	NUMBER: DAMØ19 (DAMØ15)

USED AT:	AUTHOR: Marca PROJECT: Feed Family NOTES: 1 2 3 4 5 6 7 8 9 10	DATE: 3/30/85 REV:	☒ WORKING ☒ DRAFT RECOMMENDED PUBLICATION	READER	DATE	CONTEXT: DAMØ19

Packing Instructions – The materials and procedures that package bought food so it will keep in the larder.

Packed Items – Those bought items suitable for immediate storage.

Packed Food — Food suitable for storage. (Packed food can be counted without unpacking.)

Storage Locations – What food goes where.

Stored Food — Food in the larder.

Inspection Schedule – Timetable that says when to check food for spoilage.

Withdrawn Food — Food taken out of the larder.

Spoiled Food — Food that cannot be eaten.

Good Food — Food suitable to eat.

Inventory Status Request – A request about what and how much food is on hand.

NODE: FF/A21G1	TITLE: Stock The Larder (Glossary)	NUMBER: DAMØ2Ø (DAMØ16)

Chapter

27

Writing the Specification

Recall that an SADT model is not a specification, though it contains a very organized and accurate collection of information from which to develop a specification. This chapter is made up of four workshops, designed to guide you through the steps involved in writing a specification from an SADT model. In Workshop 22, property labels are added to the A-0 and A0 diagrams. In Workshops 23 and 24, these property labels are propagated down to the first- and second-level decompositions. All these properties, together with the diagrams, become the grist from which paragraphs of the specification are written. This is done in Workshop 25. After that, the diagrams and associated text are packaged into a specification document, according to the SADT node tree for the Feed Family Model.

Go through all four workshops without a break. This will give you an accurate feel for the amount of work needed to write a specification from a validated SADT model. (After the workshops, reflect on the amount of effort it took to write the text and compare that to writing the text from scratch.) Allocate about ½ hour for the first three workshops, but don't be concerned if you take more time. Allocate about 1 hour for writing and compiling the specification in the last workshop.

Workshop

22

Annotating the Summary and Top Diagrams

Goal Annotate the A-0 and A0 diagrams with nonfunctional requirements.

Activities

1. Review Chapter 19, which discusses ways to document nonfunctional requirements using property labels, activation rules, and color. Document the annotations made in steps 2 to 5 with SADT square notes in red.
2. Place the A-0 diagram before you. Think of some nonfunctional requirements for the box. Things like frequency of activation, duration of the function, and exceptions to normal operation are examples of nonfunctional requirements for boxes.
3. Do the same for the input, output, and control arrows for the box. Volume, time of creation, time of consumption, special attributes, and storage locations are examples of nonfunctional requirements for arrows.
4. Carry the nonfunctional requirements of the A-0 diagram down to the A0 diagram. Attach them to external arrows and to their appropriate boxes. If needed, add more detail to these annotations.
5. Now concentrate on the A0 diagram itself. Annotate internal arrows and boxes with more detailed nonfunctional requirements. Concentrate on sequencing and constraining the major functions of the system.

Notes

1. The nonfunctional requirements put on the A-0 diagram will be the most general ones of the system. They provide the context for the nonfunctional requirements of all other diagrams.
2. Annotating why data is control instead of input sharpens and strengthens the story a diagram tells. This information will be very useful when you write the specification from the model.
3. Typically, an annotated A0 diagram contains the most general and important nonfunctional requirements of any diagram in the model.

Sample

1. Notice how the frequency requirement on the A-0 diagram, which describes how often the family is fed, was detailed on the A0 diagram for box 3. This detailing caused the writing of the timing requirements for all the other boxes on the A0 diagram.
2. Notice how control arrow annotations on the A0 diagram clarify the role of this information in the operation of the system.

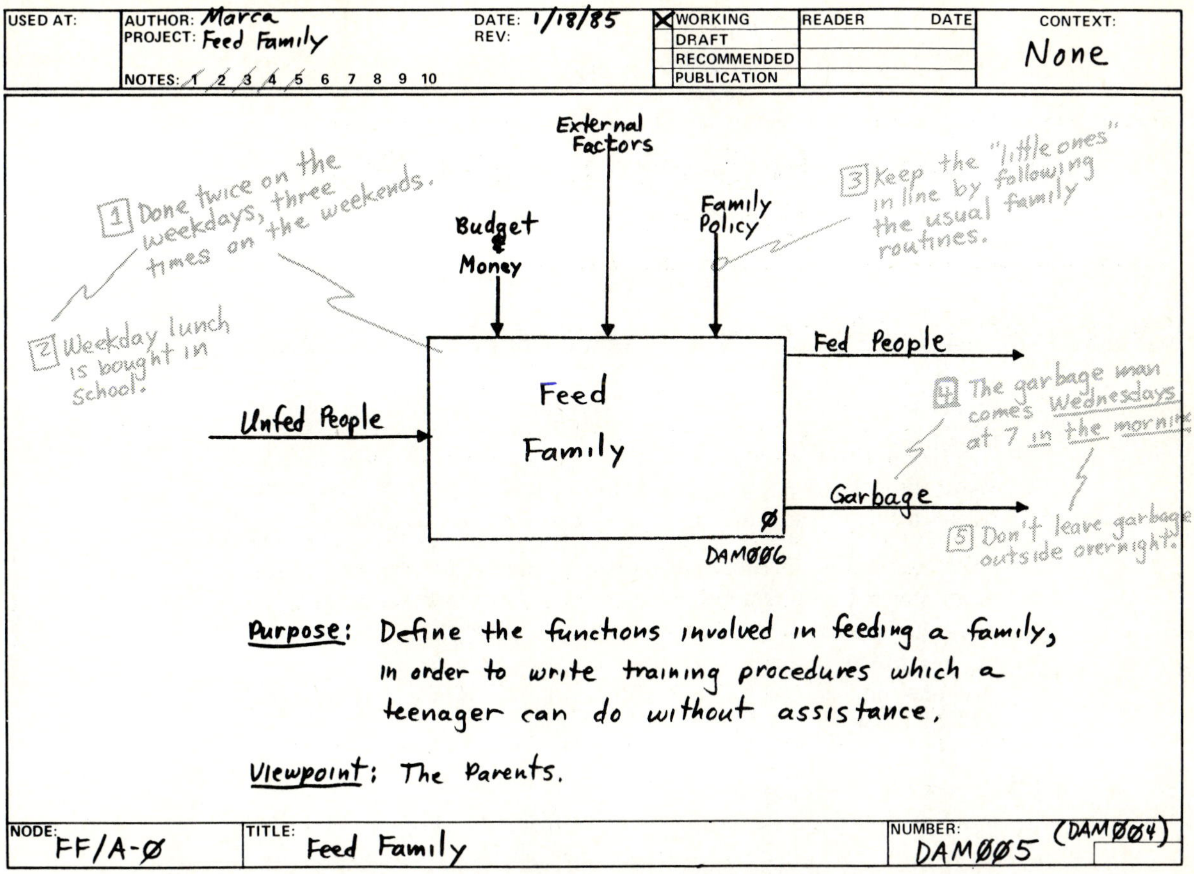
USED AT:
AUTHOR: Marca
PROJECT: Feed Family
DATE: 1/18/85
REV:
NOTES: 1 2 3 4 5 6 7 8 9 10
WORKING
DRAFT
RECOMMENDED
PUBLICATION
READER
DATE
CONTEXT:
None
External Factors
Budget & Money
Family Policy
1 Done twice on the weekdays, three times on the weekends.
2 Weekday lunch is bought in school.
3 Keep the "little ones" in line by following the usual family routines.
Unfed People
Feed Family
Ø
Fed People
Garbage
4 The garbage man comes Wednesdays at 7 in the morning
5 Don't leave garbage outside overnight.
DAMØØ6
Purpose: Define the functions involved in feeding a family, in order to write training procedures which a teenager can do without assistance.
Viewpoint: The Parents.
NODE: FF/A-Ø
TITLE: Feed Family
NUMBER: DAMØØ5
(DAMØØ4)

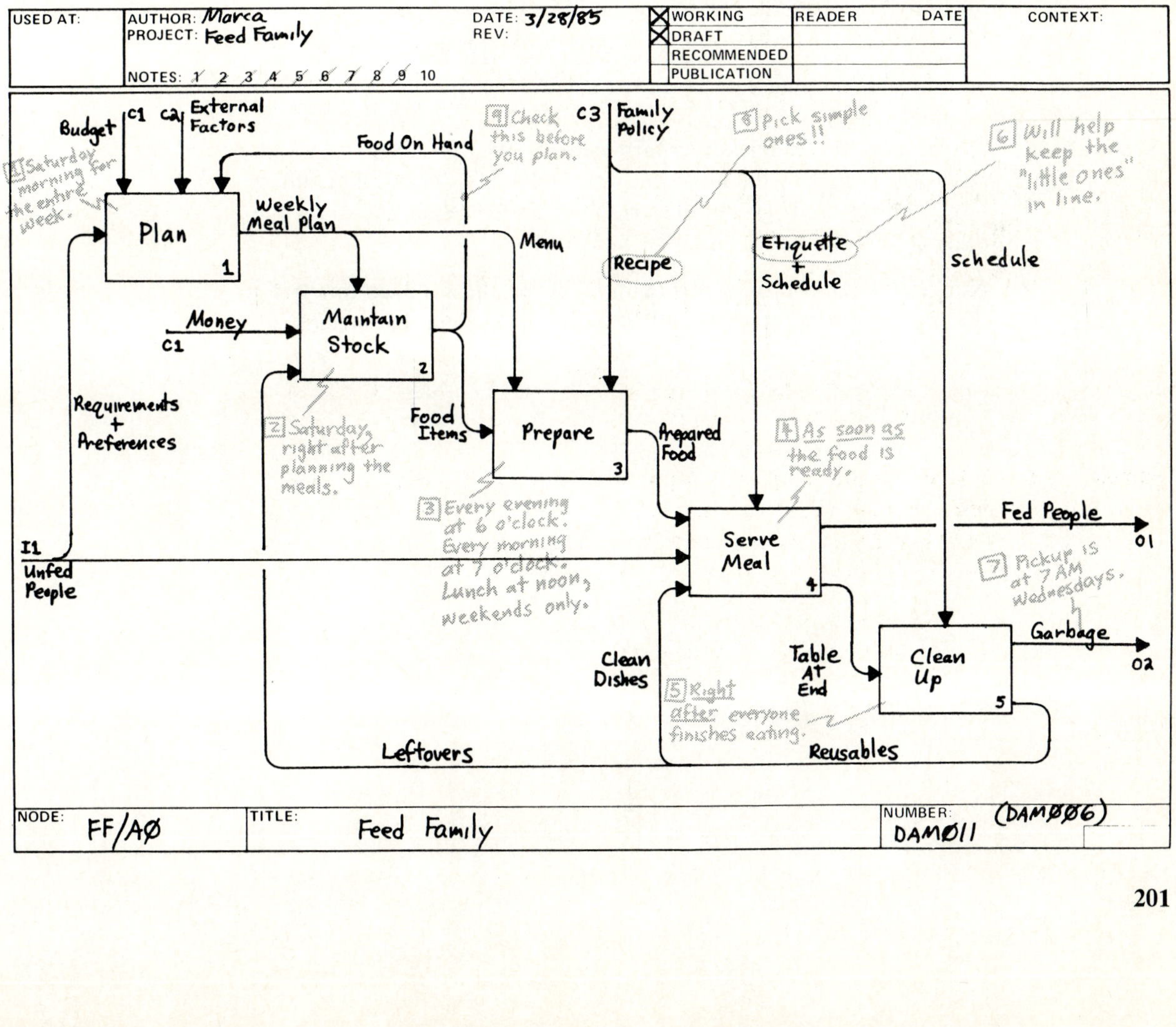
USED AT:
AUTHOR: Marca
PROJECT: Feed Family
DATE: 3/28/85
REV:
NOTES: 1 2 3 4 5 6 7 8 9 10
WORKING
DRAFT
RECOMMENDED
PUBLICATION
READER
DATE
CONTEXT:
Budget
C1
C2
External Factors
Food On Hand
9 Check this before you plan.
C3
Family Policy
8 Pick simple ones!!
6 Will help keep the "little ones" in line.
1 Saturday morning for the entire week.
Plan
1
Weekly Meal Plan
Menu
Recipe
Etiquette + Schedule
Schedule
Money
C1
Maintain Stock
2
Requirements + Preferences
2 Saturday, right after planning the meals.
Food Items
Prepare
3
Prepared Food
4 As soon as the food is ready.
3 Every evening at 6 o'clock. Every morning at 7 o'clock. Lunch at noon, weekends only.
I1
Unfed People
Serve Meal
4
Fed People
O1
7 Pickup is at 7AM Wednesdays.
Clean Dishes
Table At End
Clean Up
5
Garbage
O2
5 Right after everyone finishes eating.
Leftovers
Reusables
NODE: FF/AØ
TITLE: Feed Family
NUMBER: DAMØ11
(DAMØØ6)

Workshop

23

Annotating First-Level Decompositions

Goal Annotate the Ax diagram with nonfunctional requirements.

Activities

1. Review Chapter 19, which discusses ways to document nonfunctional requirements using property labels, activation rules, and color. Document every annotation made in this workshop with SADT square notes in red.
2. Place the A0 and Ax diagrams before you. Carry the nonfunctional requirements of the A0 diagram down to your Ax diagram.
3. Now concentrate on annotating the Ax diagram with more detailed nonfunctional requirements. Do the boxes first, external arrows second, and internal arrows last. Frequency of activation, duration, and exceptions to normal operation are appropriate for functions. Volumes, time of creation, time of consumption, special attributes, and storage locations are appropriate for arrows.

Notes

1. In general, the nonfunctional requirements on the A0 diagram will be more general than the ones you write for the Ax diagram.
2. Try to give reasons why data is control instead of input. This will often make it easier to write text from the diagram.
3. Typically, annotated Ax diagrams are the bridges between high-level and detailed nonfunctional requirements.

Sample

1. Notice how the frequency requirement of the A0 diagram, which describes when to maintain stock, was detailed on the A2 diagram at box 2. This detail triggered the writing of timing requirements for the other A2 boxes.
2. Notice how the control arrow annotations on the A2 diagram clarify the importance of information feedback during stock maintenance.

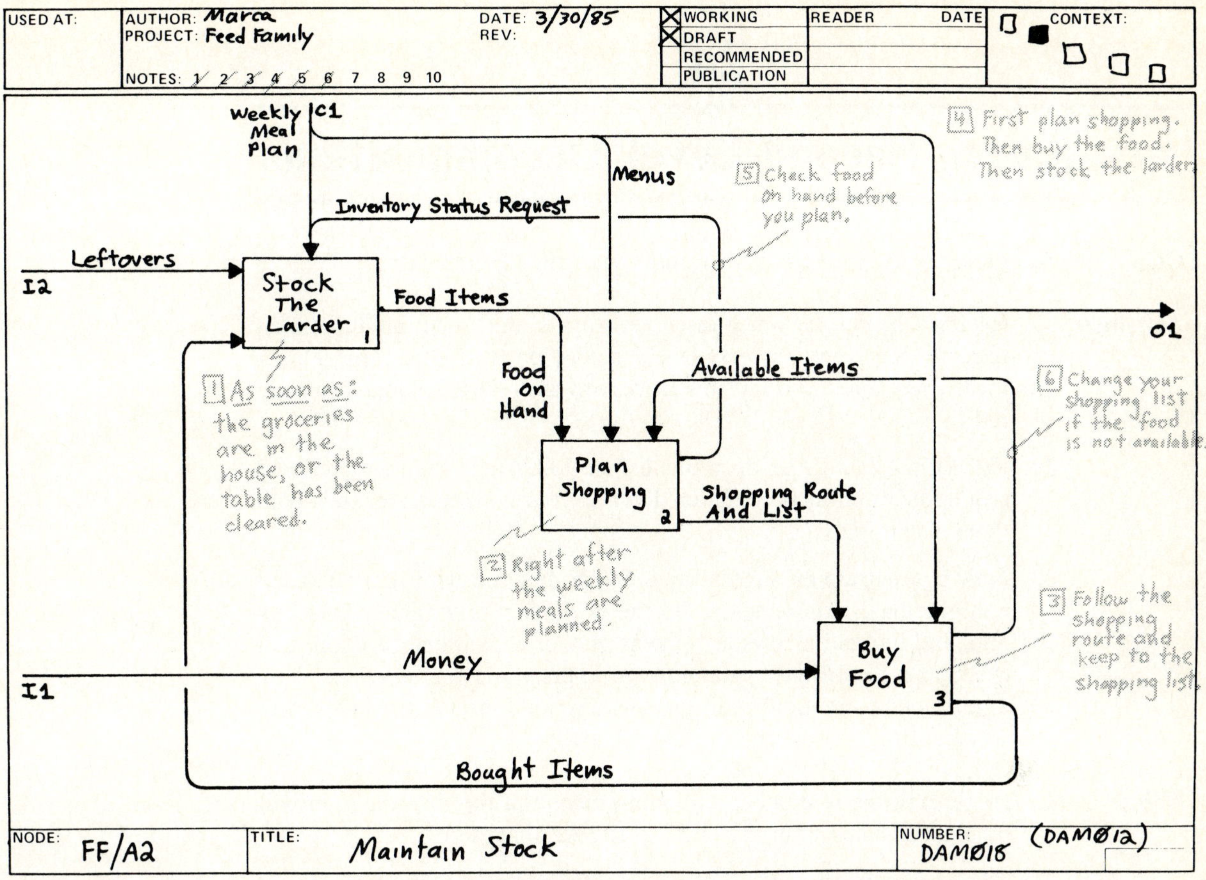
USED AT:
AUTHOR: Marca
PROJECT: Feed Family
DATE: 3/30/85
REV:
NOTES: 1 2 3 4 5 6 7 8 9 10
WORKING
DRAFT
RECOMMENDED
PUBLICATION
READER
DATE
CONTEXT:
Weekly Meal Plan
C1
Menus
Inventory Status Request
Leftovers
I2
Stock The Larder
1
Food Items
O1
Food On Hand
Available Items
Plan Shopping
2
Shopping Route And List
Buy Food
3
Money
I1
Bought Items
1 As soon as: the groceries are in the house, or the table has been cleared.
2 Right after the weekly meals are planned.
3 Follow the shopping route and keep to the shopping list.
4 First plan shopping. Then buy the food. Then stock the larder.
5 Check food on hand before you plan.
6 Change your shopping list if the food is not available.
NODE: FF/A2
TITLE: Maintain Stock
NUMBER: DAMØ18 (DAMØ12)

Workshop

24

Annotating Second-Level Decompositions

Goal Annotate the Axx diagram with nonfunctional requirements.

Activities

1. Review Chapter 19, which discusses ways to document nonfunctional requirements using property labels, activation rules, and color. Document every annotation in this workshop with SADT square notes in red.
2. Place the Ax and Axx diagrams before you. Carry the nonfunctional requirements of the Ax diagram down to your Axx diagram, and detail them accordingly.
3. Now concentrate on annotating the Axx diagram with more detailed nonfunctional requirements. Do the boxes first, external arrows second, and internal arrows last. Frequency of activation, duration, and exceptions to normal operations are appropriate for functions. Volume, time of creation, time of consumption, special attributes, and storage locations are appropriate for arrows.

Notes

1. In general, the nonfunctional requirements on the Axx diagram will be more detailed than the ones on the Ax diagram.
2. Try to give reasons why data is control instead of input. This will produce the kind of information you will need to write the text of the specification.
3. Typically, annotated Axx diagrams contain fairly detailed nonfunctional requirements.

Sample

1. The detailed nonfunctional requirements of the A21 diagram describe the necessary preventive actions required to minimize food spoilage.
2. Notice that the control arrow annotations on the A21 diagram clarify how the weekly meal plan impacts the detailed operations of stocking the larder.

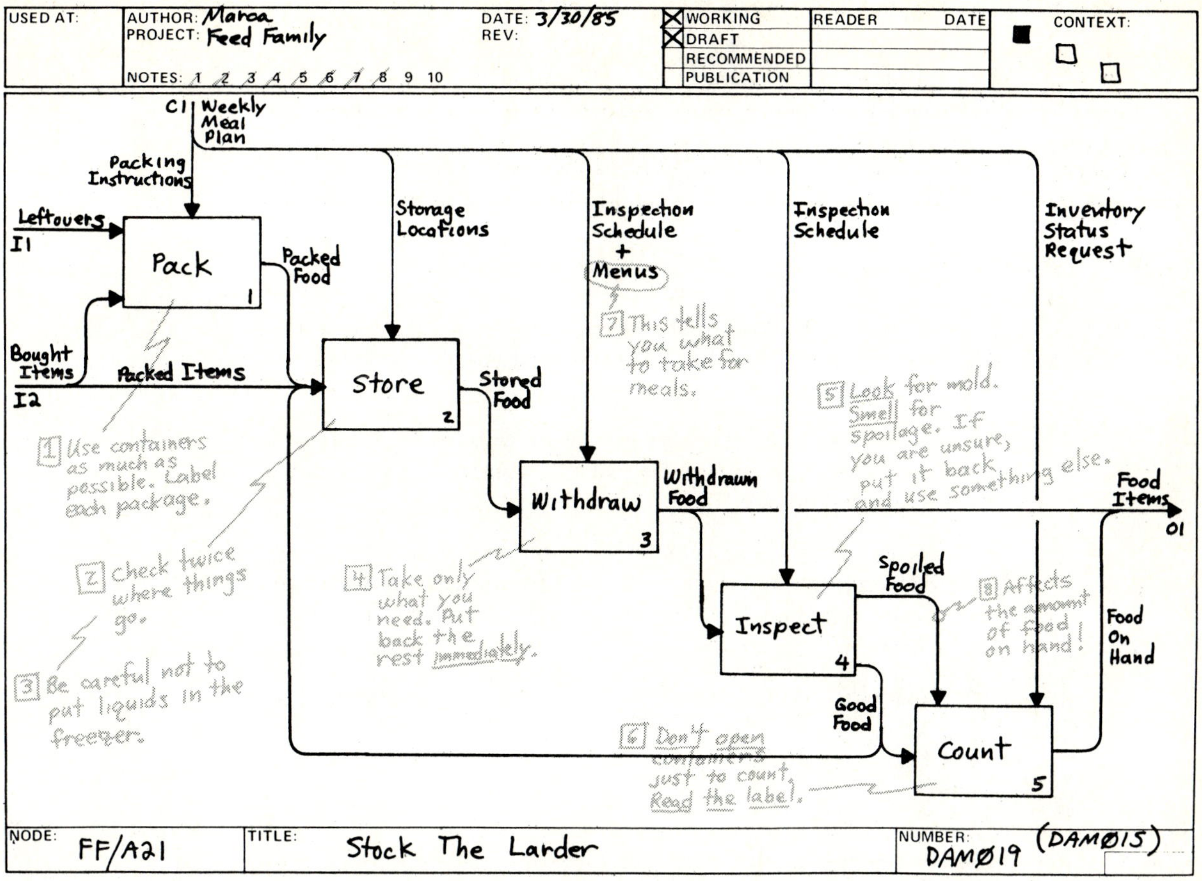

USED AT:
AUTHOR: Marca
PROJECT: Feed Family
DATE: 3/30/85
REV:
NOTES: 1 2 3 4 5 6 7 8 9 10
WORKING
DRAFT
RECOMMENDED
PUBLICATION
READER
DATE
CONTEXT:
C1 Weekly Meal Plan
Packing Instructions
Leftovers
I1
Pack
1
Packed Food
Bought Items
Packed Items
I2
Storage Locations
Store
2
Stored Food
Inspection Schedule + Menus
7 This tells you what to take for meals.
Inspection Schedule
Inventory Status Request
Withdraw
3
Withdrawn Food
Food Items
O1
1 Use containers as much as possible. Label each package.
2 Check twice where things go.
3 Be careful not to put liquids in the freezer.
4 Take only what you need. Put back the rest immediately.
5 Look for mold. Smell for spoilage. If you are unsure, put it back and use something else.
Inspect
4
Spoiled Food
8 Affects the amount of food on hand!
Food On Hand
Good Food
Count
5
6 Don't open containers just to count. Read the label.
NODE: FF/A21
TITLE: Stock The Larder
NUMBER: DAMØ19 (DAMØ15)

Workshop

25

Writing the Specification

Goal Write a description of feeding the family, according to the annotated diagrams made in Workshops 22 to 24.

Activities

1. Place your annotated Axx diagram in front of you. Keep the Ax, A0, and A-0 diagrams nearby for reference.
2. Number and name the text you are about to write with the node number and title of your Axx diagram. Remember to add the suffix /T1 to the node number.
3. Write one paragraph of text for each box on your Axx diagram, and order the paragraphs according to the box numbers. Focus each paragraph by describing the functions and nonfunctional requirements for just one box.
4. Now go back and write a paragraph about the way all the boxes of the this diagram work together. Put this paragraph before all the others on the Text Page. You have now written a section of the specification document.

Notes

1. Usually, one needs only a paragraph to describe the workings of a single SADT box, because SADT graphics compactly capture a small but well-defined part of the entire system.
2. Writing the paragraph about how all the boxes in the diagram work together is best done after each of the boxes has been described. (This is the way abstraction is done in writing.)

Sample

1. Notice how the purpose and viewpoint, taken from the A-0 diagram, summarize the specification document.
2. Notice that ordering diagrams and text according to the SADT node-numbering scheme gives a consistent organization to the writing.
3. Evaluate how small the text is for each section, and how smooth the flow of concepts is from one section to the next. This is a result of the organization and focus SADT diagrams bring to the subject.

The Feed Family Specification

Purpose The purpose of this document is to describe how to feed the family, so the teenagers of the family can do it without assistance from their parents.

Viewpoint This document was written for the teenagers by their parents.

Contents This document is a partial specification for feeding the family. It is incomplete because it does not detail many of the major feeding activities. It does, however, detail how the food inventory is maintained. The following table of contents summarizes the organization of this document:

Diagram	Title
A–0	Feed Family (Context)
A–0T1	Feed Family (Context) Text
A0	Feed Family (Top)
A0T1	Feed Family (Top) Text
A2	Maintain Stock
A2T1	Maintain Stock Text
A21	Stock the Larder
A21T1	Stock the Larder Text

Reading this document This document was written and compiled following the organization and content of the Feed Family model. All SADT models are organized in node number order. That is, the diagrams of a model appear in "depth-first tree-walk" order. This means that the most general aspects of feeding the family are presented first, and the most detailed aspects are presented last. So, to read this document for its overview, read the text associated with feeding the family (i.e., A-0 and A0). To read this document for detail, read all the text associated with maintaining food stock (i.e., A2 and A21).

A-0. Feeding the Family (Context)

All the people in the family are fed twice on weekdays (breakfast and dinner, with lunch being bought at school or work) and three times on weekend days (breakfast, lunch, and dinner). Use the money we gave you to buy the food. Be careful to stick to the budget, so you can eat during the whole week. Keep the "little ones" in line by following the usual family routines. This way, it will be harder for them to argue. And don't forget that the garbage collector comes at seven o'clock Wednesday mornings. You'll have to wake up early to put out the garbage, because it cannot stay outside overnight.

USED AT:	AUTHOR: Marca PROJECT: Feed Family NOTES: 1 2 3 4 5 6 7 8 9 10	DATE: 1/18/85 REV:	☒ WORKING ☐ DRAFT ☐ RECOMMENDED ☐ PUBLICATION	READER DATE	CONTEXT: None

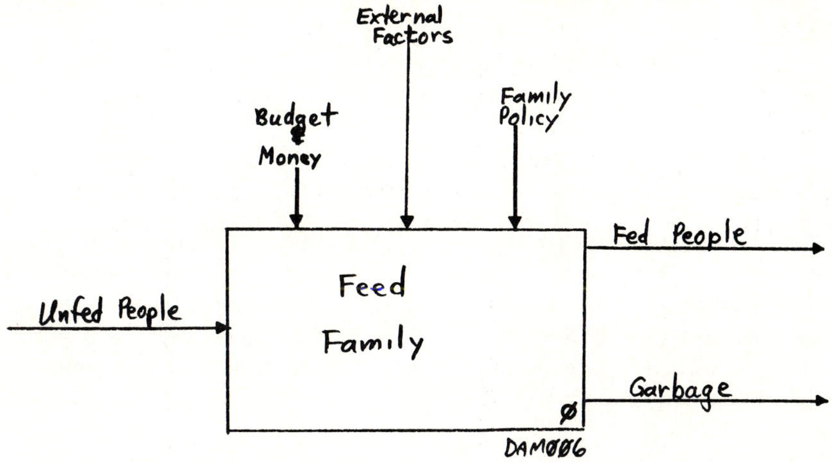

Purpose: Define the functions involved in feeding a family, in order to write training procedures which a teenager can do without assistance.

Viewpoint: The Parents.

NODE: FF/A-Ø	TITLE: Feed Family	NUMBER: DAMØØ5 (DAMØØ4)

A0. Feeding the Family (Overview)

Overview Menus for the entire week are planned every Saturday morning. Food is bought and stocked Saturday afternoon. Meals, prepared twice a day on weekdays and three times a day on weekends, are served as soon as they are cooked. The table and kitchen are cleaned right after everyone finishes eating.

Planning Plan the meals for the whole week first thing every Saturday morning. First find out what everyone wants and write it down, and then check the food on hand to see if you have enough to make all the meals. Next figure out what and how much food you need to get. But remember, you have a budget. So figure out the best way to eat what you all want with the money you have.

Stocking Right after you plan the meals, go out and buy whatever food you need. Take the weekly meal plan with you, so you can quickly get just what you want. And don't forget to take money with you when you leave the house.

Preparing Serve breakfast at seven o'clock every morning, lunch at noon every day, and dinner at six o'clock every night. Stick to these times, and you'll find it easy to feed everyone. Also, follow the menus you planned on Saturday. This way, you won't run short of food. Last, pick simple recipes to cook. If you do, there is less of a chance for something to go wrong.

Eating Serve each meal as soon as you have cooked it. This way, it will be hot and tasty. Seat all family members at the table in their usual chairs. You'll have enough to worry about without bickering over seating. Have everyone use good manners at the table and you'll enjoy the meal more.

Cleaning Right after everyone finishes eating, please clean the table, put all food scraps in the garbage can, and store the leftovers. Then, wash the dishes and clean the kitchen. (It won't take as long as you think.) This way, you'll have the rest of your time to yourselves.

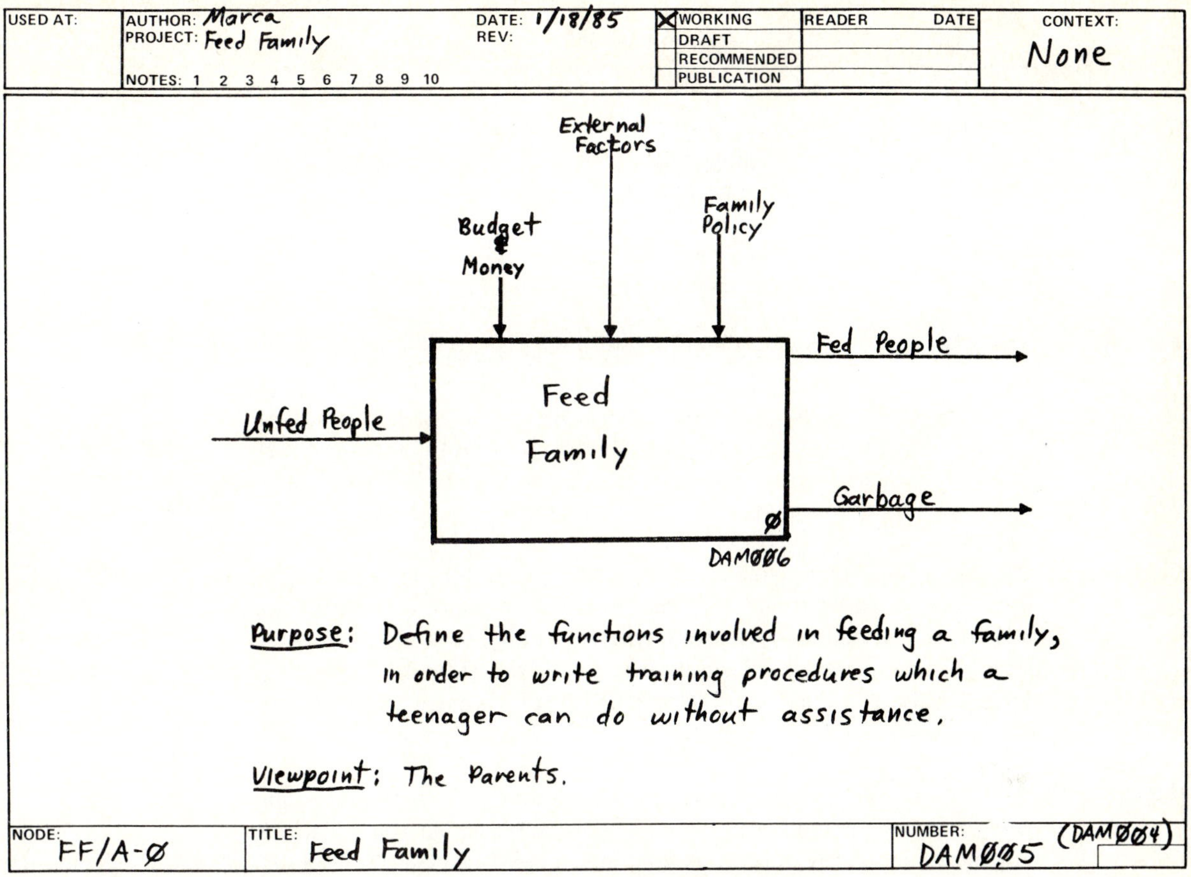

USED AT:
AUTHOR: Marca
PROJECT: Feed Family
DATE: 1/18/85
REV:
NOTES: 1 2 3 4 5 6 7 8 9 10
WORKING
DRAFT
RECOMMENDED
PUBLICATION
READER
DATE
CONTEXT:
None
External Factors
Budget & Money
Family Policy
Fed People
Unfed People
Feed Family
Garbage
Ø
DAMØØ6
Purpose: Define the functions involved in feeding a family, in order to write training procedures which a teenager can do without assistance.
Viewpoint: The Parents.
NODE: FF/A-Ø
TITLE: Feed Family
NUMBER: DAMØØ5 (DAMØØ4)

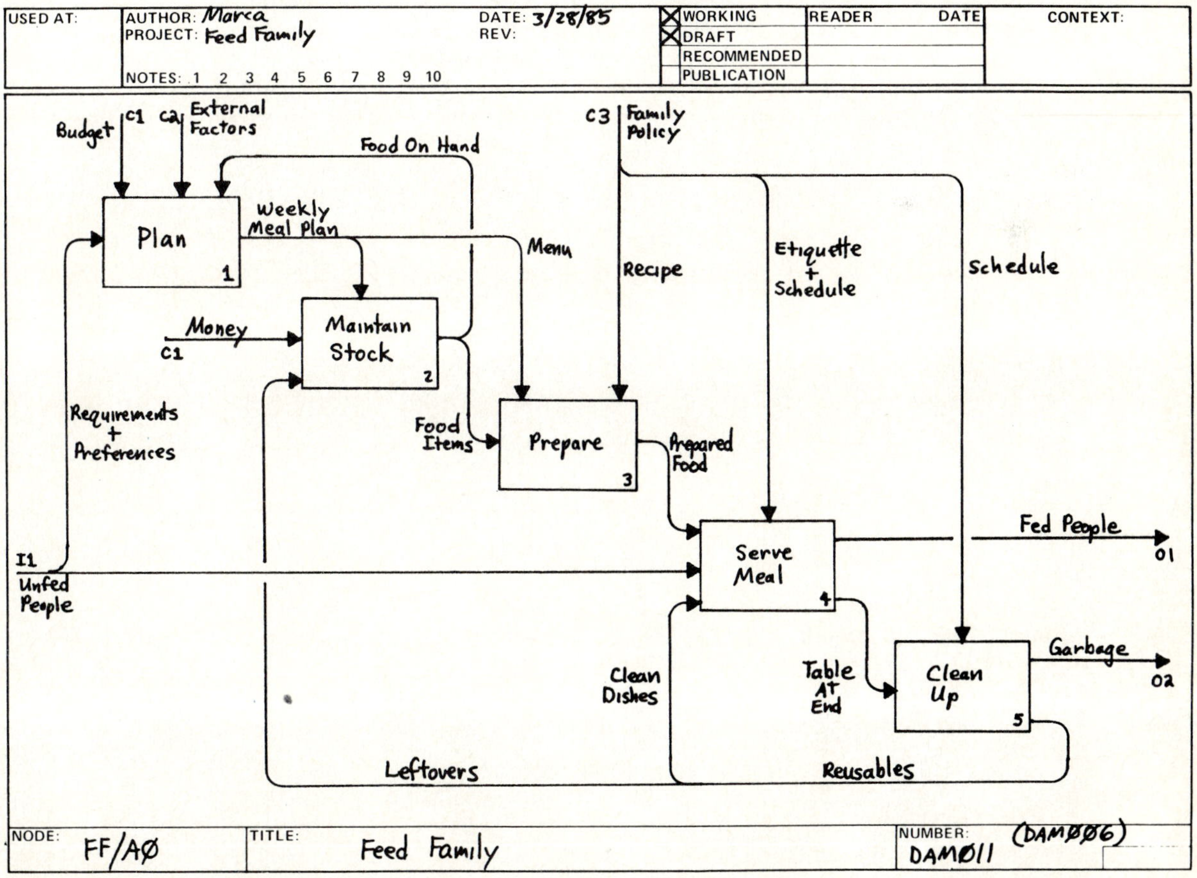

USED AT:
AUTHOR: Marca
PROJECT: Feed Family
DATE: 3/28/85
REV:
NOTES: 1 2 3 4 5 6 7 8 9 10
WORKING
DRAFT
RECOMMENDED
PUBLICATION
READER
DATE
CONTEXT:
Budget
C1
C2
External Factors
C3
Family Policy
Food On Hand
Weekly Meal Plan
Plan
1
Menu
Recipe
Etiquette + Schedule
Schedule
Money
C1
Maintain Stock
2
Requirements + Preferences
Food Items
Prepare
3
Prepared Food
Serve Meal
4
Fed People
O1
I1
Unfed People
Clean Dishes
Table At End
Clean Up
5
Garbage
O2
Leftovers
Reusables
NODE: FF/AØ
TITLE: Feed Family
NUMBER: DAMØ11 (DAMØØ6)

A2. Maintaining Stock

Overview Shopping is planned right after deciding on the weekly meals. You should also make a shopping route (where to go) and a shopping list (what to buy). When planning your shopping, check the larder to determine what and how much food is on hand. Food is bought right after the shopping plan has been made. Sometimes, when stores run out of certain foods on your list, the shopping list needs to be altered.

Stocking The word "larder" means anyplace where food is stored in the house—that is, the pantry, the refrigerator, and the freezer. The weekly meal plan tells you how to manage the larder. Two kinds of things go in the larder: items you just bought at the store and leftovers from the meal you just ate. Take things out of the larder when you want to check on what and how much food you have, or when you want to prepare a meal.

Planning Plan the shopping after you have planned the menus for the week. Check the larder to see what and how much food is on hand. Come up with a shopping list that says what to buy. Also, think about which store or stores you have to visit. If you have to go to several, make a shopping route to shorten your travels. And don't forget to check on when the stores are open.

Buying Take money, the shopping route, and the shopping list with you when you go to buy the food. Keep in mind your weekly plan when you are buying the food. This way, you can double-check yourself. Stick to the shopping list so you won't run out of money. If the grocer doesn't have what you want, change your shopping list in the store. This way you change the menus so you won't buy unnecessary food. When you get home, put the items you bought in the larder immediately. This way, they won't have a chance to spoil.

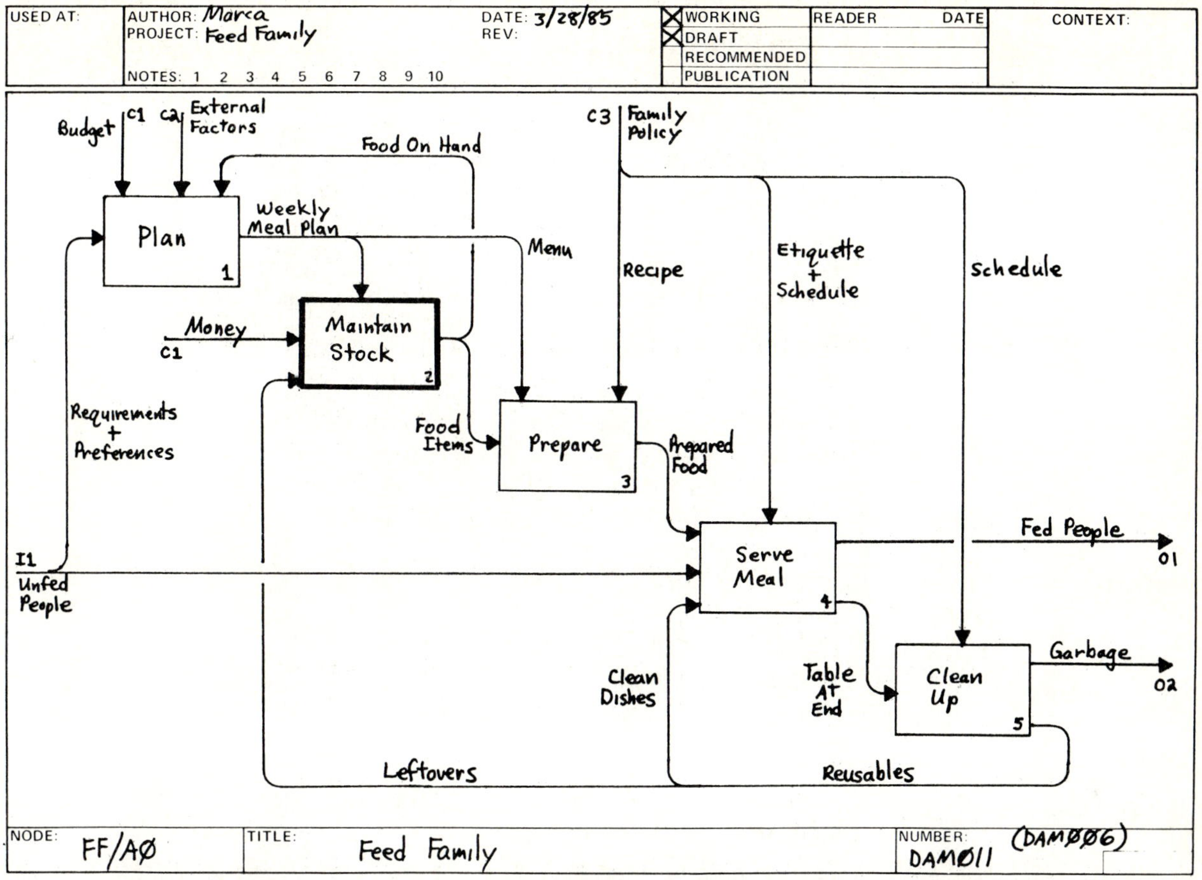
USED AT:
AUTHOR: Marca
PROJECT: Feed Family
DATE: 3/28/85
REV:
NOTES: 1 2 3 4 5 6 7 8 9 10
WORKING
DRAFT
RECOMMENDED
PUBLICATION
READER
DATE
CONTEXT:
Budget
C1
C2
External Factors
Food On Hand
C3
Family Policy
Plan
1
Weekly Meal Plan
Menu
Recipe
Etiquette + Schedule
Schedule
Money
C1
Maintain Stock
2
Requirements + Preferences
Food Items
Prepare
3
Prepared Food
I1
Unfed People
Serve Meal
4
Fed People
O1
Clean Dishes
Table At End
Clean Up
5
Garbage
O2
Leftovers
Reusables
NODE:
FF/AØ
TITLE:
Feed Family
NUMBER:
DAMØ11
(DAMØØ6)

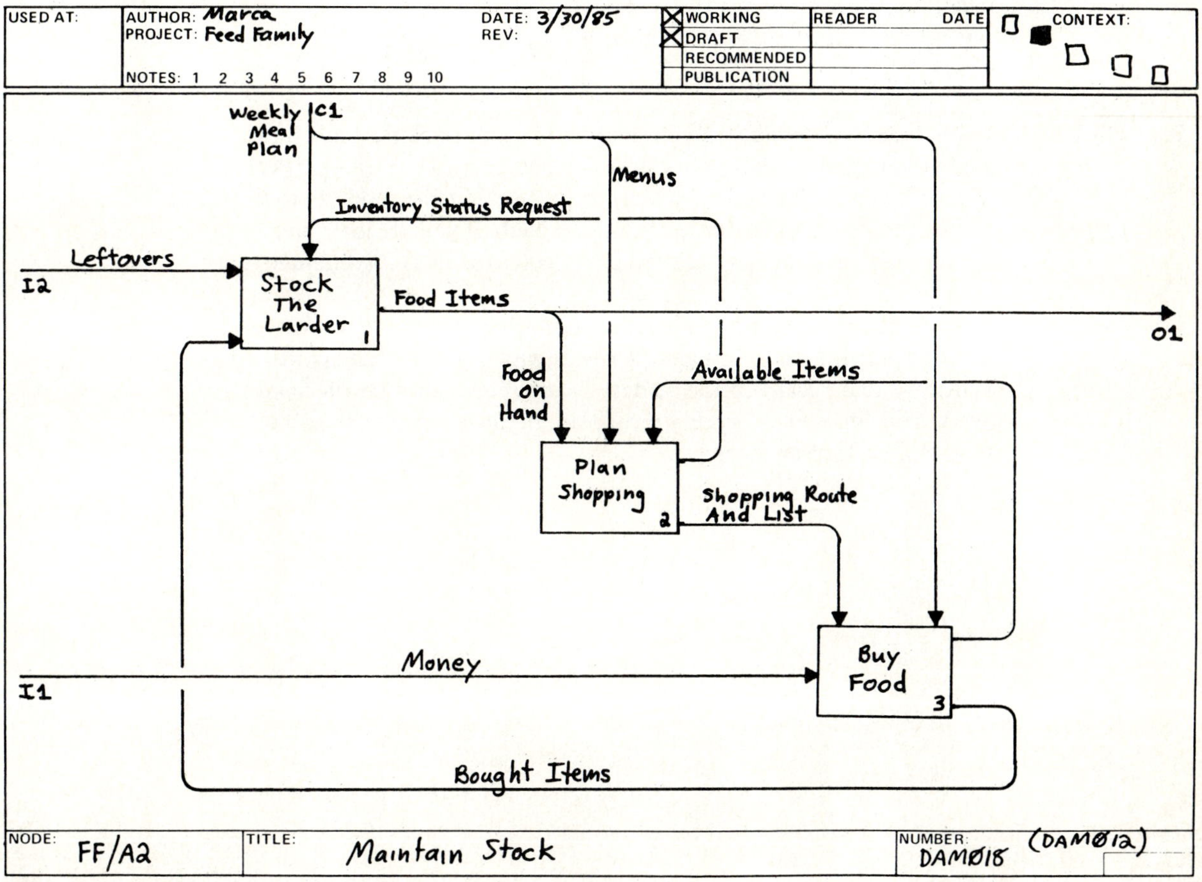
USED AT:
AUTHOR: Marca
PROJECT: Feed Family
DATE: 3/30/85
REV:
NOTES: 1 2 3 4 5 6 7 8 9 10
WORKING
DRAFT
RECOMMENDED
PUBLICATION
READER
DATE
CONTEXT:
Weekly Meal Plan
C1
Menus
Inventory Status Request
Leftovers
I2
Stock The Larder
1
Food Items
O1
Food On Hand
Available Items
Plan Shopping
2
Shopping Route And List
Money
I1
Buy Food
3
Bought Items
NODE:
FF/A2
TITLE:
Maintain Stock
NUMBER:
DAMØ18
(DAMØ12)

A21. Stocking the Larder

Overview Food is packed immediately after it is brought home. Packed food is stored in its appropriate storage location. Food is taken from storage locations just before preparing a meal or when the food must be inspected for spoilage. After inspection, good food is returned to its storage location and spoiled food is thrown out. Before planning the weekly meals, counting is done to determine what and how much food is on hand.

Packing Leftovers and most things you buy at the store will require packing before they can be stored in the larder. Use containers as much as possible, and label each container with what food it holds and how much food is in it. Try not to use plastic wrap or aluminum foil because these things take more skill to use than containers.

Storing Put packed food, or bought items that need no packing, into appropriate storage places. Check twice where things go. If you do, you will be less likely to store a particular kind of food in the wrong place. Be careful not to store liquids in the freezer, as this will surely cause their containers to break. And always make sure the refrigerator and freezer doors are shut tightly.

Withdrawing Take food out of the larder only when you want to prepare a meal or check for what and how much food is on hand. Take out only the food you need, and immediately put back the rest. This way, food won't spoil because it was accidentally left out. Again, shut the refrigerator and freezer doors tightly after you take out the food you need.

Inspecting Inspect the food carefully if you doubt it is good. Look for mold and smell it for spoilage. If you are unsure whether it can be eaten, take something else. Don't take chances with questionable food. Throw spoiled food out immediately; don't put it back. If you are just counting the food on hand, put back the good food as soon as it is counted.

Counting Remember to take account of spoiled food when you figure out what and how much food is on hand. Write down on paper what spoiled, and leave that record in the larder. Take a count of food on hand every time you make a weekly meal plan, because a week's worth of food is a lot to remember.

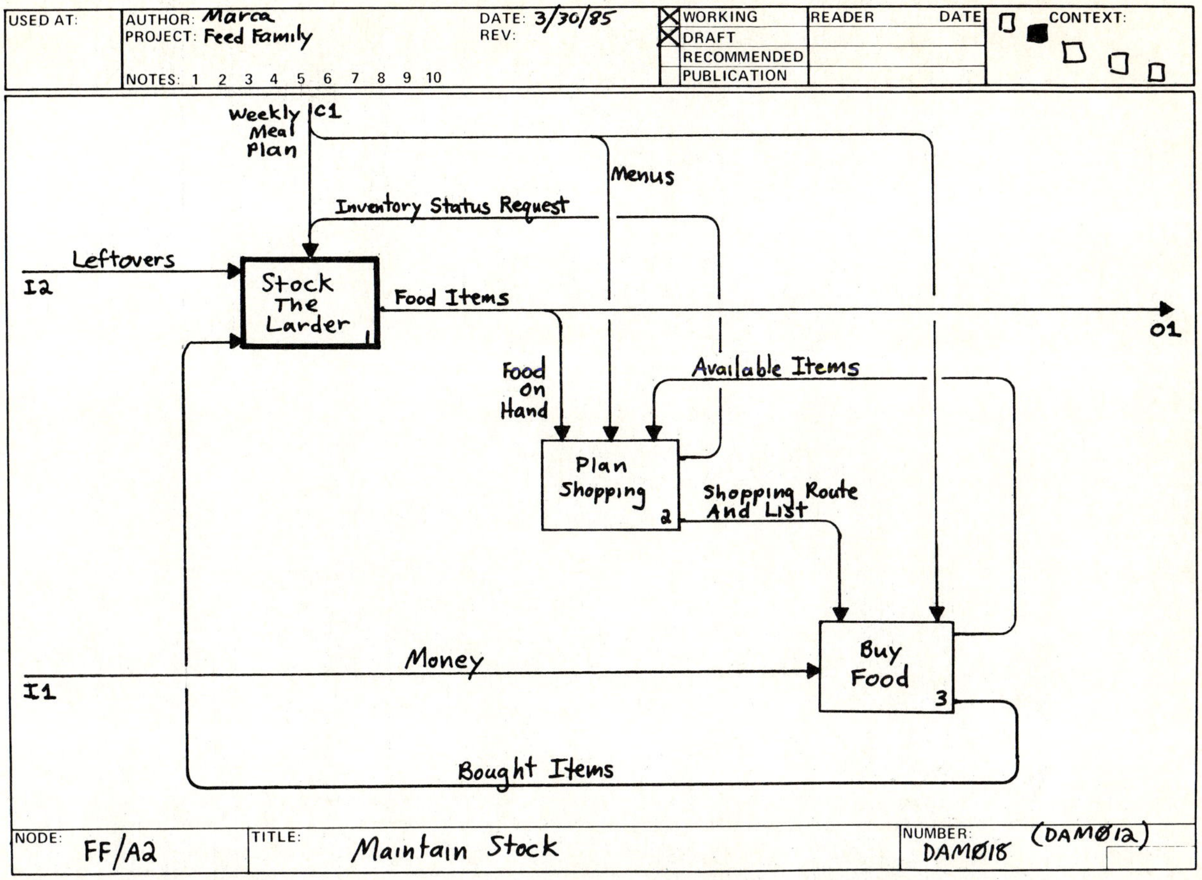
USED AT:
AUTHOR: Marca
PROJECT: Feed Family
DATE: 3/30/85
REV:
NOTES: 1 2 3 4 5 6 7 8 9 10
WORKING
DRAFT
RECOMMENDED
PUBLICATION
READER
DATE
CONTEXT:
Weekly Meal Plan
C1
Menus
Inventory Status Request
Leftovers
I2
Stock The Larder
1
Food Items
O1
Food On Hand
Available Items
Plan Shopping
2
Shopping Route And List
Money
I1
Buy Food
3
Bought Items
NODE: FF/A2
TITLE: Maintain Stock
NUMBER: DAMØ18
(DAMØ12)

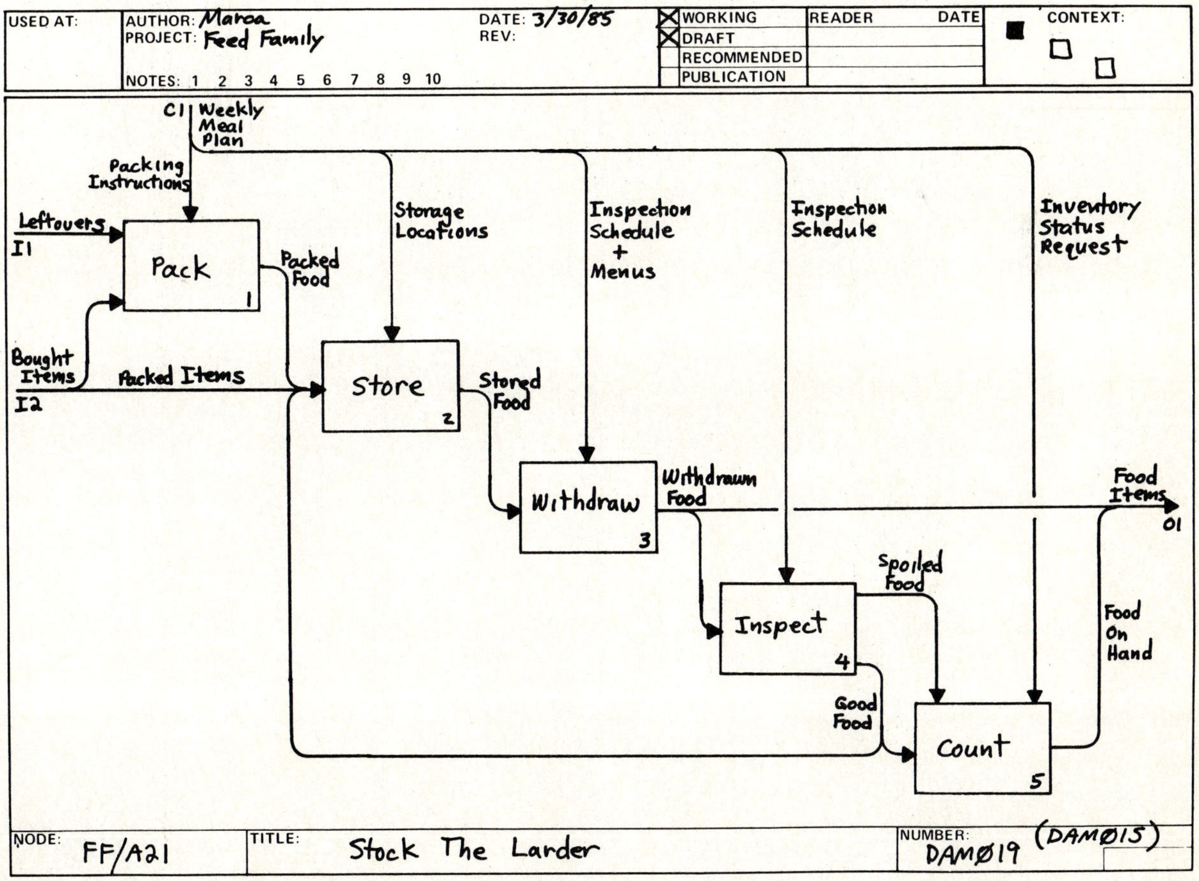
USED AT:
AUTHOR: Marca
PROJECT: Feed Family
DATE: 3/30/85
REV:
NOTES: 1 2 3 4 5 6 7 8 9 10
WORKING
DRAFT
RECOMMENDED
PUBLICATION
READER
DATE
CONTEXT:
C1
Weekly Meal Plan
Packing Instructions
Storage Locations
Inspection Schedule + Menus
Inspection Schedule
Inventory Status Request
Leftovers
I1
Pack
1
Packed Food
Bought Items
I2
Packed Items
Store
2
Stored Food
Withdraw
3
Withdrawn Food
Food Items
O1
Spoiled Food
Inspect
4
Food On Hand
Good Food
Count
5
NODE: FF/A21
TITLE: Stock The Larder
NUMBER: DAMØ19
(DAMØ15)

Part

VI

Applications of SADT

Part V presented a portion of an SADT activity model (Feed Family) in the application domain of cooperative work. This part of the book presents several other SADT models, each one a sample from a different application domain: financial management, aerospace manufacturing, data processing, and training. Each chapter presents one of these models and summarizes the application area, the background of the project that produced the model, the resources used during the project, the way the project used SADT to specify the requirements, the decomposition strategy evident in the model, the size and complexity of the model, and other important features of the model.

We have collected this information into one place for three reasons. First, each of the models presented in this part of the book is drawn in a style that differs a bit from the one we recommend and have used throughout the book. This will give you the opportunity to compare your SADT drawing style with those of five SADT experts. We strongly feel that this exposure to different styles is required to complete the teaching of how to make SADT activity models.

Second, the models that follow were specially selected to give you a feel for the effectiveness and the wide applicability of SADT. These models were picked from a very large number of well-documented SADT projects run by SofTech, Inc. (see the accompanying partial list). We know of no other structured analysis technique or methodology that has come close to matching this kind of applicability; nor do we think one will in the near future.

Third, each model presented is a representative example of the kinds of models one is likely to build in the given application domains. By presenting a spectrum of application areas, by seeing actual results, and by knowing how these results were achieved, you can begin to evaluate how SADT could be used for your particular kinds of problems. We feel strongly that an exposure to a variety of actual results will help you make sound decisions about the cost effectiveness of applying SADT in your organization.

Some Areas of SADT Application

1. Manufacturing and logistics:
   a. Integrated Computer-Aided Manufacturing
   b. Geometric modeling
   c. Process planning
   d. Manufacturing systems
   e. Logistics management
   f. Technology transfer
   g. Maintenance management
2. Management and finance
   a. Financial management
   b. Field office reporting
   c. Securities safekeeping
   d. Naval planning
   e. Delivery invoicing
   f. Order handling
   g. Discount administration
   h. Office automation
3. Training and personnel
   a. Training management
   b. Training data exchange
   c. Educational services
   d. Course development
4. Planning and development
   a. Program planning
   b. Logistics management
   c. System development
   d. Design technology
   e. Readiness measurement
   f. Technology planning
5. Avionics and defense
   a. Electronic warfare operations
   b. Simulation language
   c. Ground support equipment
   d. Tactical operations
   e. Embedded software systems
6. Communication
   a. Interoffice exchange
   b. Telecommunications
   c. Unit level switch
   d. Human-machine communications
   e. Traffic monitor
   f. Regional data switch
7. Computers
   a. Database specification
   b. Database translator
   c. Operating system enhancement
   d. Applications engineering

Chapter

28

Financial Management

28.1 Application Domain

One aspect of financial management is the monitoring of project budgets. In this particular case, SADT was used to develop the requirements for a management information system (MIS) that monitored the spending performances of government contracts. The requirements for the system were derived by describing how particular people in the financial organization, called "program element monitors" (PEM's), manually managed contractual budgets.

28.2 Project Background

The project specified, designed, and built a system that tracked the budgets for many ongoing research and development programs sponsored by the U.S. Air Force. The MIS system named RDMIS helped contract managers track the parts of an Air Force program. The capabilities of RDMIS included analysis of funding profiles, storage and retrieval of program descriptions, and report generation according to separate content and format controls. The feasibility of using RDMIS to track Air Force programs was successfully demonstrated after the system was prototyped from its SADT specification.

28.3 Project Overview

The specification phase of the project focused on the duties and tasks of a PEM, the person responsible for one or more budgetary elements. SADT was used to portray the complex budget management process in which a PEM participates. The resulting SADT activity model, presented here, describes reporting requirements, the steps necessary to establish and allocate budgets, and the steps required to monitor the progress of a program element. This model was used to design and build the prototype RDMIS. Both the successful prototype and the SADT model were then used to design and implement a full RDMIS system.

28.4 Resources Used to Make the Model

The PEM model which follows was developed in 1 year by an SADT analyst and 10 reviewers (PEM's and other members of the Air Force). Within that year, the project accomplished its primary goal, specifying RDMIS requirements. Some of the 12 months was spent educating the analyst about Air Force contract procedures, but a large percentage of the time was spent reaching consensus among the 10 members of the Air Force about what a PEM actually does. This consensus was the major reason for the success of the prototype system.

28.5 Notes about the Model

Since the PEM model had to describe the actual phases of the budget monitoring process, a sequential decomposition strategy was used. Such a strategy is often effective for describing the current procedures done by an organization. Also, the model makes use of tunneled arrows to avoid cluttering top-level diagrams with unnecessary details. Used in this way, tunneled arrows actually improve the readability of the entire model. Last, while we do not recommend shading boxes and using "squiggles" to connect labels to arrows, we have left them in the model as an example of an alternate drawing style.

A-0 Monitor Program Element

This model describes the activities performed by or affecting an Air Force Program Element Monitor (PEM) during the lifetime of a particular program element or project. A program element is the description of a mission identifying the resources needed to perform the assigned mission. The PEM is responsible for monitoring a program element. The activities are initiated by three controlling factors: the identification of a required capability may initiate a program element or project; the Planning, Programming and Budgeting System (PPBS) defines exercises and reports to be prepared, and the time cycle in which to perform them; and, inquiries will require responses to be prepared. In addition, the PEM must provide direction to the major commands involved in the program and describe it to Congress. These activities are supported by various reports and briefings from the implementing program office.

The viewpoint of this model is all activities either performed by a PEM or seen by the PEM in performing his duties.

This work may be performed by people not called PEMs. It is often done by an Action Officer, and in directorates other than Research and Development (AF/RD), that person may have a different title.

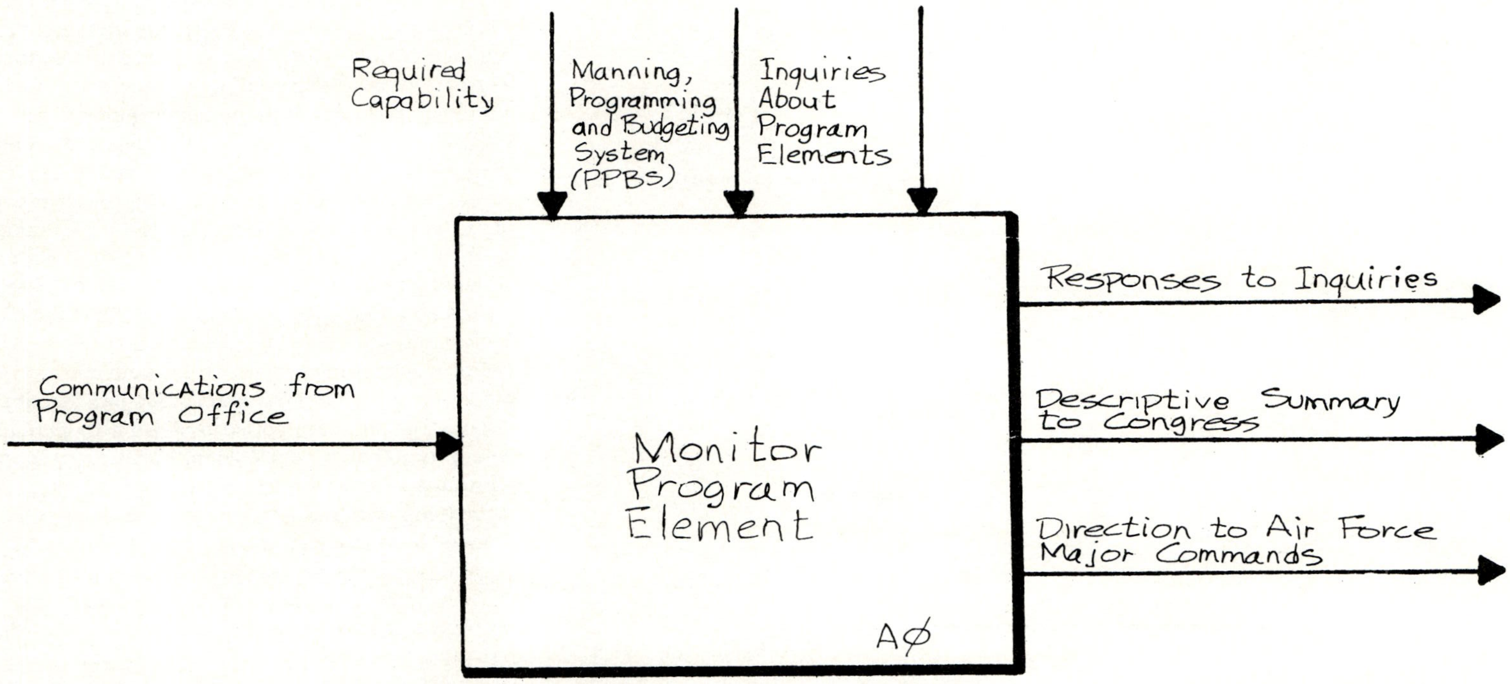

Purpose: Describe the sequence of tasks required to correctly monitor a program element.

Viewpoint: The PEM for the program.

NODE: A-Ø	TITLE: Monitor Program Element	NUMBER:

A0 Monitor Program Element

This diagram shows the monitoring activities occurring during the lifetime of a program element. The PEM and other Air Force and DoD personnel perform these activities simultaneously for several programs in different stages of development.

The process is initiated by the identification of a required capability (box 1 control 1). This can be in the form of a Required Operational Capability (ROC) received by the PEM (in most cases from the major command), or from any agency/individual perceiving a required operational capability. If program authorization is received, the system requirements are identified (box 1 output 1). However, no funds are yet appropriated. This will not occur until the next budget cycle, unless the need warrants reprioritization of other programs and their approved funding.

The PEM must then compete for funds for each program (box 2). The primary result of this competition is funding guidance given to each PEM in December (box 2 output 3). Along with this guidance, a Descriptive Summary must be prepared for Congress in January (box 2 output 2). Funds are committed for a five-year period beginning in October.

Once funding guidance is received (box 3 control 1), the PEM can provide direction to each Major (Implementing) Command for the coming year (box 3). This is provided in a Program Management Directive (PMD) (box 3 output 1).

Once a program is operational, the PEM must monitor its progress during the current year (box 4). The status of the program (box 4 output 1) then becomes a factor in the future funding and direction of the program.

One important constraint in this process is the Planning, Programming and Budgeting System (PPBS). This specifies activities, reports and timing affecting the PEM. The Five Year Defense Program (FYDP) specifies the approved Air Force plans and programs, and the Planning-Programming Guidance Memorandum (PPGM) provides Secretary of Defense (SECDEF) guidance for financial constraints and support programs.

In summary, it is important to note that the PEM is working in three time frames on each program:

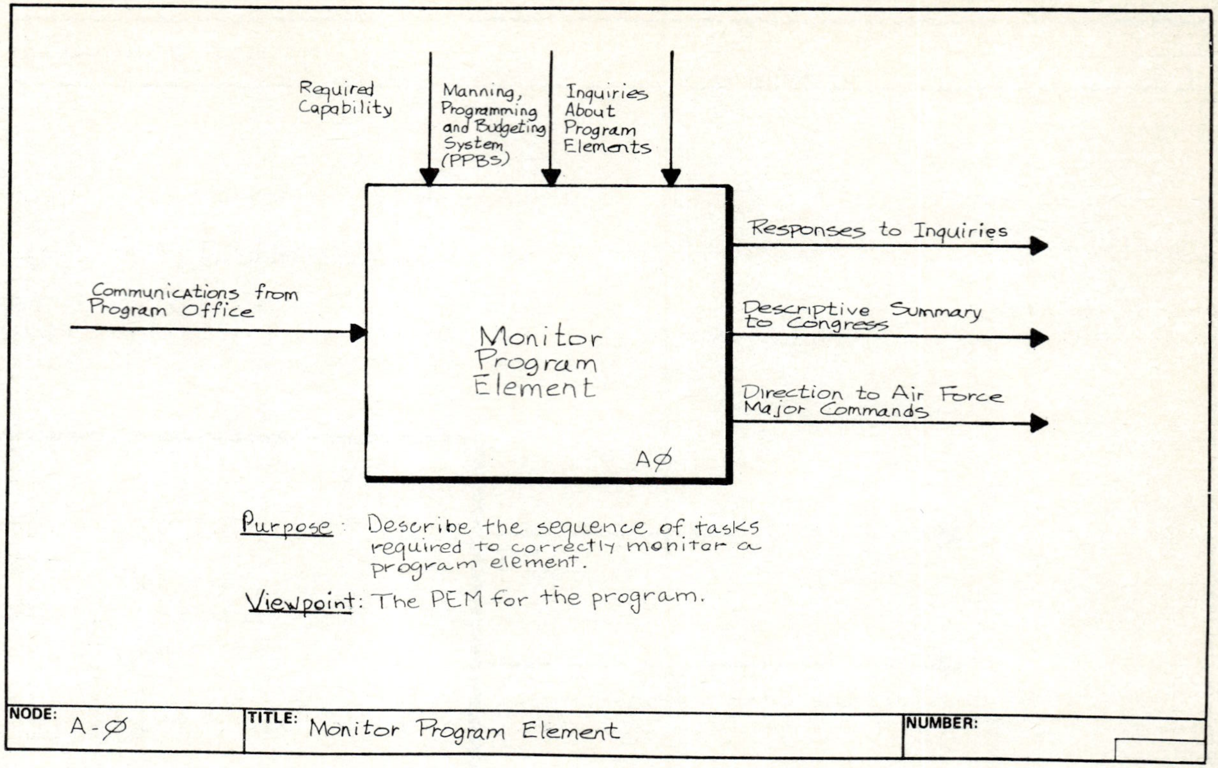

1. Current year
2. Next year
3. Five year period following next year

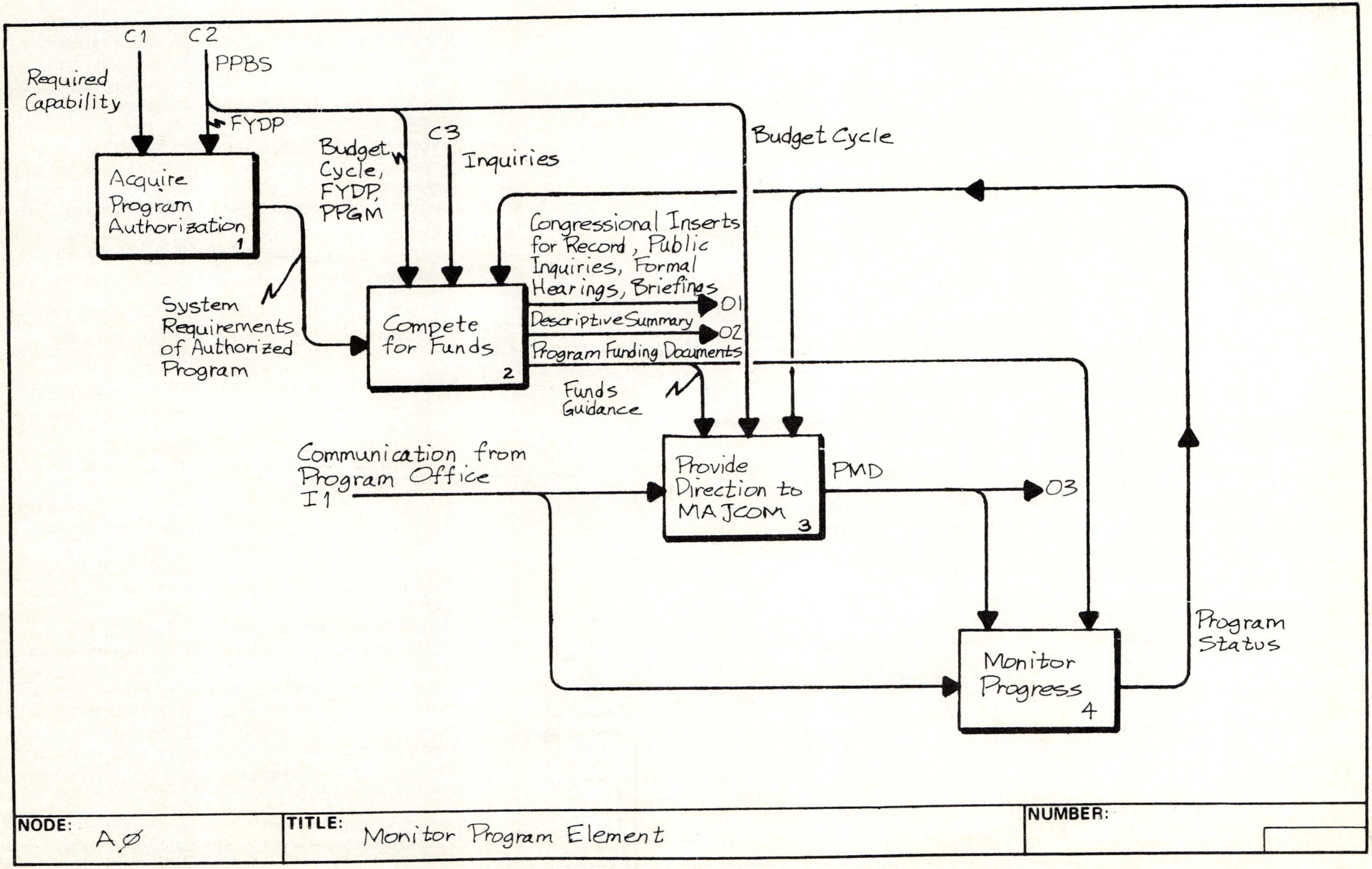
C1
C2
Required Capability
PPBS
FYDP
Acquire Program Authorization 1
Budget Cycle, FYDP, PPGM
C3
Inquiries
Budget Cycle
System Requirements of Authorized Program
Compete for Funds 2
Congressional Inserts for Record, Public Inquiries, Formal Hearings, Briefings
O1
Descriptive Summary
O2
Program Funding Documents
Funds Guidance
Communication from Program Office
I1
Provide Direction to MAJCOM 3
PMD
O3
Monitor Progress 4
Program Status
NODE: AØ
TITLE: Monitor Program Element
NUMBER:

A1 Acquire Program Authorization

A required capability can be received by the Directorate of Operational Requirements (AF/RDQ) as a ROC, or generated through the initiative of a member of the Air Staff. A division within AF/RD is designated as the focal point, with an action officer or PEM responsible for preparing a recommendation for or against validation. The recommendation is coordinated with Air Staff directorates and major commands (box 1 mechanisms 1 and 3). The recommendation (box 1 output 2) includes Budgetary Cost Information (BCI) provided by the Air Force Systems Command (AFSC) and the Air Force Logistics Command (AFLC) to support the recommendation.

The recommendation is then presented to the Requirements Review Group (RRG), which includes the following members:

AF/RDQ	Director of Operational Requirements, Chairman
AF/RDP	Director of Development and Acquisition
AF/RDR	Director of Reconnaisance and Electronic Warfare
AF/XOO	Director of Operations
AF/XOX	Director of Plans
AF/LGY	Director of Maintenance and Engineering
AF/PRP	Director of Programs

The RRG reviews the recommendation (box 2) and determines whether to validate it or not. It may not be validated for a number of reasons: the requirement may not be cost effective, or the requirement is unsubstantiated. If the program is validated the requirements of the system must be obtained. This requires the cooperation of the Air Staff, PEM, operating, developing and supporting commands.

If the program is small enough or has high level interest, then it may bypass validation by the RRG. In this case, it is said to have functional validation (box 1 output 1).

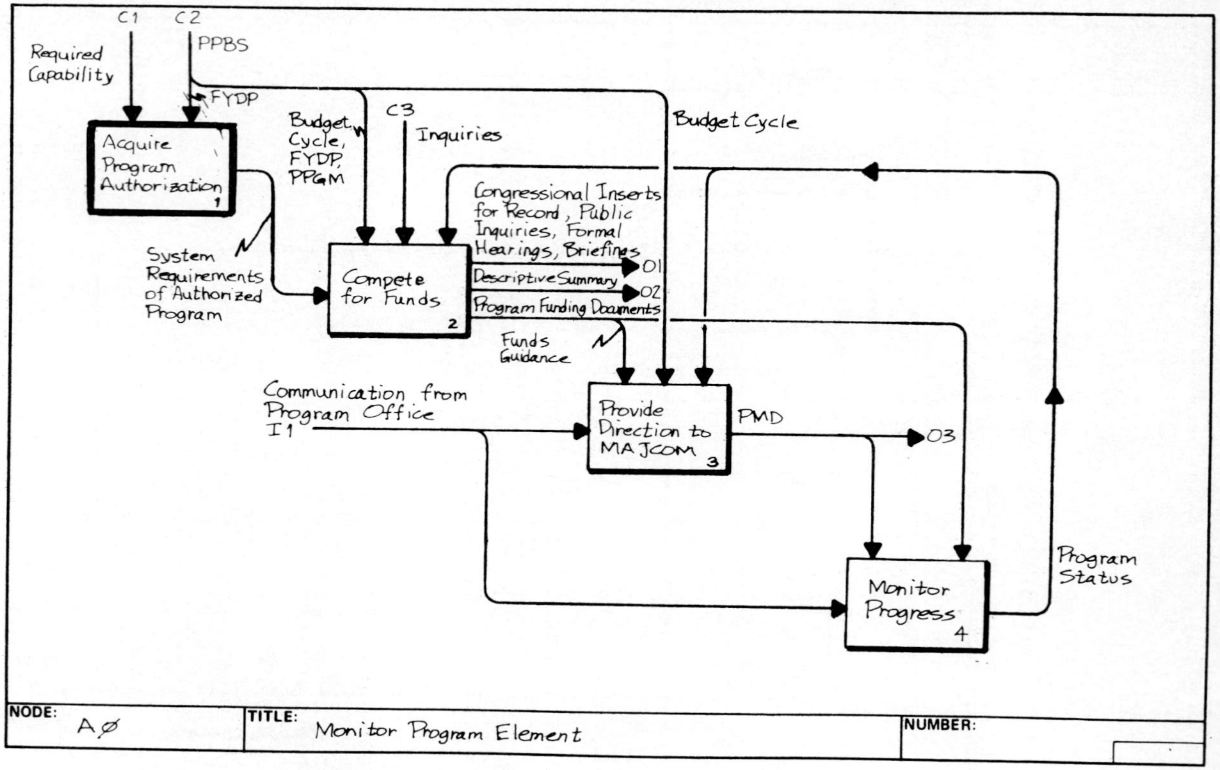

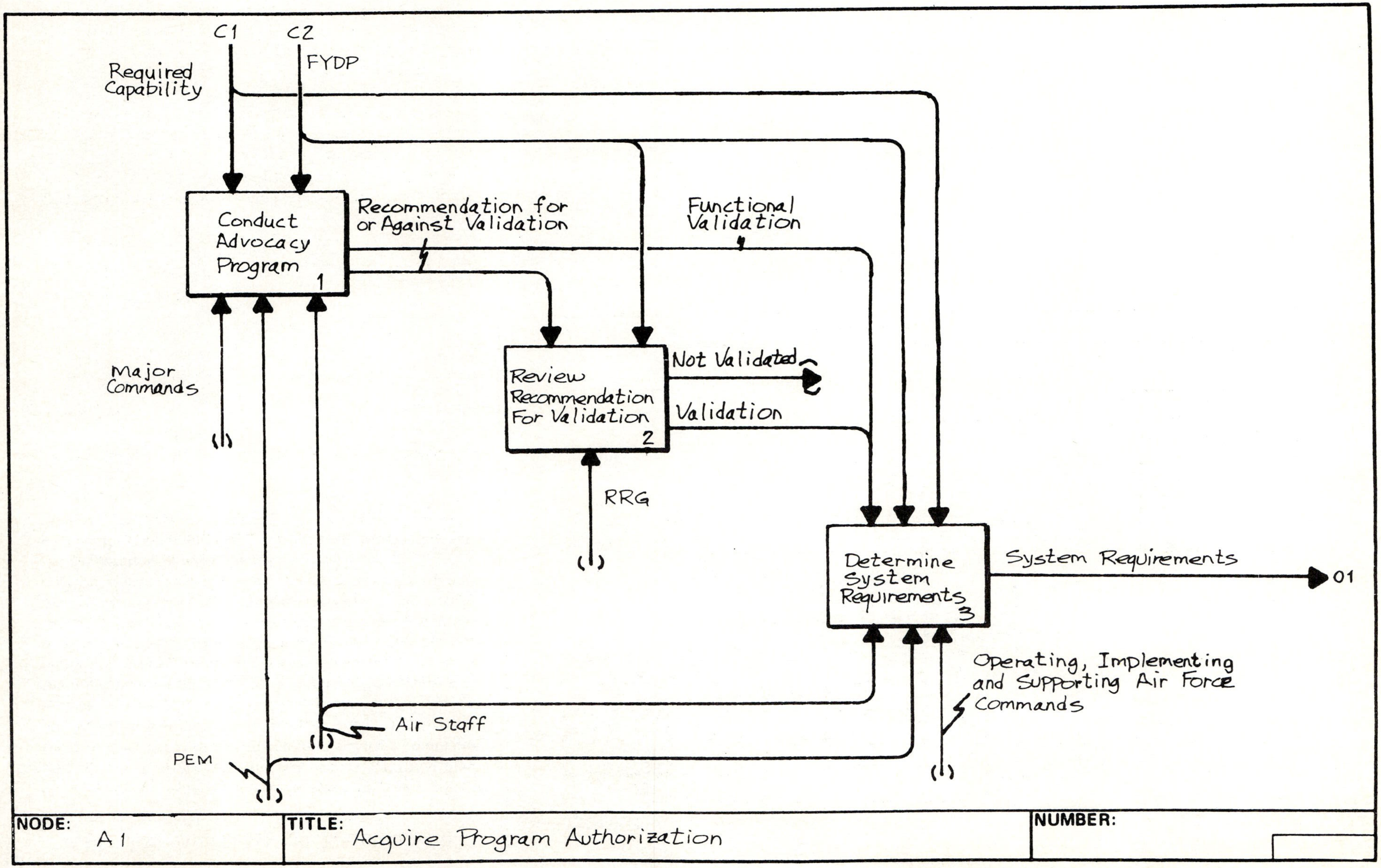
C1
C2
Required Capability
FYDP
Conduct Advocacy Program 1
Recommendation for or Against Validation
Functional Validation
Review Recommendation For Validation 2
Not Validated
Validation
Major Commands
RRG
Determine System Requirements 3
System Requirements
01
Operating, Implementing and Supporting Air Force Commands
Air Staff
PEM
NODE: A1
TITLE: Acquire Program Authorization
NUMBER:

A13 Determine System Requirements

Once the program is validated, the PEM or action officer prepares a PMD specifying the requirements of the system (box 1). This PMD is sent to the implementing command for further action (box 1 output 1). In the case of a small system, no action is required. For a Class V modification, the implementing command is directed to prepare cost and feasibility information in the form of a modification proposal and analysis (MPA). This is coordinated through the various affected commands and Air Staff. The procedure for the acquisition of a major system involves the selection of a proposal (box 3 input 1) for the new defense system and the preparation of a Decision Coordinating Paper (DCP). The DCP provides summary management information and program rationale.

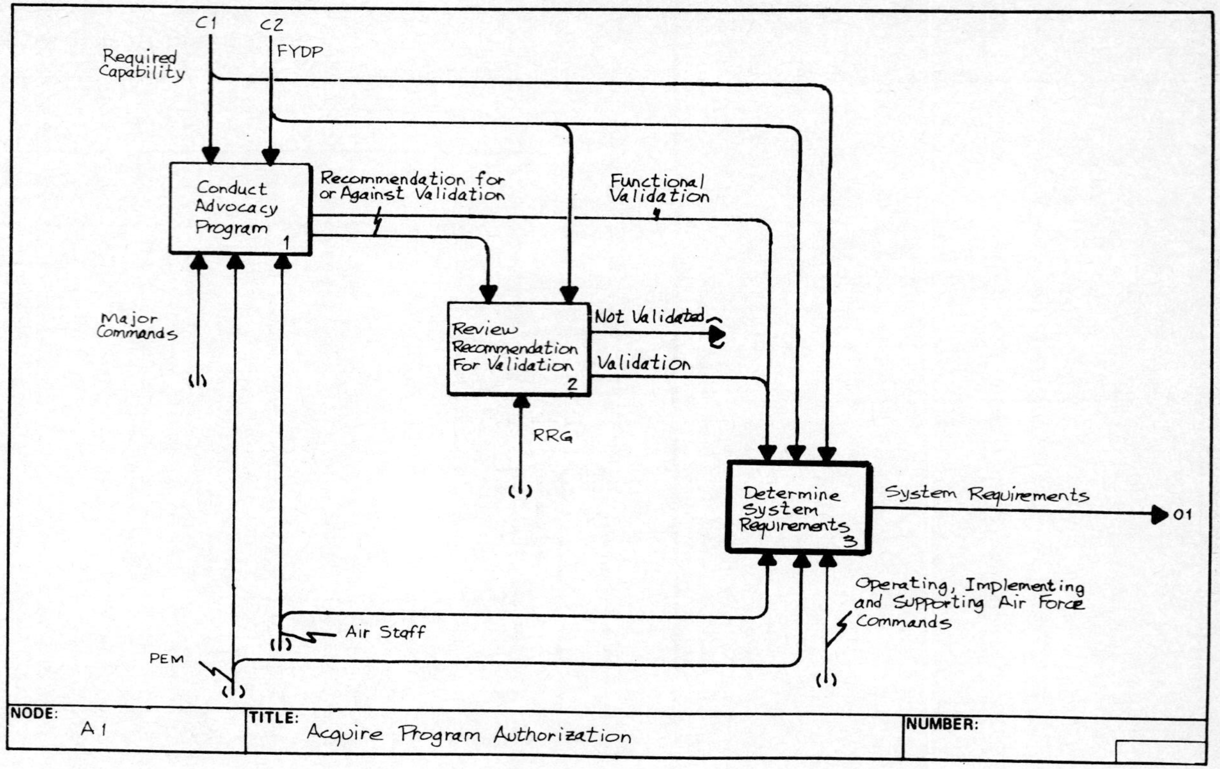

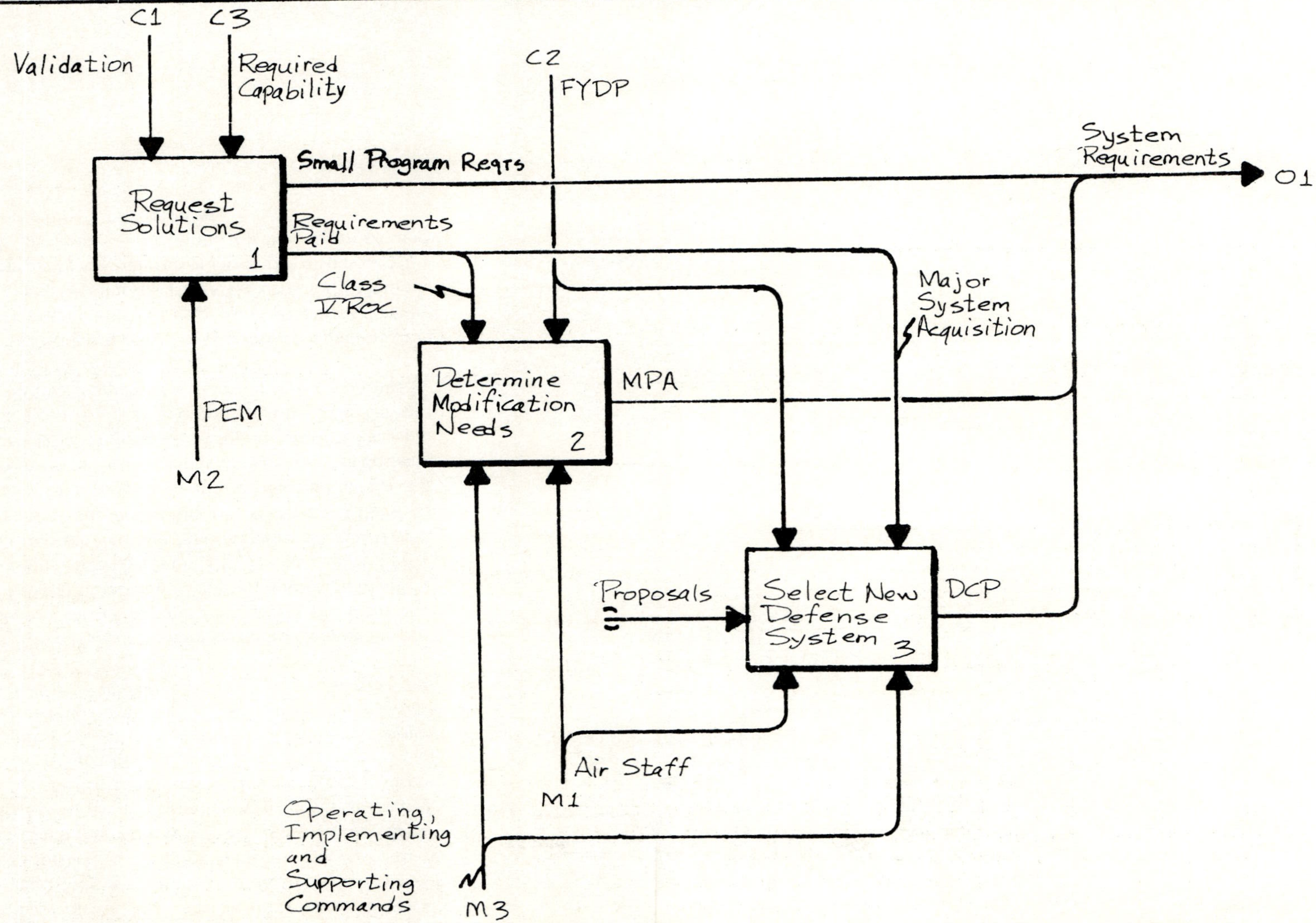

NODE: A13 | TITLE: Determine System Requirements | NUMBER:

A2 Compete for Funds

The first phase of the funds competition occurs between November and May when the Air Force budget is developed in the form of a Program Objective Memorandum (POM). Through interactions between the Air Force directors and PEMs an Air Force budget is determined (box 1) Programs may be denied funding at this point (box 1 output 1). The POM is submitted to the Office of the Secretary of Defense (OSD) where it is further reviewed and budget guidance issued (box 2 output 1). *The agencies and individuals involved in these activities are detailed on the next diagram.*

The Air Force Budget for submission to Congress is prepared based on the POM and guidance in the form of a Tentative Program Decision Memorandum (TPDM), Program Decision Memorandum (PDM) pair, or a PDM, Amended Program Decision Memorandum (APDM) pair. It may be either one depending on whether a TPDM or PDM is issued first. A program may be denied funds at this point in the process (box 3 output 2). The funding decision will be received in the form of a Program Budget Decision (PBD) (box 4 control 1). In addition, the Five Year Defense Program (FYDP) is updated at this point (box 3 output 1), and other funding documents are received.

Finally, the program must be described to Congress during their examination of the budget (box 4). This is done by providing Congressional Inserts for Records and Public Inquiries as responses to inquiries, a Descriptive Summary for an overall program description, formal hearings and briefings to members of Congress.

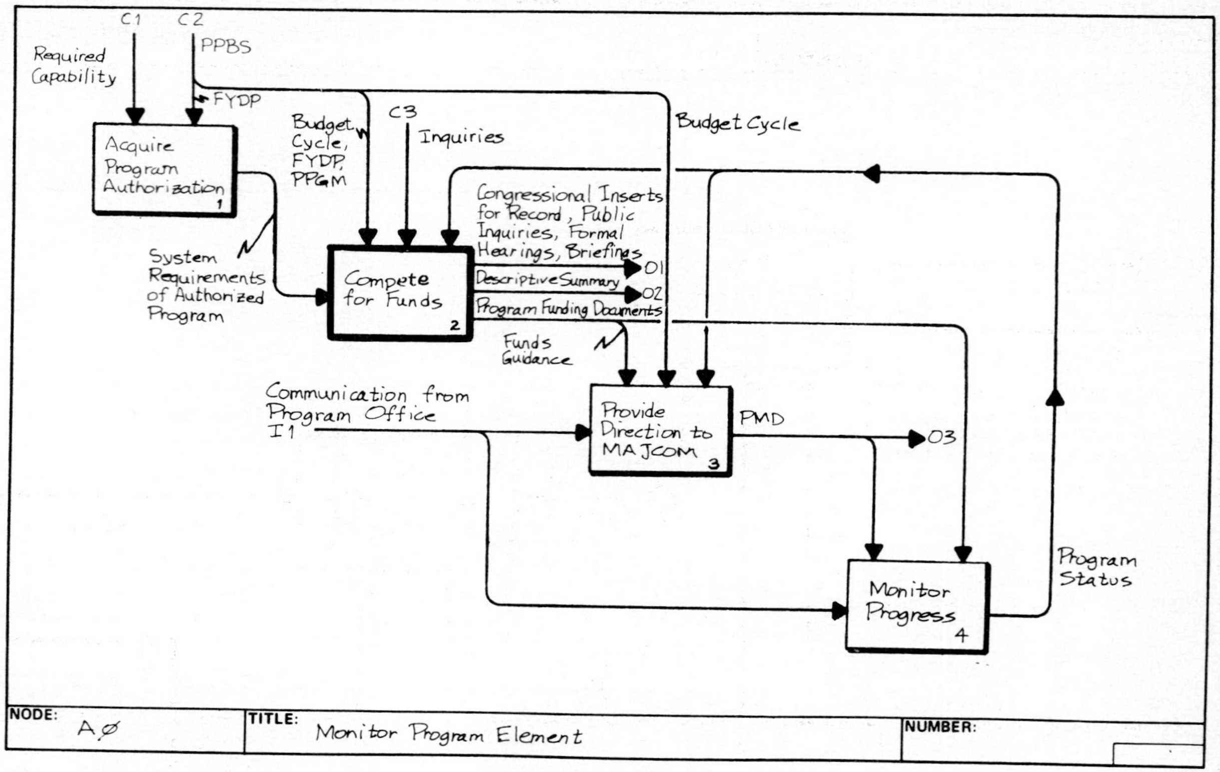

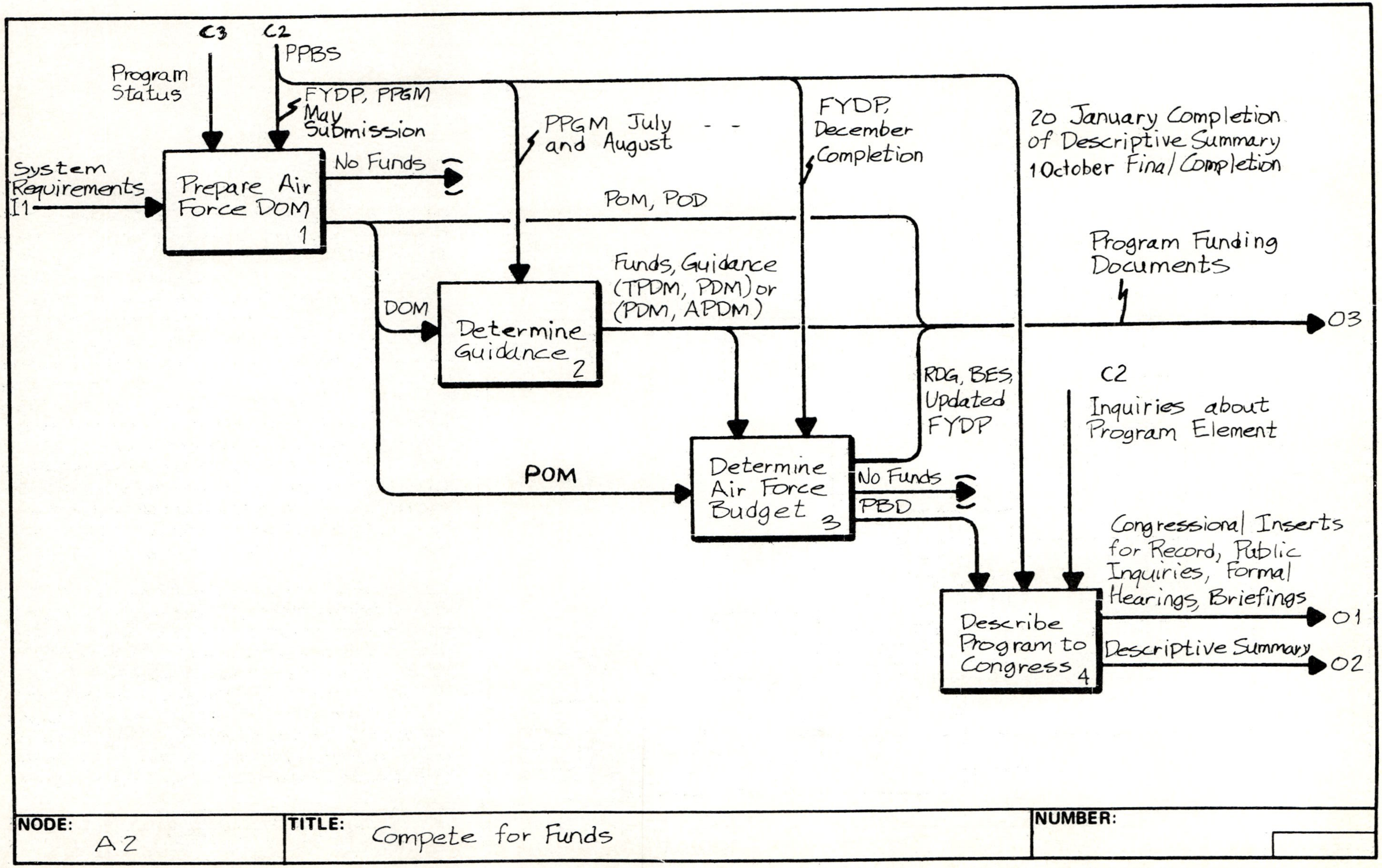
C3
C2
PPBS
Program Status
System Requirements
I1
FYDP, PPGM May Submission
No Funds
Prepare Air Force DOM 1
PPGM July and August
POM, POD
FYDP, December Completion
20 January Completion of Descriptive Summary 10ctober Final Completion
DOM
Determine Guidance 2
Funds, Guidance (TPDM, PDM) or (PDM, APDM)
Program Funding Documents
O3
RDG, BES, Updated FYDP
C2
Inquiries about Program Element
POM
Determine Air Force Budget 3
No Funds
PBD
Congressional Inserts for Record, Public Inquiries, Formal Hearings, Briefings
O1
Describe Program to Congress 4
Descriptive Summary
O2
NODE: A2
TITLE: Compete for Funds
NUMBER:

A21 Prepare Air Force Program Objective Memorandum (POM)

In November (box 1 control 2), the PEM begins preparation of the Program Objective Directive (POD) to support the Air Force directors in preparing the initial version of the budget. The POD is reviewed by the Air Force directors and the Directorate of Planning, Programming and Analysis (AF/RDX) (box 2) until it meets their satisfaction. Then either the program element is denied funds (box 2 output 1) or the POD is used (box 3 input 1) to help prepare the Program Objective Memorandum (box 3). The POM specifies resource requirements within fiscal guidelines.

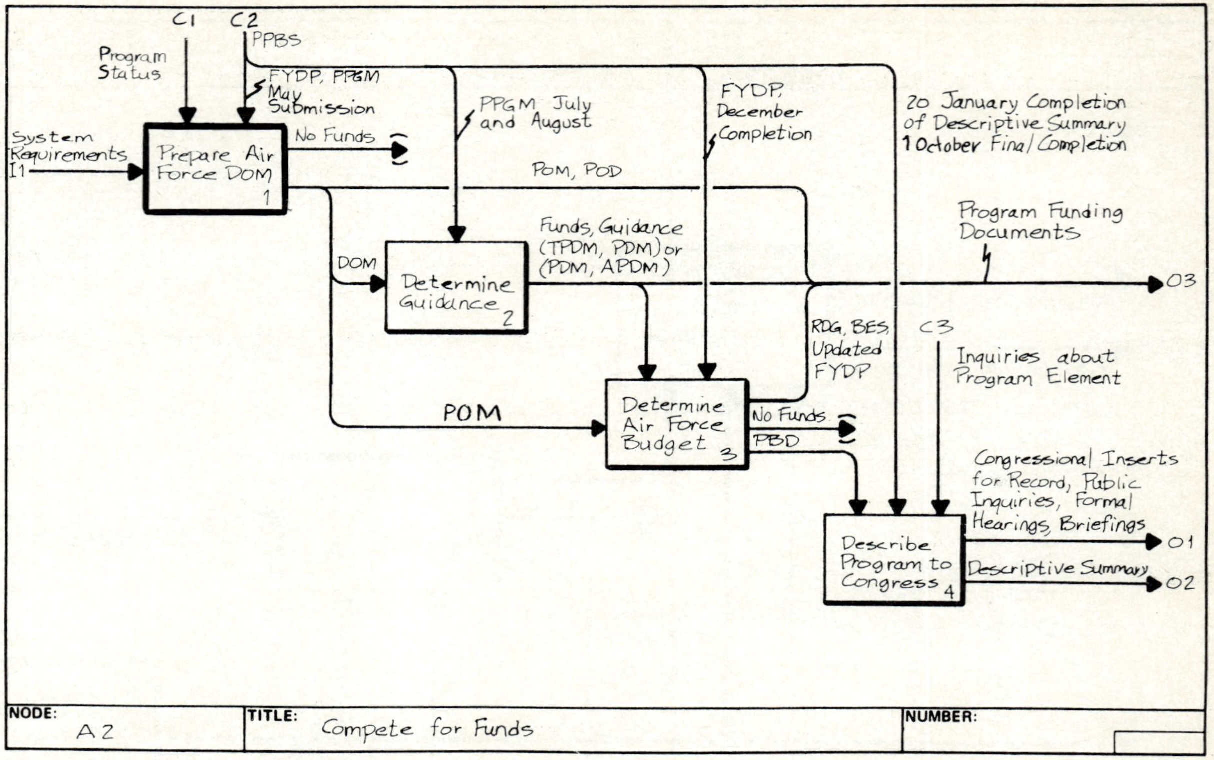

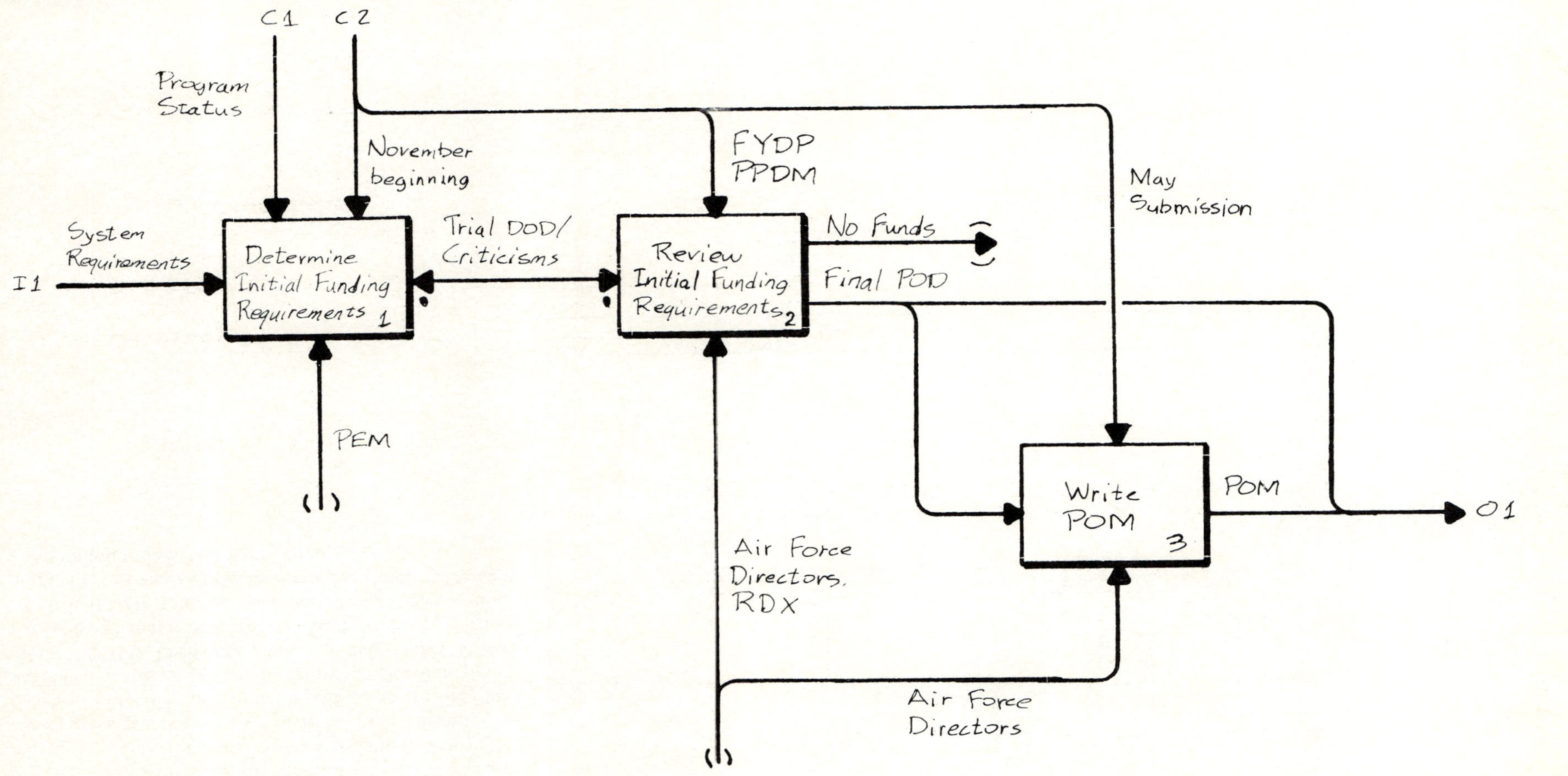

NODE: A21 | TITLE: Prepare Air Force Program Objective Memorandum (POM) | NUMBER:

A22 Determine Guidance

Once the Program Objective Memorandum (POM) is prepared by the Air Force, OSD reviews it under the constraints imposed by the Planning-Programming Guidance Memorandum (PPGM). OSD prepares issue papers analyzing the POM and these are submitted to the Air Force where the PEM responds through reclamas. The POM, issue papers and reclamas are reviewed by the Secretary of Defense (SECDEF) and initial modifications to the POM are issued in a series of Tentative Program Decision Memoranda (TPDM). Then final guidance is determined through resolution of issues between the Air Force, supported by PEMs and OSD. Guidance is reported in a PDM, or if the original report was a PDM, an Amended Program Decision Memorandum (APDM) is prepared.

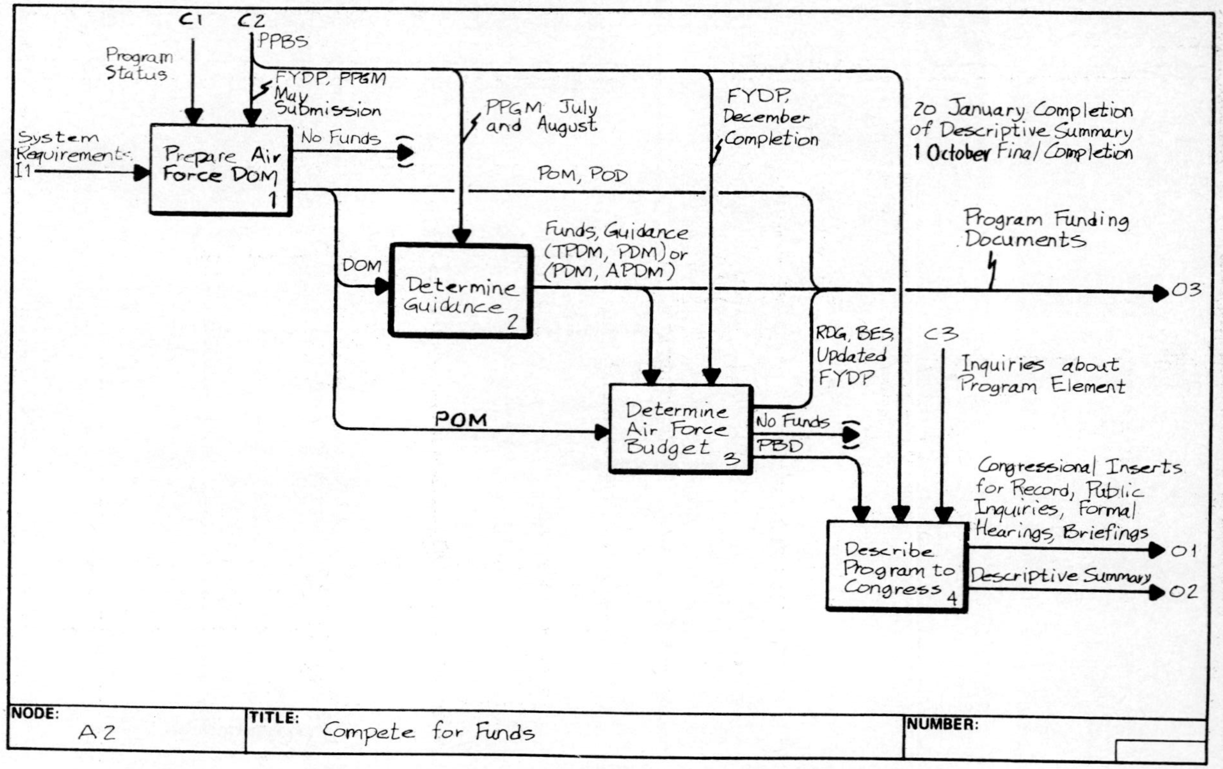

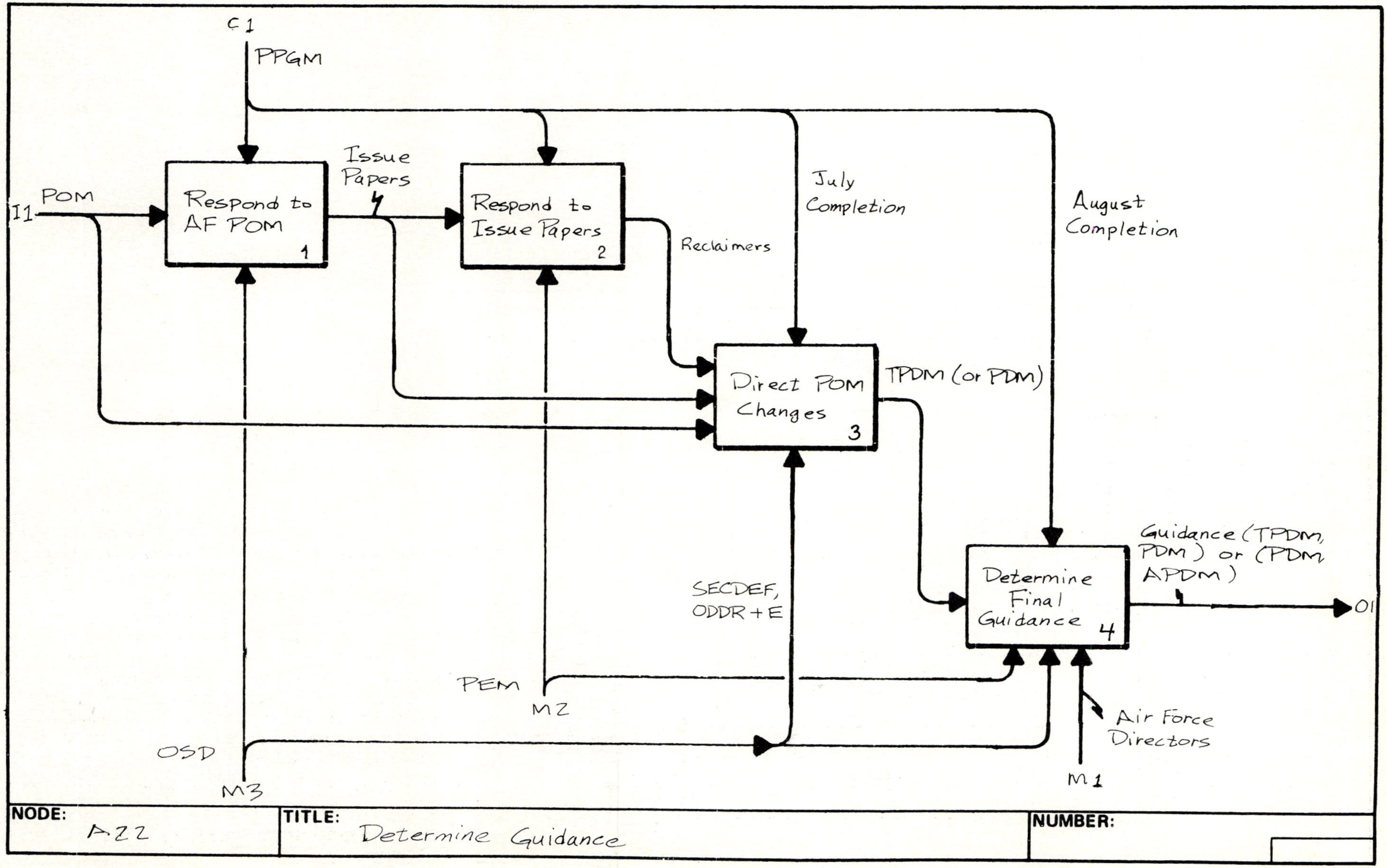
C1
PPGM
I1
POM
Respond to AF POM 1
Issue Papers
Respond to Issue Papers 2
Reclaimers
July Completion
August Completion
Direct POM Changes 3
TPDM (or PDM)
Determine Final Guidance 4
Guidance (TPDM, PDM) or (PDM, APDM)
O1
SECDEF, ODDR + E
PEM
M2
OSD
M3
Air Force Directors
M1
NODE: A22
TITLE: Determine Guidance
NUMBER:

A224 Determine Final Guidance

The PEM reviews the TPDM and comments on the changes to the POM. OSD then identifies remaining issues yet to be resolved. These issues are resolved by the end of August in meetings held between OSD and Air Force directors. Then the PDMs (or APDMs if a PDM was originally written) are written to provide guidance and establish program decisions.

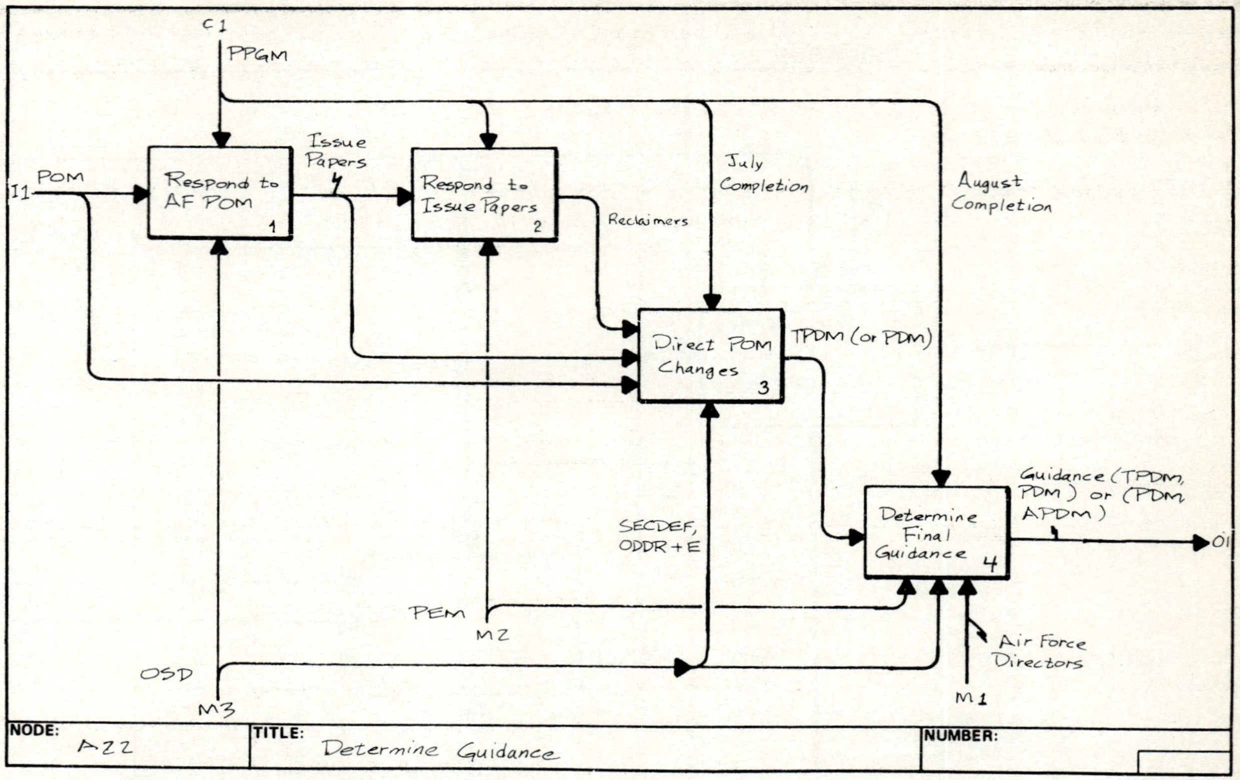

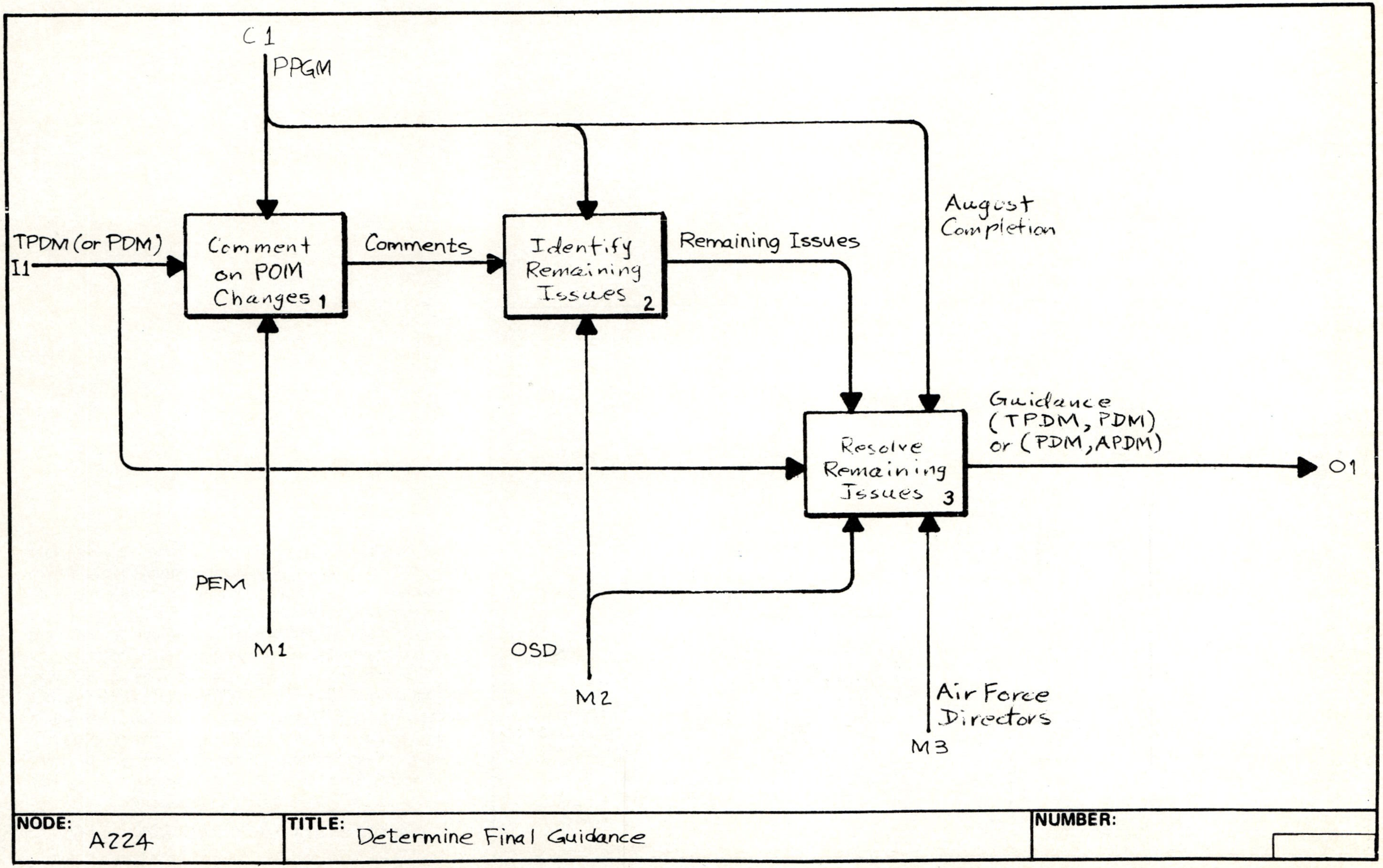

NODE: A224 | TITLE: Determine Final Guidance | NUMBER:

A23 Determine Air Force Budget

The final Air Force budget submitted to the Office of the Secretary of Defense (OSD) is the Budget Estimate Submission (BES) (box 2 output 1). It is based on an RD5 report that is prepared by the PEM (box 1). The RD5 report contains budget information for five years with significant detail for the first two years (box 1 output 1).

The budget is reviewed (box 3) by OSD and the Office of Management and Budget (OMB) and the final budget is completed in December for inclusion in the President's budget. The FYDP is updated in the process (box 3 output 2). Changes to the budget are specified in a series of documents called Program Budget Decisions (PBDs) (box 3 output 3).

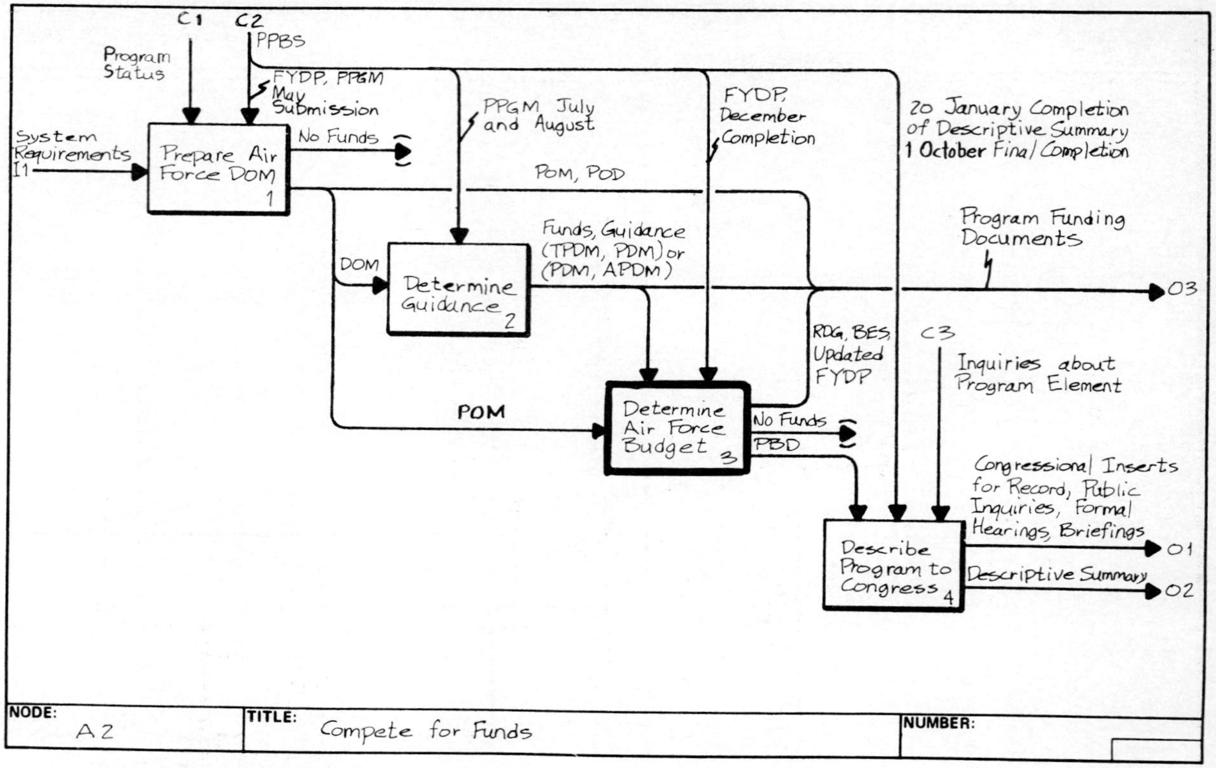

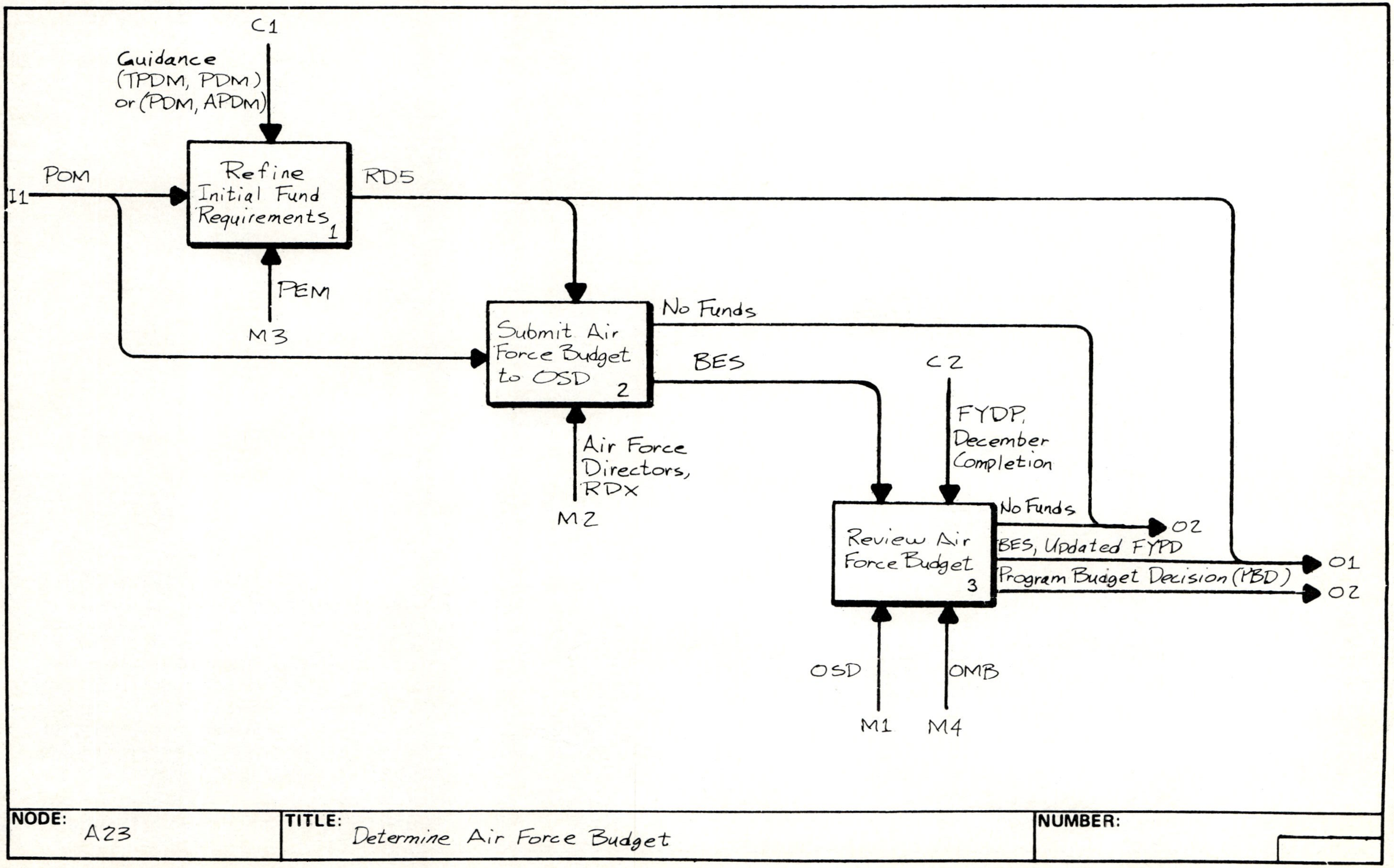

NODE: A23 | TITLE: Determine Air Force Budget | NUMBER:

A24 Describe Program to Congress

It is the responsibility of the Deputy Chief of Staff for Research and Development (DCS/R+D) to advocate the program to Congress (box 3). Support is provided by the PEM (box 1 mechanism 1) who prepares a Program Element Review (PER) (box 1 output 1). He also writes a Descriptive Summary (box 2 output 1) which provides a description of the program to Congress. The Descriptive Summary, along with other information received (box 3 output 1), provoke inquiries (box control 1) from Congress about the program element.

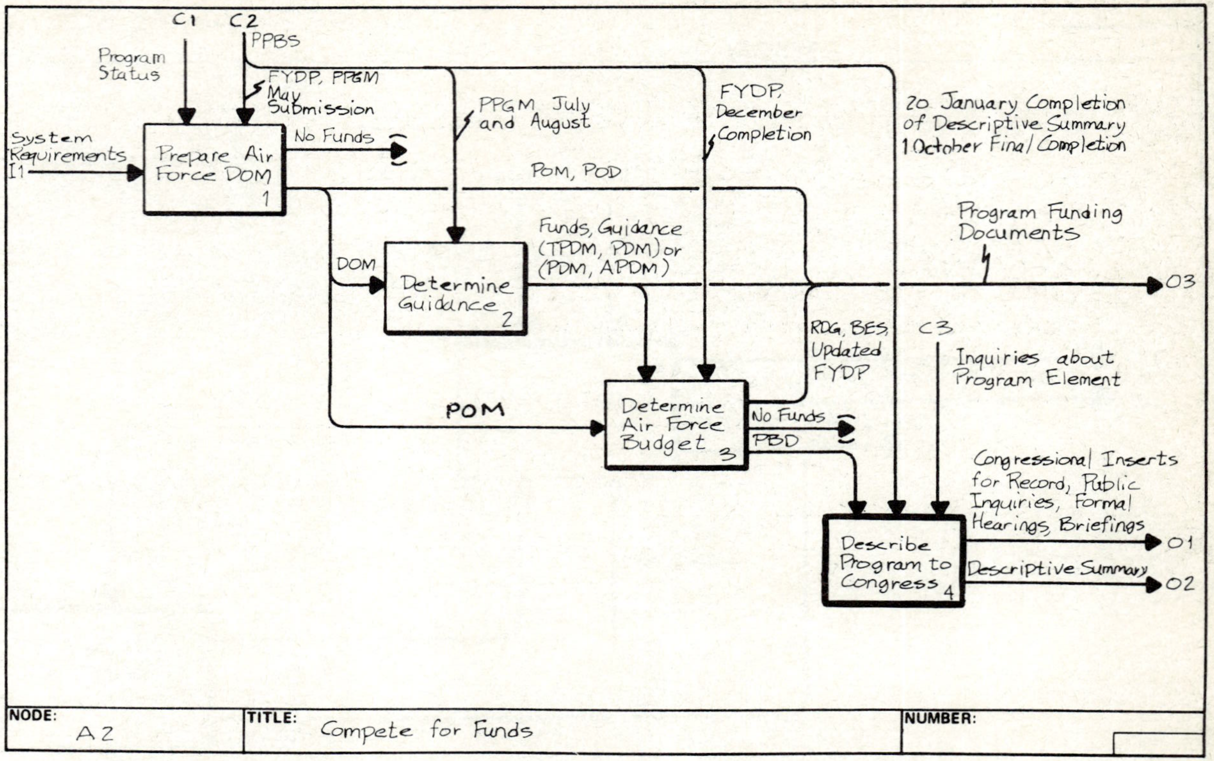

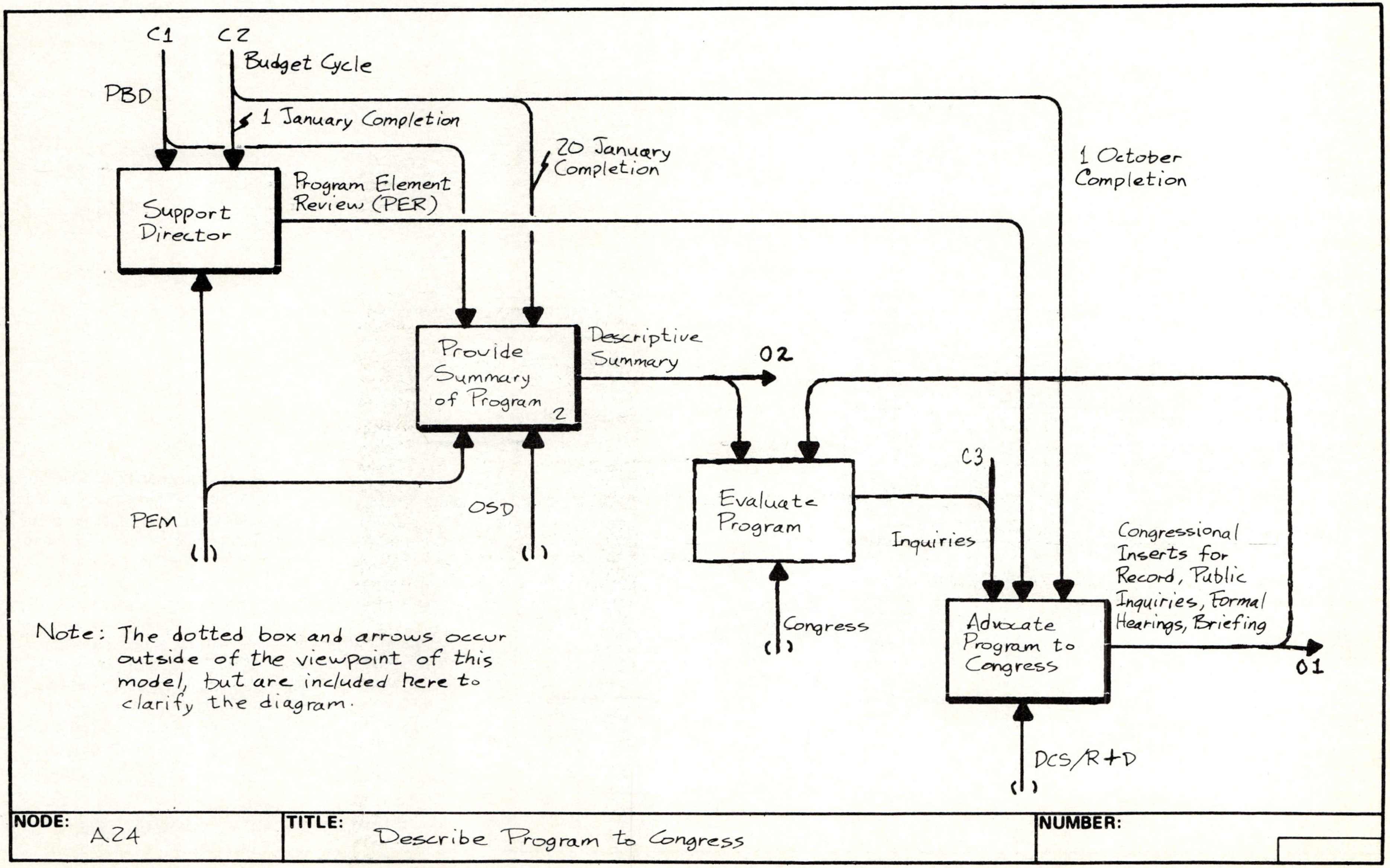
C1
C2
Budget Cycle
PBD
1 January Completion
20 January Completion
1 October Completion
Support Director
Program Element Review (PER)
Provide Summary of Program
2
Descriptive Summary
O2
Evaluate Program
C3
Inquiries
PEM
OSD
Congress
Advocate Program to Congress
Congressional Inserts for Record, Public Inquiries, Formal Hearings, Briefing
O1
DCS/R+D
Note: The dotted box and arrows occur outside of the viewpoint of this model, but are included here to clarify the diagram.
NODE: A24
TITLE: Describe Program to Congress
NUMBER:

A3 Provide Direction to Major Command

The Program Management Directive (PMD) is a primary document prepared by the PEM (box 3). It provides direction to the major commands involved with the program. The direction provides funding guidelines and broad technical guidance. It is usually prepared each year by the first of October, but can be written whenever a change occurs that requires new direction (box 3 control 3).

To prepare the PMD, milestones and funding to achieve them are determined first (boxes 1 and 2). This is an iterative process requiring cooperation from the using command, program office and PEM. Information from the program office, such as the Program Management Plan and Procurement Plan, respond to the PMD and provide input for subsequent activities.

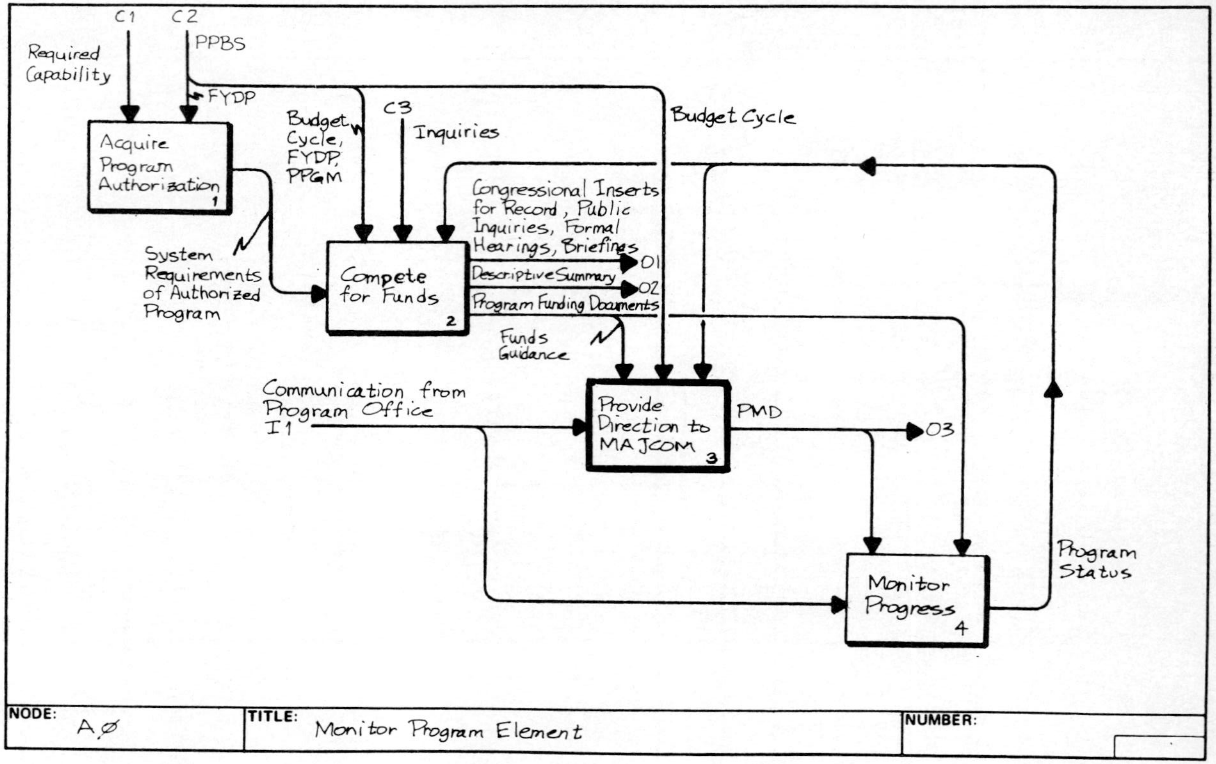

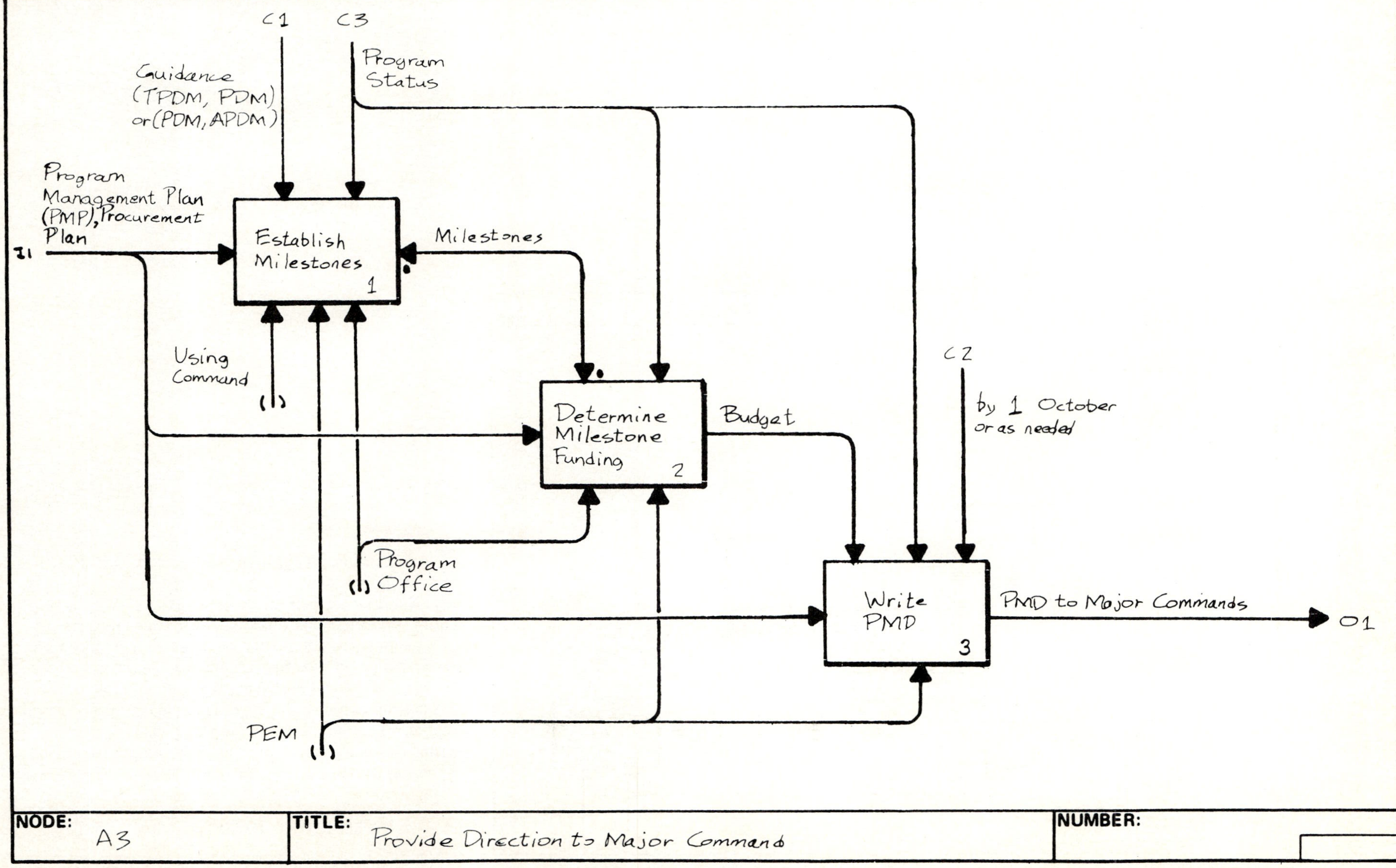

NODE: A3 **TITLE:** Provide Direction to Major Command **NUMBER:**

A4 Monitor Progress

The program will be in one of two states at all times. The procurement phase (box 2) occurs before a contract is let to perform the work. The funding, schedule, PMD requirements, D+F, Procurement Plan and Source Selection Plan are all required to monitor the procurement. Once the procurement is let the program is in the contract phase (box 3). The program status can be determined by evaluating the cost, schedule and technical performance progress to date (box 4). These activities are supported by a variety of reports and briefings prepared by the program office. These will be prepared as often as the PEM desires.

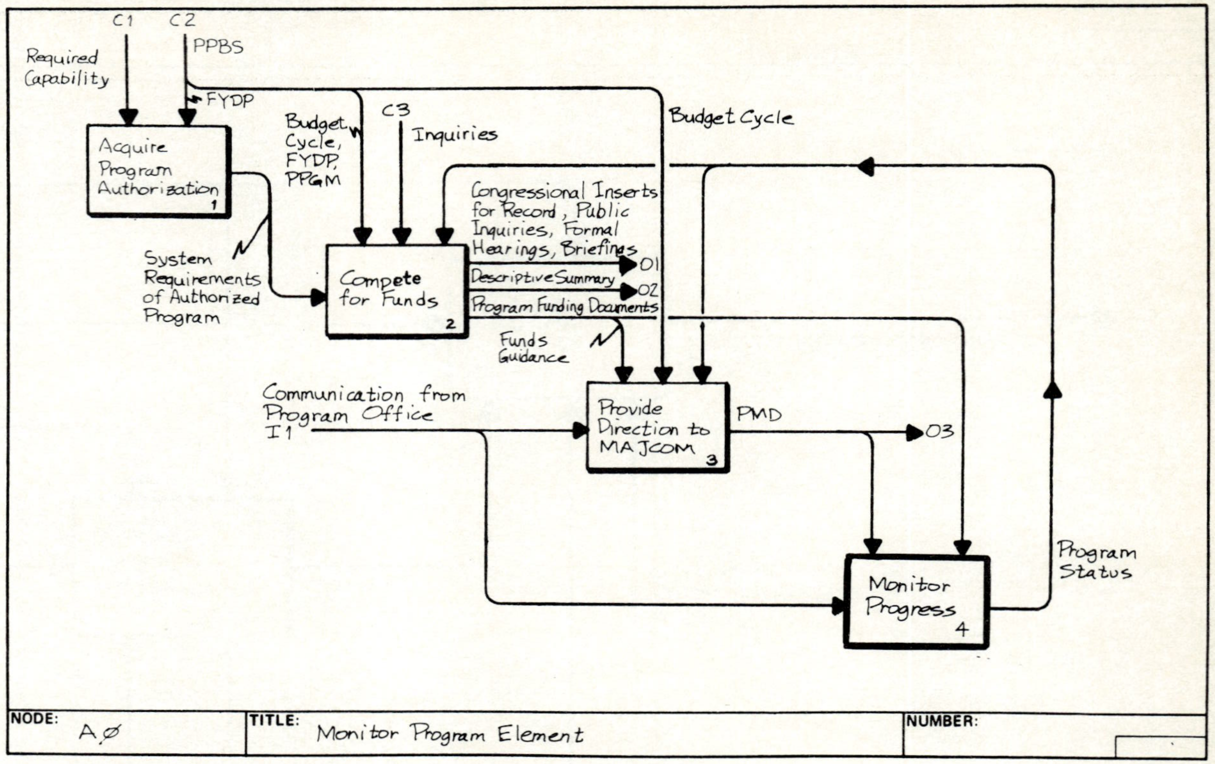

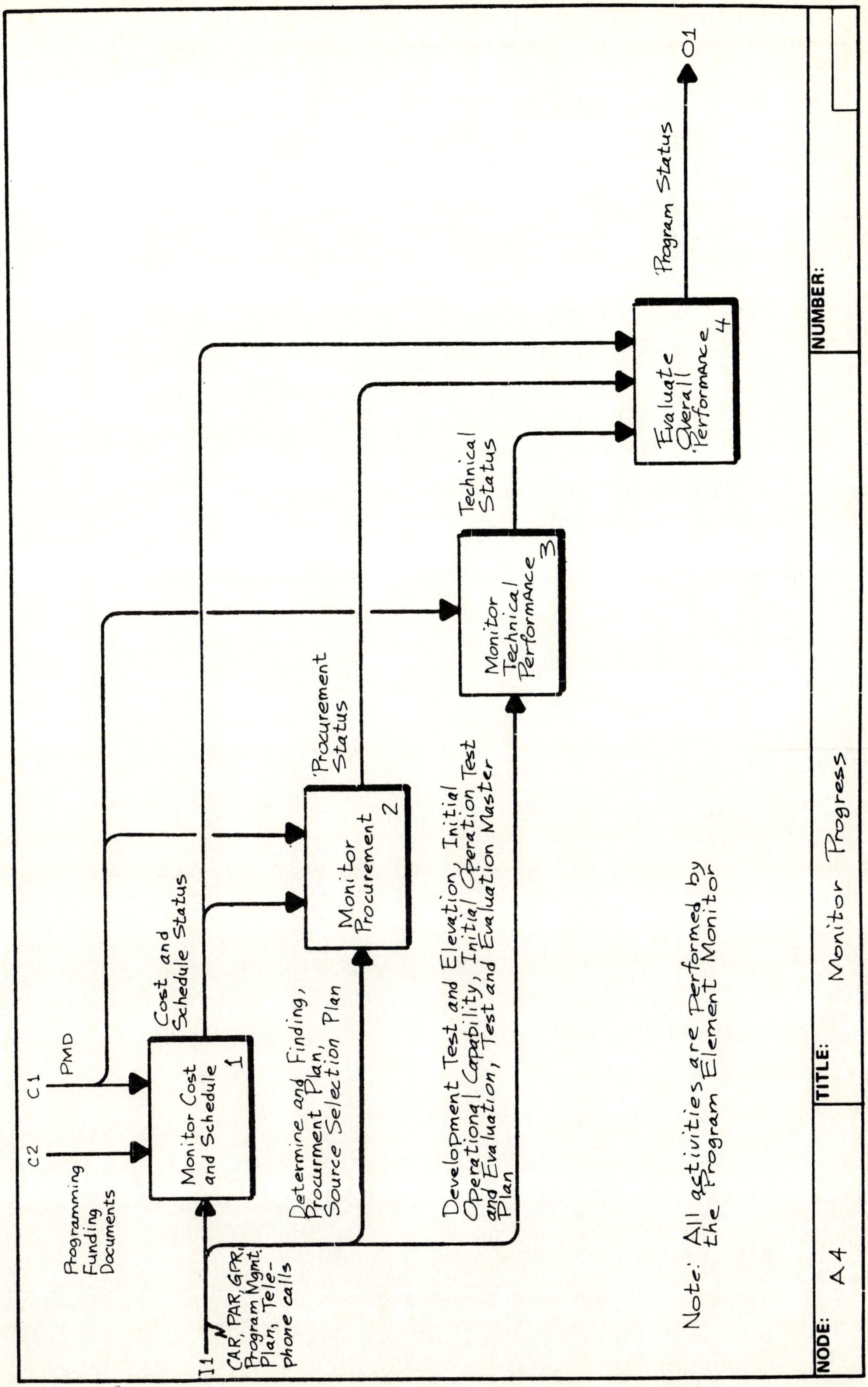

Programming Funding Documents
C2
C1
PMD
Monitor Cost and Schedule 1
Cost and Schedule Status
I1
CAR, PAR, GPR, Program Mgmt. Plan, Tele-phone calls
Determine and Finding, Procurment Plan, Source Selection Plan
Monitor Procurement 2
Procurement Status
Development Test and Elevation, Initial Operational Capability, Initial Operation Test and Evaluation, Test and Evaluation Master Plan
Monitor Technical Performance 3
Technical Status
Evaluate Overall Performance 4
Program Status
O1
Note: All activities are performed by the Program Element Monitor
NODE: A 4
TITLE: Monitor Progress
NUMBER:

Diagram Index for the Monitor Program Element Model

- **A-0** Monitor Program Element
  - **A0** Monitor Program Element
    - **A1** Acquire Program Authorization
      - **A13** Determine System Requirements
    - **A2** Compete for Funds
      - **A21** Prepare Air Force Program Objective Memorandum
      - **A22** Determine Guidance
        - **A224** Determine Final Guidance
      - **A23** Determine Air Force Budget
      - **A24** Describe Program to Congress
    - **A3** Provide Direction to Major Command
    - **A4** Monitor Progress

Chapter 29

Aerospace Manufacturing

29.1 Application Domain

Aerospace manufacturing is concerned with the efficient mass production of aircraft for military use. Many large American companies exist solely to propose, design, and built aircraft for the U.S. Air Force. There are also many other firms which have a large proportion of their business in the aerospace manufacturing arena. Coordinating the efforts of these diverse manufacturers was becoming more and more difficult, especially in the late 1960s and early 1970s. So, the U.S. Air Force decided to use existing computer technology to ease this burden.

29.2 Project Background

The U.S. Air Force Integrated Computer-Aided Manufacturing (ICAM) Program was established to apply computer technology to boost aerospace manufacturing effectiveness. Central to the ICAM effort was the development of an architecture for manufacturing. This architecture was designed to provide the basis for automated support software systems for the batch manufacturing process, with initial emphasis on sheet metal fabrication and subassembly. The goal was to define a hierarchical structure of the entire process, including machines, materials, procedures, people, and organizations. Once this was accomplished, the architecture could then be used to specify integrated modernization and computerization projects. This goal set by the ICAM project is indeed the largest SADT modeling effort ever undertaken.

29.3 Project Overview

The modeling part of this ICAM project used SADT to capture the manufacturing experience of all aerospace and industrial participants. Multiple models of different aspects of manufacturing were developed based on extensive interviews with experts at the participating companies and also based on documents provided by them. Extensive review and rework of these models led to the synthesis of a composite manufacturing architecture model.

29.4 Resources Used to Make the Model

A core modeling team comprising eight authors and many support staff members produced the intermediate multiple models and final composite model. Over 70 experts from over 15 aerospace companies and the U.S. Air Force were interviewed. These people participated in various reviews of kits, models, and project status. The whole effort took 5 full years to complete.

29.5 Notes about the Model

The goal of describing the current sheet metal fabrication technology required a sequential decomposition strategy. This made easier the job of describing existing manufacturing processes having many mutually constraining activities. Especially noteworthy is the high quality of the diagrams for a model of this size. The complete ICAM model contains around 90 diagrams and 250 pages of written specifications. Almost all the model is decomposed down to three levels, and some of it is decomposed to six levels. For example, node A333433 describes how to set up a machine on the shop floor. Because of its size, we elected to present only the first two levels of the model.

A-0 Manufacture Product (Context)

Diagram A-0 is simply Box 3 of Diagram A-1, separated from all the rest of that diagram to emphasize its inputs, controls, and outputs. The monolithic content of this activity is expanded and expounded in the next diagram, A-0.

We might look upon A-0 as the trunk of a tree, and all the succeeding diagrams as its branches, unfolding to display the detailed complexity of aerospace manufacturing. The preceding diagrams (A-3, A-2, A-1) are the roots of the tree, establishing the complex environment in which manufacturing resides.

The primary input is "Procureable Items"—the things that are bought or given to the manufacturer (e.g., sheet metal or engines), to be incorporated in the output or used in its manufacture (drills). Implicit also is the sum total of previous knowledge which will be used in decision making.

The primary control is the product design, and the primary outputs are the product itself, support systems, parts and prototypes.

Other controls are the manufacturing requirements set forth in the original plan (see Diagram A-2), and the management directives and corresponding reports. To unclutter the subsequent diagrams, the "Directives/Reports" arrows are not drawn, but must be understood to be present. The parentheses at the arrow head show this implicit existence.

The primary output is the product and other manufactured items (parts, kits, prototypes). Other outputs are information useful to planning, field support, and design.

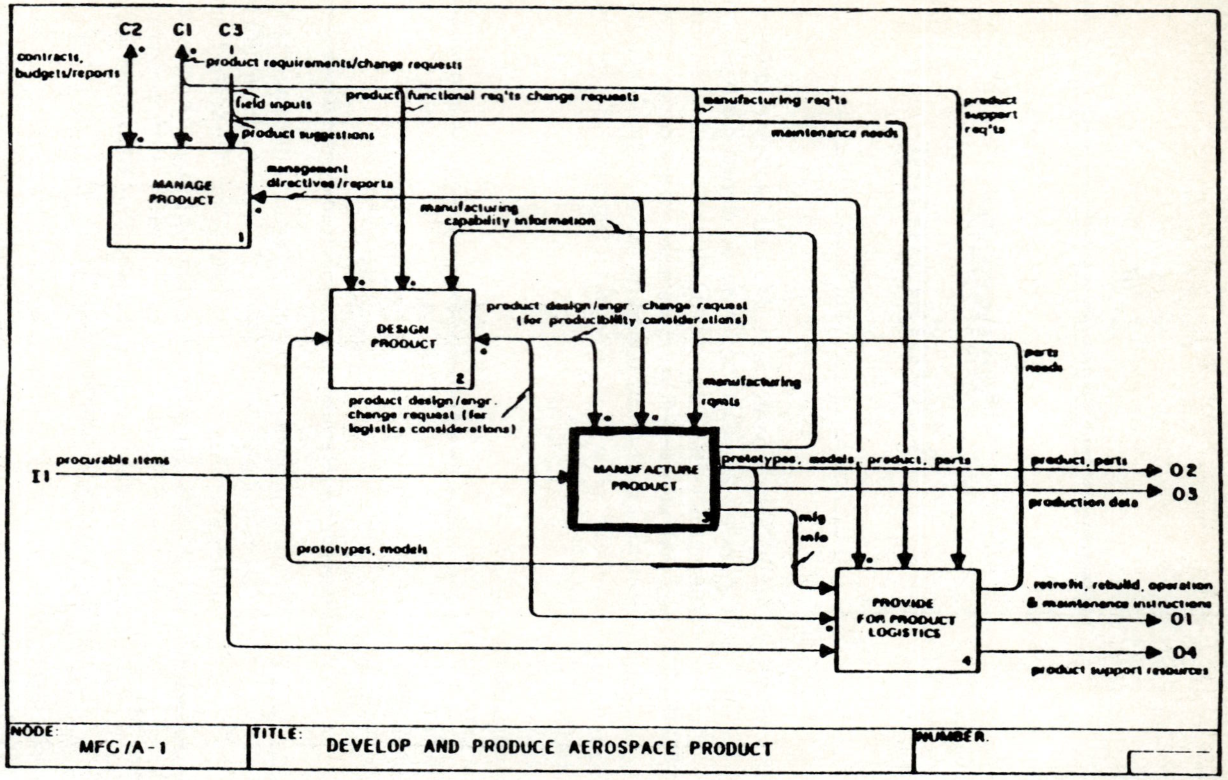

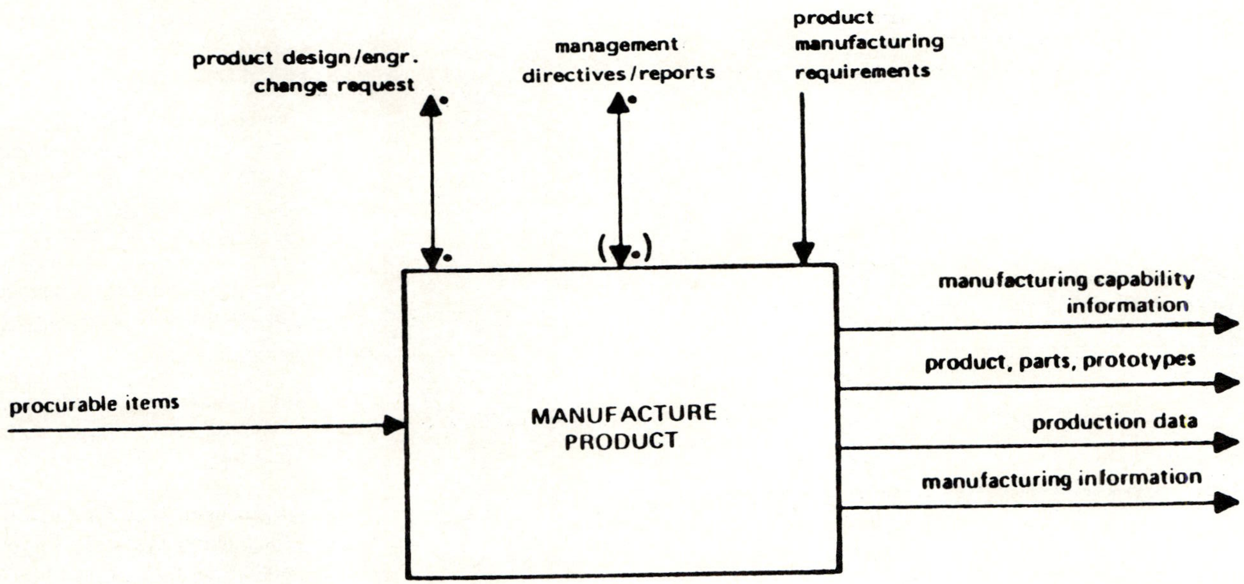

Purpose: To provide an industry wide acceptable architecture of aerospace manufacturing which encompasses all aspects of the manufacturing process.

Viewpoint: Aerospace manufacturer

NODE: MFG/A-0	TITLE: MANUFACTURE PRODUCT (CONTEXT)	NUMBER:

A0 Manufacture Product

A0 is the source diagram from which all subsequent diagrams in the architecture of aerospace manufacturing arise. It has the same inputs, controls, and outputs as does the one-block Diagram A-0.

There are two "planning" activities shown on A0; Box 1 "Plan for Manufacture" and Box 3 "Plan Production". The first relates to the strategy of producing the total product—the major subdivision of the vehicle structure, the basic method of manufacture, and the trade-offs to optimize facilities requirements, cost, and time schedules. The second relates to the strategy of producing the individual parts—the route sheets, operation sheets, and the list of machine tools, forms and cutters, fixtures and gauges required.

There are two "provisioning" activities shown on A0; Box 4 "Provide Production Resources"—facilities, equipment, tools, and people; and Box 5 "Obtain Manufacturing Materials"—those items which will ultimately be included in the delivered product.

There is an administrative activity; Box 2 "Make and Administer Schedules and Budgets". This activity produces those schedules and budgets which provide proper coordination between the separate activities of Boxes 3, 4, 5, and 6.

Finally there is "Produce Product". This is where the form and character of materials are altered and the pieces assembled. Because of the complexity of this diagram, four diagrams follow, each extracts and explains one flow of information or materials through those six activities.

NOTE: The four following diagrams, A0F1, A0F2, A0F3, and A0F4 present simplified versions which emphasize various aspects of this A0 diagram. The same activity boxes appear on all four diagrams.

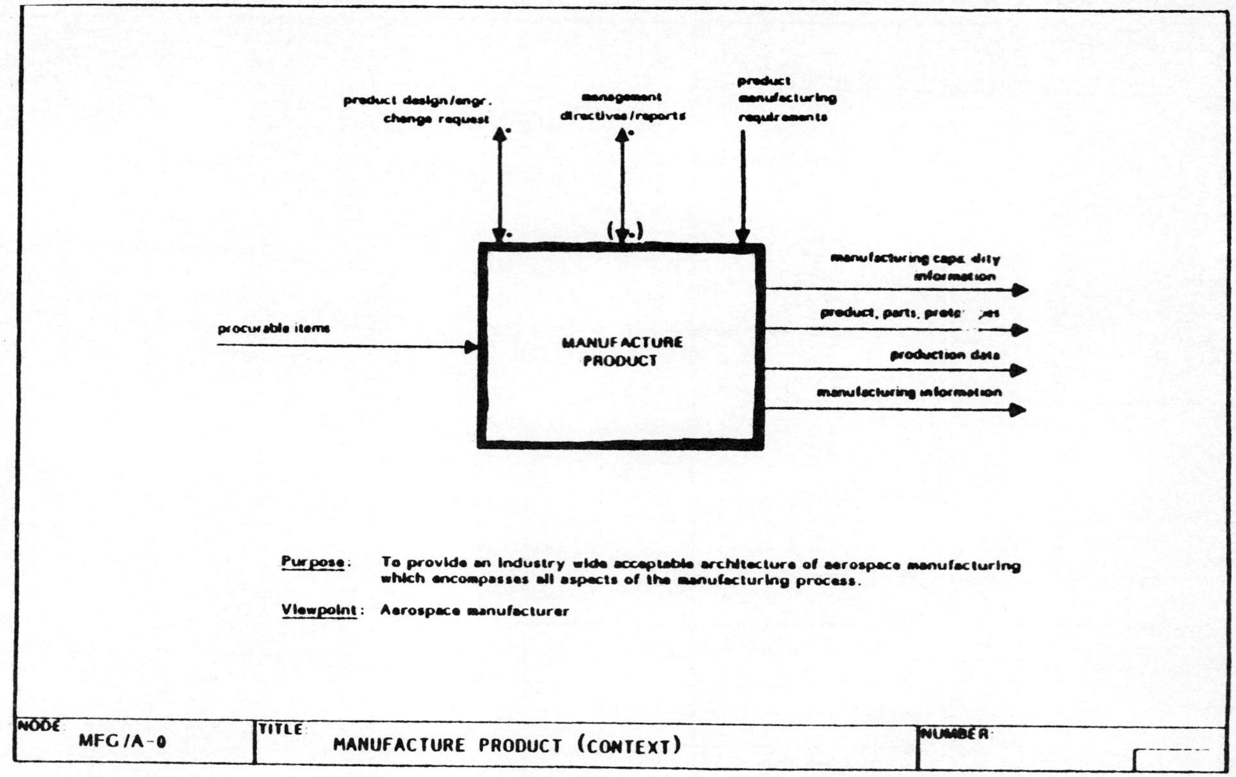

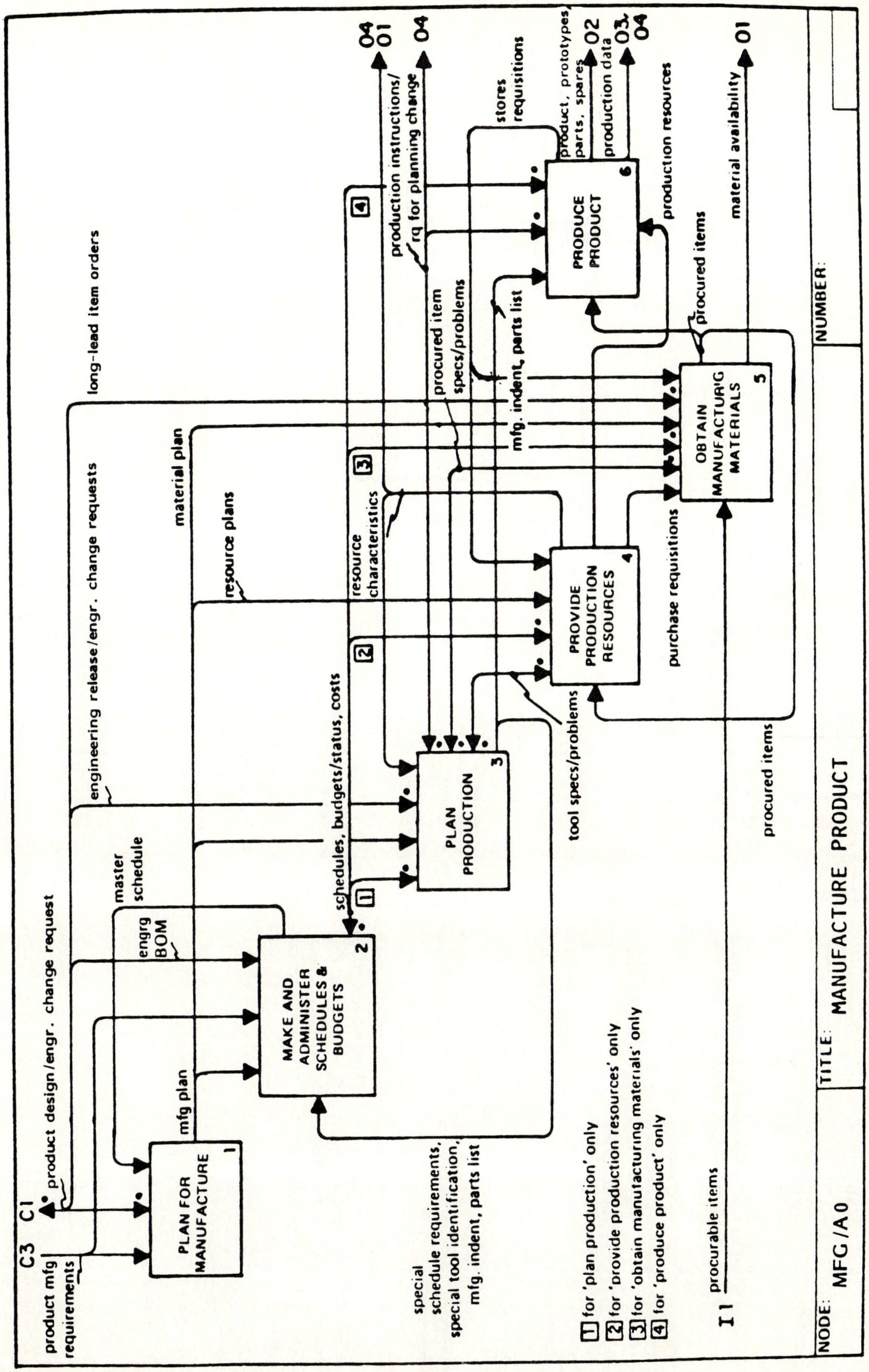
C3
C1
product mfg requirements
product design/engr. change request
PLAN FOR MANUFACTURE
1
mfg plan
engrg BOM
master schedule
MAKE AND ADMINISTER SCHEDULES & BUDGETS
2
schedules, budgets/status, costs
engineering release/engr. change requests
PLAN PRODUCTION
3
long-lead item orders
material plan
resource plans
resource characteristics
procured item specs/problems
mfg. indent, parts list
tool specs/problems
PROVIDE PRODUCTION RESOURCES
4
purchase requisitions
OBTAIN MANUFACTURING MATERIALS
5
procured items
production instructions/ rq for planning change
stores requisitions
PRODUCE PRODUCT
6
product, prototypes, parts, spares
production data
production resources
material availability
O4
O1
O2
O3
special schedule requirements, special tool identification, mfg. indent, parts list
1 for 'plan production' only
2 for 'provide production resources' only
3 for 'obtain manufacturing materials' only
4 for 'produce product' only
I1
procurable items
NODE: MFG/A0
TITLE: MANUFACTURE PRODUCT
NUMBER:

A1 Plan for Manufacture

The "Plan for Manufacture" is made as follows:

Beginning with the product design as released from the Design activity, and with the product manufacturing requirements established by the needs statements, and with the sum total of past experience and know-how and the capabilities of the manufacturer, a manufacturing plan is developed. This is an iterative process of assumption, evaluation and comparison with requirements.

Box 1 develops the grand strategy of the product manufacture, breaking it down by major subassemblies and then successively dependent subassemblies. Manufacturing methods, source of material, the plant capabilities (present or purchasable), costs and delivery times are assessed. Box 2 evaluates each strategy, and feeds back the evaluation. Box 1 can then introduce variations to improve (or optimize) the plan. The ultimate decision on a strategy occurs when Box 1 determines that the plan under consideration adequately meets the design and product manufacturing requirements.

When a plan is selected, a detailed manufacturing plan and support activity plan are developed. No attempt is made at this stage to carry these work statements to the ultimate level of detail. A later activity, A3, Plan for Production, performs that task. However, Activity A1 determines the boundaries within which product planning is carried on.

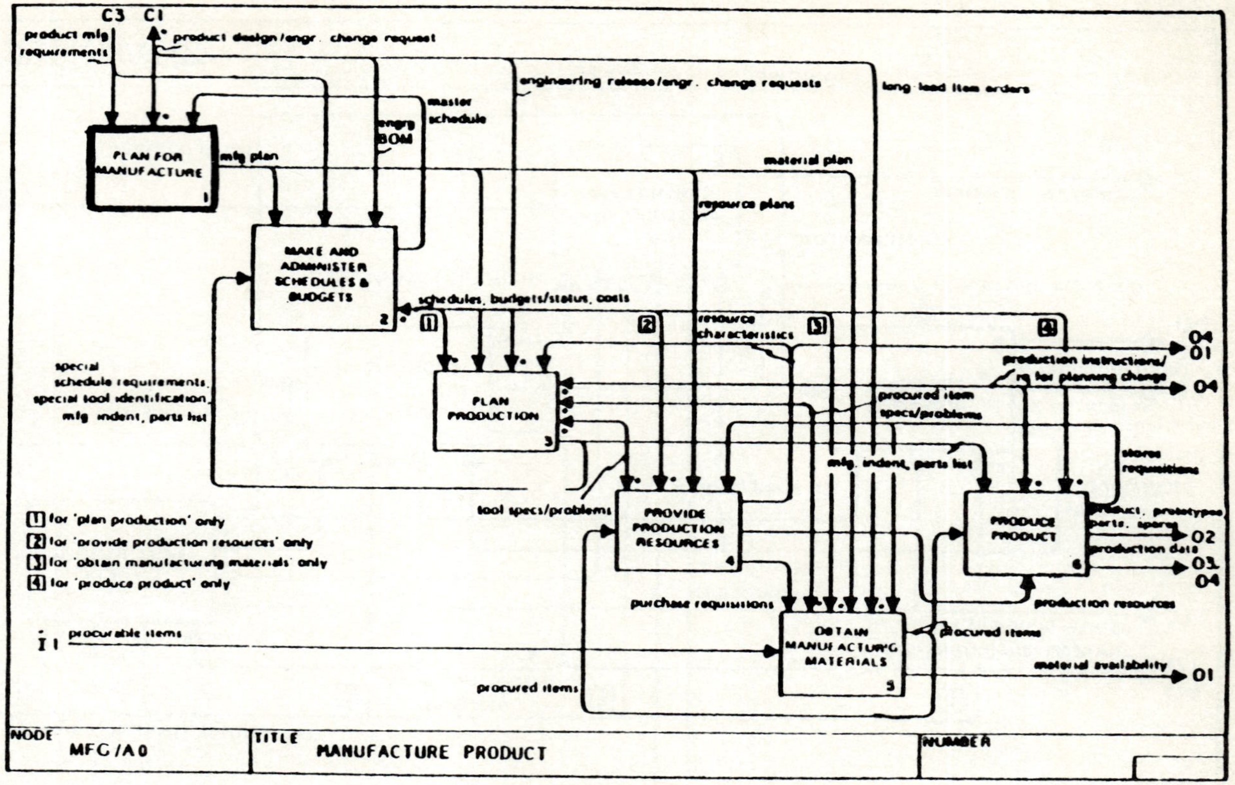

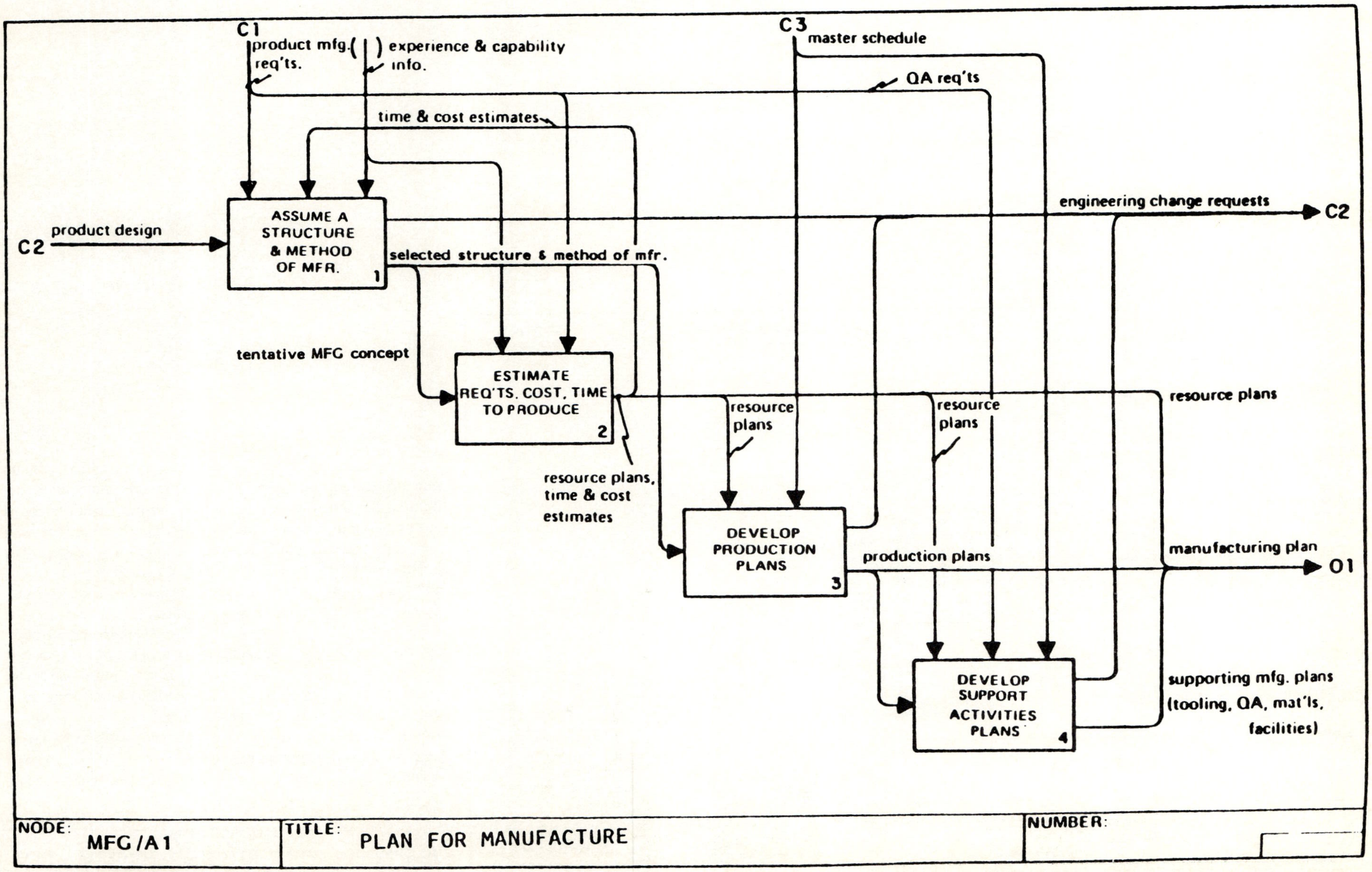
C1
product mfg. req'ts.
experience & capability info.
time & cost estimates
C3
master schedule
QA req'ts
C2
product design
ASSUME A STRUCTURE & METHOD OF MFR.
1
selected structure & method of mfr.
engineering change requests
C2
tentative MFG concept
ESTIMATE REQ'TS. COST, TIME TO PRODUCE
2
resource plans, time & cost estimates
resource plans
resource plans
resource plans
DEVELOP PRODUCTION PLANS
3
production plans
manufacturing plan
O1
DEVELOP SUPPORT ACTIVITIES PLANS
4
supporting mfg. plans (tooling, QA, mat'ls, facilities)
NODE:
MFG/A1
TITLE:
PLAN FOR MANUFACTURE
NUMBER:

A11 Assume a Structure and Method of Manufacture

This activity and the following (A22) perform the interactive convergence on a selected manufacturing concept. This activity performs the evaluation. The time and cost estimate feedback from A12 allows this activity to refine the concept.

In addition to the product design and the product manufacturing input, this activity draws very heavily in all its phases, on the backlog of experience and know-how of the creative personnel.

As the product design is broken down into successively finer subassembly levels, make or buy decisions are made. For these things to be made, manufacturing methods are hypothesized and defined. Similarly, experience and know-how are called upon in devising an assembly method. In the process, producibility considerations requiring (or making desirable) engineering changes may be identified, and change requests generated.

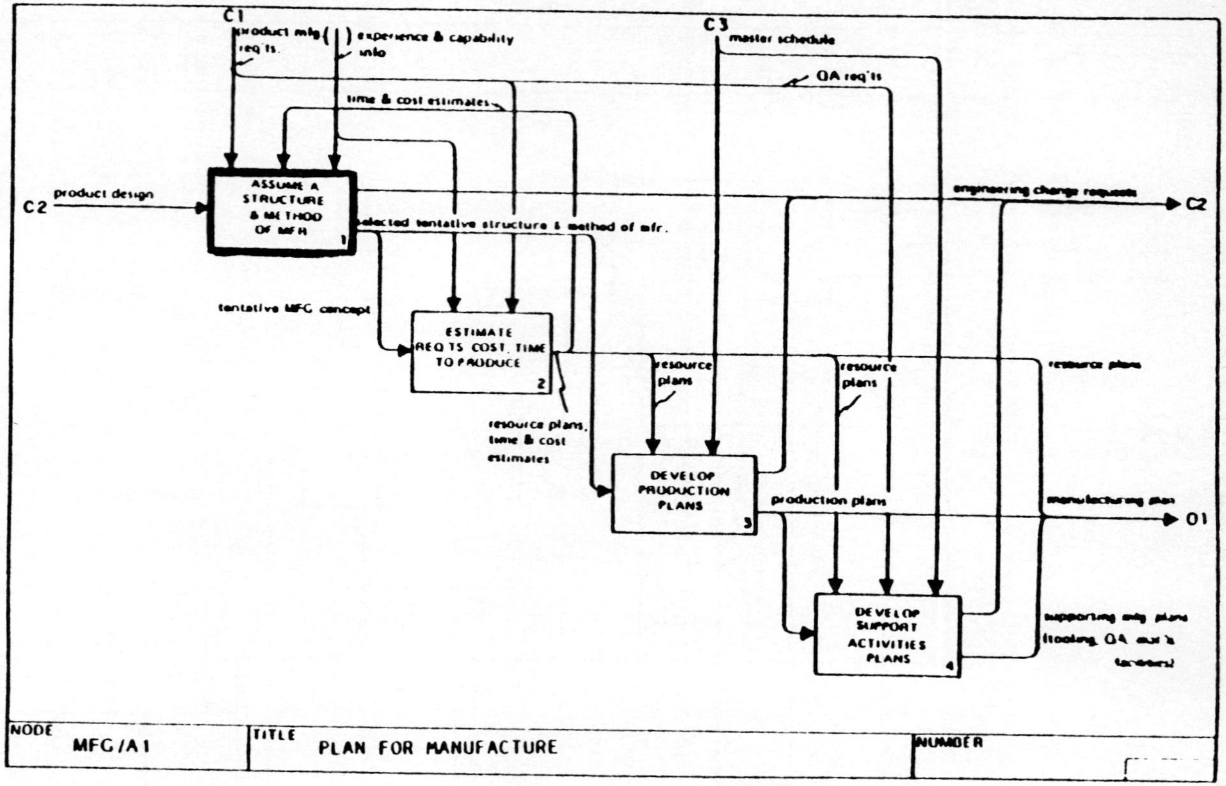

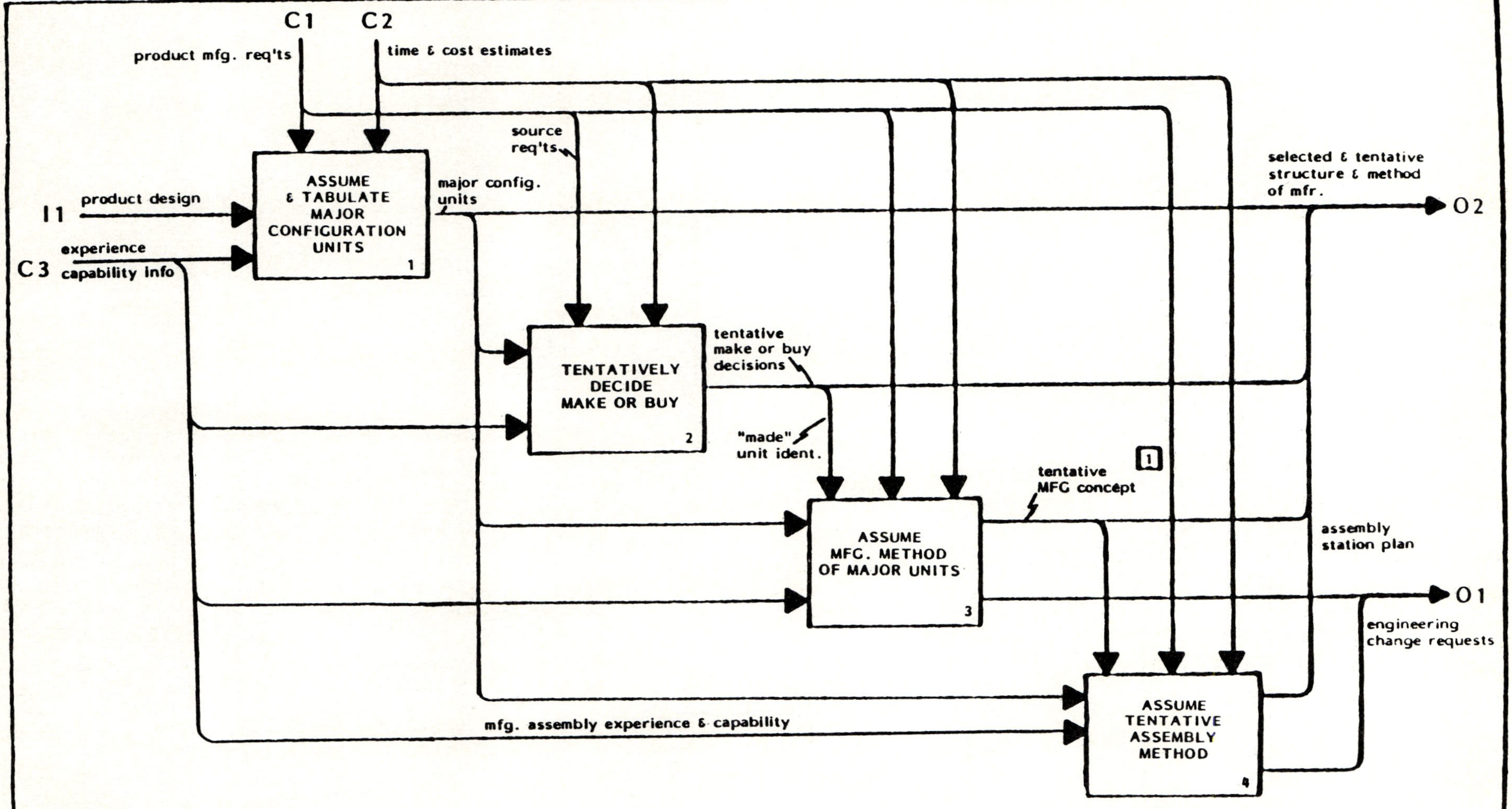

[1] est. structure - number & level of subass'ys & parts in each major unit

NODE: MFG/A11

TITLE: ASSUME A STRUCTURE AND METHOD OF MANUFACTURE

NUMBER:

A12 Estimate Requirements, Time, and Cost to Produce

This activity evaluates the tentative manufacturing plans submitted by the preceding Activity A11, and returns its evaluation to the preceding activity.

Evaluation is divided into these areas of expertise:

- Facilities (plant space and services), equipment (machine tools and test equipment) and manpower. The need for each is estimated, and compared to what is available. If inadequate, an estimate is made of necessary acquisitions, their cost and their availability.
- Cost of materials, including acquisition, in raw material or finished form; fabrication: inspection; assembly; and test.
- Time needed to start up production and rate at which finished products can be delivered.

Again, the accumulated experience of the manufacturing concern in making similar products in the past will play a major factor in this activity.

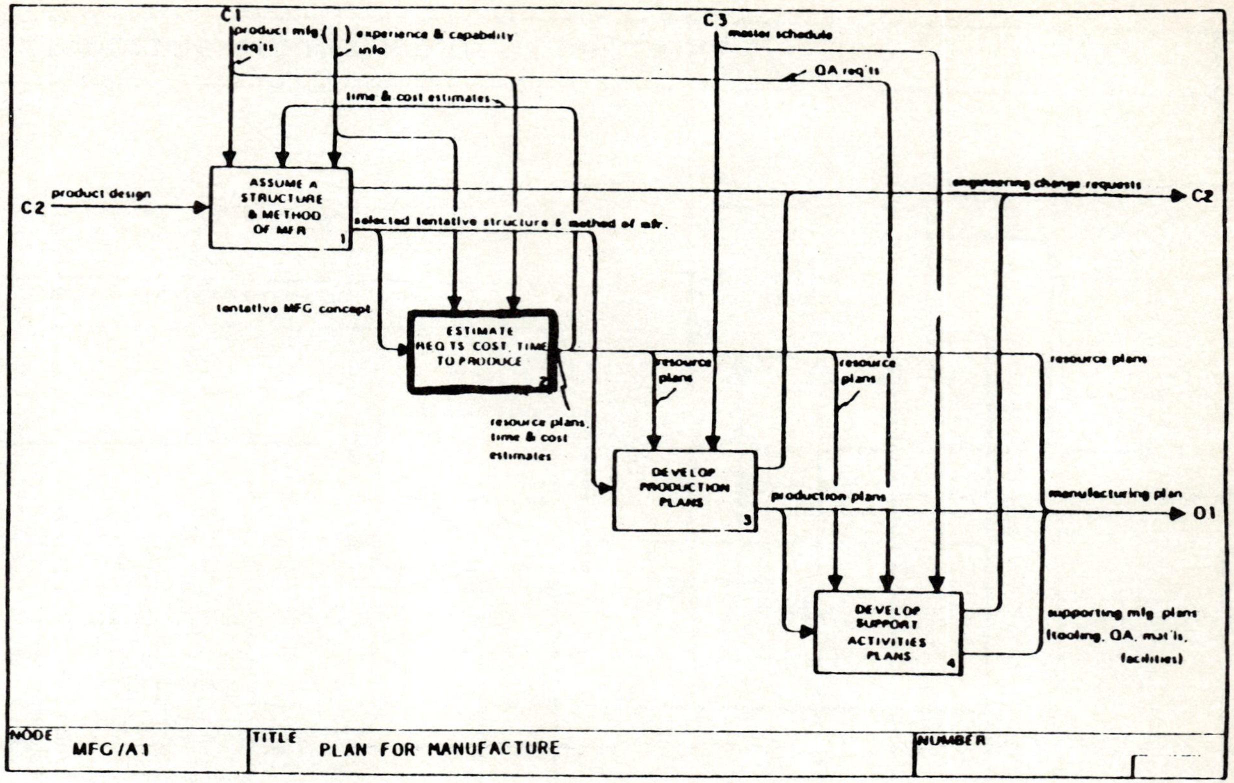

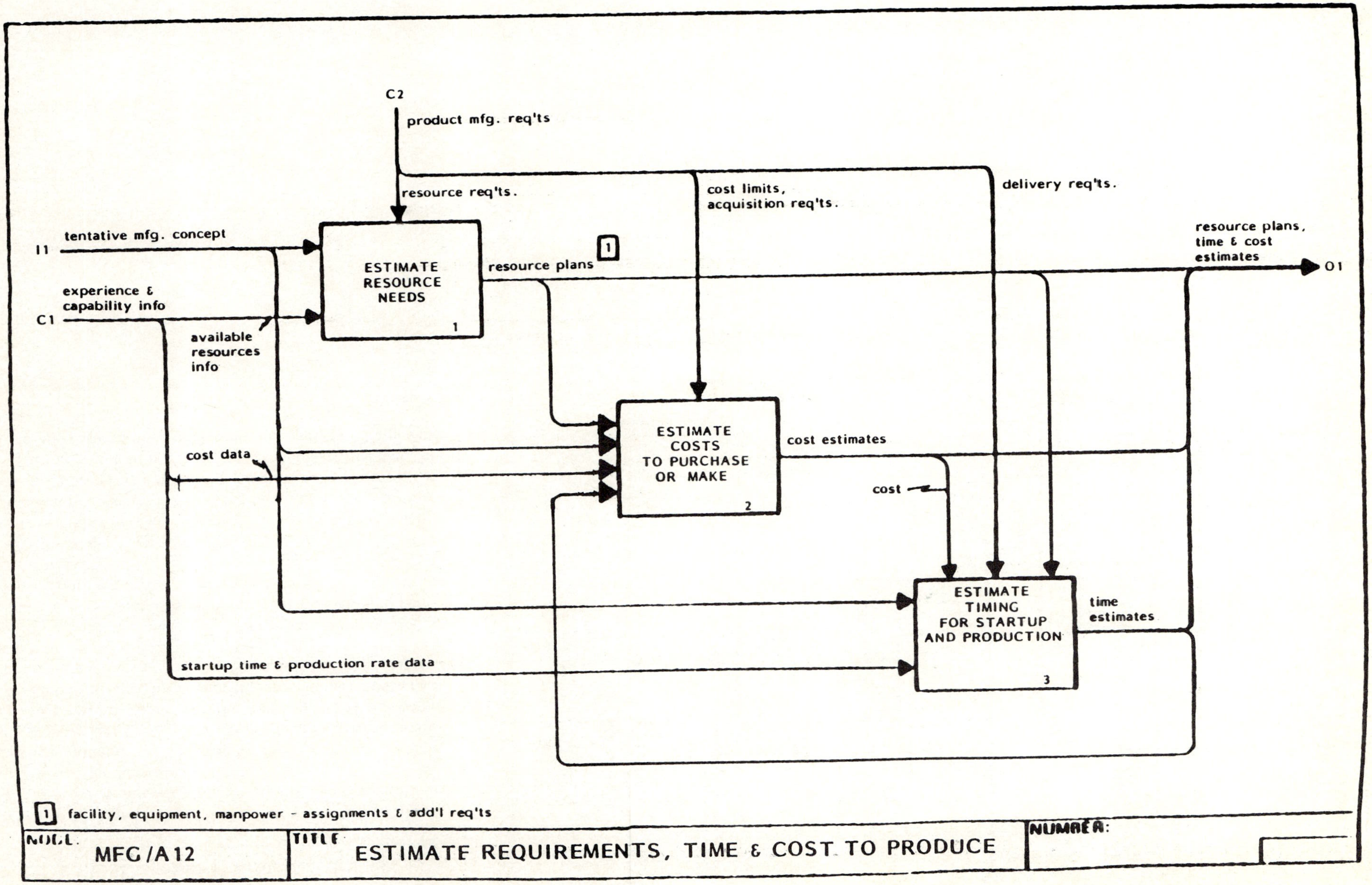
C2
product mfg. req'ts
resource req'ts.
cost limits,
acquisition req'ts.
delivery req'ts.
I1
tentative mfg. concept
experience &
capability info
C1
ESTIMATE
RESOURCE
NEEDS
1
resource plans
1
available
resources
info
cost data
ESTIMATE
COSTS
TO PURCHASE
OR MAKE
2
cost estimates
cost
resource plans,
time & cost
estimates
O1
ESTIMATE
TIMING
FOR STARTUP
AND PRODUCTION
3
time
estimates
startup time & production rate data
1 facility, equipment, manpower - assignments & add'l req'ts
NODE:
MFG/A12
TITLE
ESTIMATE REQUIREMENTS, TIME & COST TO PRODUCE
NUMBER:

A13 Develop Production Plans

When a manufacturing plan has been selected (Activity A11) it is transmitted for further expansion and exposition to this and the following activity.

The manufacturing flow plan forms the basis upon which this activity is structured, particularly so in aerospace manufacture in which assembly is the dominant activity. Assembly can be portrayed by a dendritic (branching) pattern, like the leaves and limbs on a tree. The major branches join the trunk in a succession corresponding to major or "line" assembly operations, and smaller twigs join the branches in similar succession corresponding to batch assembly operations. Each twig carries a number of leaves, corresponding to detail parts.

At this stage of planning the complete list of details and first assemblies may not be available, but estimates based on experience can be made, and the batch and line assembly and installation operations can be quite fully planned.

The major output is a fully stated manufacturing plan. In addition, requests for engineering changes may be generated.

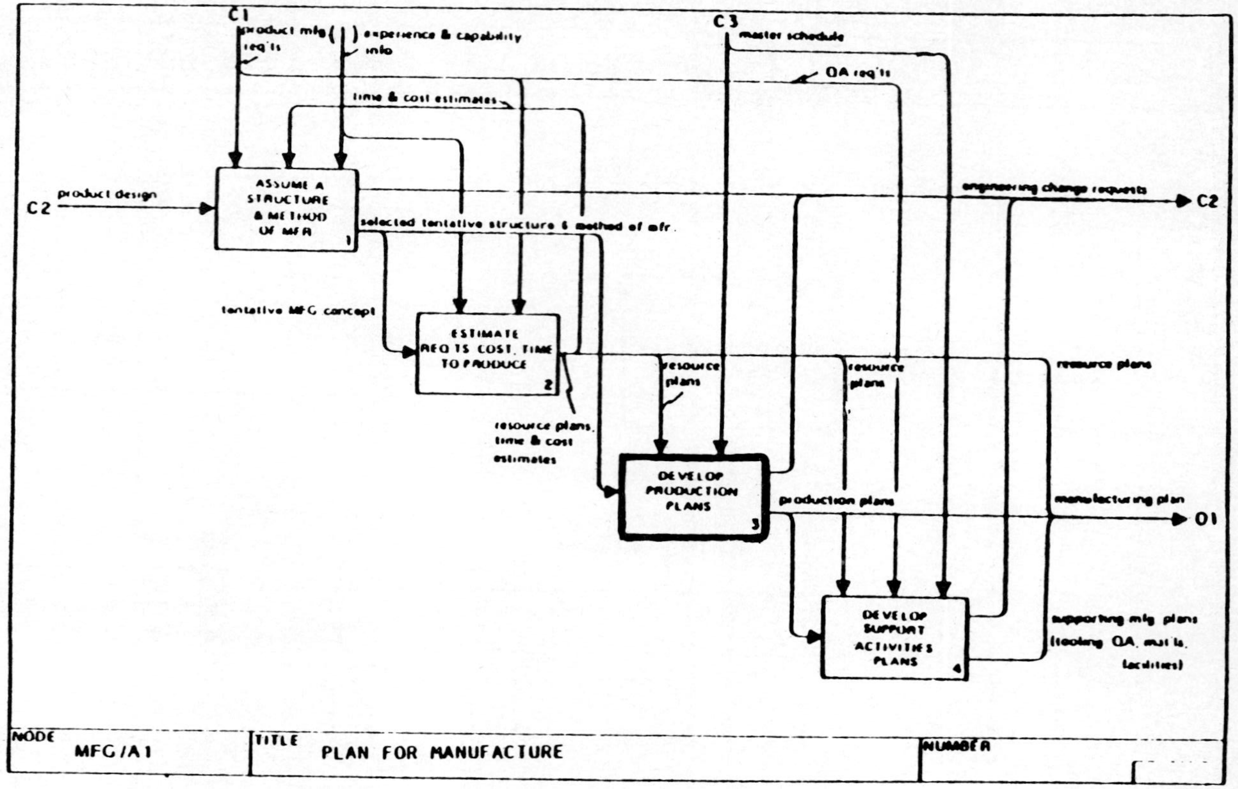

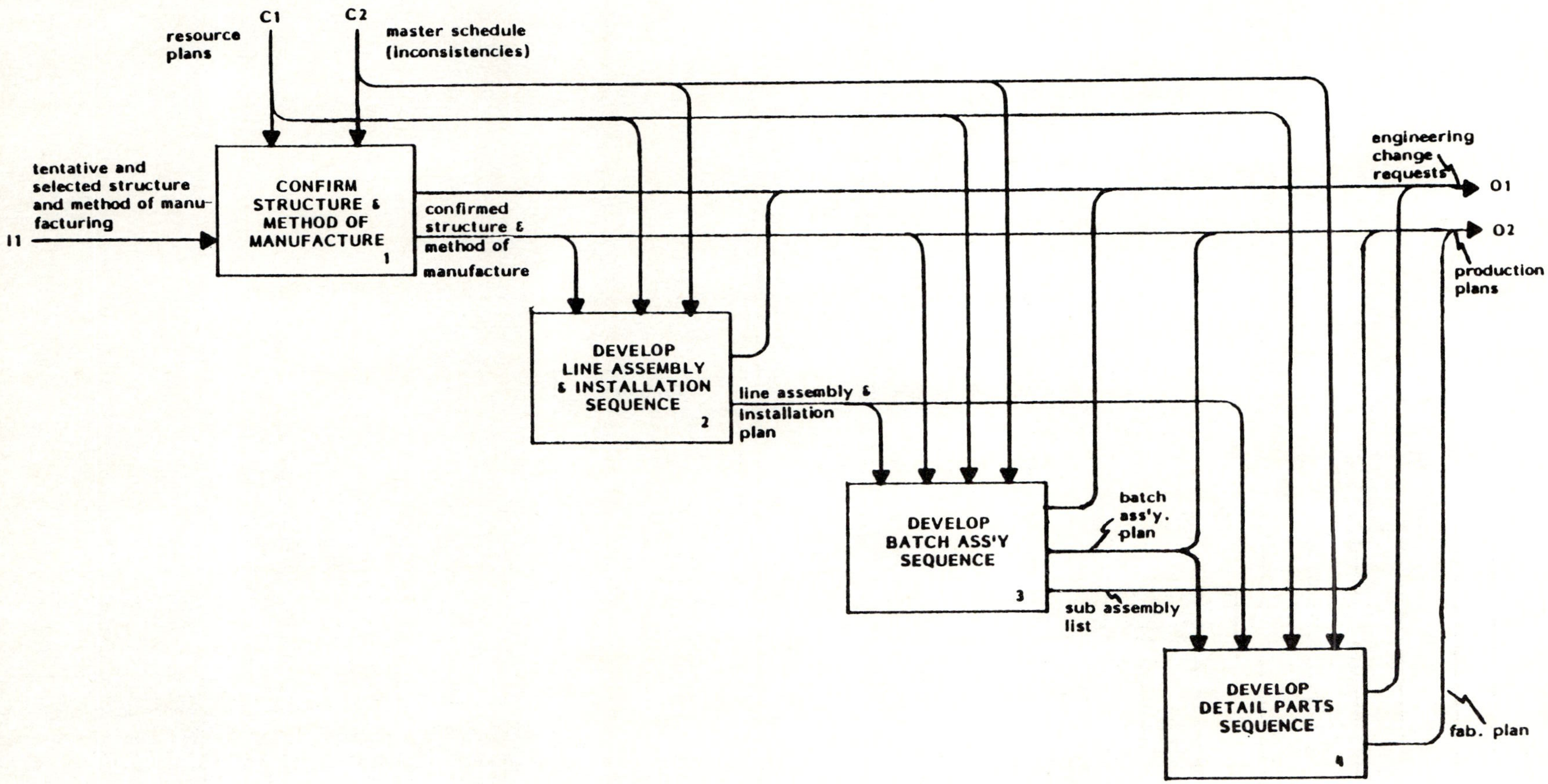

NODE: MFG/A13

TITLE: DEVELOP PRODUCTION PLANS

NUMBER:

A14 Develop Support Activities Plans

This activity does for manufacturing support what the previous activity (A13) did for the main manufacturing plan. The difference lies in the fact that Activity A13 dealt with making the product, which was ultimately to be shipped out, whereas Activity A14 deals with services in support of that operation, but which do not appear as tangible components of the product.

The input is the main manufacturing plan (from Activity A13); the outputs are plans for quality assurance, materials management, and tooling. Again, needs for engineering changes may be identified.

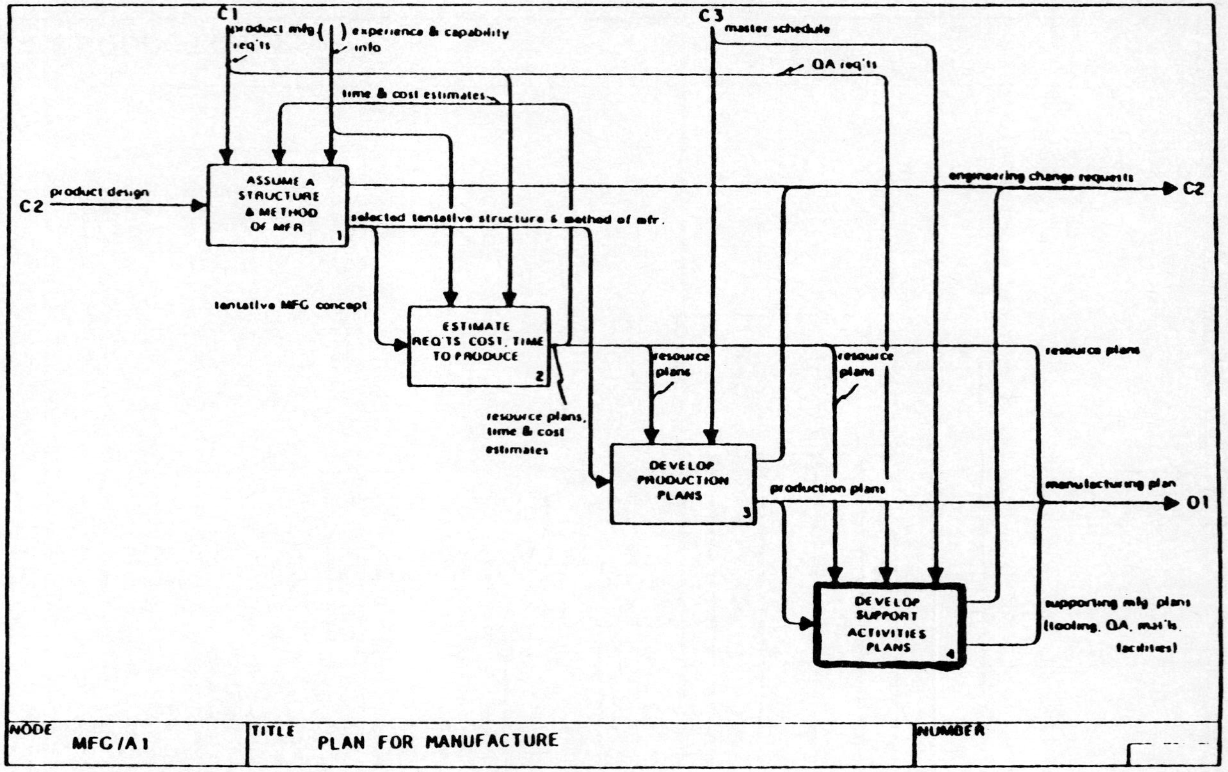

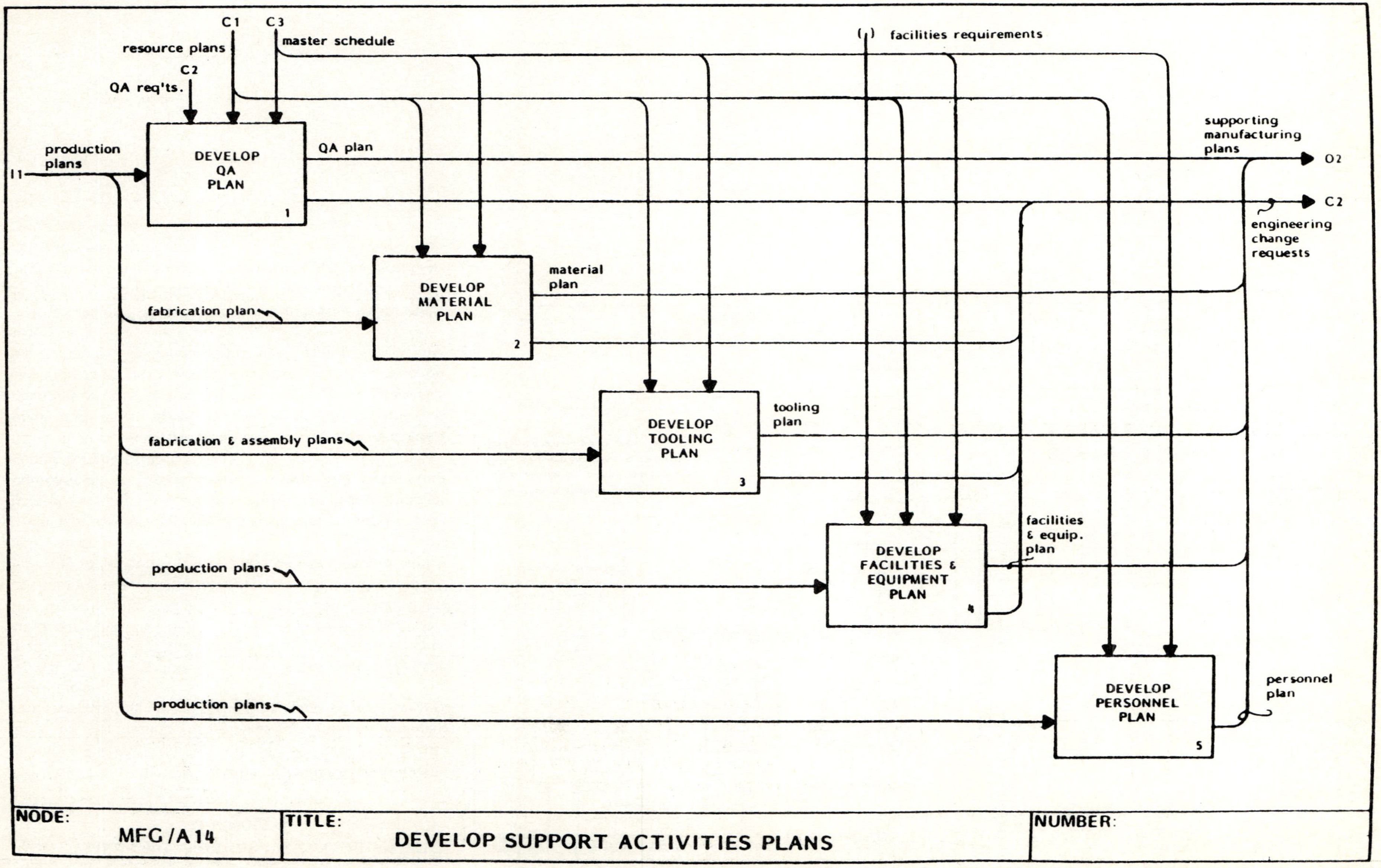

NODE: MFG/A14	TITLE: DEVELOP SUPPORT ACTIVITIES PLANS	NUMBER:

A2 Make and Administer Schedules and Budgets

Working from the production schedule requirements (mainly delivery dates), the WBS (Work Breakdown Structure), and the manufacturing plans (preliminary versions), a master schedule showing the time spans of major events is produced.

This master schedule is expanded using (initially) the engineering parts list and (finally) the manufacturing indentured parts list and tool identifications to produce coordinating schedules which give the (negotiated) need dates for those items which interface between A3, A4, A5, and A6 (e.g., the manufacturing instructions, facilities and equipment, manpower, materials).

The manufacturing plans and schedules are used to estimate costs, which are combined with the product budget requirements to produce the budgets and accounting scheme for the project.

Throughout the project, actual costs and times are monitored, and when necessary, the need for revisions to schedules or budgets are identified and fed back. Schedules and budgets are also revised whenever revised production requirements (C2) or revised manufacturing plans (C1) arrive.

Note that only those schedules which coordinate other activities are handled here. There is further detailed scheduling for parts planning, resource providing, material control, and production control.

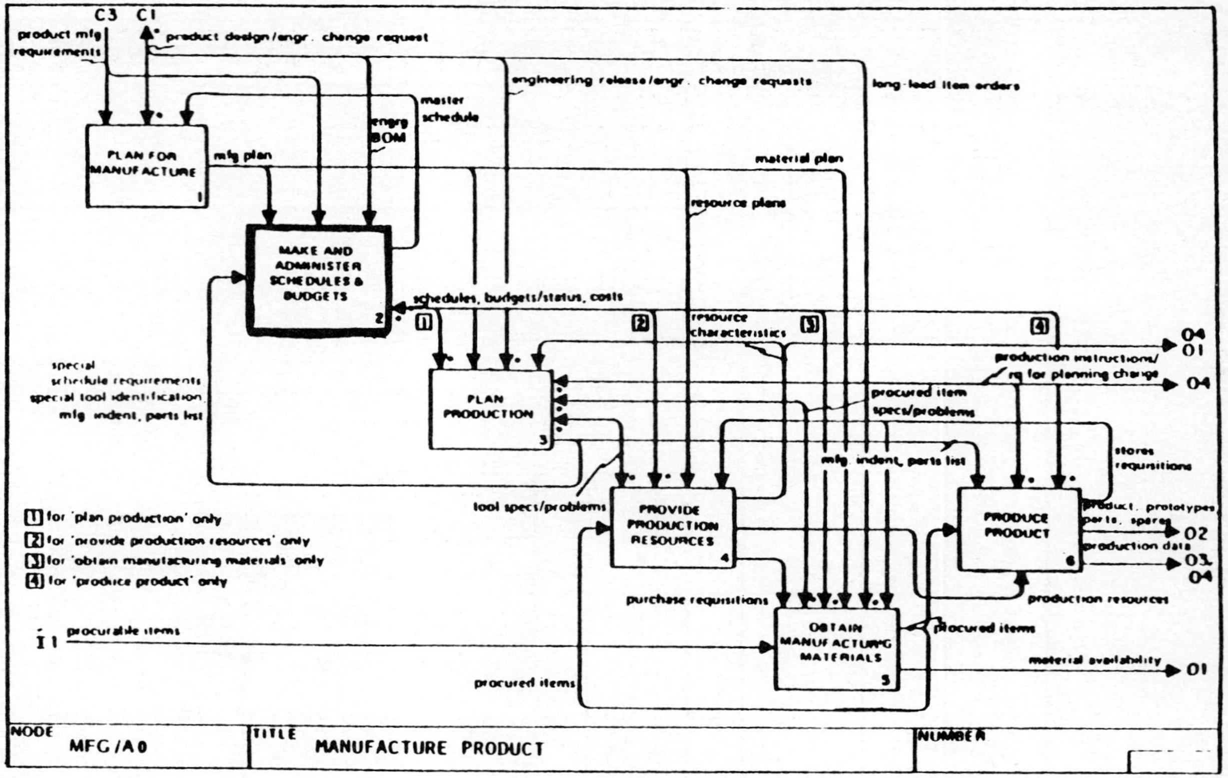

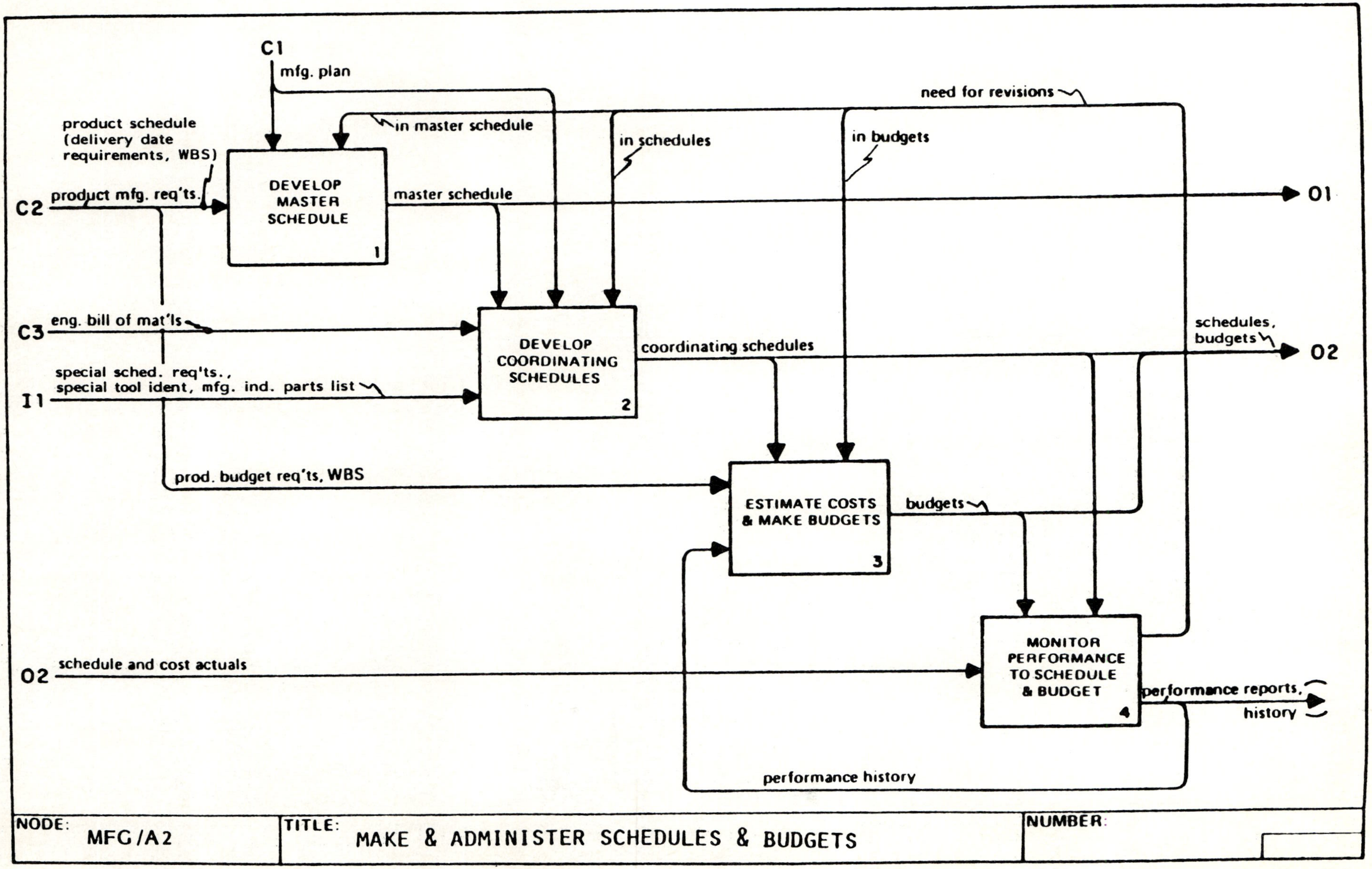
C1
mfg. plan
product schedule
(delivery date
requirements, WBS)
C2
product mfg. req'ts.
DEVELOP
MASTER
SCHEDULE
1
in master schedule
master schedule
need for revisions
in schedules
in budgets
O1
C3
eng. bill of mat'ls
special sched. req'ts.,
special tool ident, mfg. ind. parts list
I1
DEVELOP
COORDINATING
SCHEDULES
2
coordinating schedules
schedules,
budgets
O2
prod. budget req'ts, WBS
ESTIMATE COSTS
& MAKE BUDGETS
3
budgets
O2
schedule and cost actuals
MONITOR
PERFORMANCE
TO SCHEDULE
& BUDGET
4
performance reports,
history
performance history
NODE:
MFG/A2
TITLE:
MAKE & ADMINISTER SCHEDULES & BUDGETS
NUMBER:

A21 Develop Master Schedule

The work breakdown structure, from I1, is converted in "Apply Time Spans" (Box 1), into the estimated times for the WBS items. These in turn are integrated into preliminary versions of the master schedule (Box 2). The next step is to compare the preliminary master schedule with the delivery schedule to identify and to then resolve any conflicts (Box 3 and Box 4, respectively). Any re-estimated or negotiated times are fed back from this process to Box 1. The output of Box 4, completed master schedule, is then sent to management for their approval.

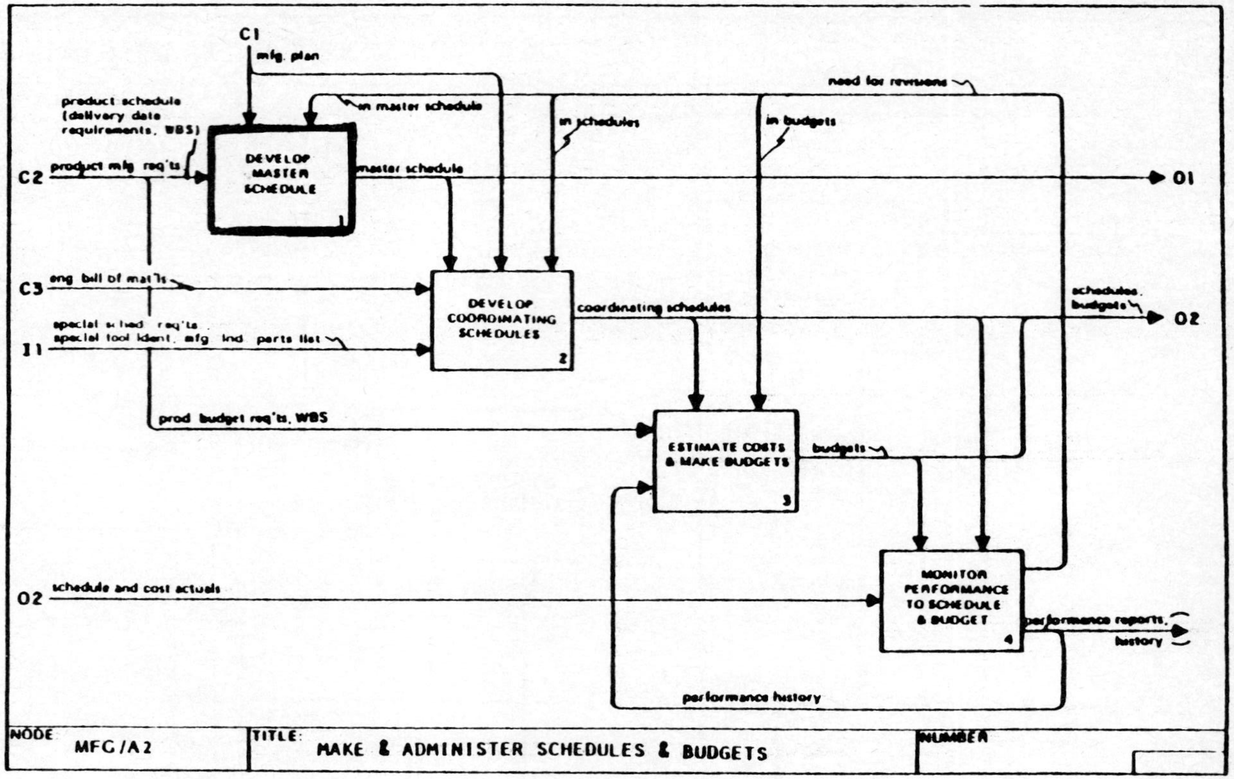

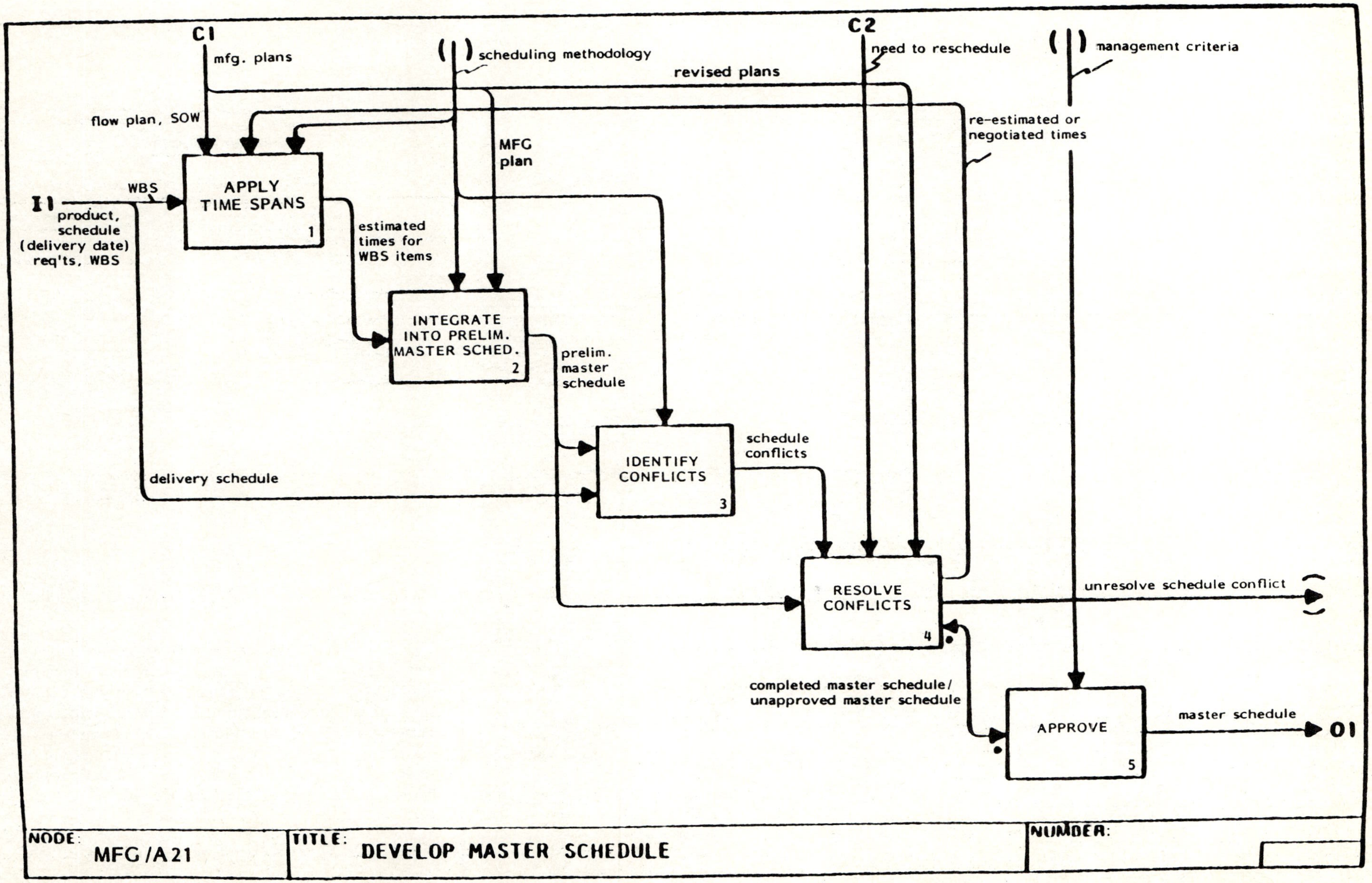
C1
mfg. plans
flow plan, SOW
scheduling methodology
revised plans
C2
need to reschedule
management criteria
re-estimated or negotiated times
MFG plan
I1
WBS
product, schedule (delivery date) req'ts, WBS
APPLY TIME SPANS
1
estimated times for WBS items
INTEGRATE INTO PRELIM. MASTER SCHED.
2
prelim. master schedule
IDENTIFY CONFLICTS
3
delivery schedule
schedule conflicts
RESOLVE CONFLICTS
4
unresolve schedule conflict
completed master schedule/ unapproved master schedule
APPROVE
5
master schedule
O1
NODE: MFG/A21
TITLE: DEVELOP MASTER SCHEDULE
NUMBER:

A22 Develop Coordinating Schedules

The engineering bill of materials and the manufacturing indentured parts list are converted, in Box 1, to the set-backs and times used in the preparation of both the prototype schedule and the production schedule. These set-backs and flow times, together with the appropriate parts of the manufacturing plan, transform the master schedule into coordinating schedules for the prototype, production, and production support activities.

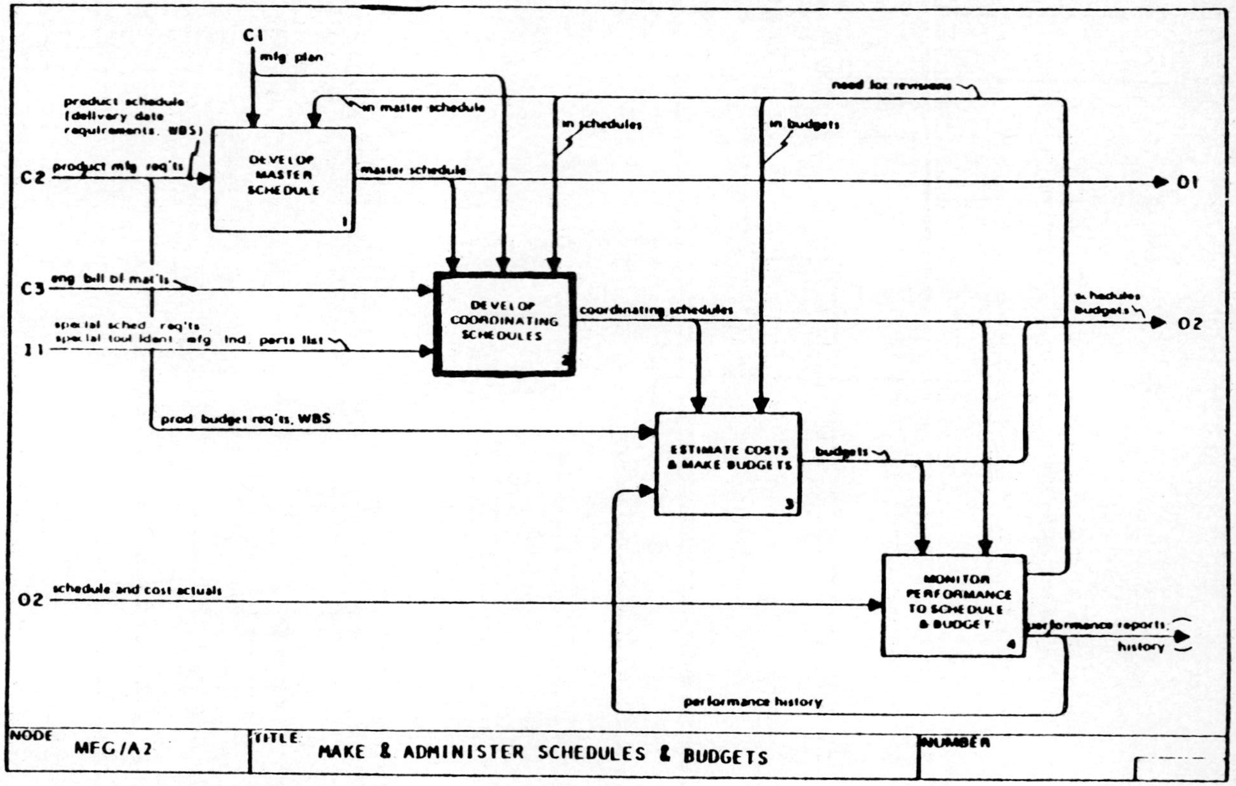

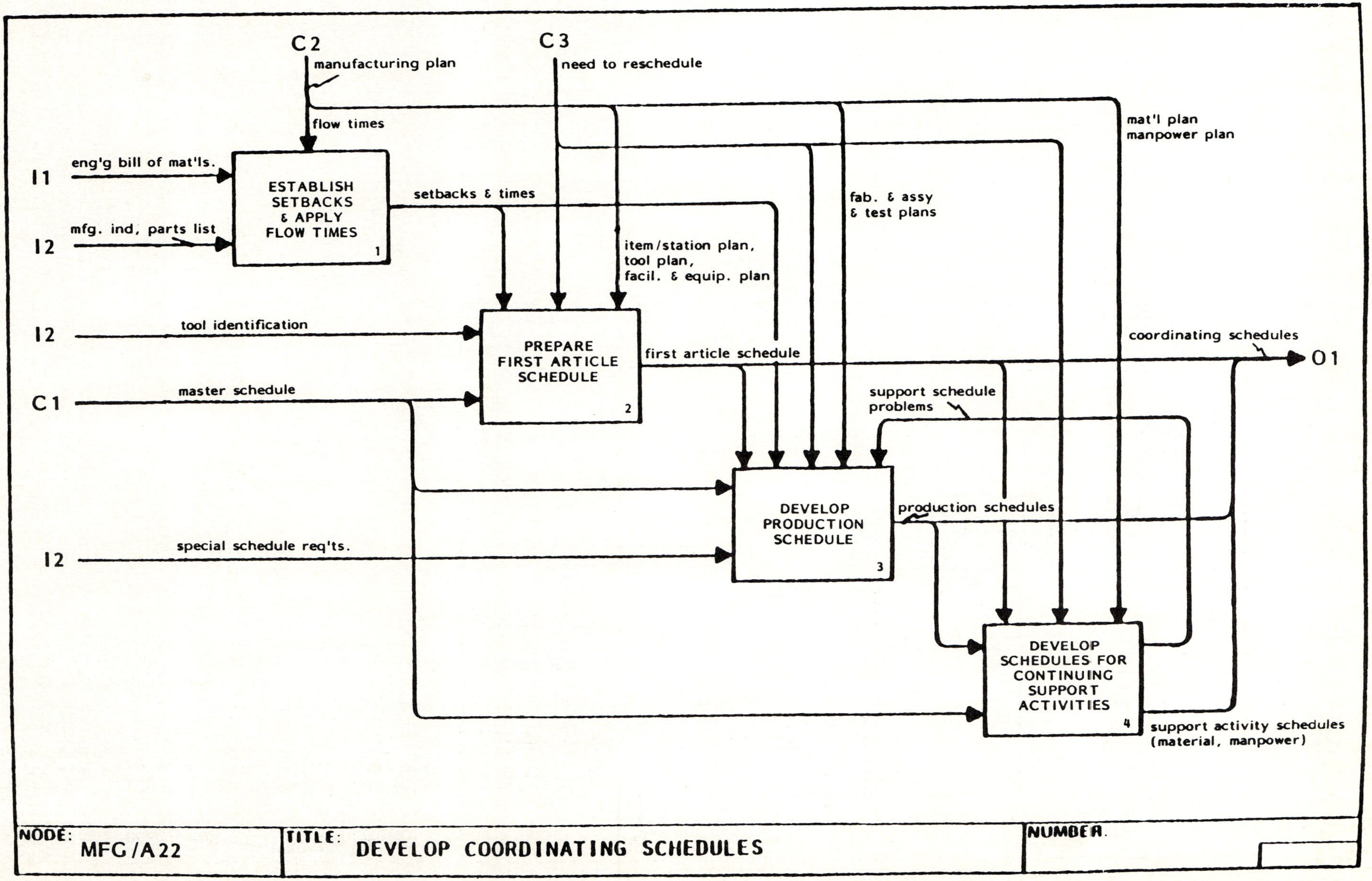
C2
manufacturing plan
flow times
C3
need to reschedule
I1
eng'g bill of mat'ls.
I2
mfg. ind, parts list
ESTABLISH SETBACKS & APPLY FLOW TIMES
1
setbacks & times
mat'l plan
manpower plan
fab. & assy
& test plans
item/station plan,
tool plan,
facil. & equip. plan
I2
tool identification
C1
master schedule
PREPARE FIRST ARTICLE SCHEDULE
2
first article schedule
coordinating schedules
O1
support schedule
problems
DEVELOP PRODUCTION SCHEDULE
3
production schedules
I2
special schedule req'ts.
DEVELOP SCHEDULES FOR CONTINUING SUPPORT ACTIVITIES
4
support activity schedules
(material, manpower)
NODE: MFG/A22
TITLE: DEVELOP COORDINATING SCHEDULES
NUMBER:

A23 Estimate Costs and Make Budgets

For planning purposes, the product budget for the manufacturing activity is subdivided into planning packages based on the Work Breakdown Structure (Box 1). Simultaneously, accounts are assigned identifying contract lot releases (Box 2). Working budgets are then developed as Cost Account Packages, which are assigned structured subaccounts and released (Box 3 and 4). Finally, functional (departmental) budgets are issued, permitting work to commence.

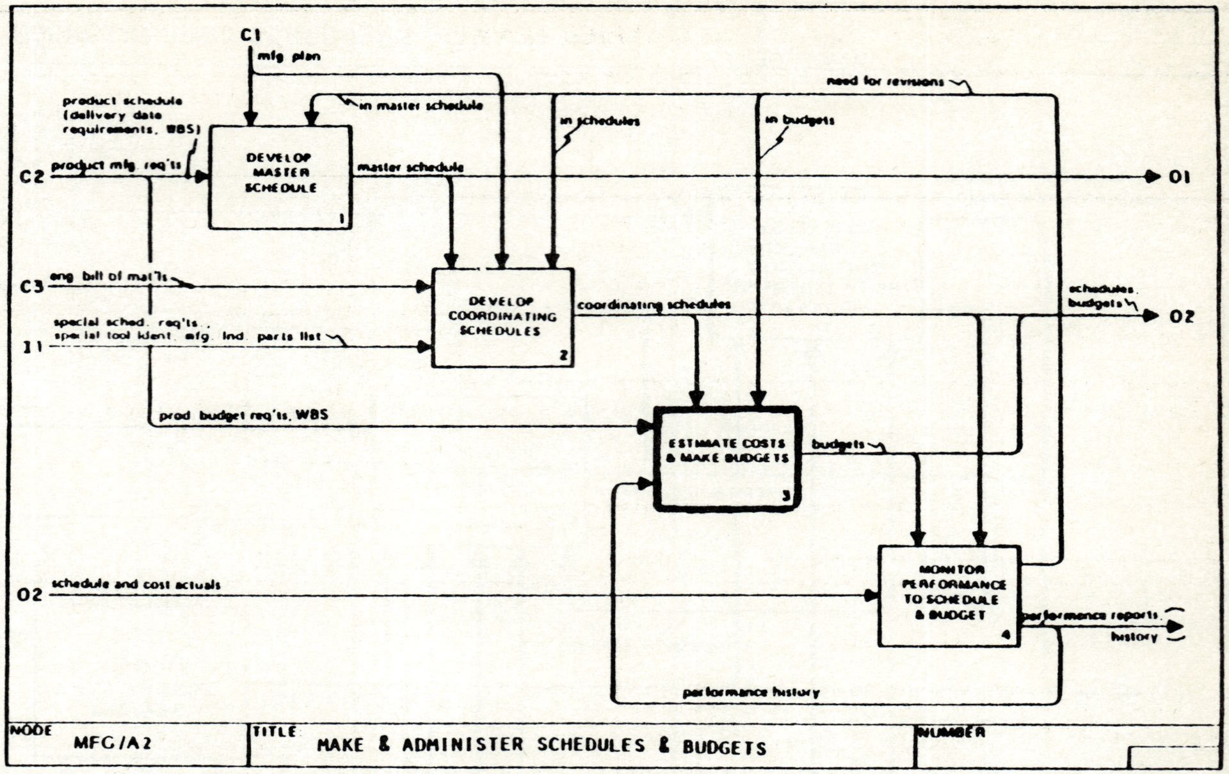

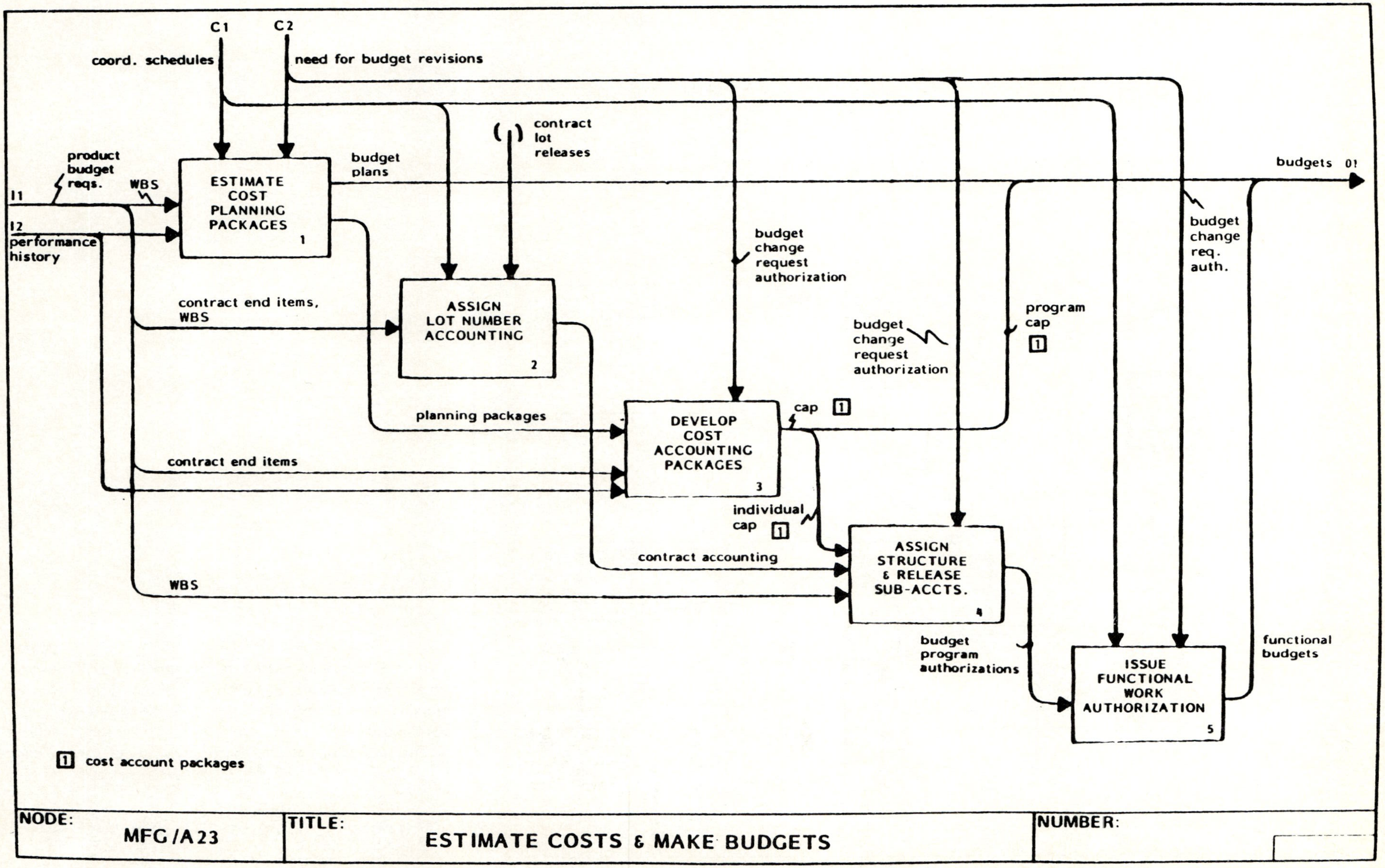
C1
C2
coord. schedules
need for budget revisions
contract
lot
releases
product
budget
reqs.
WBS
I1
I2
performance
history
ESTIMATE
COST
PLANNING
PACKAGES
1
budget
plans
budgets O1
ASSIGN
LOT NUMBER
ACCOUNTING
2
contract end items,
WBS
budget
change
request
authorization
budget
change
req.
auth.
program
cap
1
budget
change
request
authorization
planning packages
DEVELOP
COST
ACCOUNTING
PACKAGES
3
cap 1
contract end items
individual
cap 1
ASSIGN
STRUCTURE
& RELEASE
SUB-ACCTS.
4
contract accounting
WBS
budget
program
authorizations
ISSUE
FUNCTIONAL
WORK
AUTHORIZATION
5
functional
budgets
1 cost account packages
NODE:
MFG/A23
TITLE:
ESTIMATE COSTS & MAKE BUDGETS
NUMBER:

A24 Monitor Performance to Schedule and Budget

This activity involves the development and monitoring of display methods (Box 1) for the various production schedules and budgets. Box 2 records the actual schedule and cost data which then is evaluated in Box 3. This evaluation may result in requests for changes in the schedule and budget.

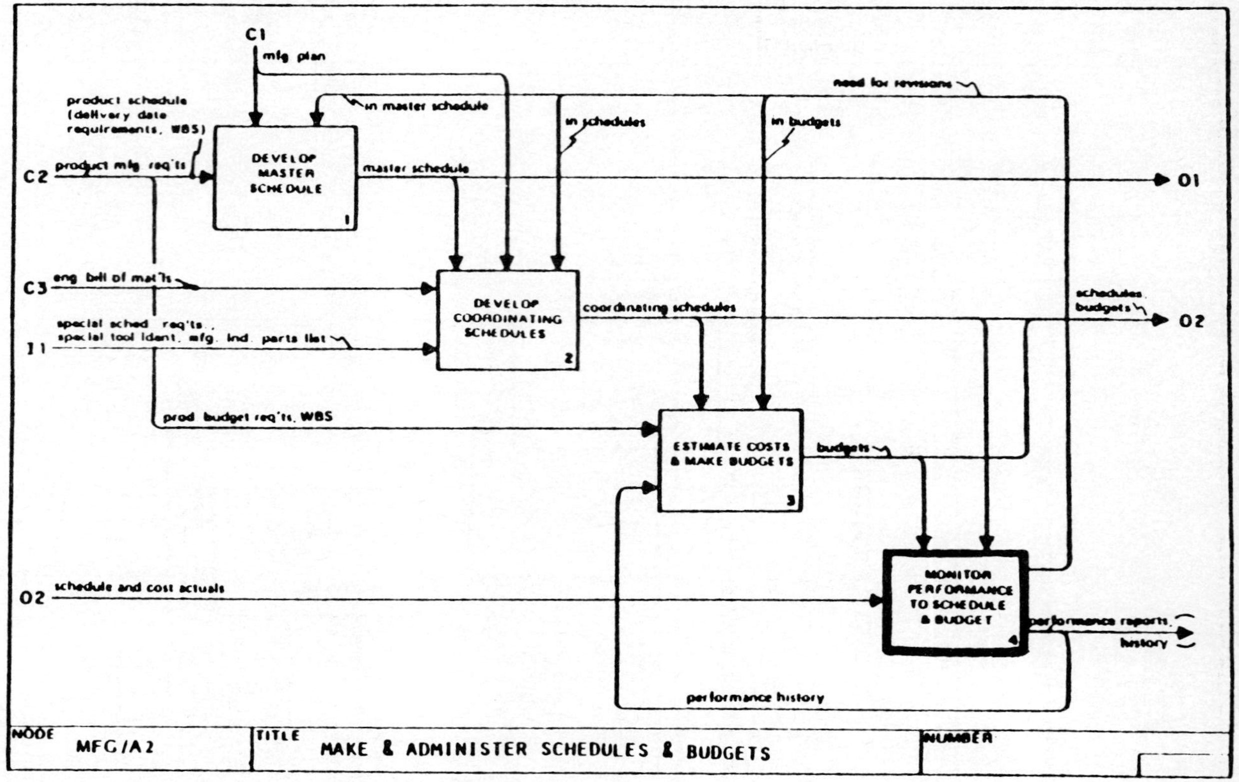

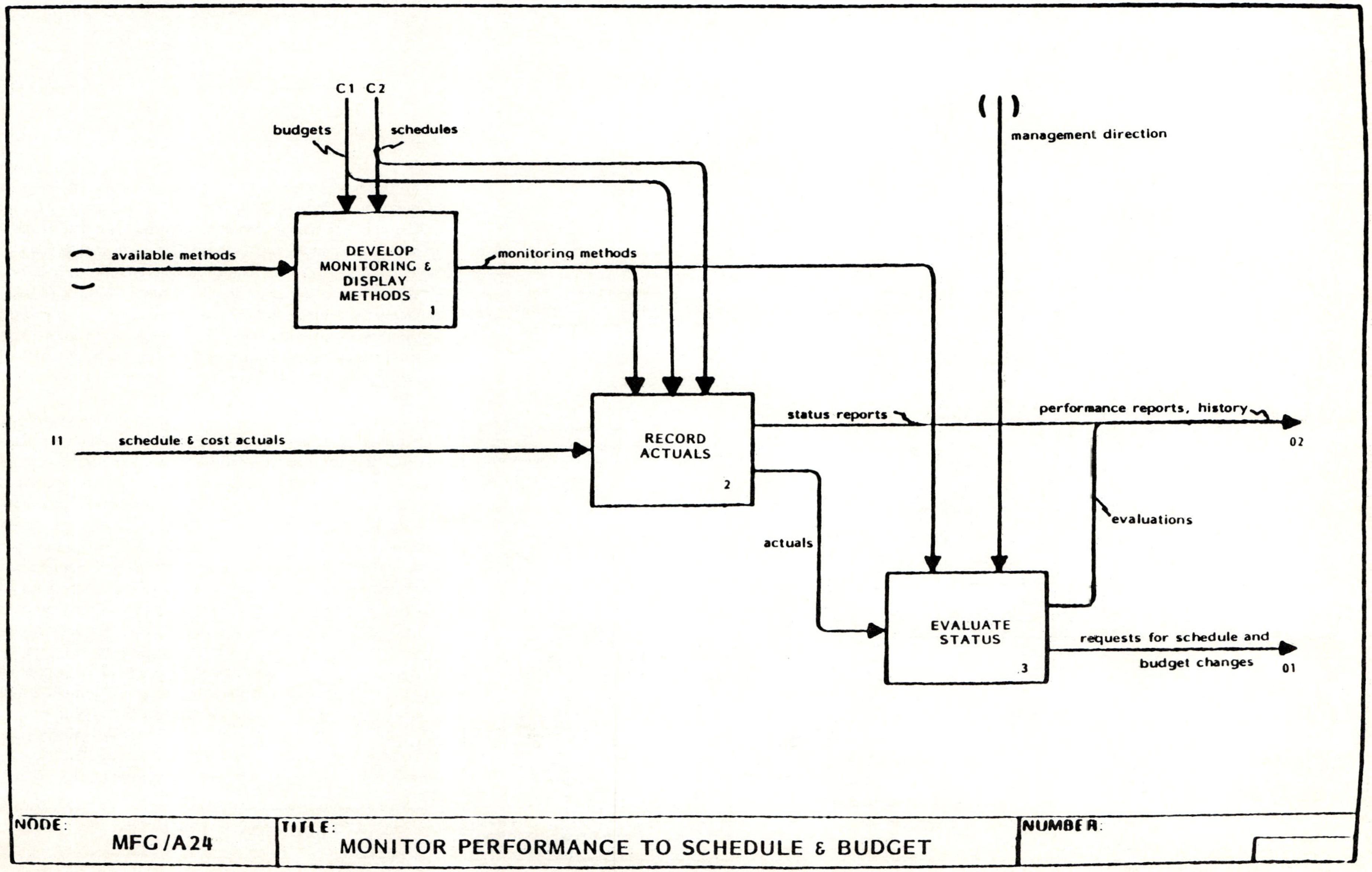
C1 C2
budgets
schedules
available methods
DEVELOP MONITORING & DISPLAY METHODS
1
monitoring methods
management direction
I1
schedule & cost actuals
RECORD ACTUALS
2
status reports
performance reports, history
02
actuals
evaluations
EVALUATE STATUS
3
requests for schedule and budget changes
01
NODE:
MFG/A24
TITLE:
MONITOR PERFORMANCE TO SCHEDULE & BUDGET
NUMBER:

A3 Plan Production

The activity "Plan Production" has the overall task of converting engineering designs into detailed manufacturing methods which conform to the manufacturing plans.

Coordinating schedules (which give engineering release dates, planning needs dates, tool spec need dates, etc.) are expanded into detailed schedules for production planning which allow appropriate assignments to be made (Box 1).

For each engineering item (or changed item), a detailed manufacturing method is devised (Box 2). This involves identifying the manufacturing parts to the extent that they are planned to exist temporarily in different structures or in different forms from the engineering versions (for example, groups of parts may be put together into subassemblies not identified by engineering or holes may be omitted in detail parts and drilled after assembly to assure alignment), and identifying the tools to be used. Production plans and resource characteristics are necessary considerations in this task. This task is often performed by manufacturing engineers.

Detailed instructions for making each part are an elaboration of the detailed manufacturing method. In the process of making them (Box 3), specifications for tools and for procured items are produced. All these must be validated (Box 4) to be sure that the items produced or acquired will conform to the engineering design.

Throughout this whole process, there may be feedback to engineering in the form of engineering change requests (C3 output). Feedback to planning from tooling, procurement, and production comes in as "Requests for Change in Planning" (C3).

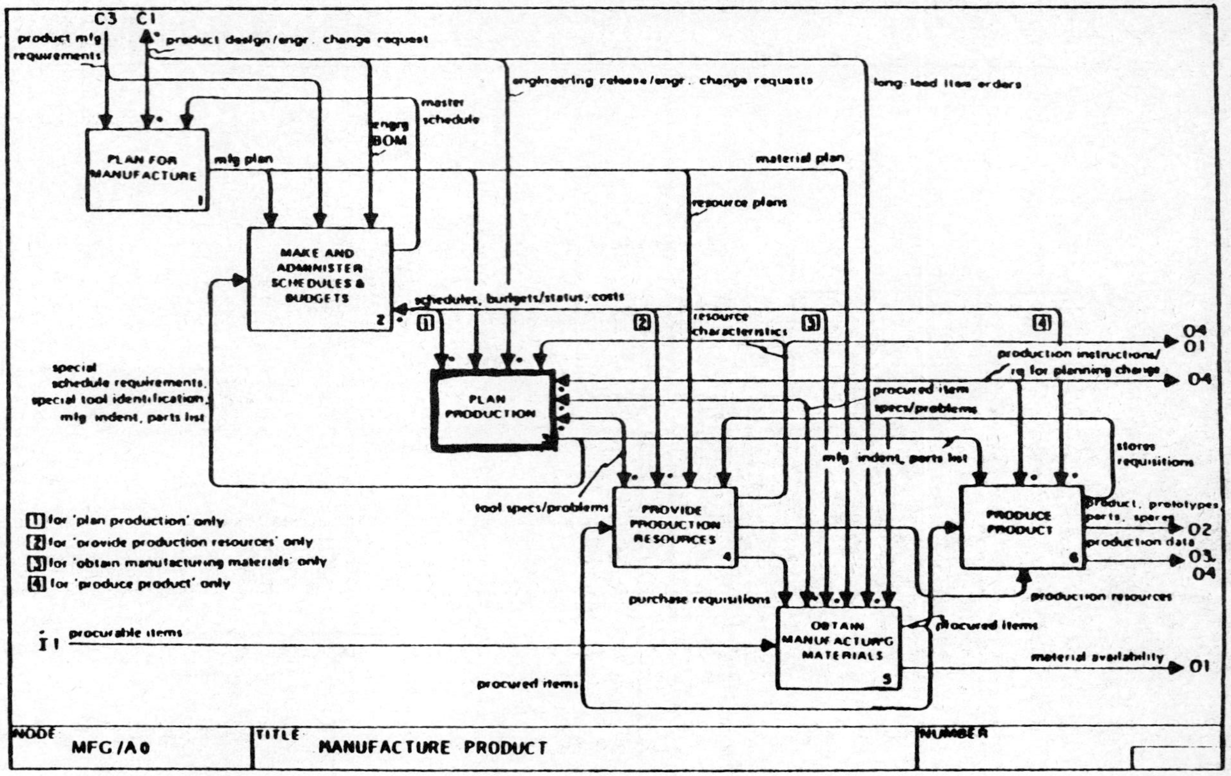

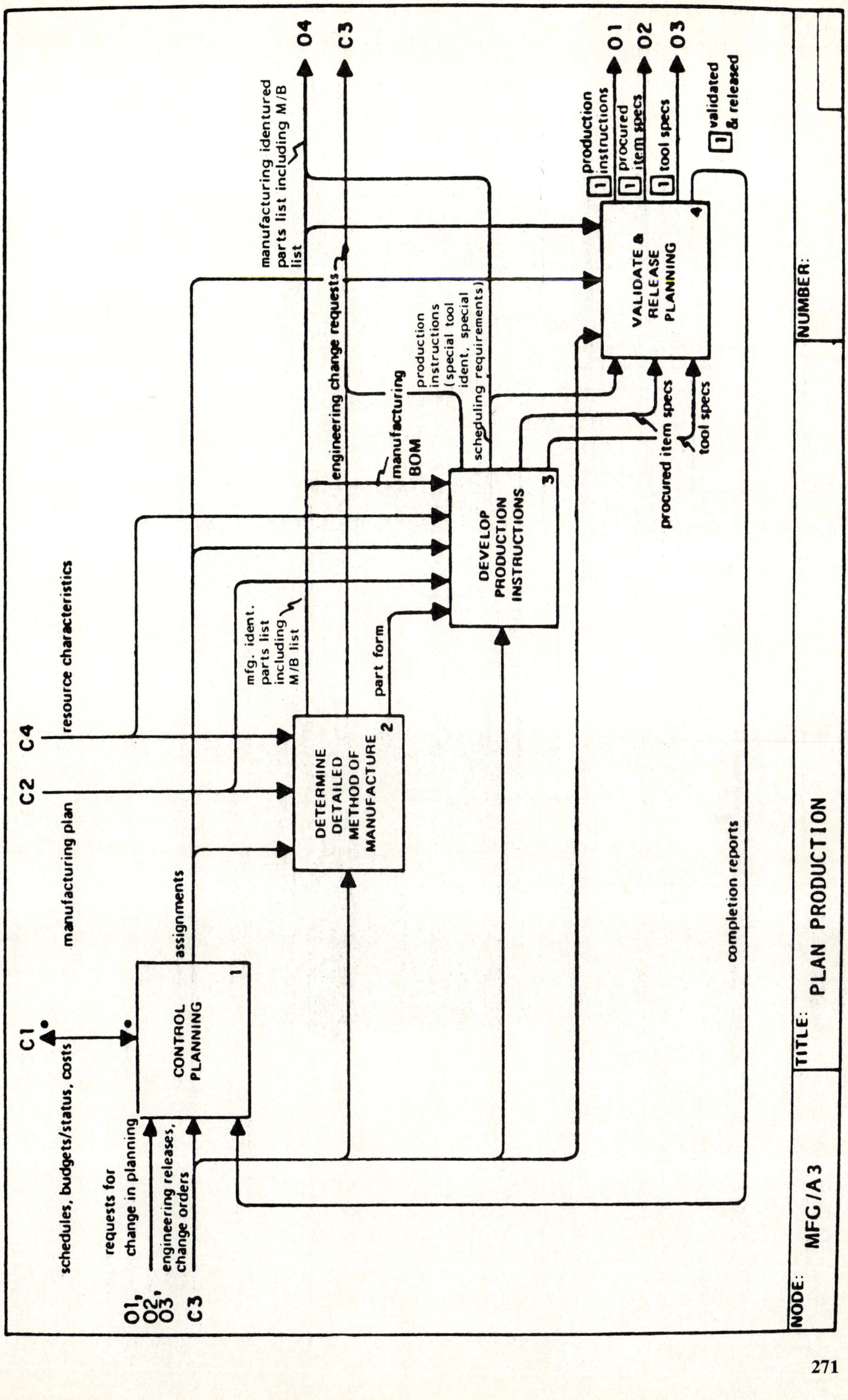
C1
schedules, budgets/status, costs
requests for change in planning
engineering releases, change orders
O1, O2, O3
C3
CONTROL PLANNING
1
assignments
manufacturing plan
C2
C4
resource characteristics
DETERMINE DETAILED METHOD OF MANUFACTURE
2
mfg. ident. parts list including M/B list
part form
DEVELOP PRODUCTION INSTRUCTIONS
3
manufacturing BOM
engineering change requests
production instructions (special tool ident, special scheduling requirements)
procured item specs
tool specs
manufacturing identured parts list including M/B list
VALIDATE & RELEASE PLANNING
4
1 production instructions
1 procured item specs
1 tool specs
1 validated & released
completion reports
O4
C3
O1
O2
O3
NODE:
MFG/A3
TITLE:
PLAN PRODUCTION
NUMBER:

A32 Determine Detailed Method of Manufacture

This activity converts, in Box 1, the engineering release into the manufacturing indentured parts list. Using this parts list, together with the manufacturing plan, Box 2 outputs a make or buy decision for each part. For those parts that are to be made, Box 3 outputs the precise manufacturing form of the part or sub-part. Engineering change requests may be initiated by any of the activities of this diagram.

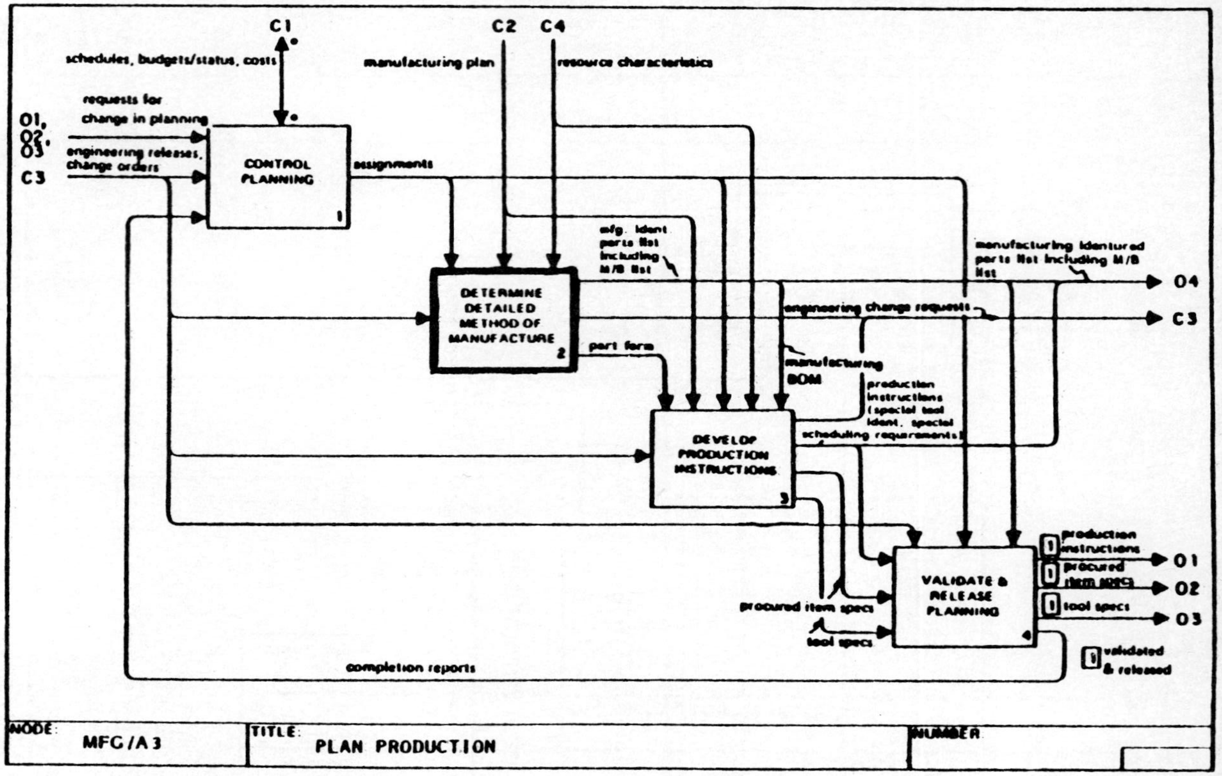

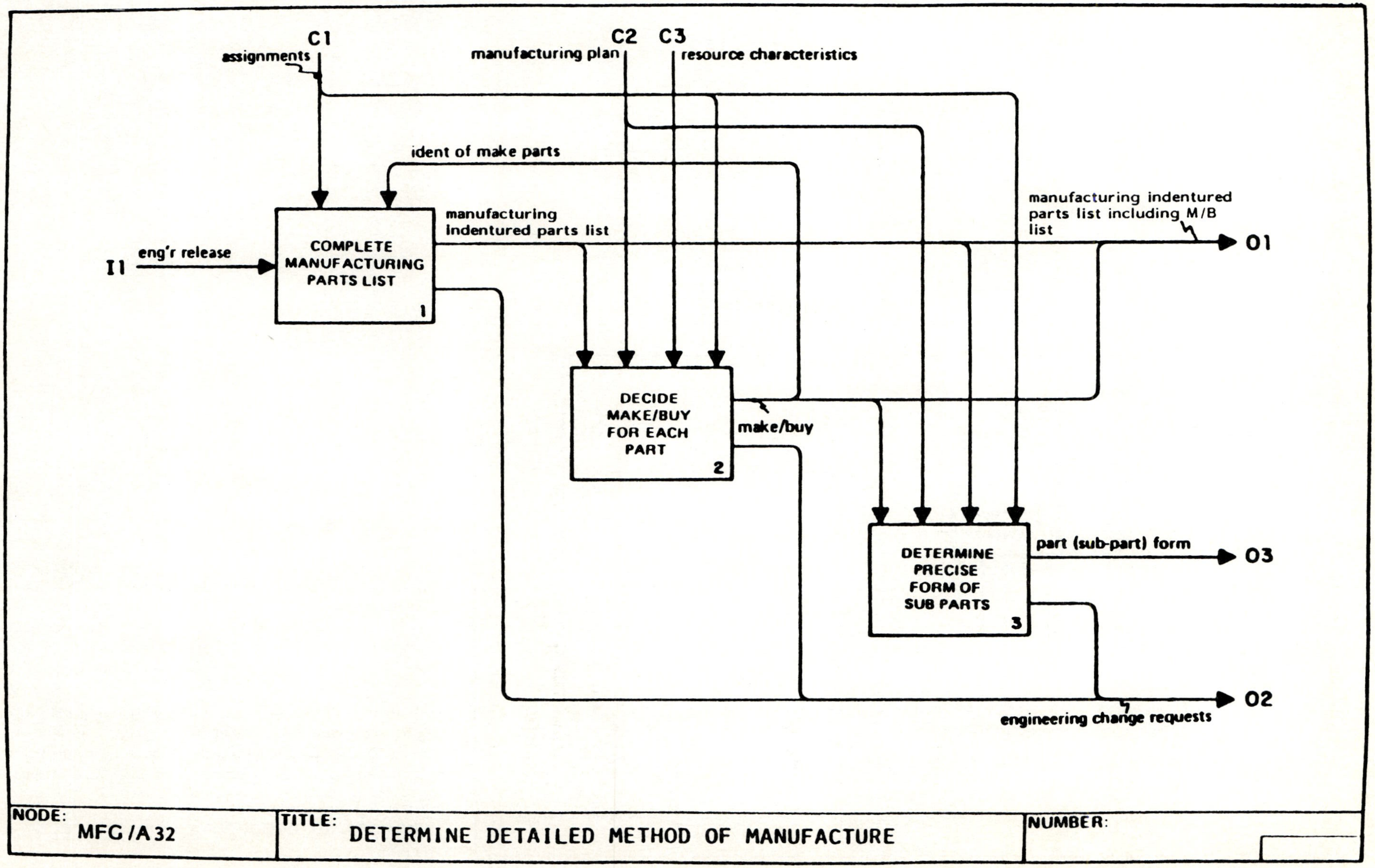
C1
assignments
C2
manufacturing plan
C3
resource characteristics
ident of make parts
I1
eng'r release
COMPLETE MANUFACTURING PARTS LIST
1
manufacturing indentured parts list
DECIDE MAKE/BUY FOR EACH PART
2
make/buy
manufacturing indentured parts list including M/B list
O1
DETERMINE PRECISE FORM OF SUB PARTS
3
part (sub-part) form
O3
engineering change requests
O2
NODE: MFG/A32
TITLE: DETERMINE DETAILED METHOD OF MANUFACTURE
NUMBER:

A33 Develop Production Instructions

This activity splits the planning of the production instructions into four separate areas; Plan Line Assemblies and Installations (Box 1), Plan Batch Assemblies (Box 2), Plan Detailed Parts (Box 3), and Plan for Procured Items (Box 4).

Each of these activities converts the engineering release and change orders into an appropriate set of production instructions. Procured item specs are outputs from either the detailed parts plan or the procured items plan. Engineering change requests can be originated from any of these activities.

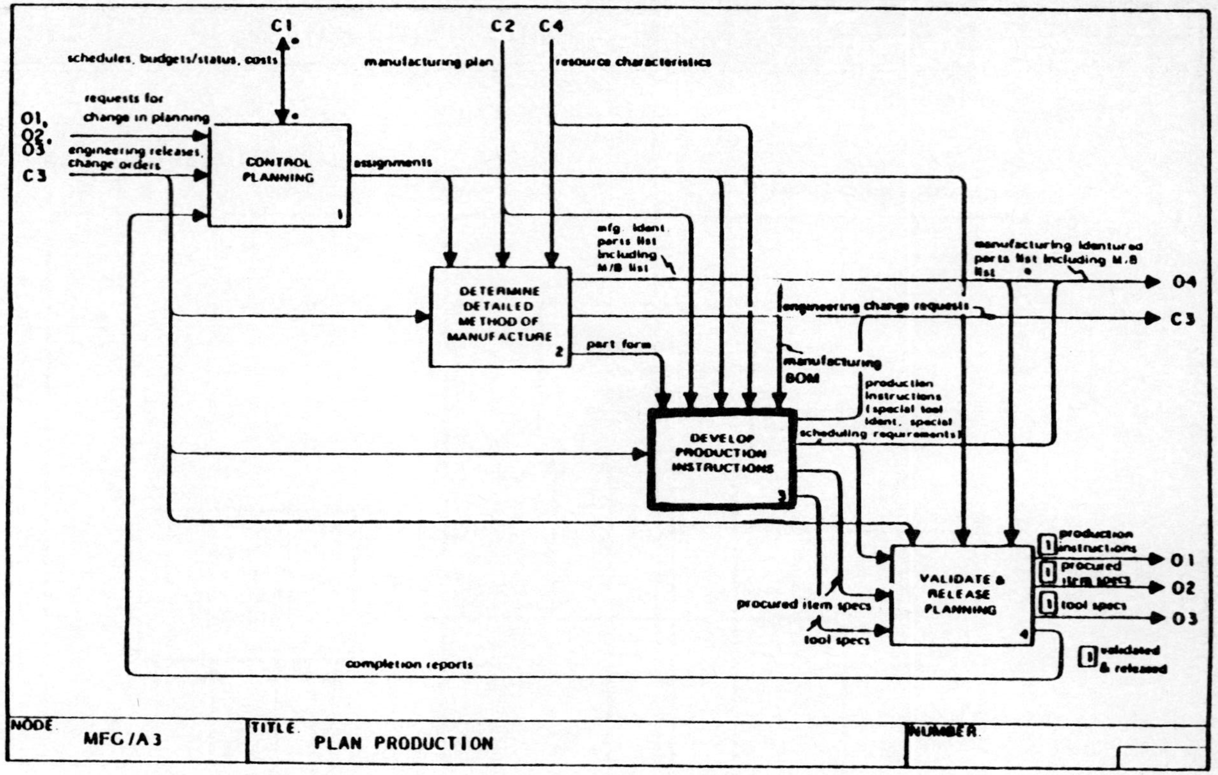

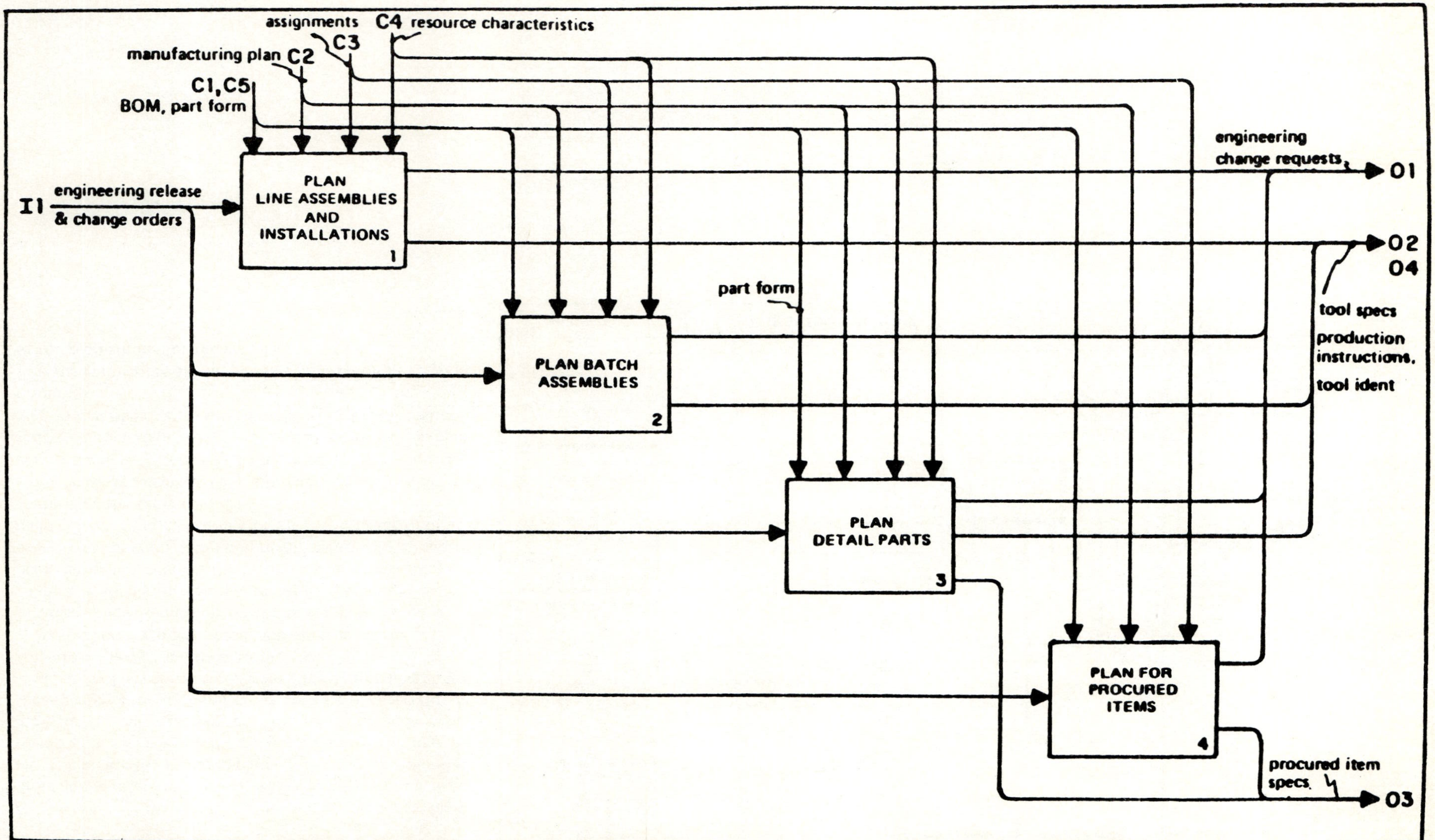

NODE: MFG/A33 | TITLE: DEVELOP PRODUCTION INSTRUCTIONS | NUMBER:

A4 Provide Production Resources

Production resources come in four major categories: facilities, equipment, tools, and people. The primary controls for all of them are the resource plans (C3—from the manufacturing plan) which specify what they are to be and the methods for getting them; and the schedules (C2) which specify when they are required. In the case of tools (Box 3), detailed tool specifications (from production planning—A3) are needed in addition to the tool plans.

Interaction between these activities is fairly limited. Facilities place limitations on the space, power, and other support available for equipment, while the equipment imposes needs for these on the facilities. Since most tools are used in conjunction with equipment, the interface requirements of the equipment must be known in order to design or procure the tools (Box 3).

Purchase requisitions (03) ultimately result in the procured items (I1) from which the resources are made. Resource characteristics (01) are provided to production planning so that plans can conform to the available (or planned) resources.

Tools and people are supplied to production in response to the requisitions for them (C4).

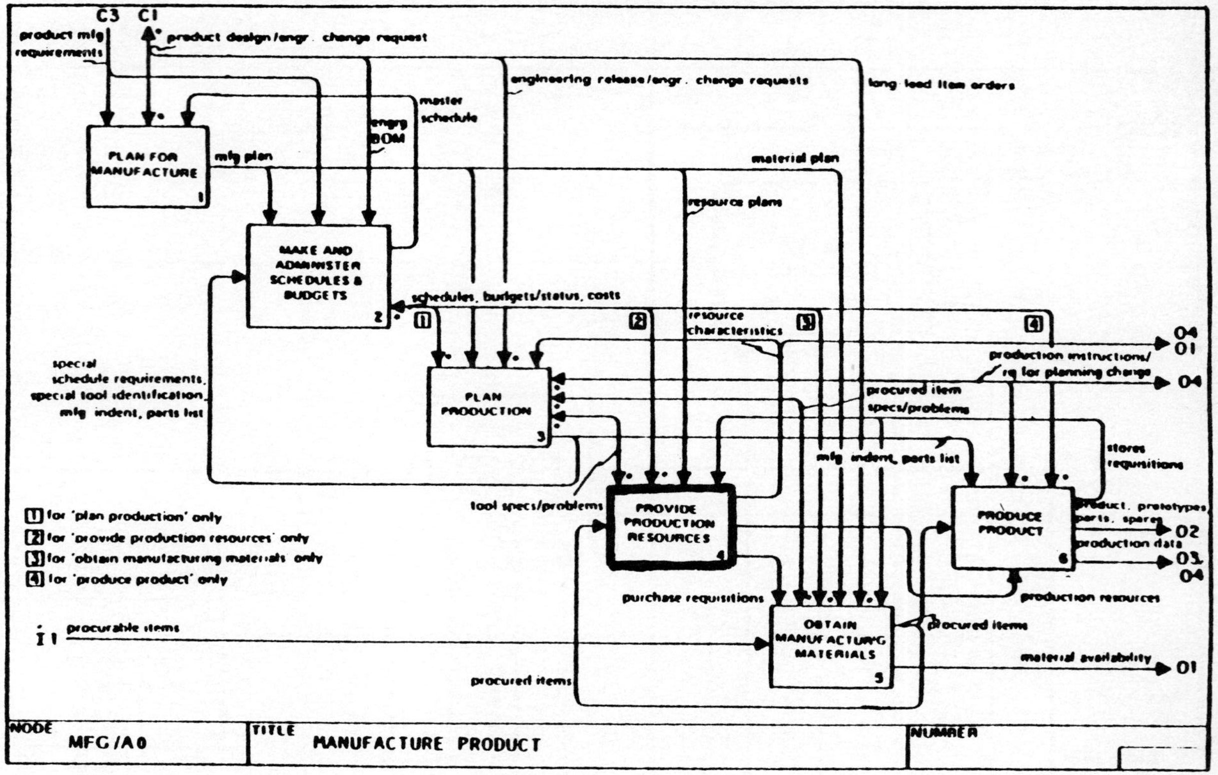

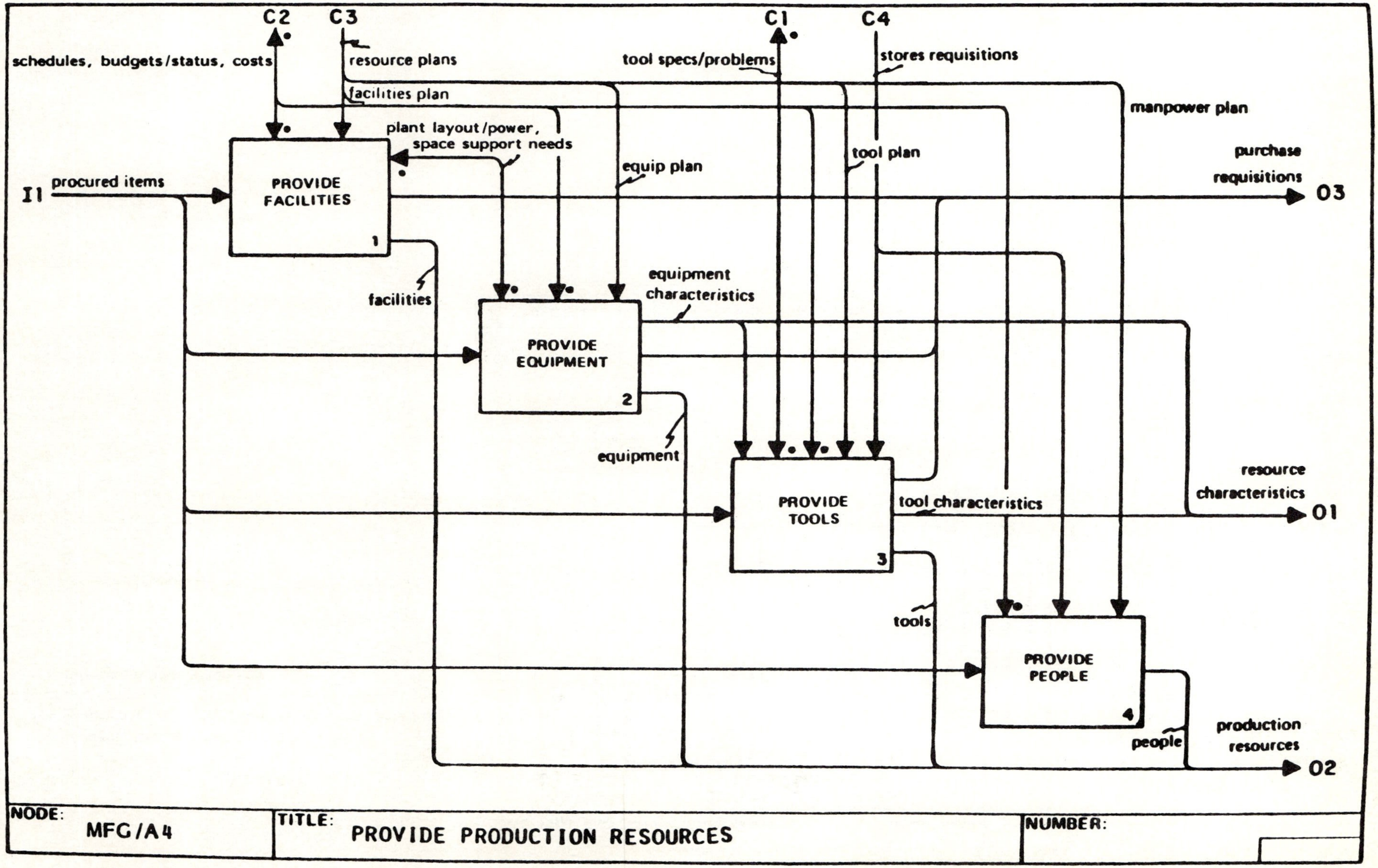

C2
C3
C1
C4
schedules, budgets/status, costs
resource plans
facilities plan
tool specs/problems
stores requisitions
manpower plan
plant layout/power,
space support needs
equip plan
tool plan
I1
procured items
PROVIDE FACILITIES
1
purchase requisitions
O3
facilities
equipment characteristics
PROVIDE EQUIPMENT
2
equipment
PROVIDE TOOLS
3
tool characteristics
resource characteristics
O1
tools
PROVIDE PEOPLE
4
people
production resources
O2
NODE: MFG/A4
TITLE: PROVIDE PRODUCTION RESOURCES
NUMBER:

A41 Provide Facilities

Provide facilities encompasses planning, building, modifying, operating and maintaining facilities within the prescribed budgets to house the work identified.

The facility plan (C2) and the schedule, work orders and tracking information (C1) act as controls in that they identify facility requirements and schedules.

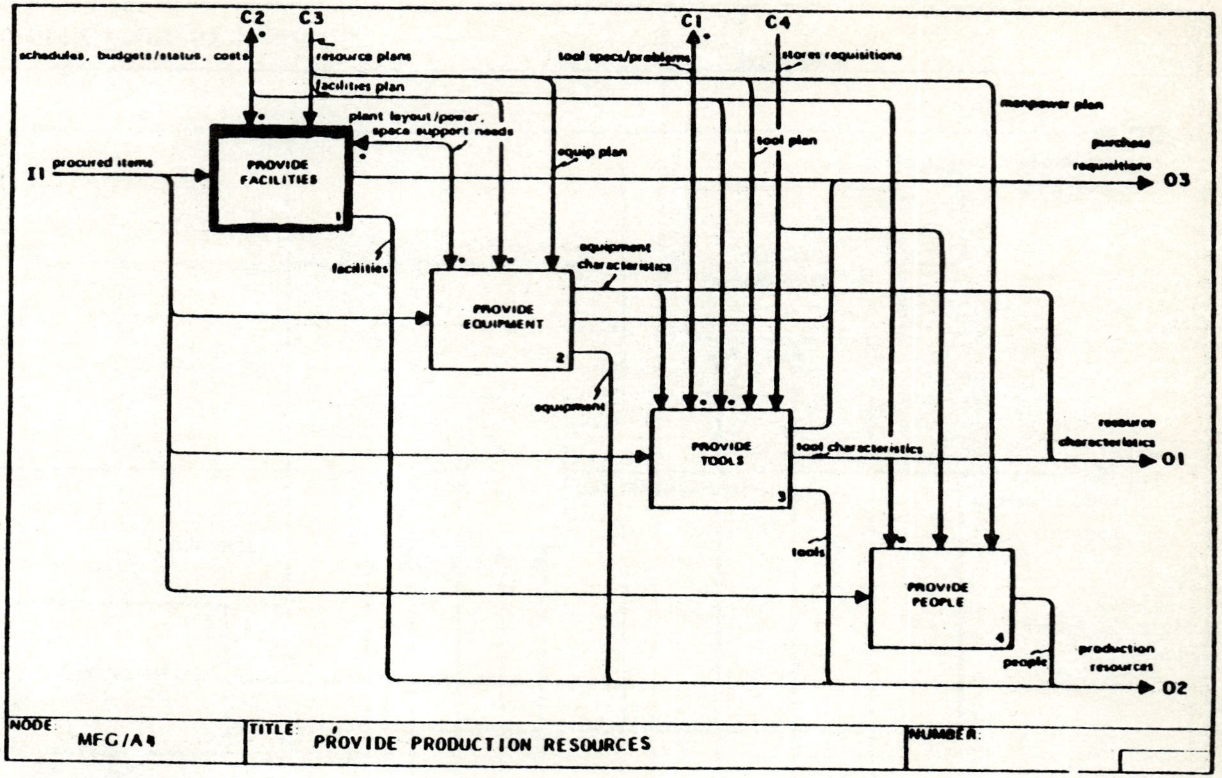

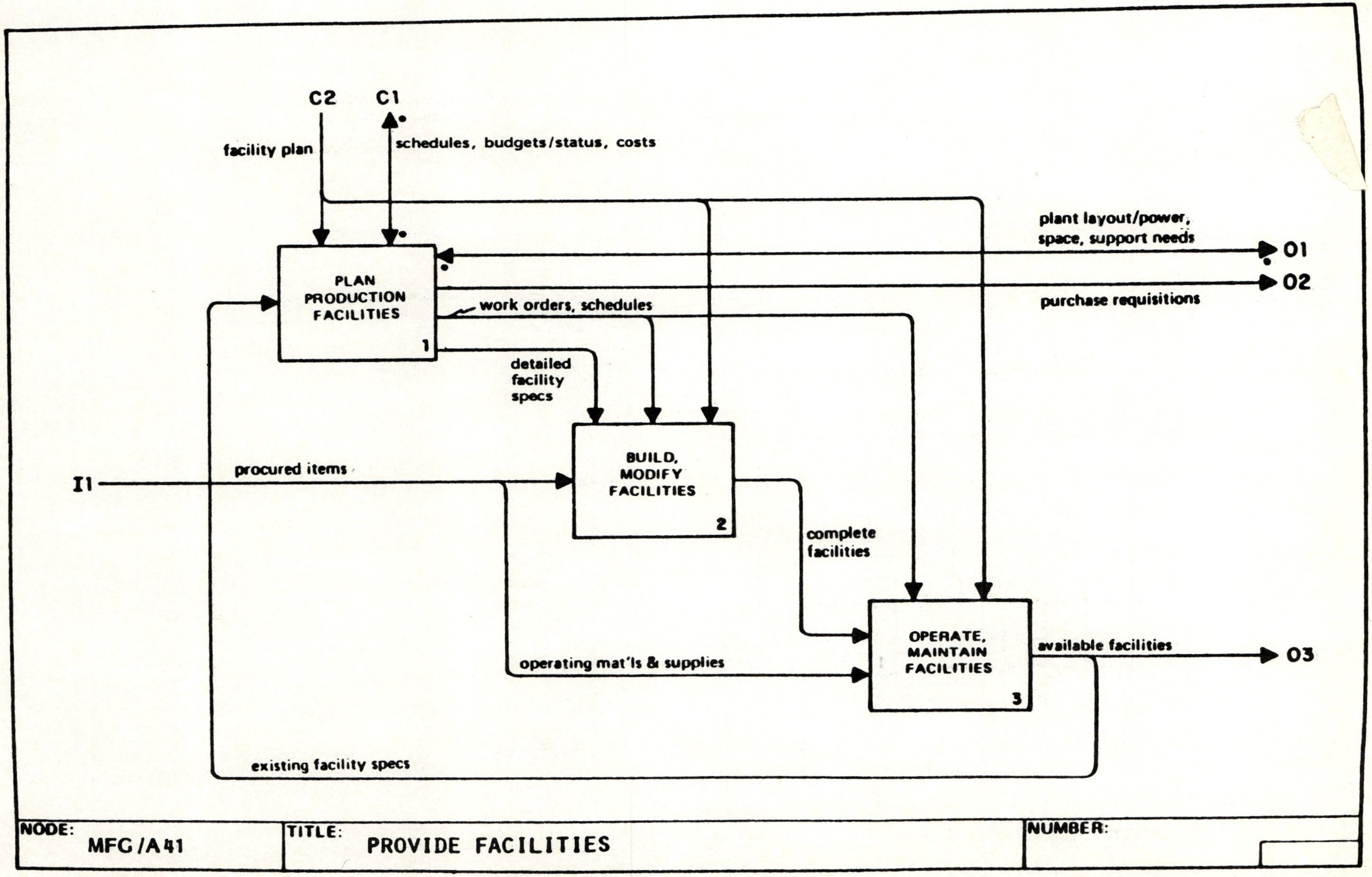

NODE: MFG/A41 | TITLE: PROVIDE FACILITIES | NUMBER:

A42 Provide Equipment

This function develops equipment plans (Box 1), issues material requisitions (Box 2), and, after installation, controls (Box 3) and maintains the equipment (Box 4).

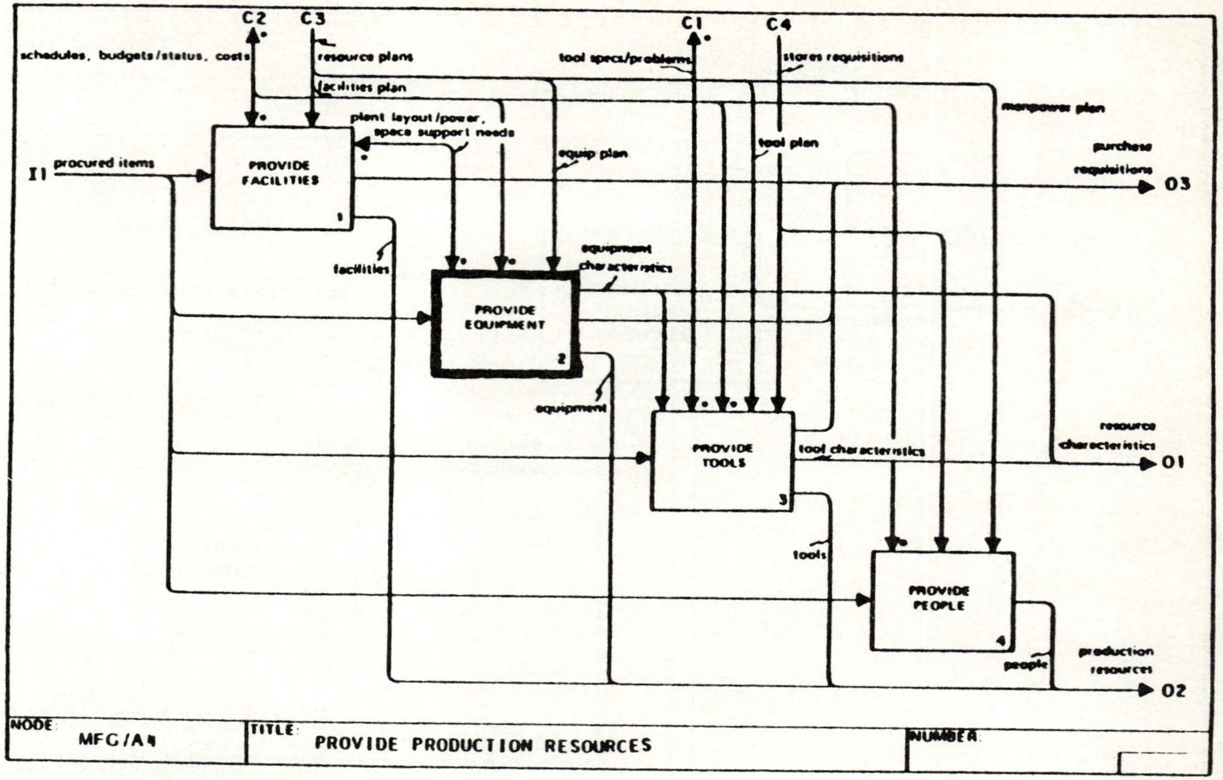

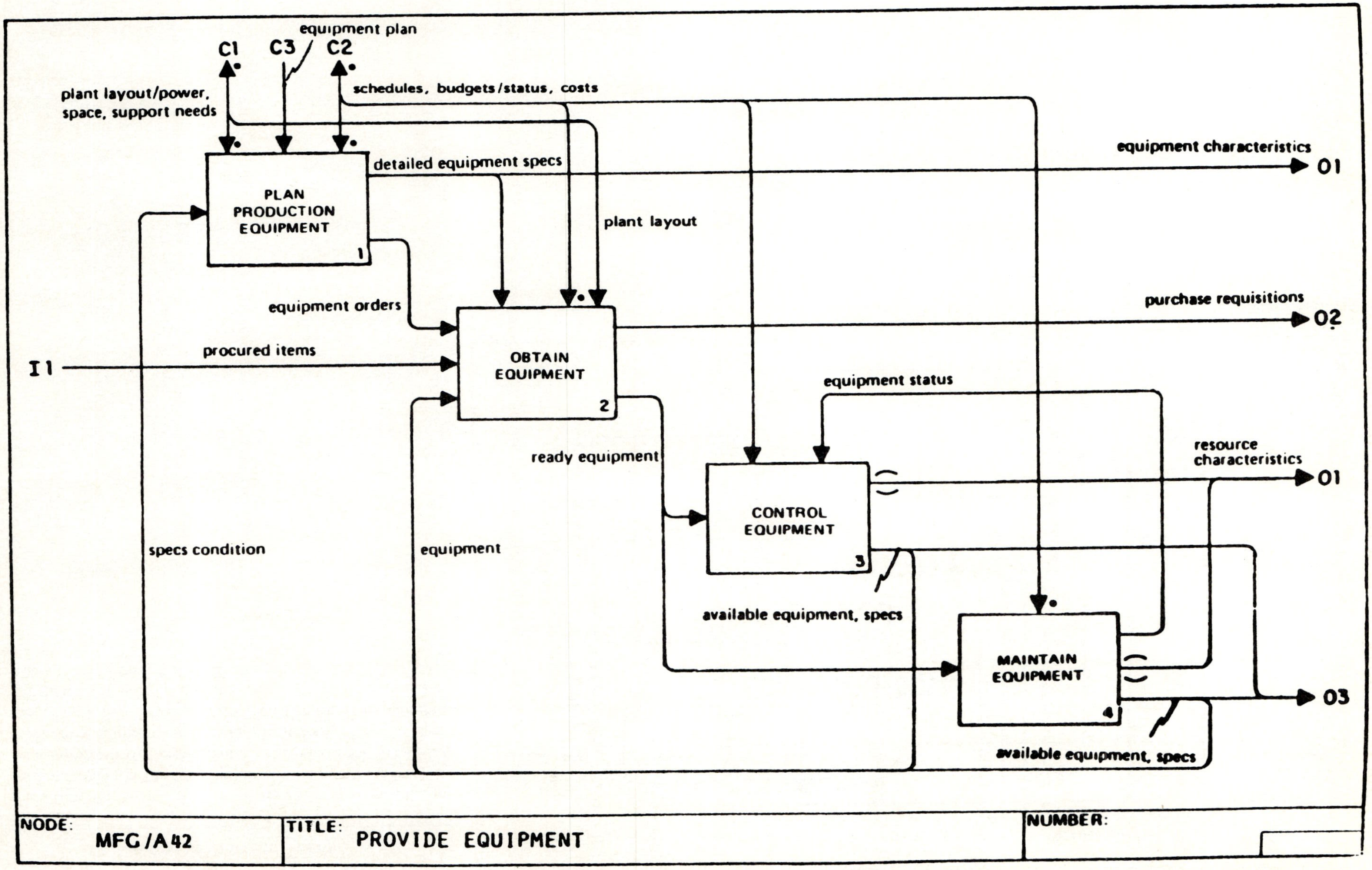

C1
C3
C2
equipment plan
plant layout/power,
space, support needs
schedules, budgets/status, costs
PLAN
PRODUCTION
EQUIPMENT
1
detailed equipment specs
equipment characteristics
O1
plant layout
equipment orders
I1
procured items
OBTAIN
EQUIPMENT
2
purchase requisitions
O2
equipment status
ready equipment
CONTROL
EQUIPMENT
3
resource
characteristics
O1
specs condition
equipment
available equipment, specs
MAINTAIN
EQUIPMENT
4
O3
available equipment, specs
NODE:
MFG/A42
TITLE:
PROVIDE EQUIPMENT
NUMBER:

A43 Provide Tools

This function is the process of design, planning, making, and managing of the tools required to support the work to be accomplished. Requirements originate in the Tool Plan (C4) and are used in the design of the tools (Box 2). These designs are then planned (Box 3) which give the manufacturing instructions for making tools (Box 4). Tools are then managed and distributed for use (Box 5).

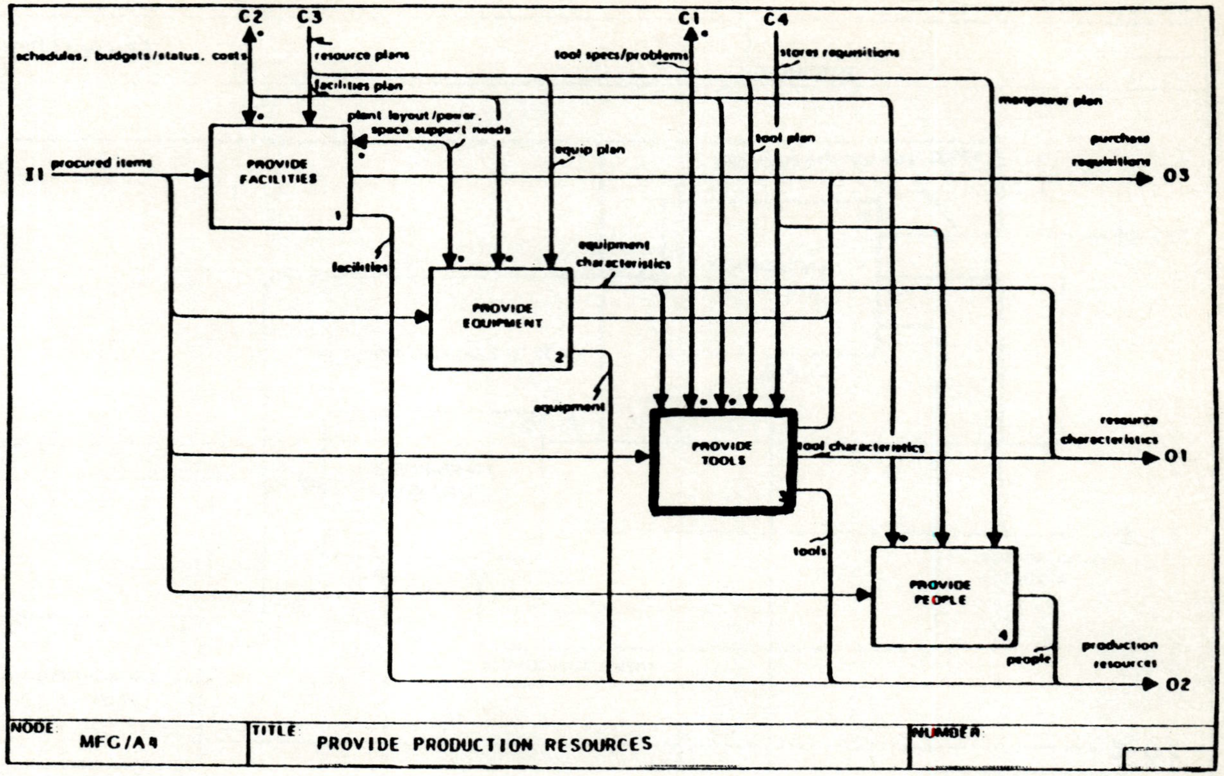

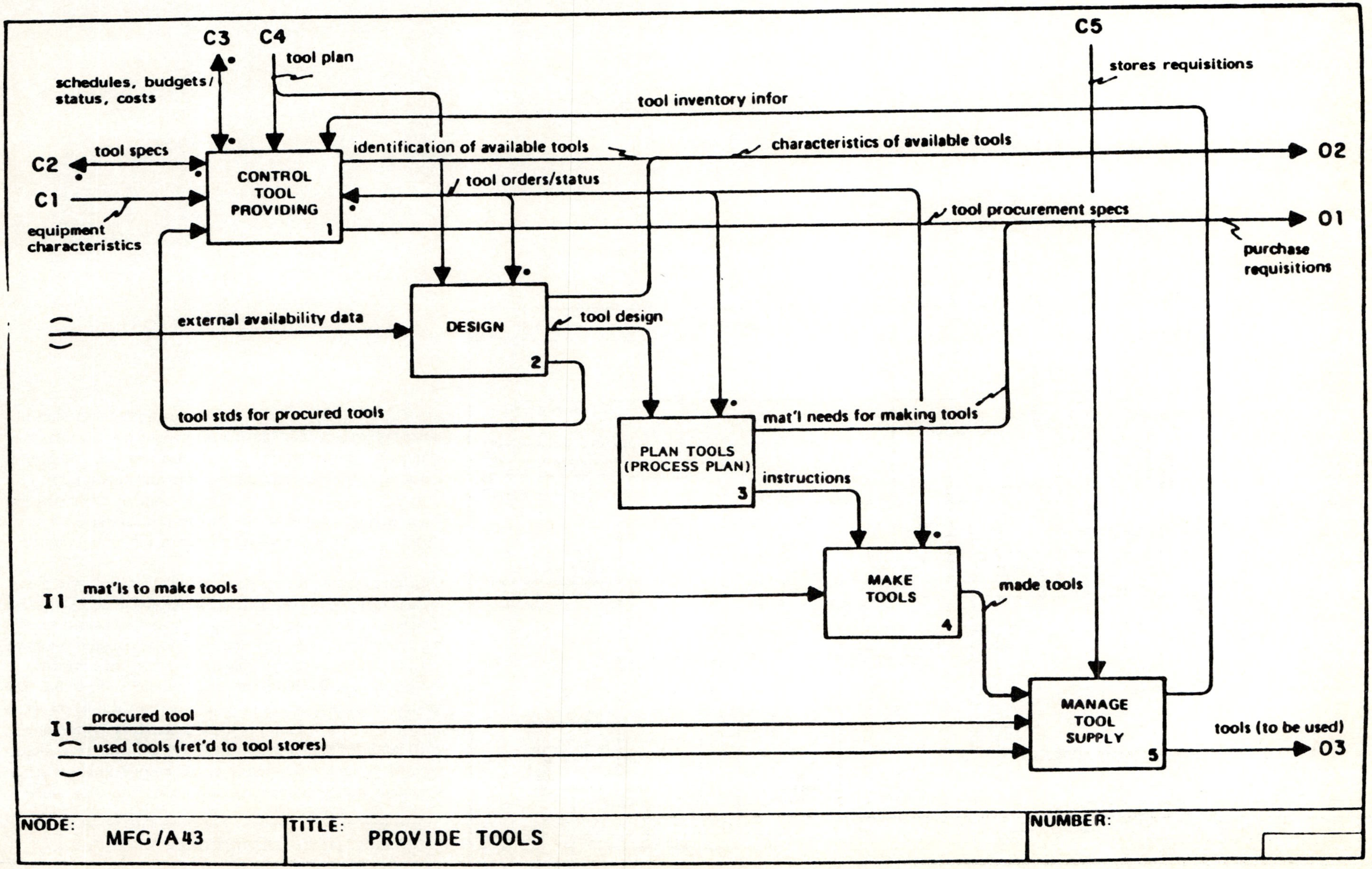

NODE: MFG/A43	TITLE: PROVIDE TOOLS	NUMBER:

A5 Obtain Manufacturing Materials

Starting with the overall material plan (C4), the materials needs schedule and budget (C3) and the advanced material orders (C5), detailed materials requirements planning (Box 1) can occur. As detailed procured item specs are available (C2), inventory (on hand and on order) can be allocated, and requisitions generated for additional procurements.

Procurement (Box 2) actually obtains the items (I1) from vendors, subcontractors, the customer (GFE, etc.), reporting any problems in availability (to specification or to schedule). Received material is inspected (Box 3) and if accepted, stored until issued (01) in response to a material requisition (C6).

Material records (inventory, on-order, rejection) are critical in knowing what is or will be available so that all material needs can be met.

In addition, information on material availability information (types and lead times for obtaining as well as what is in inventory) is provided (to engineering) so that designers can take this into account.

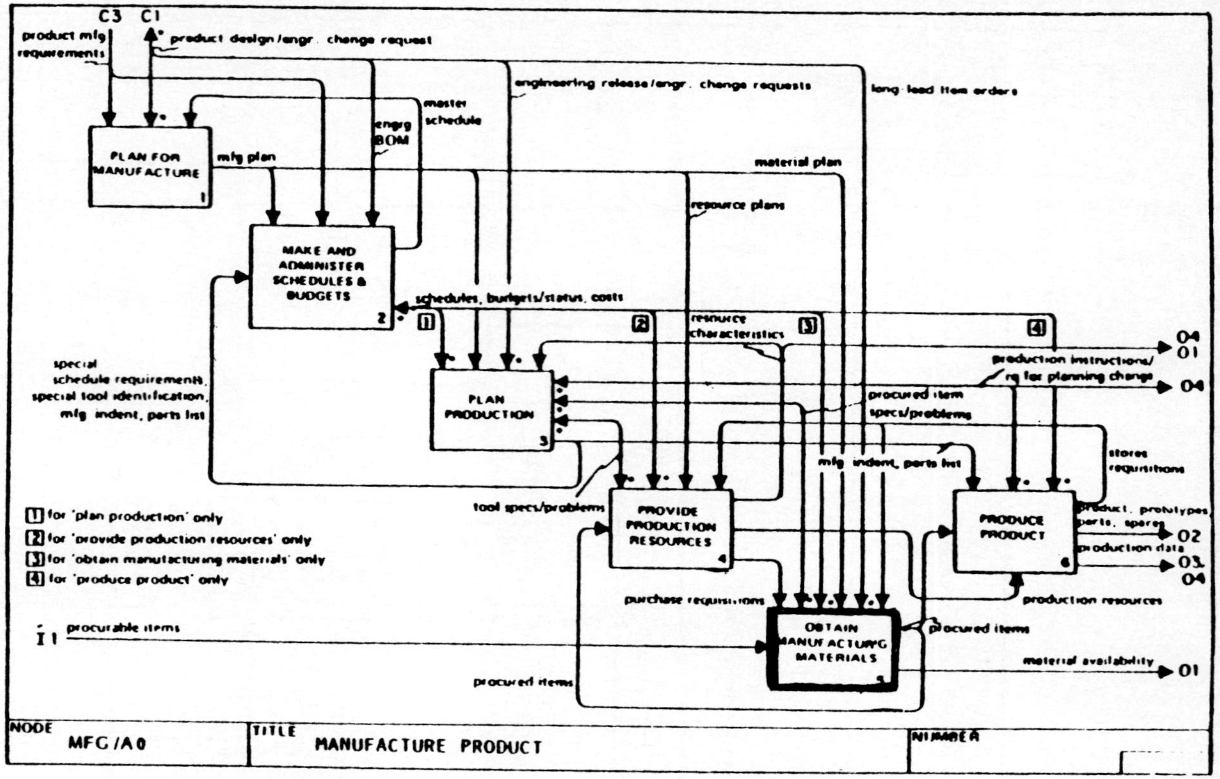

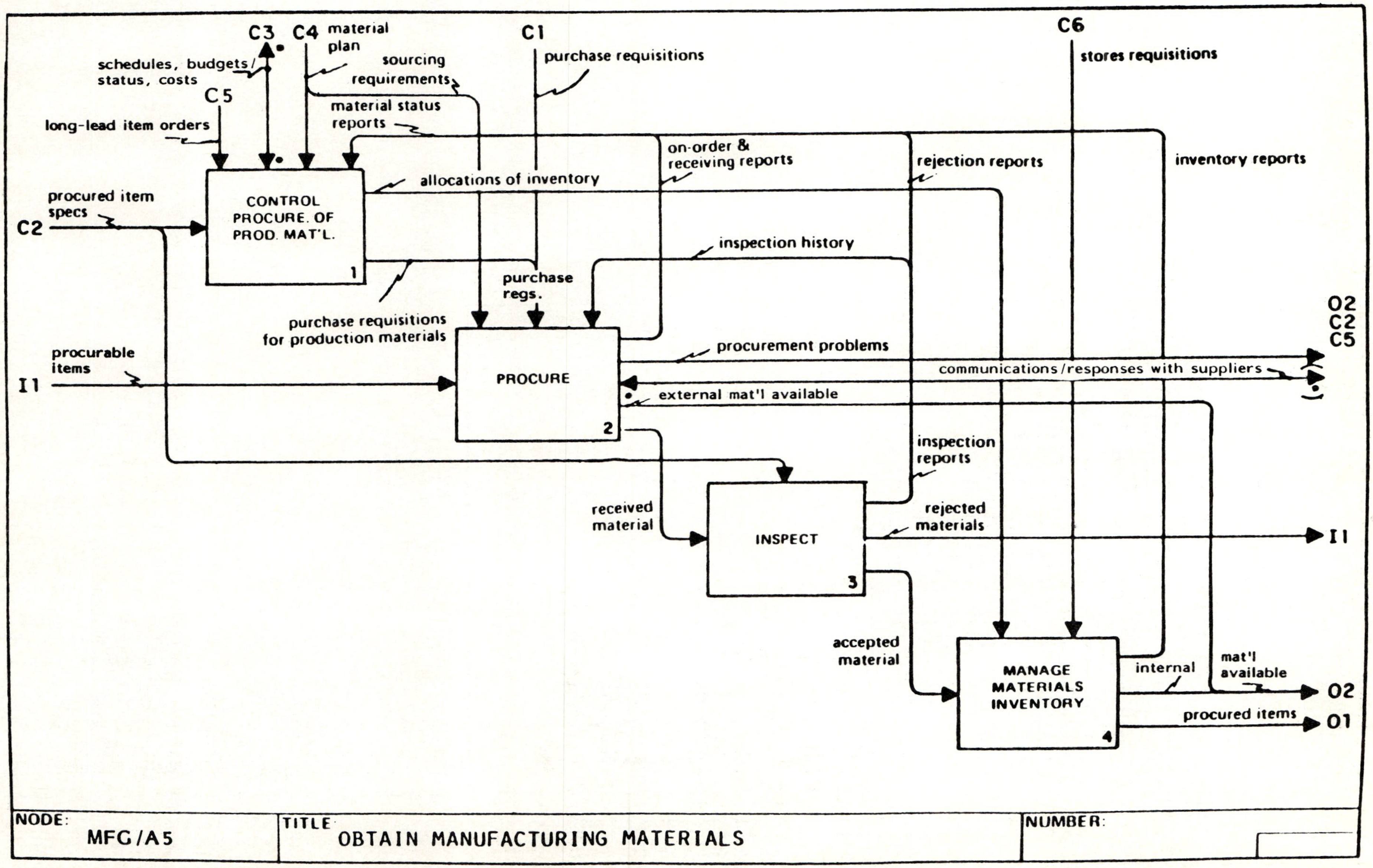
C3
C4
material plan
C1
purchase requisitions
C6
stores requisitions
schedules, budgets/status, costs
sourcing requirements
C5
material status reports
long-lead item orders
on-order & receiving reports
rejection reports
inventory reports
allocations of inventory
procured item specs
C2
CONTROL PROCURE. OF PROD. MAT'L.
1
inspection history
purchase regs.
purchase requisitions for production materials
O2
C2
C5
procurable items
procurement problems
I1
PROCURE
2
communications/responses with suppliers
external mat'l available
inspection reports
received material
rejected materials
INSPECT
3
I1
accepted material
MANAGE MATERIALS INVENTORY
4
internal
mat'l available
O2
procured items
O1
NODE: MFG/A5
TITLE: OBTAIN MANUFACTURING MATERIALS
NUMBER:

A51 Control Procurement of Production Materials

This function is responsible for procurement of the required production materials. Total material requirements (Box 1) are identified from the advanced orders (Long Lead), the material plan (C3) and scheduled requirements (C2). An inventory reconciliation (Box 2) is accomplished and additional procurement needs are identified (A51.203). An ordering approach is identified (Box 3) and material requisitions (Box 4) are written.

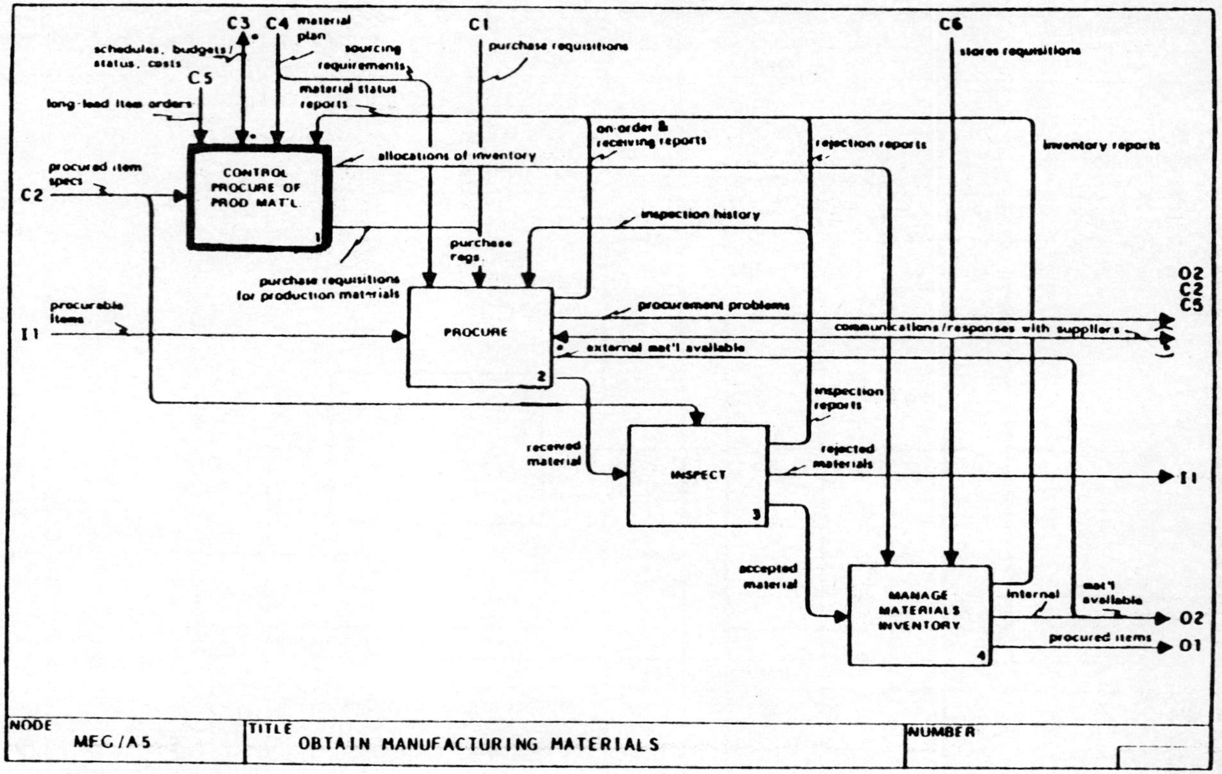

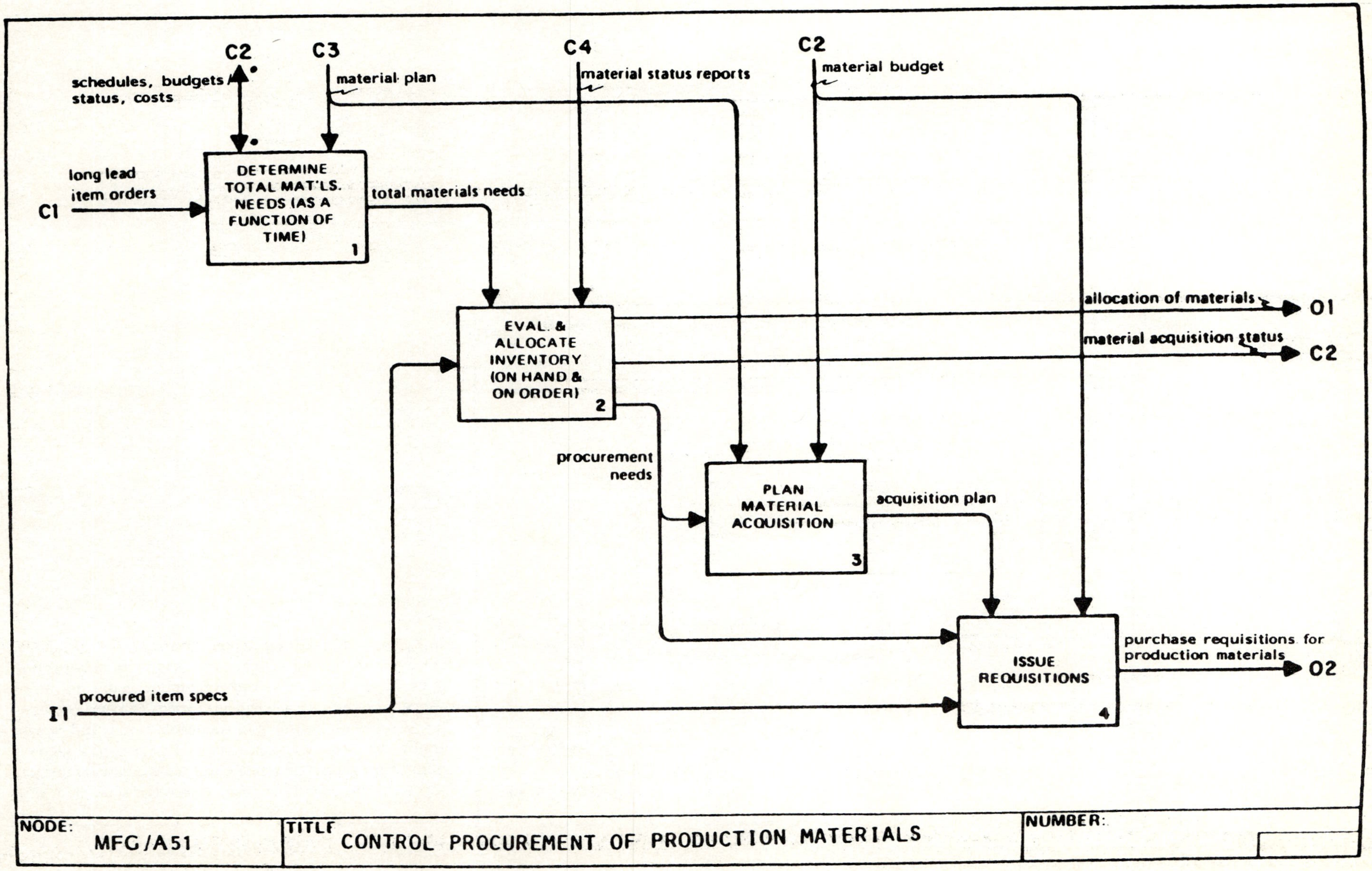

C2
C3
C4
C2
schedules, budgets
status, costs
material plan
material status reports
material budget
long lead
item orders
C1
DETERMINE TOTAL MAT'LS. NEEDS (AS A FUNCTION OF TIME)
1
total materials needs
EVAL. & ALLOCATE INVENTORY (ON HAND & ON ORDER)
2
allocation of materials
O1
material acquisition status
C2
procurement
needs
PLAN MATERIAL ACQUISITION
3
acquisition plan
ISSUE REQUISITIONS
4
purchase requisitions for
production materials
O2
I1
procured item specs
NODE:
MFG/A51
TITLE
CONTROL PROCUREMENT OF PRODUCTION MATERIALS
NUMBER:

A52 Procure

This function is responsible for developing orders, tracking, receiving purchased parts and materials, and closing orders. It collates (Box 1) data from vendor listing sources requirements, purchase requisition, and inspection history. The collected package goes to the place order (Box 2) function. From this function purchase orders go to the selected vendor(s) for work. They are tracked (Box 3) and complete work is received (Box 4). The orders are then closed (Box 5).

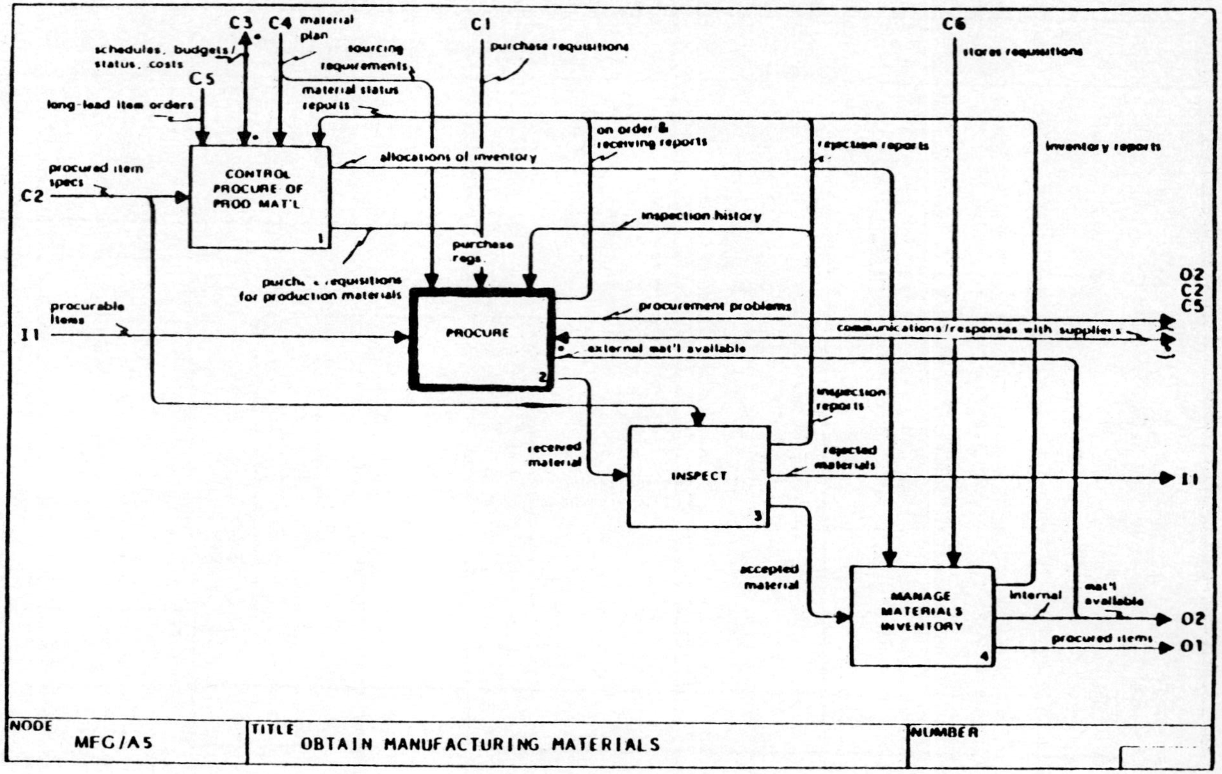

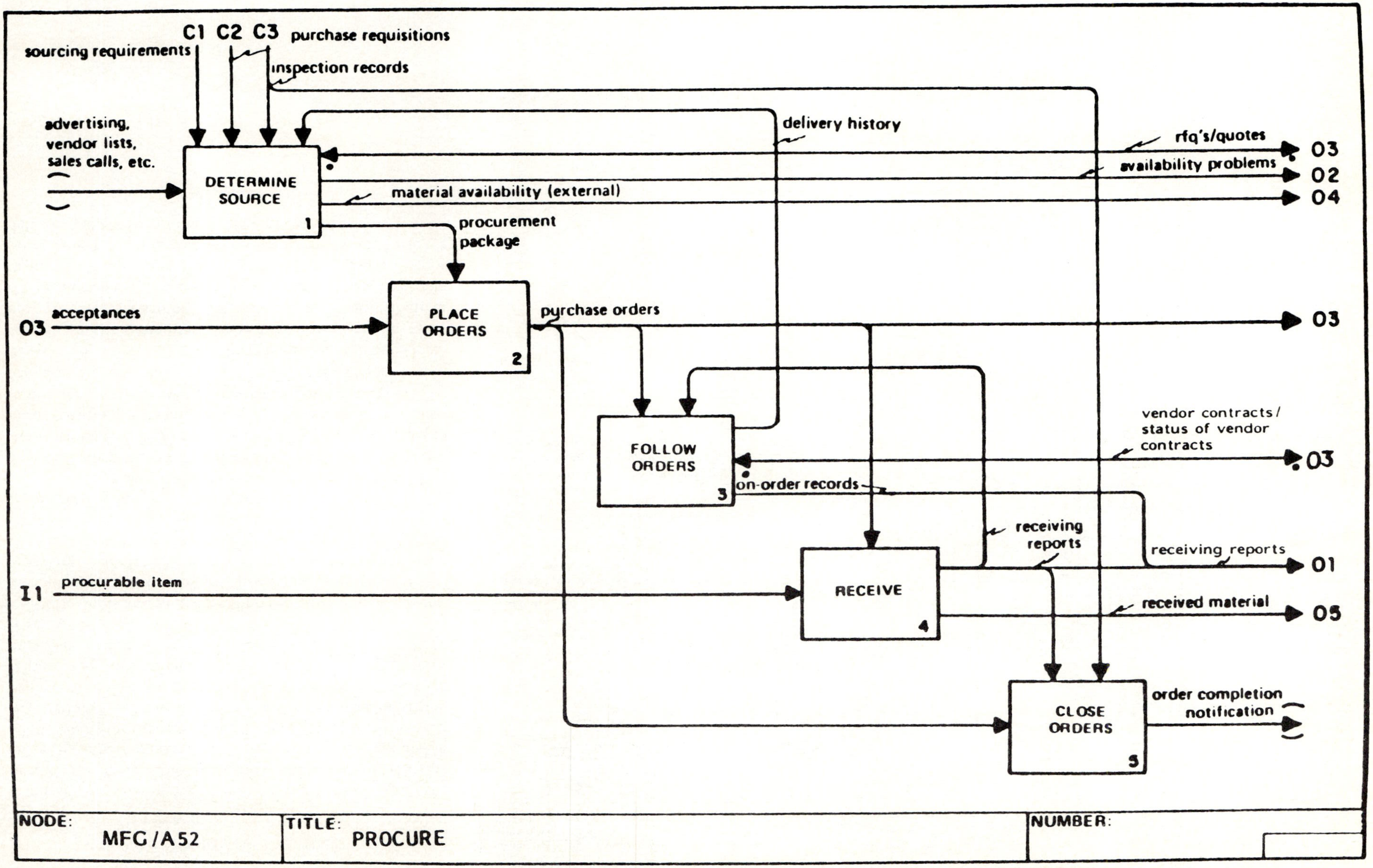
C1 C2 C3
sourcing requirements
purchase requisitions
inspection records
advertising, vendor lists, sales calls, etc.
DETERMINE SOURCE 1
delivery history
rfq's/quotes
O3
availability problems
O2
material availability (external)
O4
procurement package
O3
acceptances
PLACE ORDERS 2
purchase orders
O3
FOLLOW ORDERS 3
vendor contracts/ status of vendor contracts
O3
on-order records
receiving reports
receiving reports
O1
I1
procurable item
RECEIVE 4
received material
O5
CLOSE ORDERS 5
order completion notification
NODE: MFG/A52
TITLE: PROCURE
NUMBER:

A6 Produce Product

Using the manufacturing indentured parts list (C1) and the production instructions (C2), production control (Box 1) can determine the proper times for performing production tasks (taking into account load leveling requirements as well as scheduled due dates). This results in the issue of kitting orders (for collecting needed items) and work packages (to initiate required fabrication, assembly, test). The work packages include quantities, dates, and the instructions (from C1) for the tasks.

All the tools and materials (finished parts and subassemblies plus procured materials) necessary for a task are collected into kits (Box 2). As necessary, requisitions are made to obtain tools from tool stores or materials from procured item stores. In addition, Box 2 prepares spare parts and repair retrofit kits to be delivered to the customer.

Actual physical production (shop floor) is done by Box 3, converting the kits into finished items in accordance with the work packages. In the process, difficulties encountered may result in requests for planning changes (C2). Finished items are returned to production stores (if subparts) or forwarded to test (if completed items). Test, checkout and delivery (Box 4) result in products ready for delivery. Occasionally, testing requires disassembly, resulting in parts needing further assembly. Both Boxes 3 and 4 contribute to the production technical data which describes each specific instance of a product (inspection and test results, lot or serial control of information on components, etc.). Box 5 delivers the tested products, kits, parts, etc.

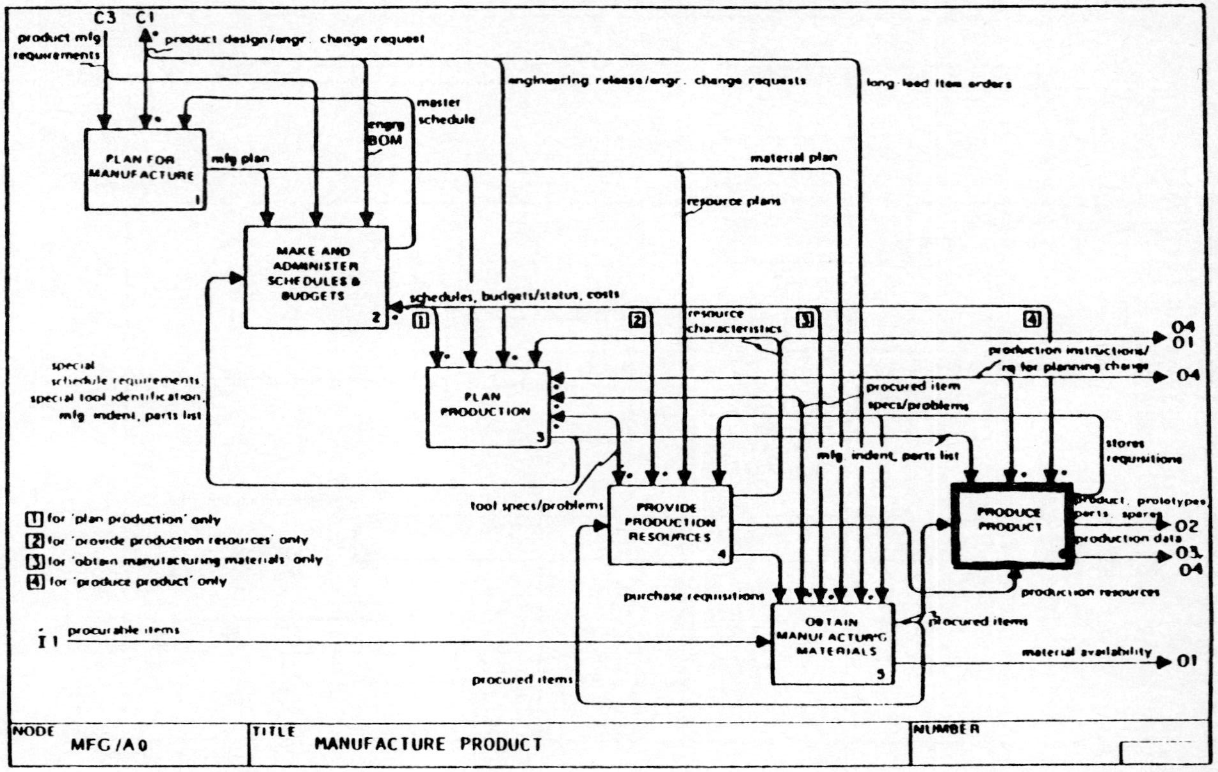

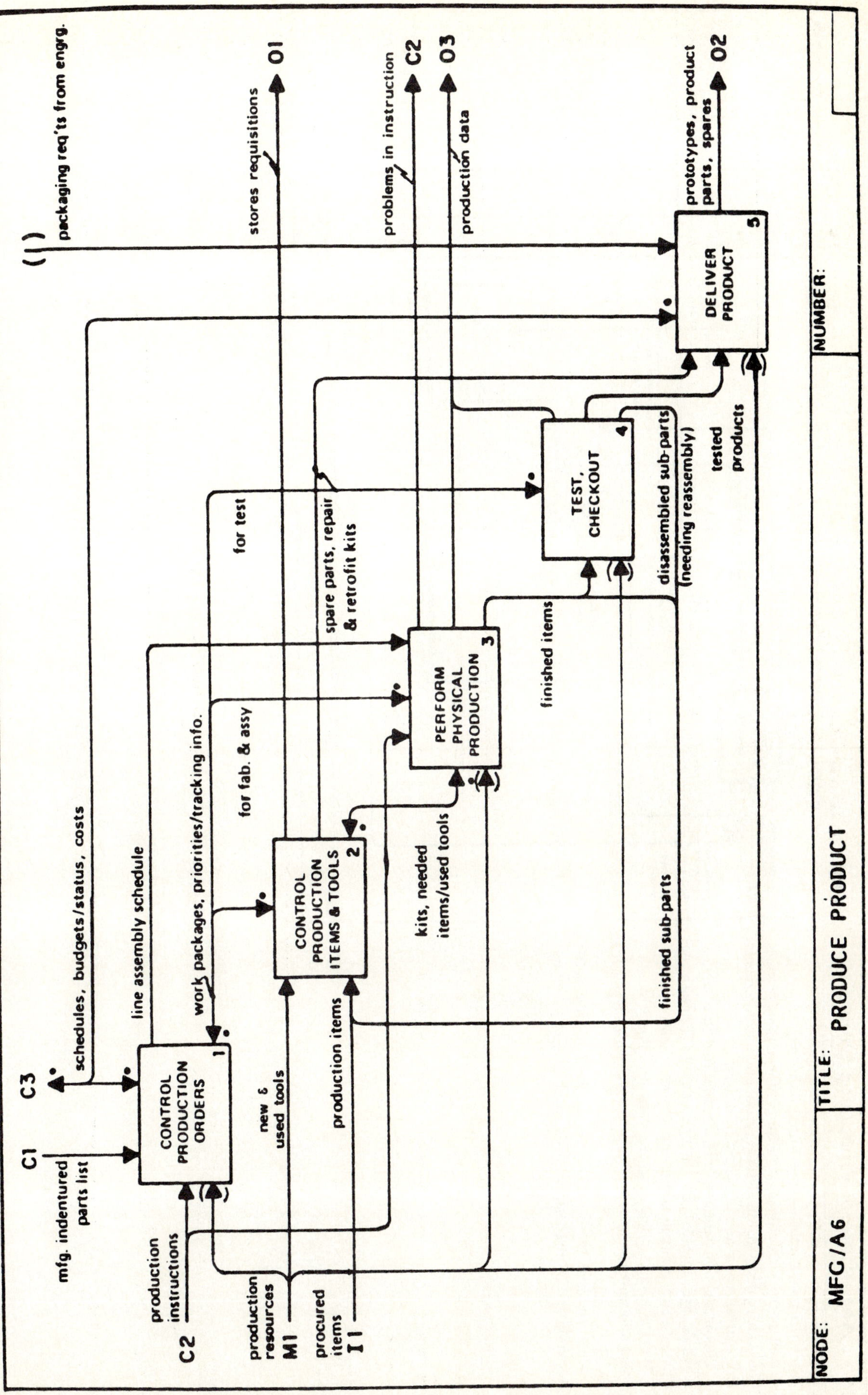

C1
mfg. indentured parts list
C3
schedules, budgets/status, costs
packaging req'ts from engrg.
C2
production instructions
production resources
M1
procured items
I1
CONTROL PRODUCTION ORDERS
1
line assembly schedule
work packages, priorities/tracking info.
new & used tools
production items
CONTROL PRODUCTION ITEMS & TOOLS
2
for fab. & assy
for test
stores requisitions
O1
kits, needed items/used tools
spare parts, repair & retrofit kits
PERFORM PHYSICAL PRODUCTION
3
problems in instruction
C2
production data
O3
finished items
finished sub-parts
TEST, CHECKOUT
4
disassembled sub-parts (needing reassembly)
tested products
DELIVER PRODUCT
5
prototypes, product parts, spares
O2
NODE:
MFG/A6
TITLE:
PRODUCE PRODUCT
NUMBER:

A61 Control Production Orders

Using the manufacturing indentured parts lists, schedules, as well as stock records and tracking information, production control develops the detailed production requirements and spares, repair and retrofit ship orders.

The production requirements incorporated in a release schedule are adjusted for changes due to Engineering Changes, procurement status and shortage items.

Work Packages released to the shop are subject to additional adjustments at the work center level where priorities are set.

Once released these orders are followed up for status reporting and cost reporting for feedback to Boxes 2 and 4.

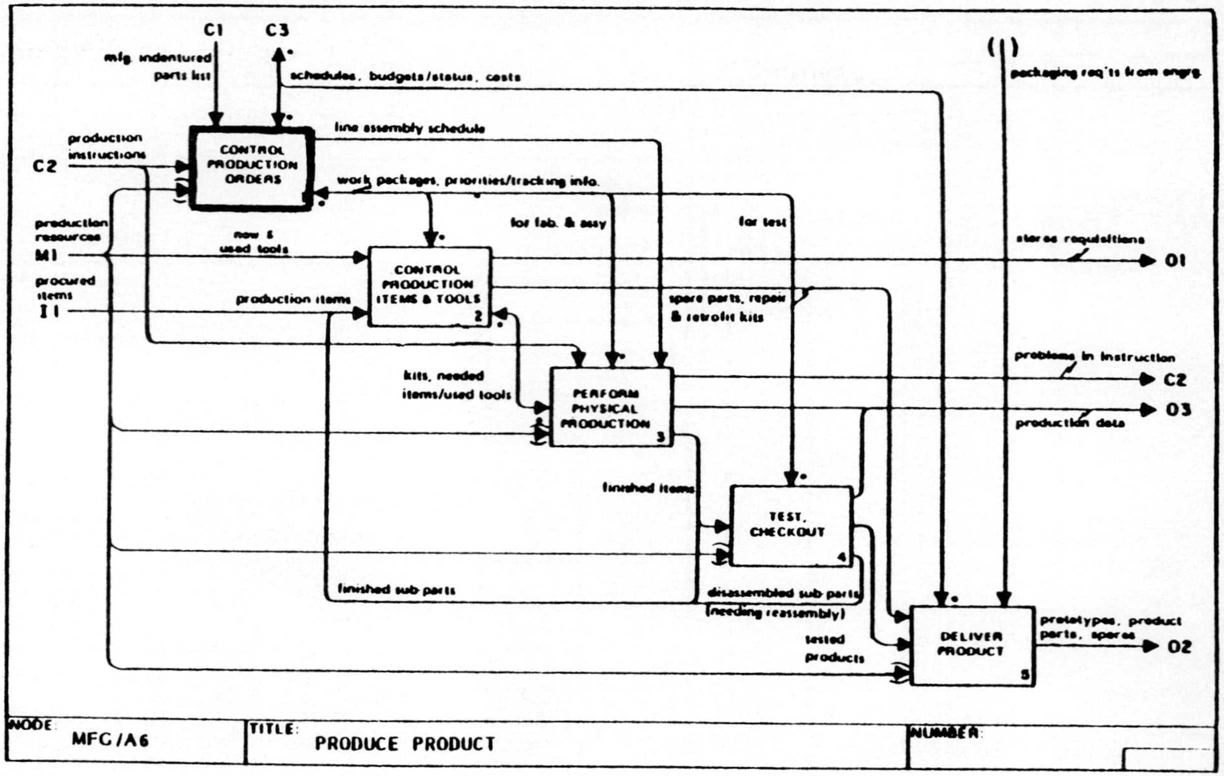

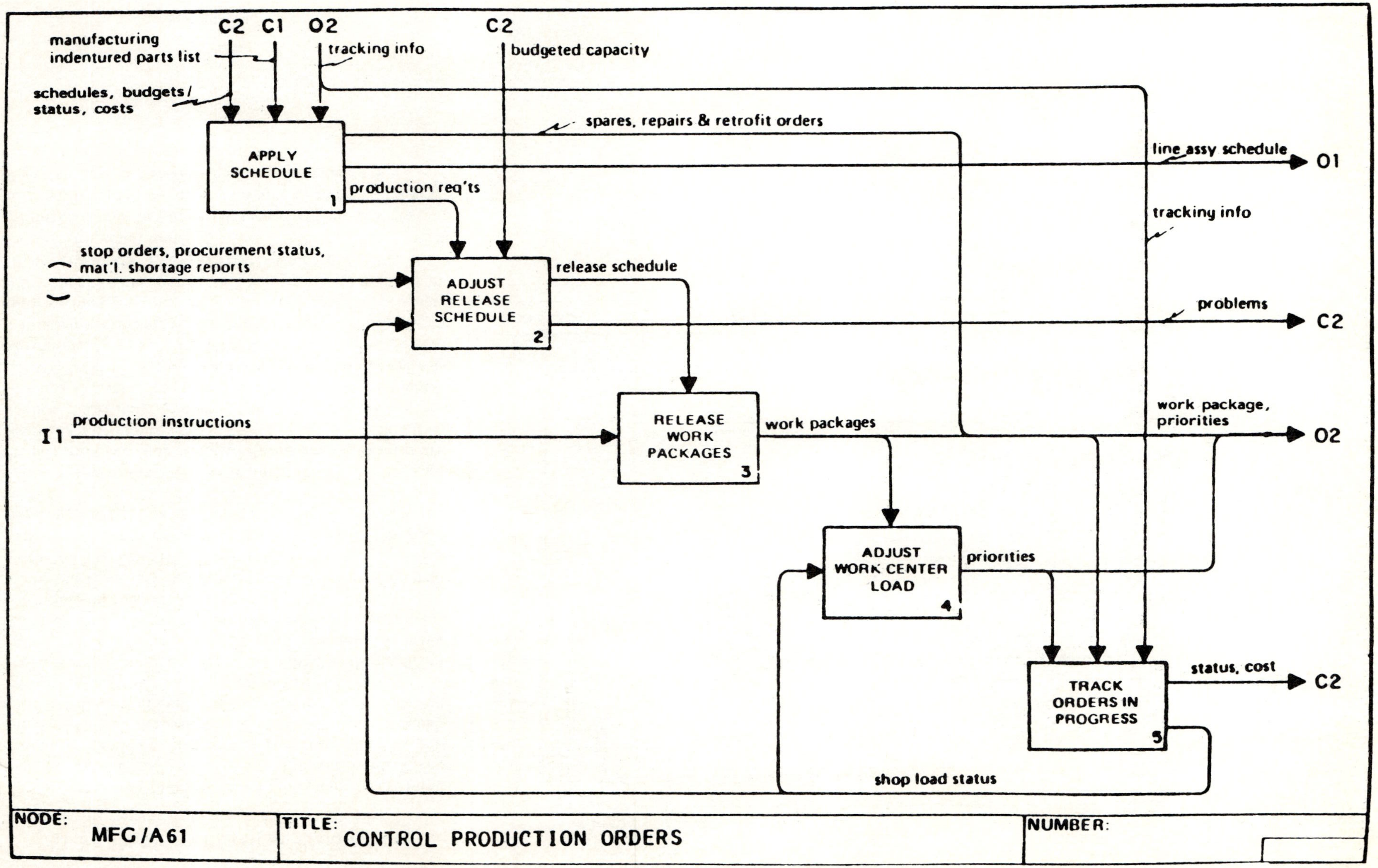

NODE: MFG/A61	TITLE: CONTROL PRODUCTION ORDERS	NUMBER:

A63 Perform Physical Production

The functions performed by the activities depicted are those which convert purchased items into the end product. The consecutive activities depicted convert raw material into details; details and subassemblies into further subassemblies and assemblies, and finally build subassemblies, details and assemblies into major assemblies and installations.

The three activities are not shown linked as the results of each stage are passed back into storage prior to being released into the next stage at production. (This looping is not visible on this diagram but may be seen at A6).

The work package requirements are the central control on each of the activities. The status of each activity is fed back to "Control Production Orders" (A61). Scrapped parts, production time, and operator assignment is also fed back.

In addition to the materials required to be converted into the final product, tools required during these processes are also shown as inputs.

These tools may accompany the materials, or may be obtained at a later time. On completion of the production activity, (or if tools are worn during production), they will be returned to storage for refurbishment or replacement.

Problems regarding that method of manufacturing stated in the planning instruction, or problems regarding producibility found in the design that have been overlooked in "Plan Production" (A3), will be transmitted to "Plan Production" for resolution.

Finished items may not all be passed to a storage area but may be forwarded for inspection and/or testing, and may be put through a checkout phase prior to delivery.

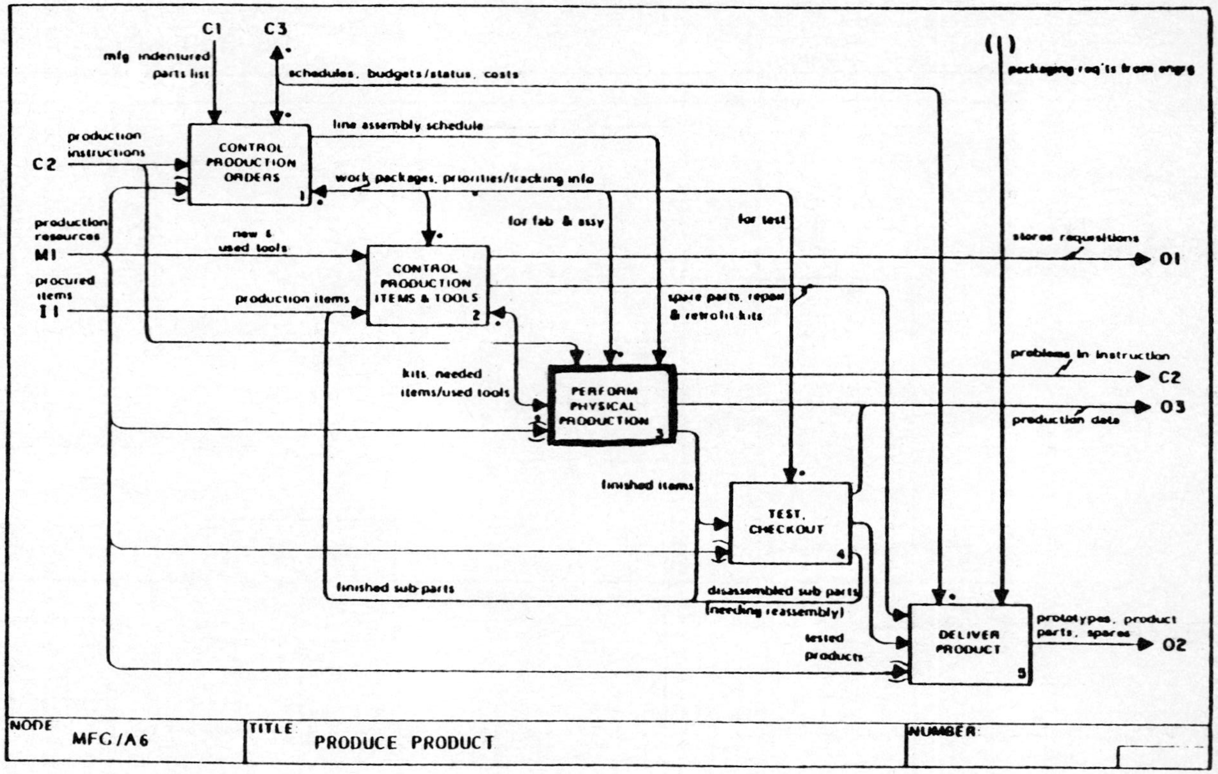

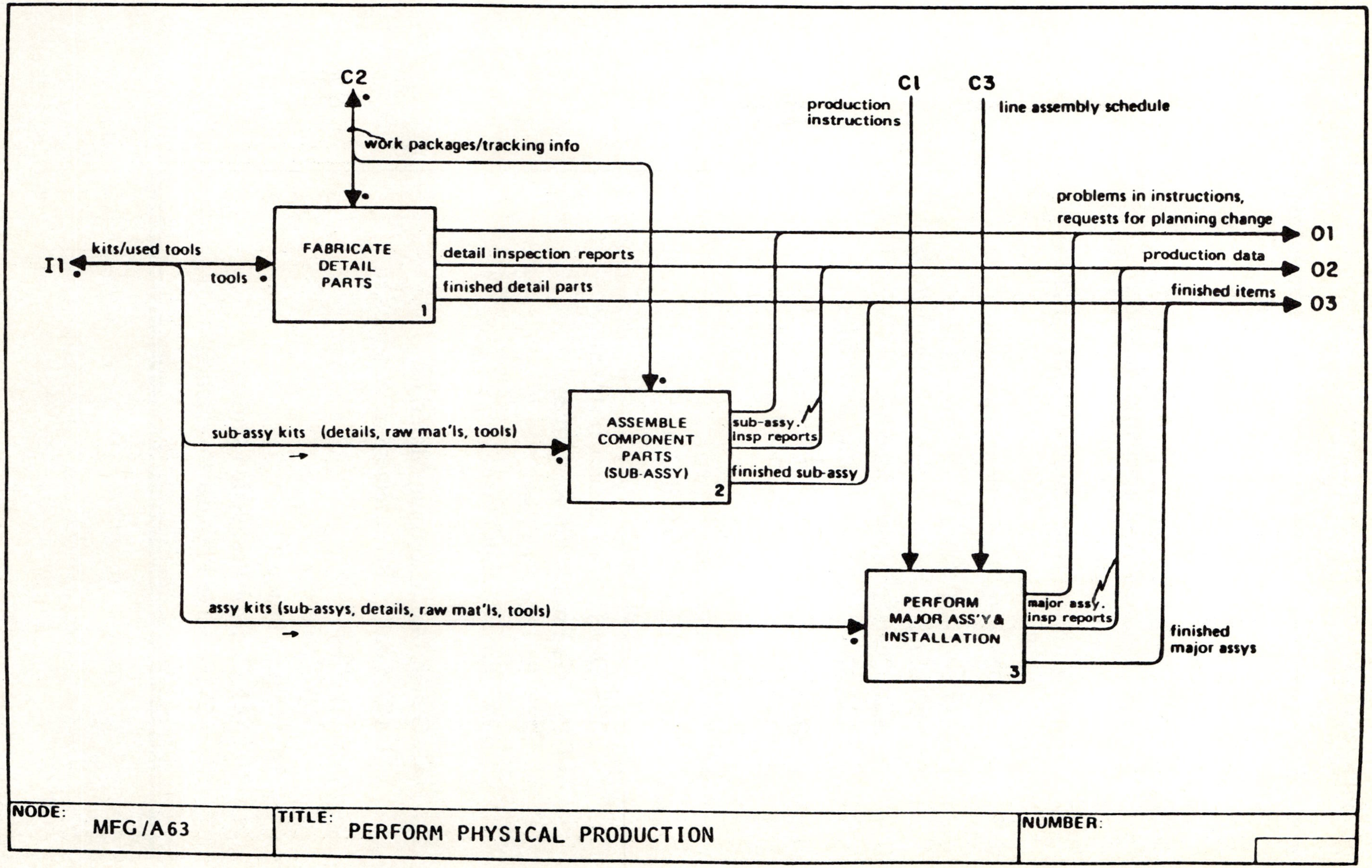

NODE: MFG/A63 | TITLE: PERFORM PHYSICAL PRODUCTION | NUMBER:

Diagram Index for the Integrated Computer-Aided Manufacturing Model

A-0 Manufacture Product

- **A0** Manufacture Product
- **A1** Plan for Manufacture
  - **A11** Assume a Structure and Method of Manufacture
  - **A12** Estimate Requirements, Time, and Cost to Produce
  - **A13** Develop Production Plans
  - **A14** Develop Support Activities Plans
- **A2** Make and Administer Schedules and Budgets
  - **A21** Develop Master Schedule
  - **A22** Develop Coordinating Schedules
  - **A23** Estimate Costs and Make Budgets
  - **A24** Monitor Performance to Schedule and Budget
- **A3** Plan Production
  - **A32** Determine Detailed Method of Manufacture
  - **A33** Develop Production Instructions
- **A4** Provide Production Resources
  - **A41** Provide Facilities
  - **A42** Provide Equipment
  - **A43** Provide Tools
- **A5** Obtain Manufacturing Materials
  - **A51** Control Procurement of Production Materials
  - **A52** Procure
- **A6** Produce Product
  - **A61** Control Production Orders
  - **A63** Perform Physical Production

Chapter

30

Data Processing

30.1
Application Domain

Data processing commonly deals with the manipulation of data into information from which people make day-to-day business decisions and from which an organization can manage itself. One of the biggest stumbling blocks to developing high-quality data-processing systems is the lack of understanding of how a company works.

30.1
Project Background

Large American oil companies have complicated billing processes: Their products are sold over a large geographic area, and prices and taxes vary from state to state. Their customer base is very large and constantly changing. Their sales records come in from regional areas and hence must be properly processed to reflect regional trends. A data-processing system for this kind of application must reliably process information from all these perspectives.

30.3
Project Overview

One large American oil company used SADT to describe its current central billing process so it could be improved. As in most data-processing systems, both people and computers process sales records to produce invoices. The goal of the SADT project was to identify the interface points between the people and the computers. These defined interface points would then be used to plan how to improve the use of computers inside the central billing organization.

30.4
Resources Used to Make the Model

Three SADT authors interviewed 10 experts in the central billing organization. Another 10 people from various parts of the company reviewed diagrams and approved the final model. The project lasted a total of 4 months.

30.5
Notes about the Model

The top- and first-level diagrams exhibit a functional decomposition strategy. These diagrams group the activities of central billing by functional relatedness. Lower levels of the model were made by a sequential decomposition strategy in order to emphasize how bills are processed. Sometimes the A-0 diagram does not sufficiently describe how a system fits into its environment. So, a higher-level context diagram is drawn to show that fit. The context diagram for an A-0 diagram is numbered A-1, and the Central Billing model is introduced by such a diagram, which explains the information sources and destinations for central billing. Also, the model uses two-way arrows to simplify the description of human-machine dialogs, and it uses questions to describe the workings of a box. We do not recommend the latter technique, but have included it to illustrate an alternate style of system description.

A-1 Do Company Billing (to establish CBS context only)

This sketch of company billing operations is drawn only to identify certain assumptions about Central Billing. It is not intended to be complete, except with regard to the data arrows connected to "Do Central Billing."

The assumptions made are two. First, delivery documents will be batched and other documents segregated outside the area of Central Billing. Second, re-issuing invoices and special (i.e., manual) handling are also not the concern of the Central Billing System.

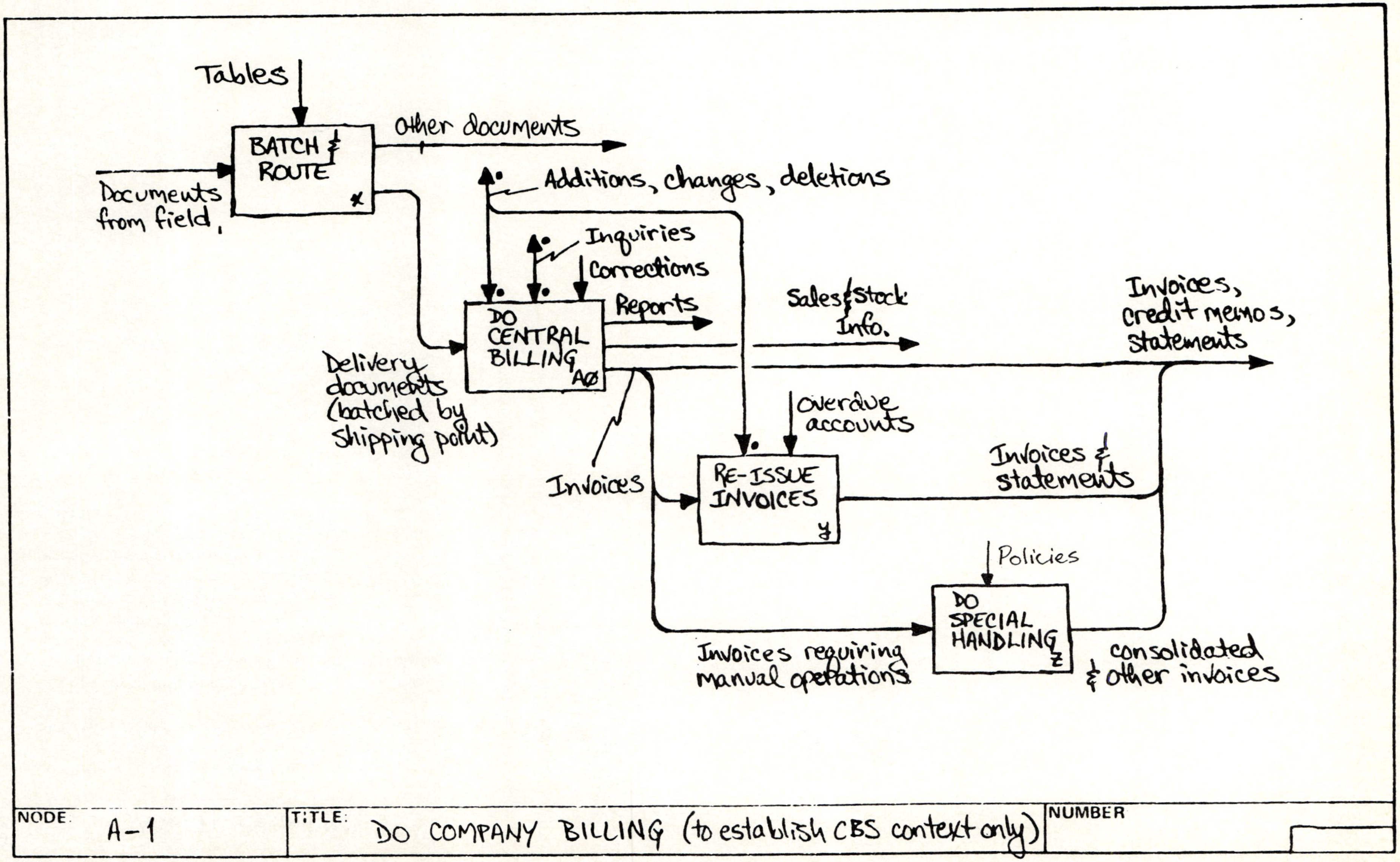

NODE: A-1

TITLE: DO COMPANY BILLING (to establish CBS context only)

NUMBER:

A-0 Do Central Billing

As the title says, this diagram forms a title page for the CBS Functional Activity Model. It also establishes the interface constraints (data arrows) to be tracked through the rest of the model.

The Central Billing System performs all processing necessary to produce invoices, stock and sales information, and certain reports, when given as input the delivery documents provided daily by shipping points.

It is worth emphasizing that this is a "functional" model, drawn (we hope) without regard to any particular manual or automatic implementation.

As the box numbers show, details will be found in the A0 diagram.

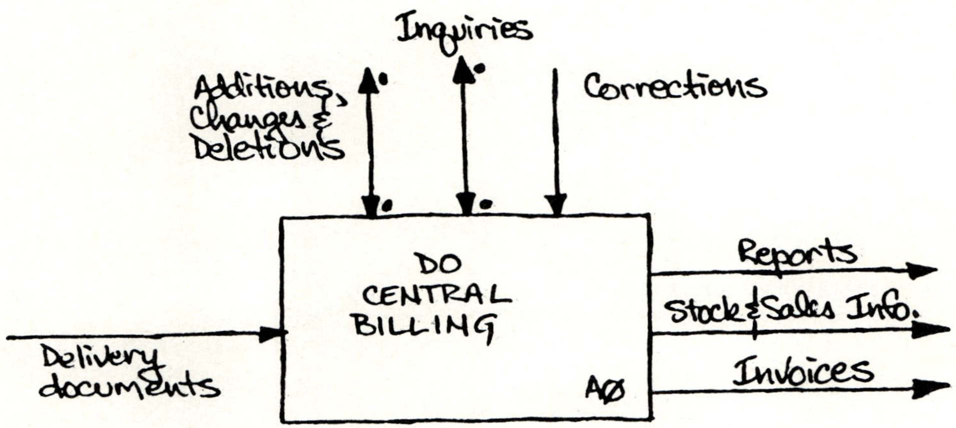

Note: This model is intended to show functional requirements, that is, what functions the Central Billing System is to perform, without regard to who or what performs the functions.

NODE: A-Ø	TITLE: DO CENTRAL BILLING (title page)	NUMBER:

A0 Do Central Billing

(Note: the text, in this space, that accompanies a diagram is a series of observations about the diagram. Diagrams are as complete as possible. Text is not.)

Central Billing is composed of three major subsystems. The first (box 1) provides reference information. The files (box 1, output 2) feed back to show that they can be updated. The second subsystem (box 2) validates delivery data. Corrections may be entered. The third subsystem (box 3) creates invoices, and has an interface to the stock and sales systems. Note that "reports" are composed of three different sets of data, one from each of the subsystems.

The main stream of processing is that delivery documents (box 2, input 1) are validated. The delivery transactions (box 2, output 2 to box 3, input 1) are then used to produce invoices (box 3, output 3). For this diagram and others, this "main stream" is shown on a FEO which immediately follows the diagram.

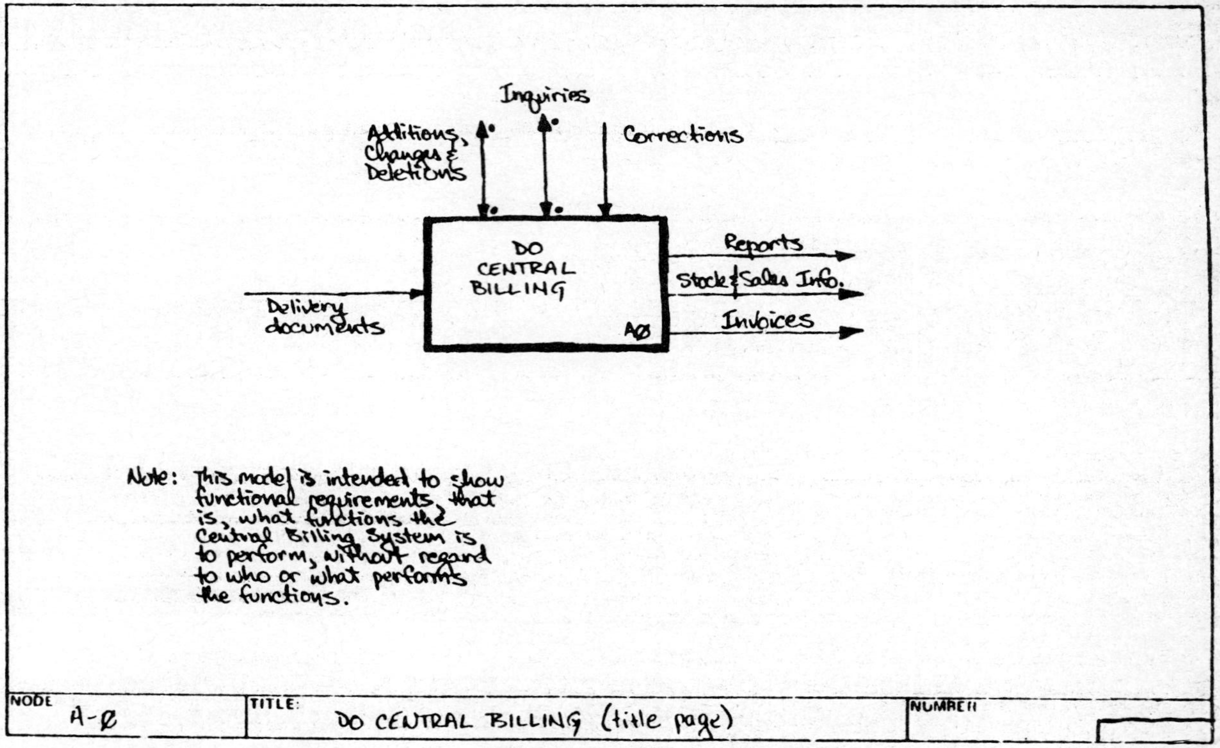

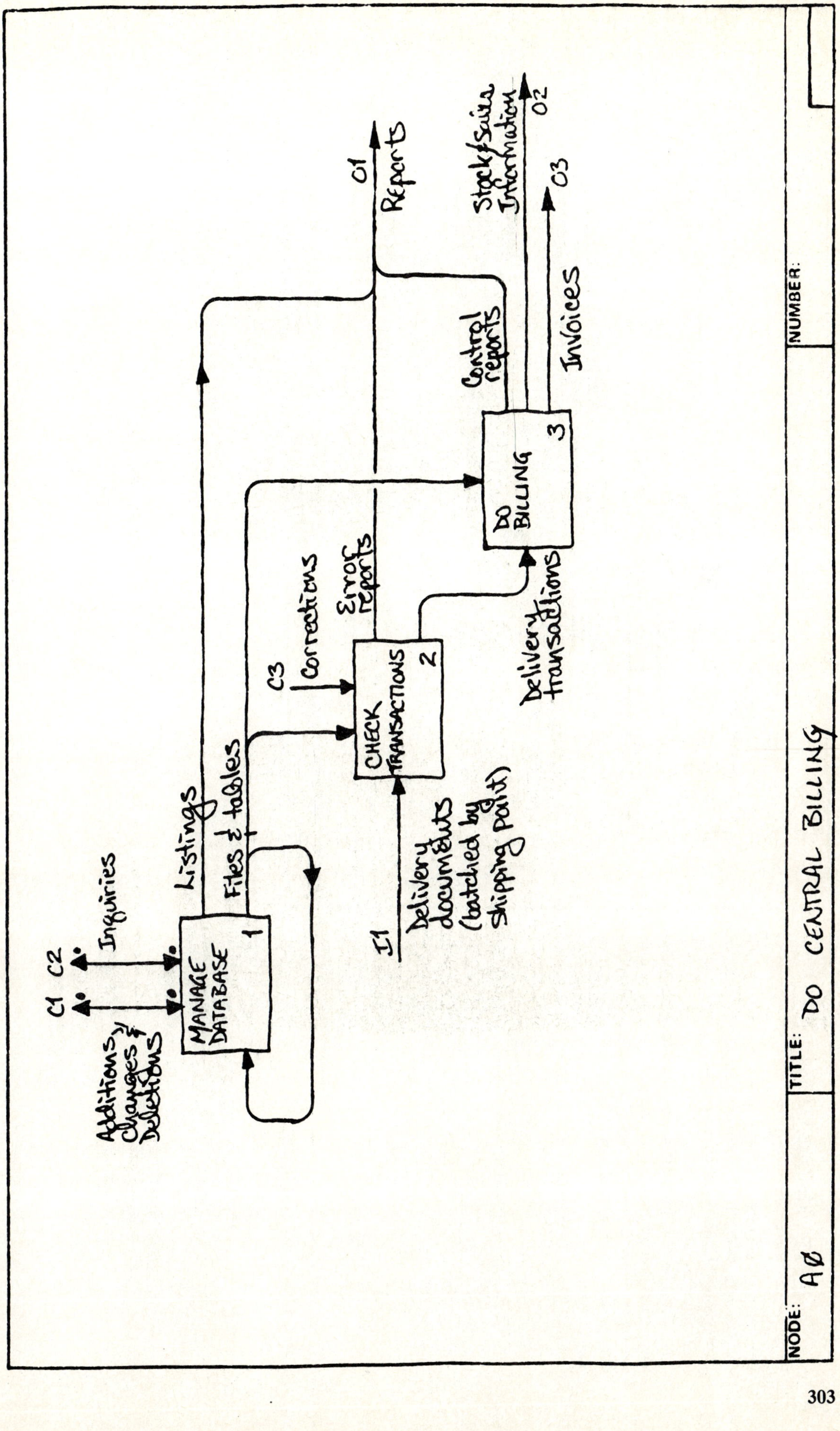
Additions, Changes & Deletions
C1
C2
Inquiries
MANAGE DATABASE
1
Listings
Files & tables
C3
Corrections
CHECK TRANSACTIONS
2
I1
Delivery documents (batched by shipping point)
Error reports
Delivery transactions
DO BILLING
3
C1
Reports
Control reports
Stock & sales Information
O2
O3
Invoices
NODE: AØ
TITLE: DO CENTRAL BILLING
NUMBER:

A1 Manage Database

The Central Billing database is composed of tables (which are relatively stable), customer records, and price records. Throughout this model, all references to a "customer file" mean this Central Billing file.

Note that box 1 is a trivial box and is not detailed. Maintenance and prompt updating of customer and price records are important requirements. The system must also respond to inquiries (box 4) and may produce specific responses or mass listings.

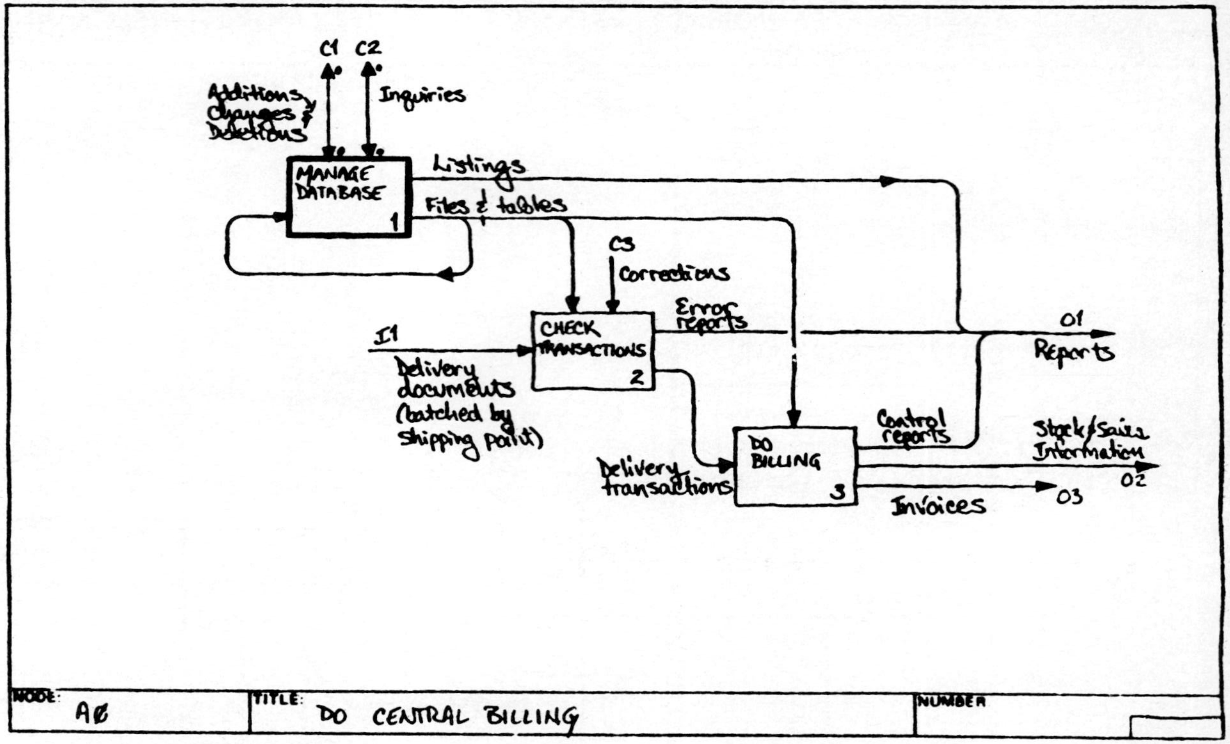

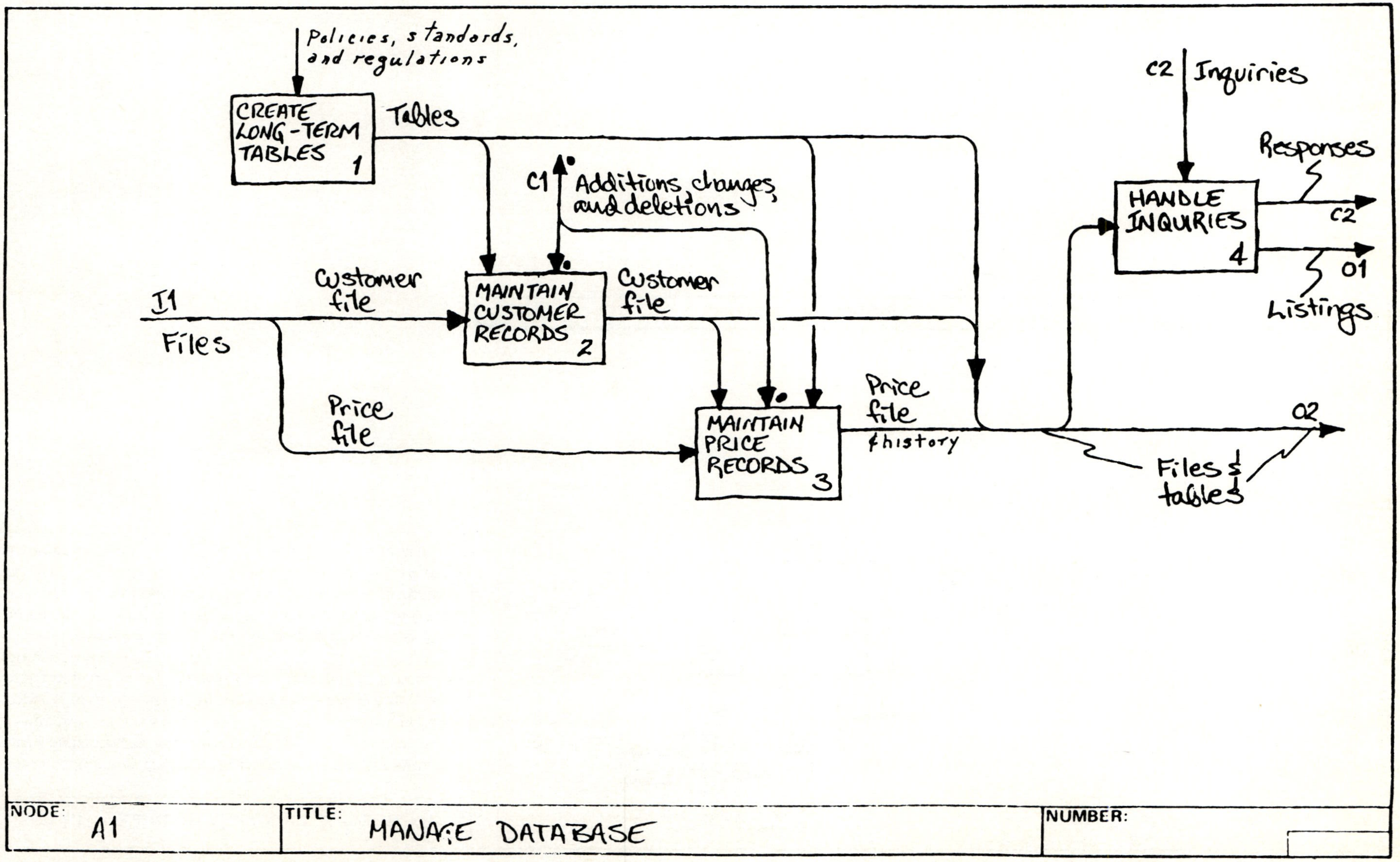

Policies, standards, and regulations
CREATE LONG-TERM TABLES 1
Tables
C1
Additions, changes, and deletions
I1
Files
Customer file
MAINTAIN CUSTOMER RECORDS 2
Customer file
Price file
MAINTAIN PRICE RECORDS 3
Price file & history
C2
Inquiries
HANDLE INQUIRIES 4
Responses
C2
O1
Listings
O2
Files & tables
NODE: A1
TITLE: MANAGE DATABASE
NUMBER:

A12 Maintain Customer Records

Before making any changes to customer records, the system must confirm that the request is an approved change (box 1) and that all necessary data has been provided in the proper format (box 2). If these conditions are not met, an error notification is produced (box 1, output 1 or box 2, output 1).

Modifications to the database are of two types. First, we can change a field in an existing record (box 3). Second, we can add or delete (rarely) all information about a customer (box 4). These activities modify an existing version of the customer file (input 1 on boxes 3 and 4) to create a new version (output 1 on boxes 3 and 4).

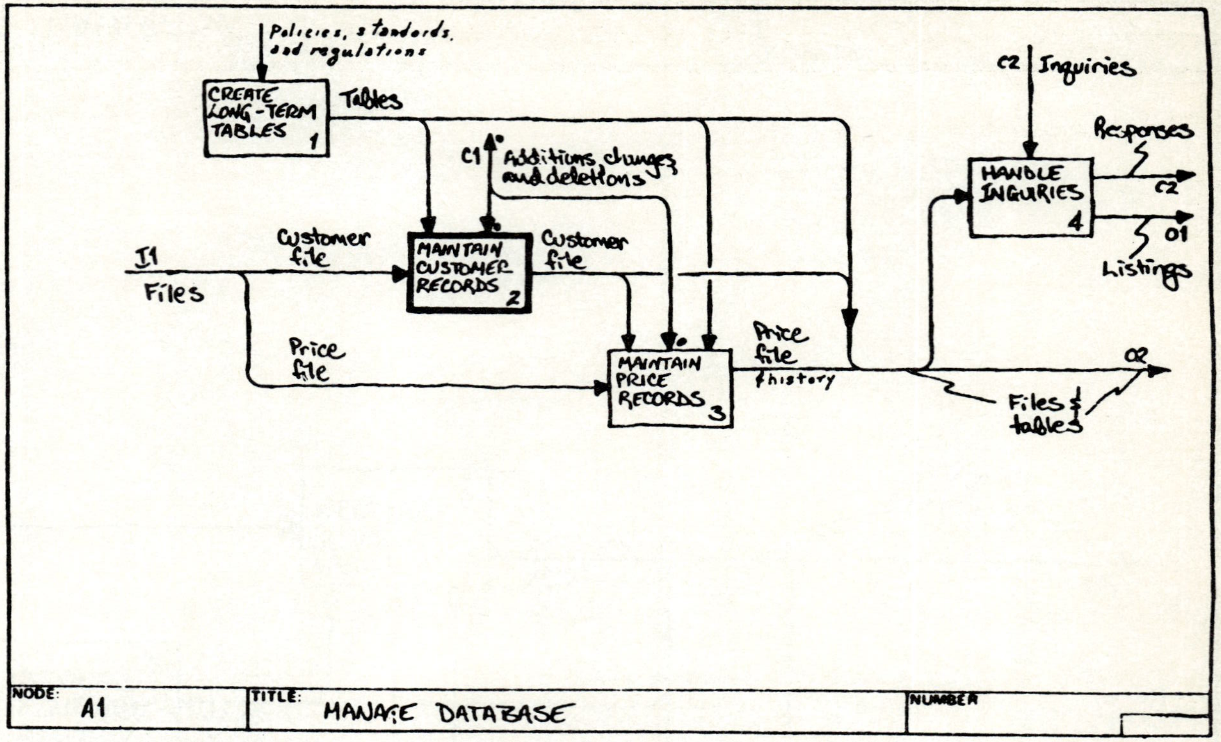

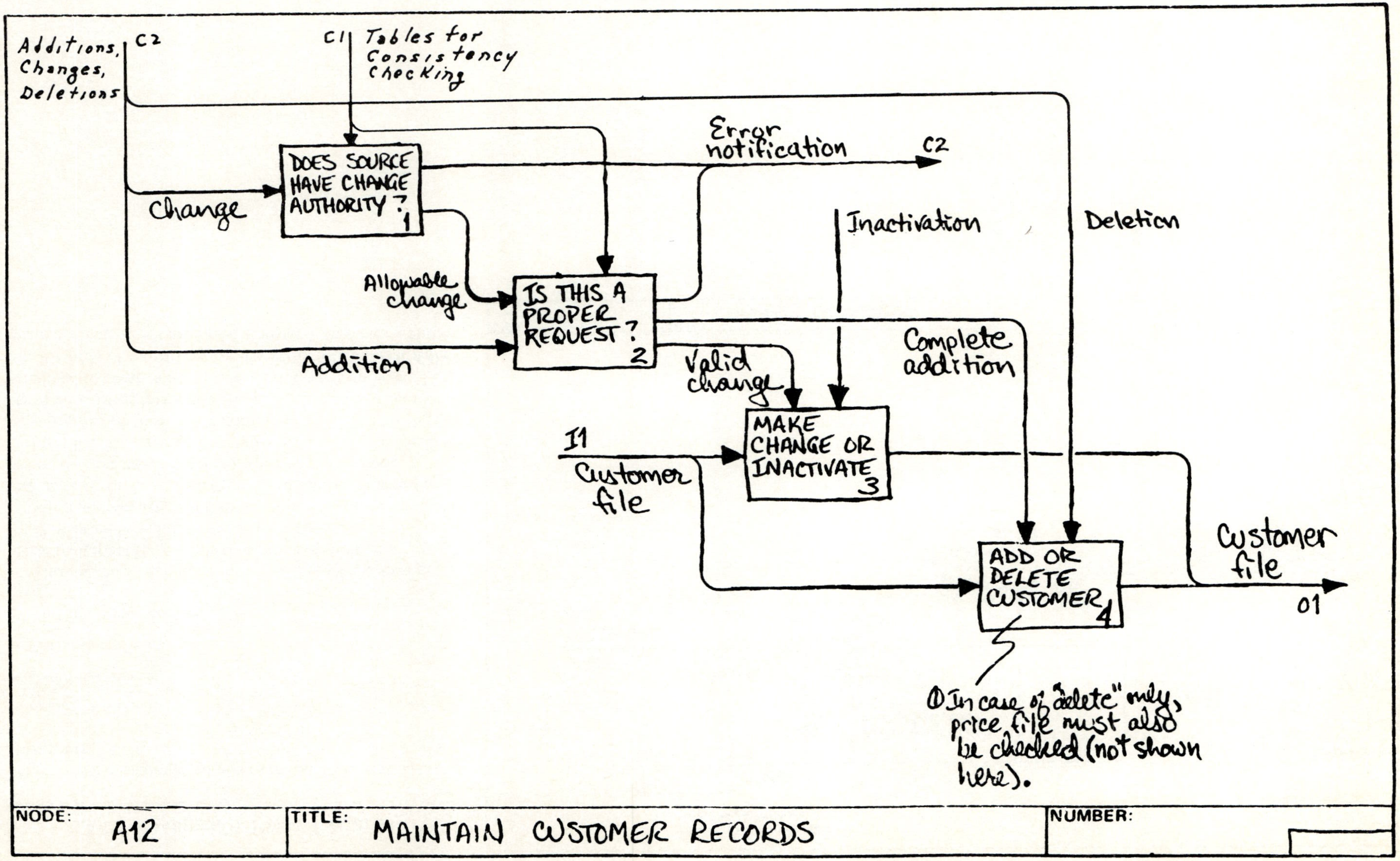
Additions, Changes, Deletions
C2
C1
Tables for Consistency Checking
DOES SOURCE HAVE CHANGE AUTHORITY ? 1
Change
Allowable change
IS THIS A PROPER REQUEST ? 2
Addition
Error notification
C2
Inactivation
Deletion
Complete addition
Valid change
MAKE CHANGE OR INACTIVATE 3
I1
Customer file
ADD OR DELETE CUSTOMER 4
Customer file
01
① In case of "delete" only, price file must also be checked (not shown here).
NODE: A12
TITLE: MAINTAIN CUSTOMER RECORDS
NUMBER:

A13 Maintain Price Records

The price file is composed of several different sections (box 1 output):

- Posted prices,
- Individual prices,
- Price adjustments,
- And line item charges.

Thus, when a change is requested, the system must determine its impact (box 1), then use the requirements for each type of price to validate the request (box 2).

Valid updates are held (box 2, output 3) and the accounting department is notified (box 2, output 2). Upon approval (control 2 on boxes 3 and 4), the change is made.

Modifications are of two types: changing a field in an existing record (box 3), or adding or deleting a price (box 4). The existing price file (input 1 on boxes 3 and 4) is modified to create a new price file (output 2 on boxes 3 and 4). Records of old prices are never discarded, even if they were erroneous. All are saved (output 1 on boxes 3 and 4).

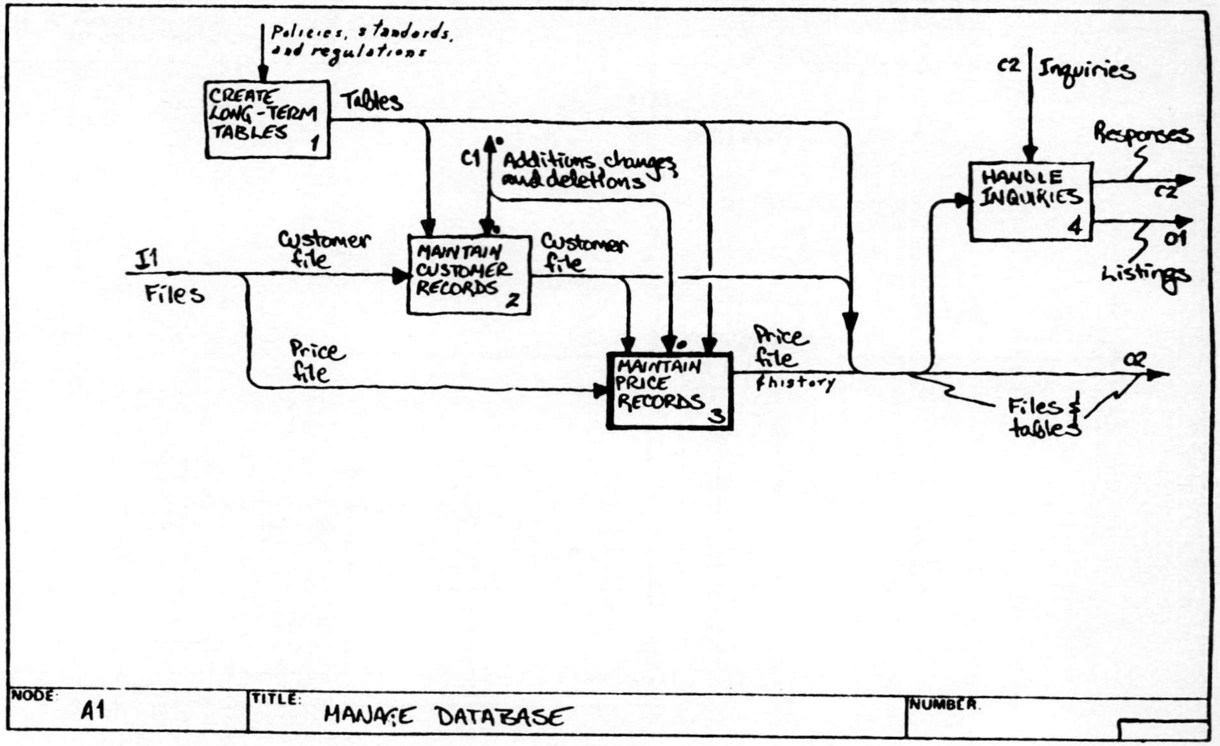

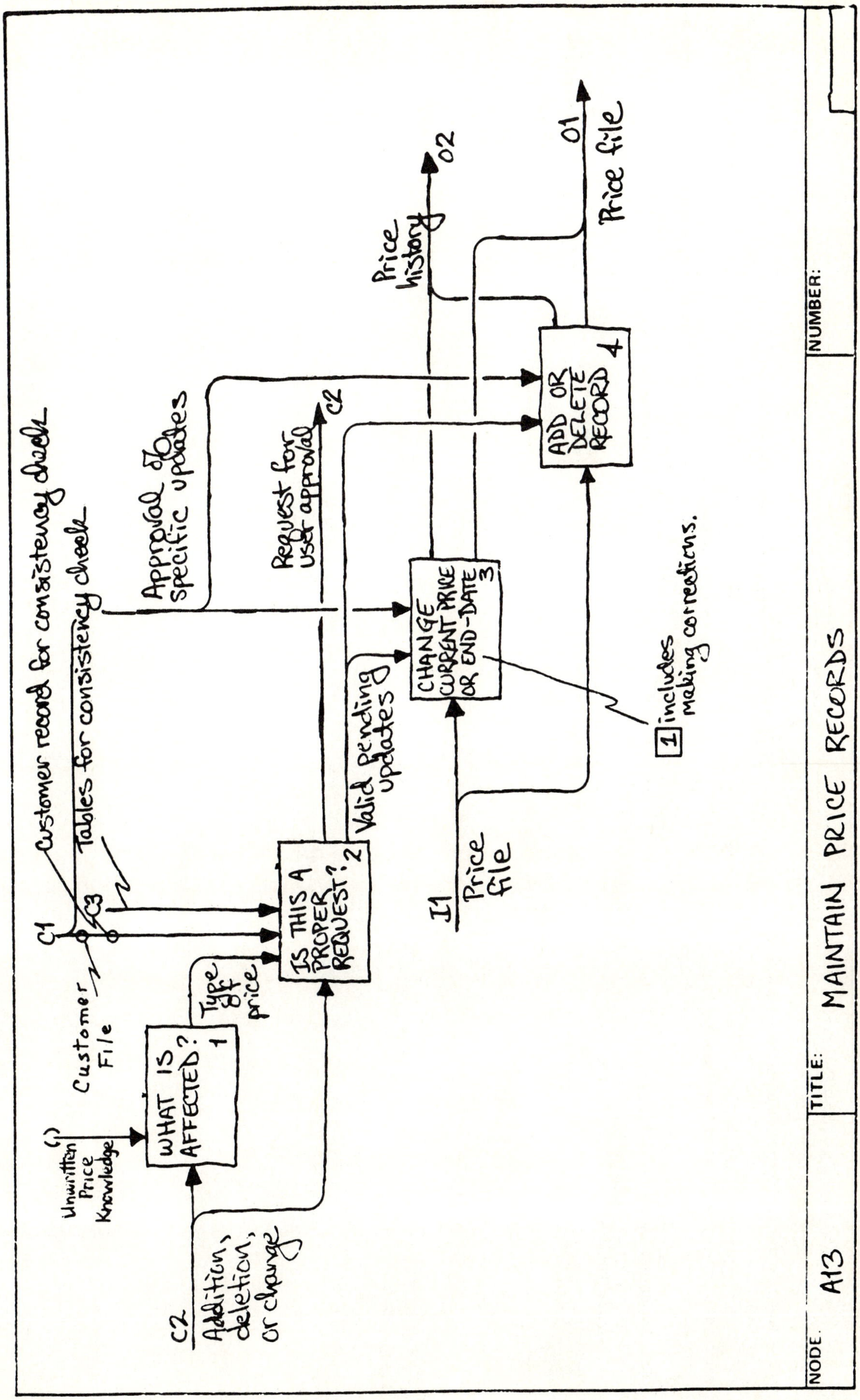

NODE: A13
TITLE: MAINTAIN PRICE RECORDS
NUMBER:
C2 Addition, deletion, or change
Unwritten Price Knowledge
WHAT IS AFFECTED? 1
Customer File
Type of price
C1
C3
Customer record for consistency check
Tables for consistency check
IS THIS A PROPER REQUEST? 2
Request for user approval
C2
Approval of specific updates
Valid pending updates
I1 Price file
CHANGE CURRENT PRICE OR END-DATE 3
1 includes making corrections.
ADD OR DELETE RECORD 4
Price history
O2
O1
Price file

A14 Handle Inquiries

The system must use the information in the files and tables (input 1 on all boxes) to report information of two kinds. First, it must report all information about a specific customer (box 1, output 1) or a specific price (box 2, output 1). For customer records, the customer number and marketing office must be supplied (box 1, control 1). For price information, selection—i.e., sorting—criteria must be provided.

Second, the system must give mass listings of sections of the price file (box 2, output 2) or of entire tables (box 3, output 1).

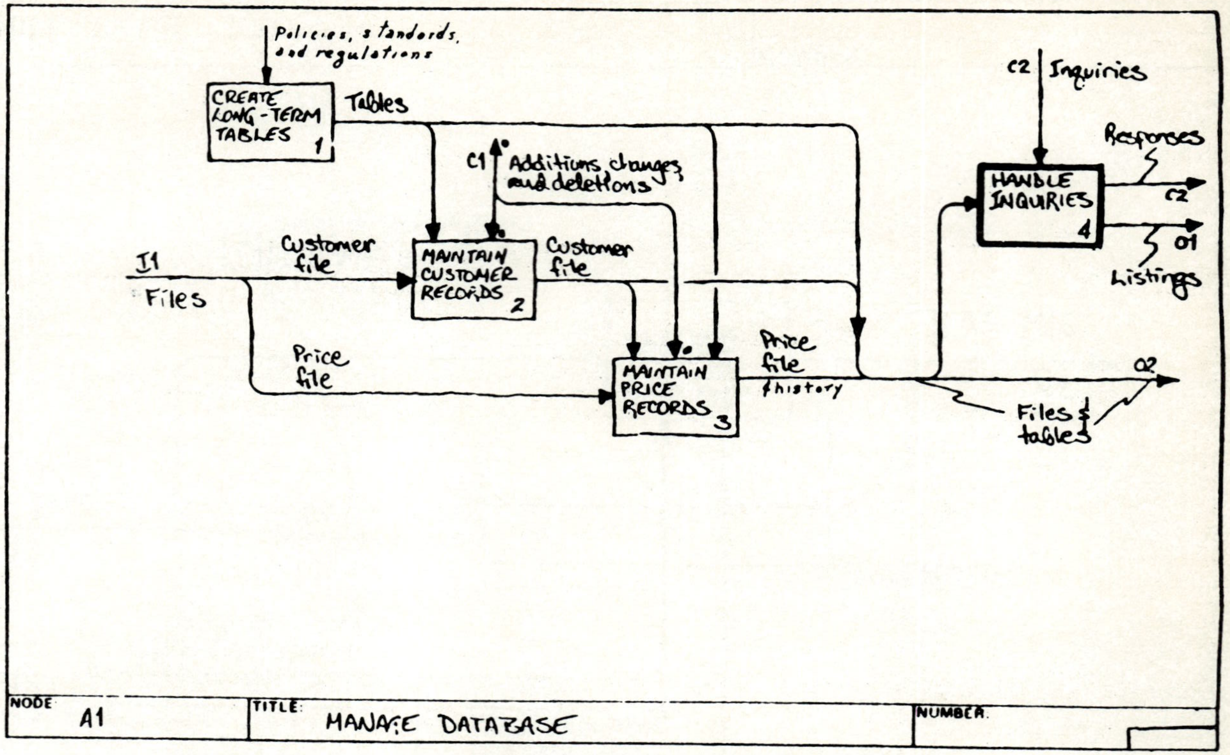

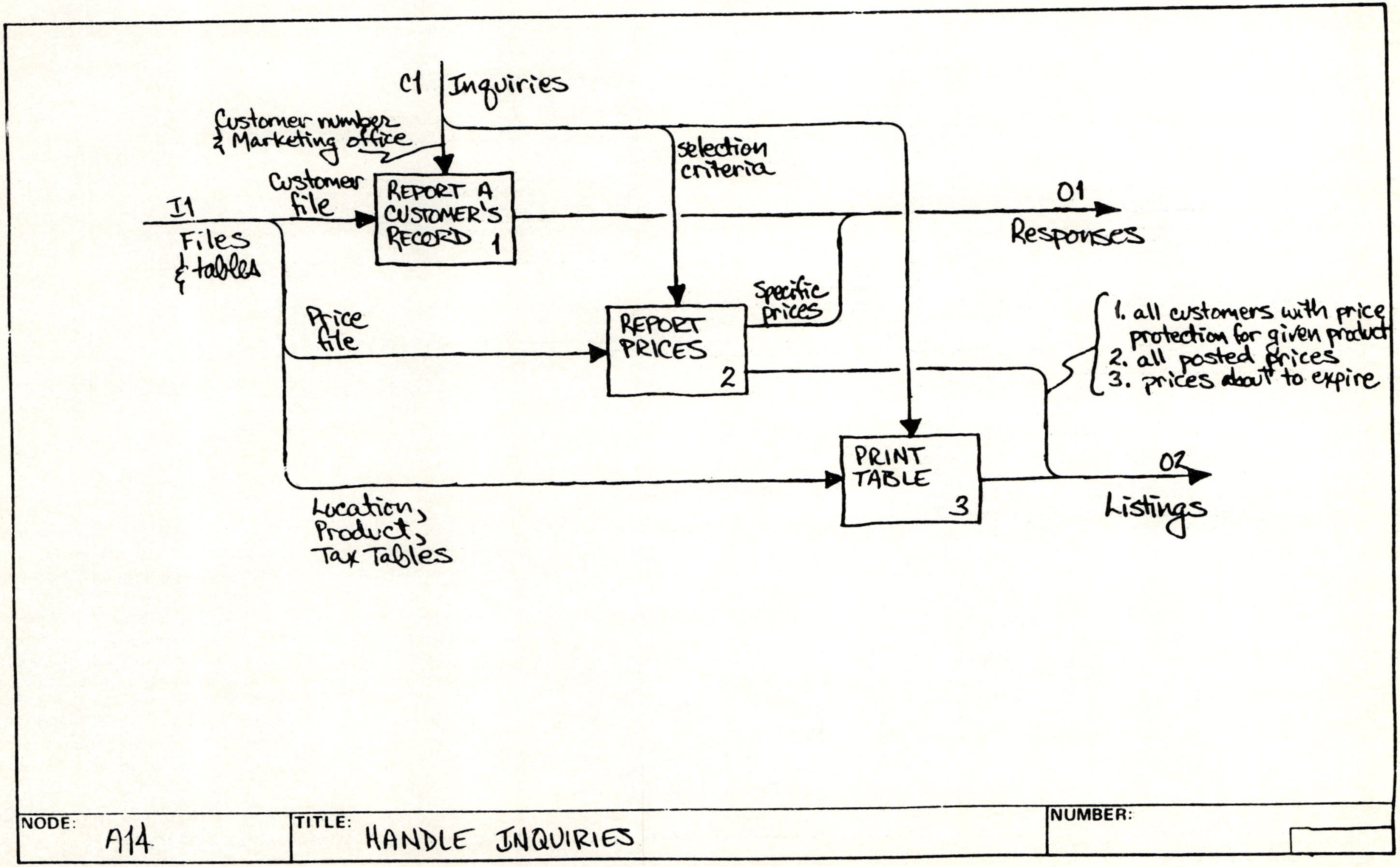
C1 Inquiries
Customer number & Marketing office
Customer file
I1 Files & tables
REPORT A CUSTOMER'S RECORD 1
selection criteria
O1 Responses
Price file
REPORT PRICES 2
Specific prices
1. all customers with price protection for given product
2. all posted prices
3. prices about to expire
PRINT TABLE 3
O2 Listings
Location, Product, Tax Tables
NODE: A14
TITLE: HANDLE INQUIRIES
NUMBER:

A2 Check Transactions

Delivery documents arrive (box 1 input 1) batched by shipping point and contained in a "batch pocket" which has certain control information written on it. The first task of data validation is to insure that the batch controls are valid, and to hold any batch with invalid controls (box 1). Next, each document in the batch must be checked for validity and completeness (box 2). Finally, document totals must be compared to batch controls (box 3), and any errors (whether originating in box 2 or box 3) must be corrected. The error reports (box 3 output 2) summarize the errors discovered and held for correction. Valid delivery transactions (box 3 output 1) are passed to the billing subsystem in batches.

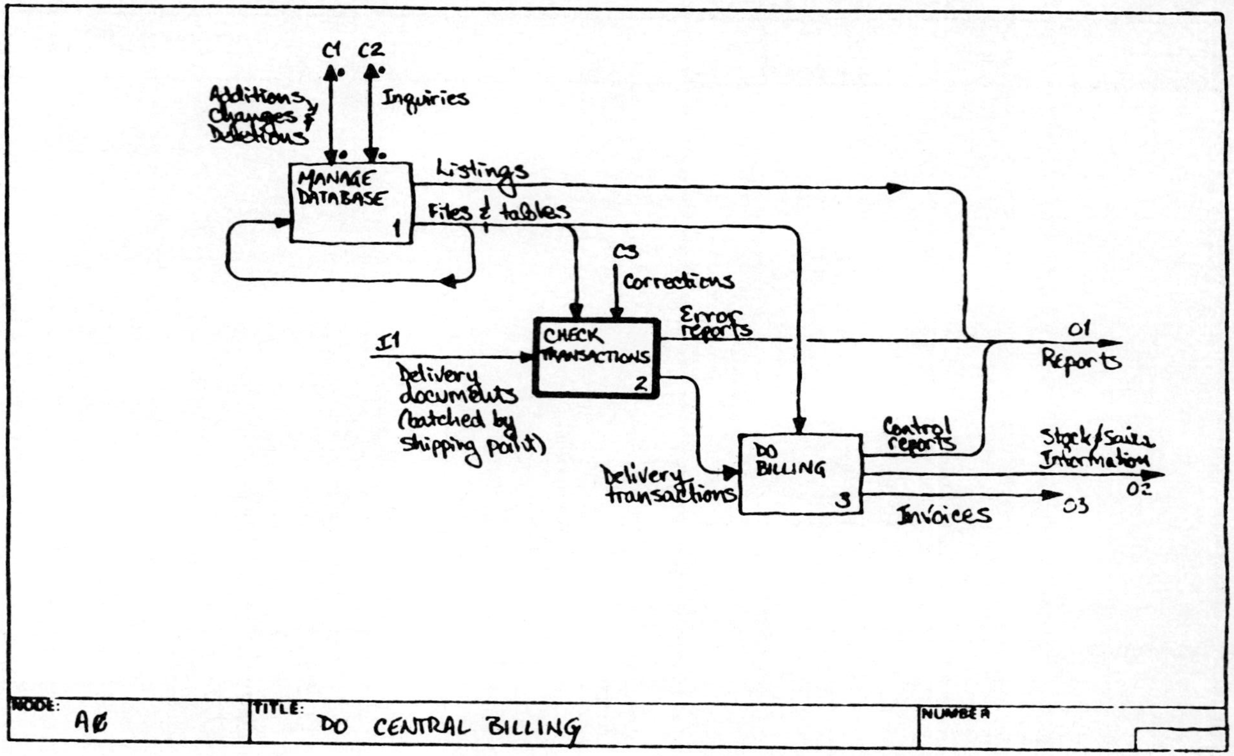

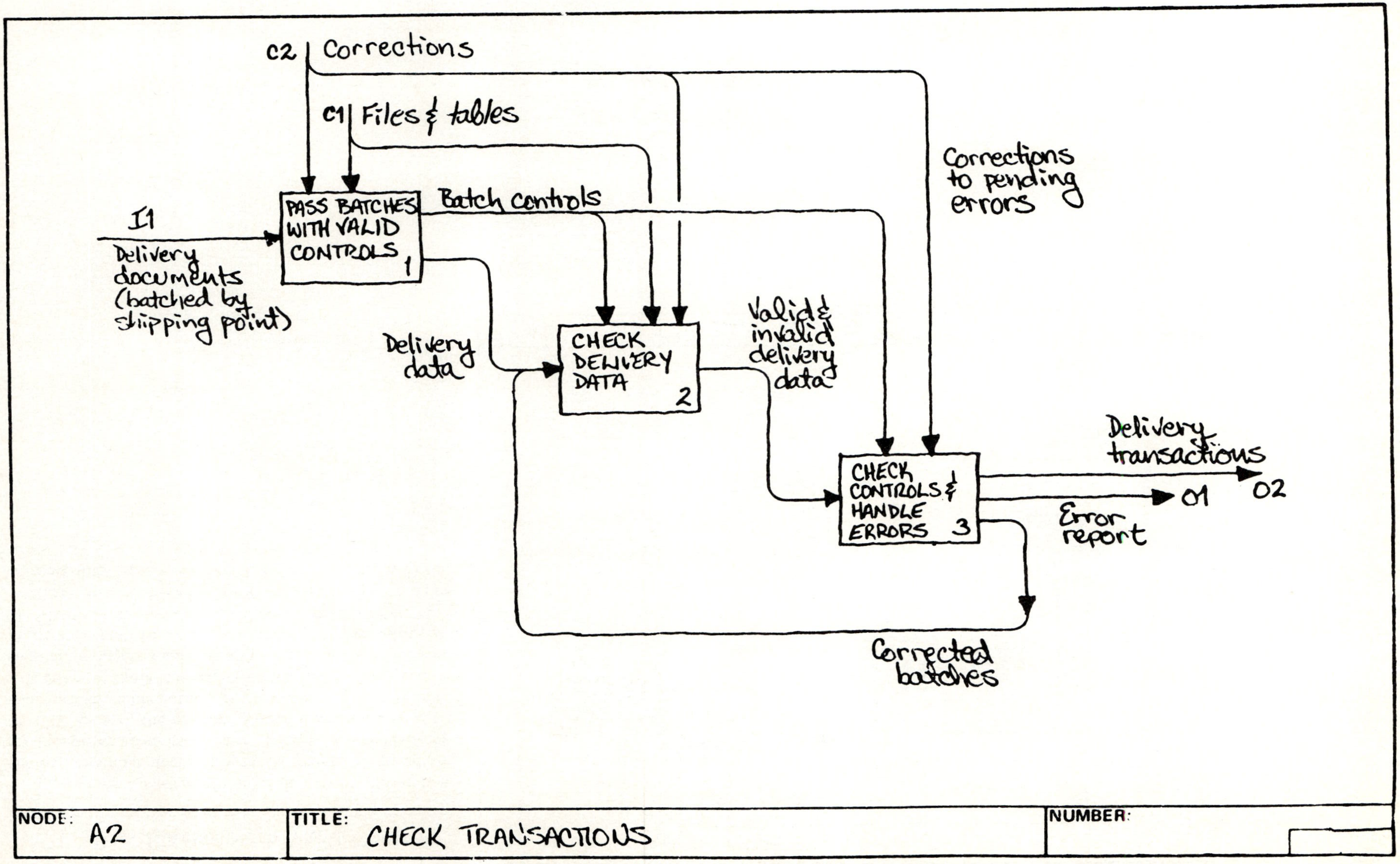

C2 Corrections
C1 Files & tables
I1
Delivery documents (batched by shipping point)
PASS BATCHES WITH VALID CONTROLS 1
Batch controls
Delivery data
CHECK DELIVERY DATA 2
Valid & invalid delivery data
Corrections to pending errors
CHECK CONTROLS & HANDLE ERRORS 3
Delivery transactions
O2
Error report
O1
Corrected batches
NODE: A2
TITLE: CHECK TRANSACTIONS
NUMBER:

A21 Pass Batches with Valid Controls

Two questions are asked of the batch control information, before examining documents in the batch. First, all the product group codes listed on the pocket must be allowable (box 1). Second, the product groups listed must be those which are available at the indicated shipping point (box 2).

If either of these tests fails (output 1 on boxes 1 and 2), an attempt may be made to fix the control information (box 3). If this cannot be done, the entire batch is set aside (box 4). After any fix, the data is retested (box 3, output 1).

If the controls are valid (box 2, output 2), then the delivery data may be entered (box 4, input 1 to output 1).

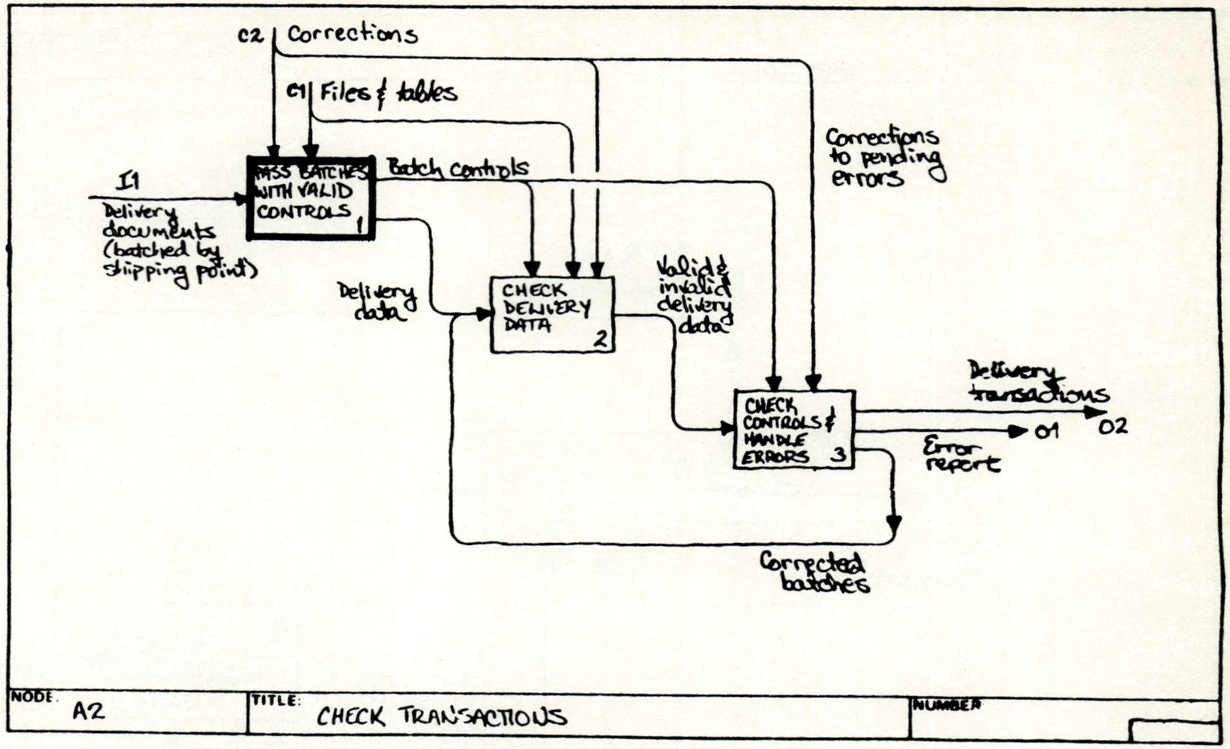

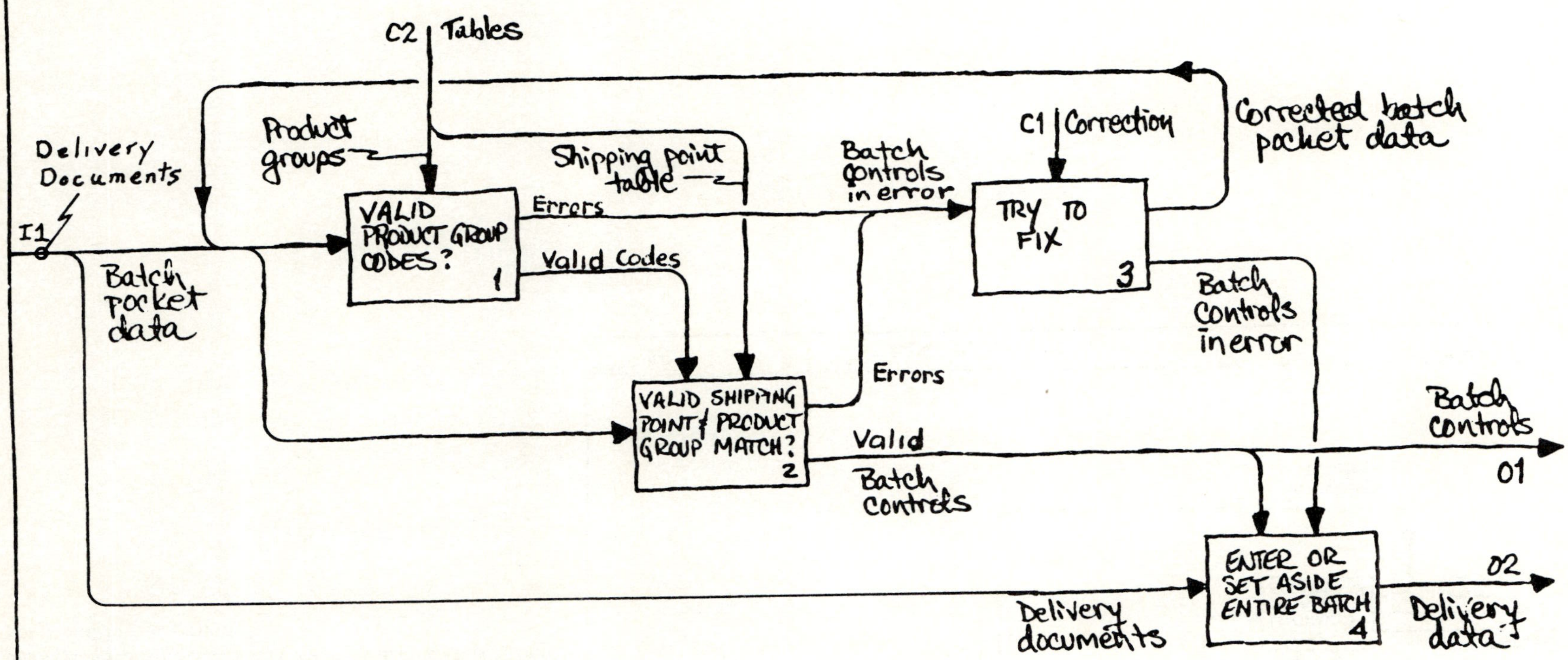

NODE: A21 | TITLE: PASS BATCHES WITH VALID CONTROLS | NUMBER:

A22 Check Delivery Data

Each document within a batch is examined (box 1) and if no errors are discovered, is passed along (box 2, output 2). If any data is in error, an attempt may be made to fix it (box 4) and it will be retested (box 4, output 1).

Correct data may be incomplete, due to the need for information not on the delivery ticket.

Each ticket is tested (box 3) to determine the need for such information.

Delivery data which is both correct and complete is forwarded as valid (box 3, output 2). Data which is either incorrect or incomplete is forwarded as invalid (box 3, output 1 and box 4, output 2).

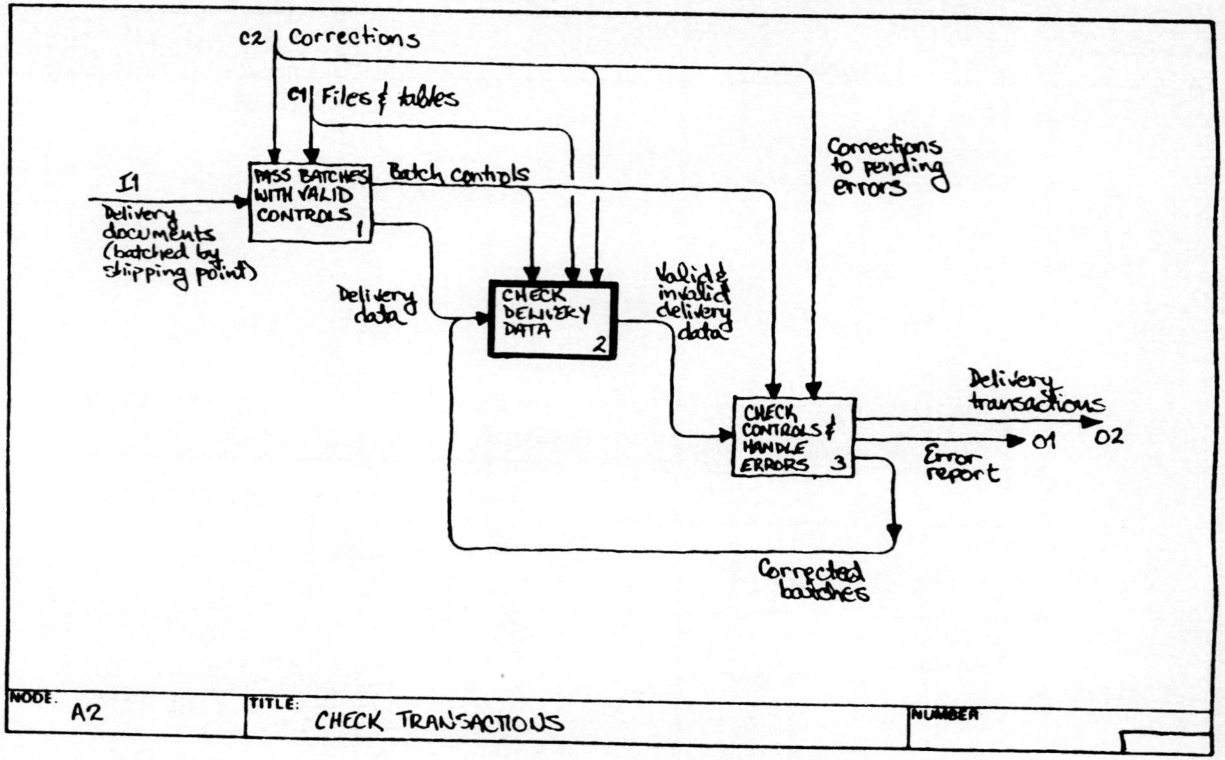

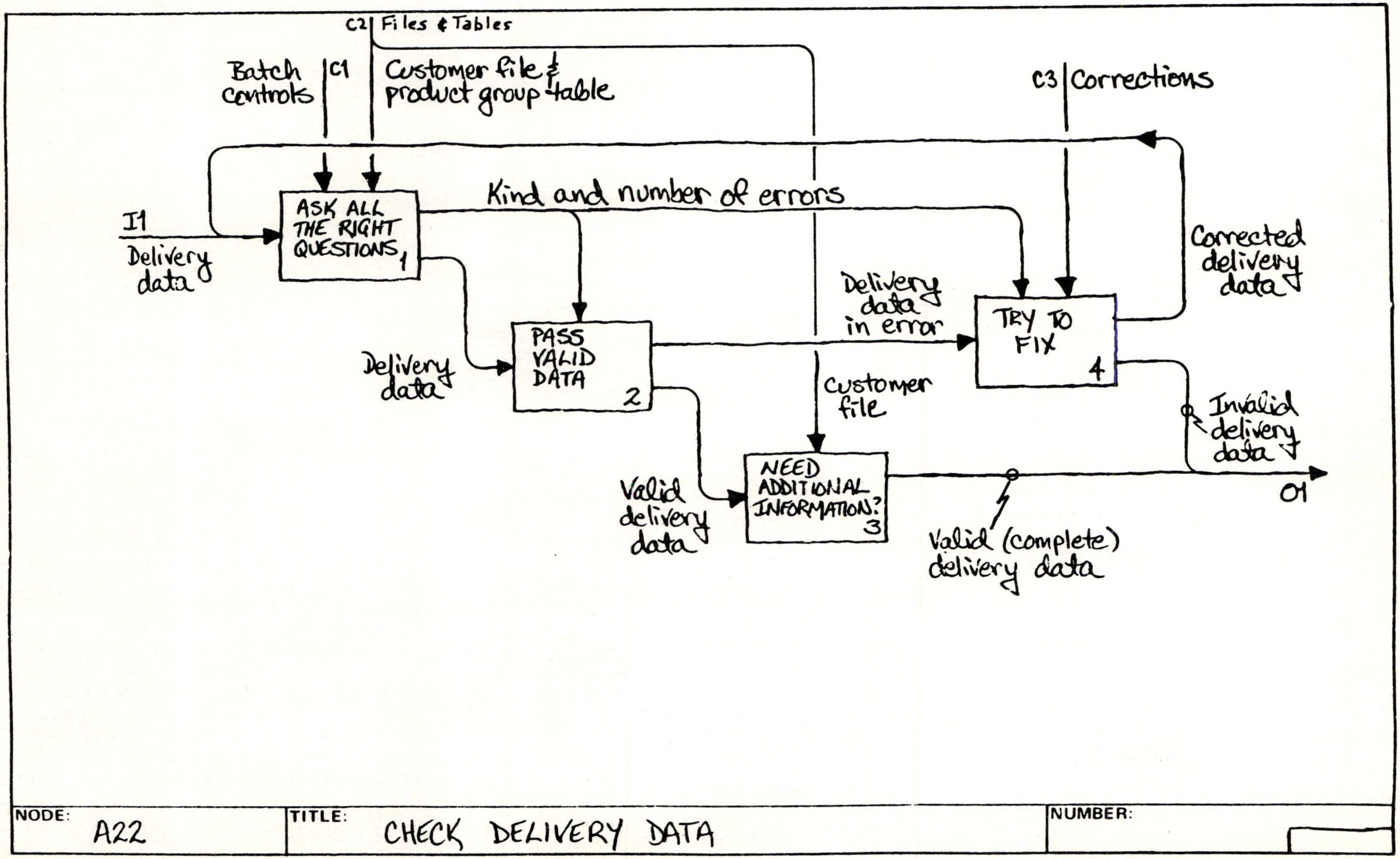
C2 Files & Tables
Batch controls
C1
Customer file & product group table
C3 Corrections
I1
Delivery data
ASK ALL THE RIGHT QUESTIONS 1
Kind and number of errors
PASS VALID DATA 2
Delivery data
Delivery data in error
TRY TO FIX 4
Corrected delivery data
Customer file
NEED ADDITIONAL INFORMATION? 3
Valid delivery data
Invalid delivery data
O1
Valid (complete) delivery data
NODE: A22
TITLE: CHECK DELIVERY DATA
NUMBER:

A23 Check Controls and Handle Errors

After all documents in a batch have been checked, two other batch controls are validated (box 1). These are total number of tickets in the batch, and total net product volume (box 1, control 1).

If all batch controls and delivery data are acceptable, the batch is submitted for billing (box 2).

If either the batch is out-of-balance, or delivery data is invalid, the batch or the data (with batch header) is saved in a pending errors file (box 3, output 2) and a notification issued. When a correction is entered (box 4, control 1), these batches are retested (box 4, output 1).

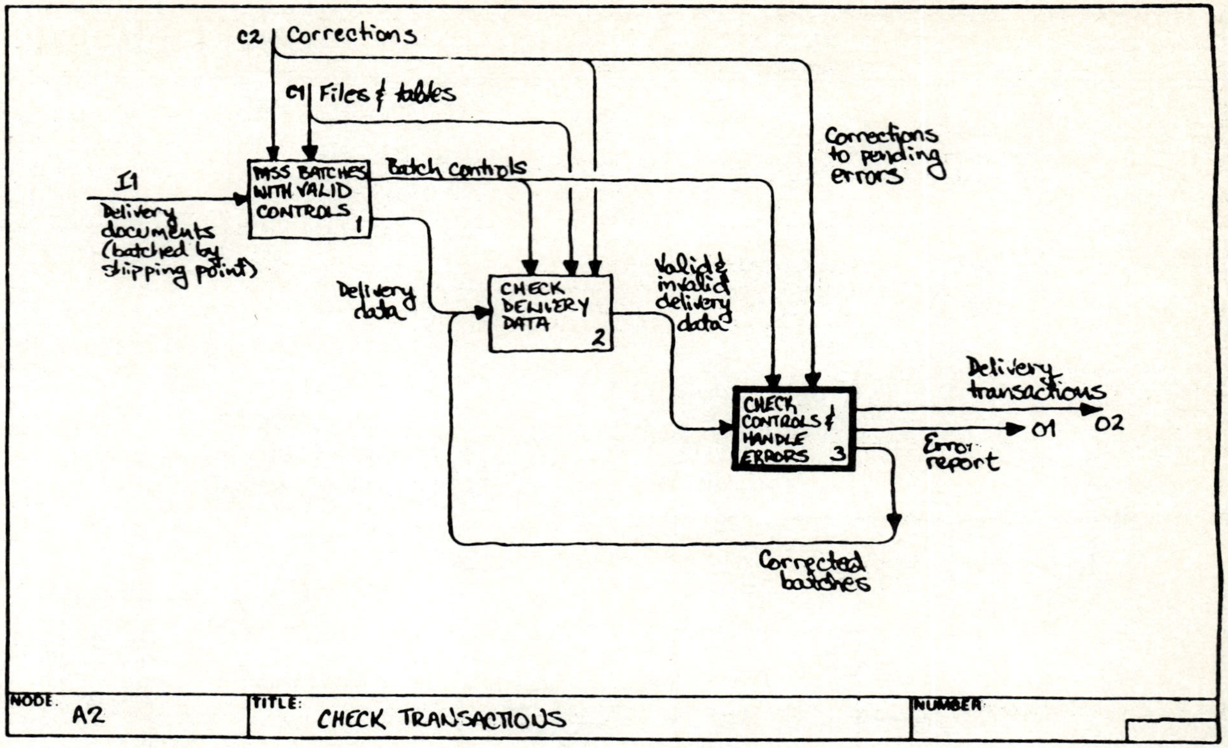

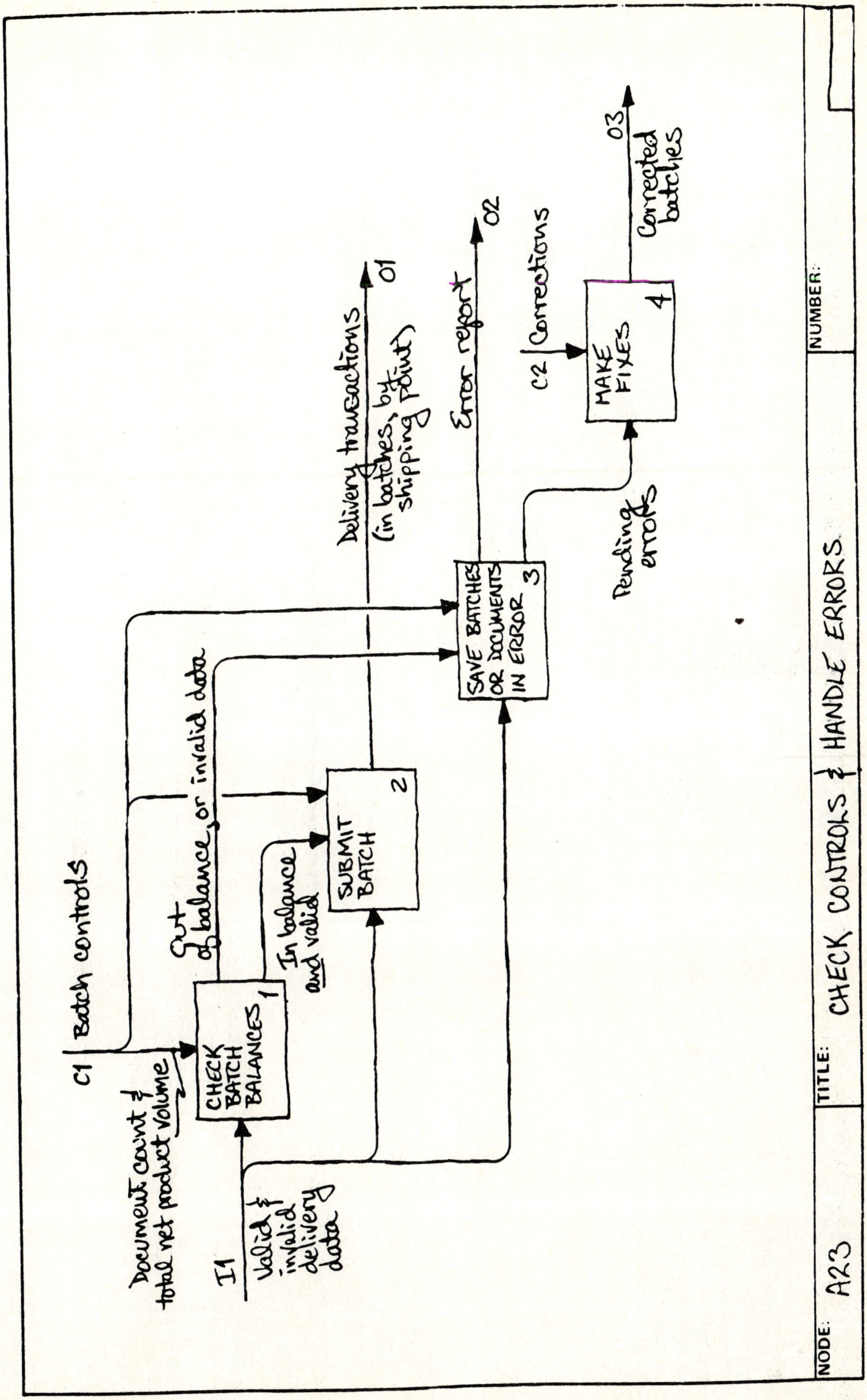

C1 Batch controls
Document count & total net product volume
I1 Valid & invalid delivery data
CHECK BATCH BALANCES 1
Out of balance, or invalid data
In balance and valid
SUBMIT BATCH 2
Delivery transactions (in batches, by shipping point)
O1
SAVE BATCHES OR DOCUMENTS IN ERROR 3
Error report
O2
Pending errors
C2 Corrections
MAKE FIXES 4
Corrected batches
O3
NODE: A23
TITLE: CHECK CONTROLS & HANDLE ERRORS
NUMBER:

A3 Do Billing

The billing subsystem breaks down into five relatively straightforward tasks. Box 1 represents certain "overhead" functions. Note that there is no link between box 2 and box 3 (because product taxes are not applied to prices). Also note that the only input to box 5 is labeled "line items" (because, by the time we reach box 5, we need only know that a set of line items has been calculated which must be printed on an invoice). These line items consist of a transaction price (box 2, output 2), product taxes (box 3, output 2), and a sales tax line item (box 4, output 2).

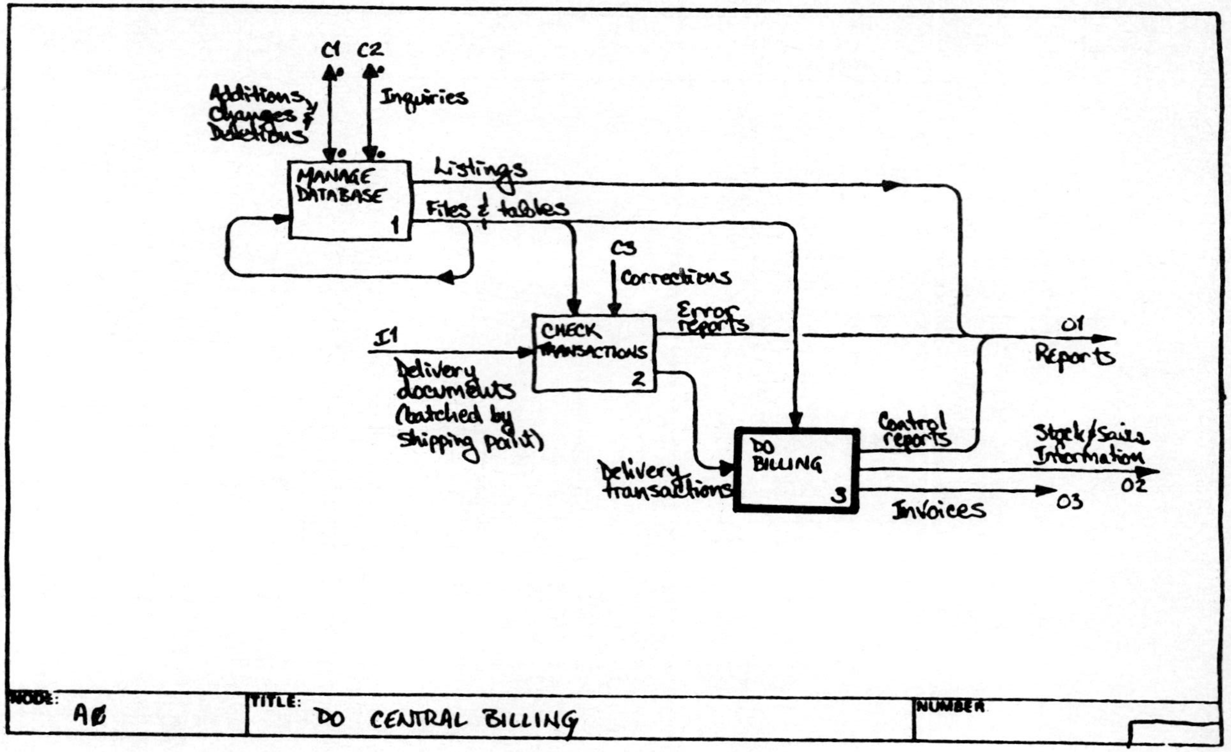

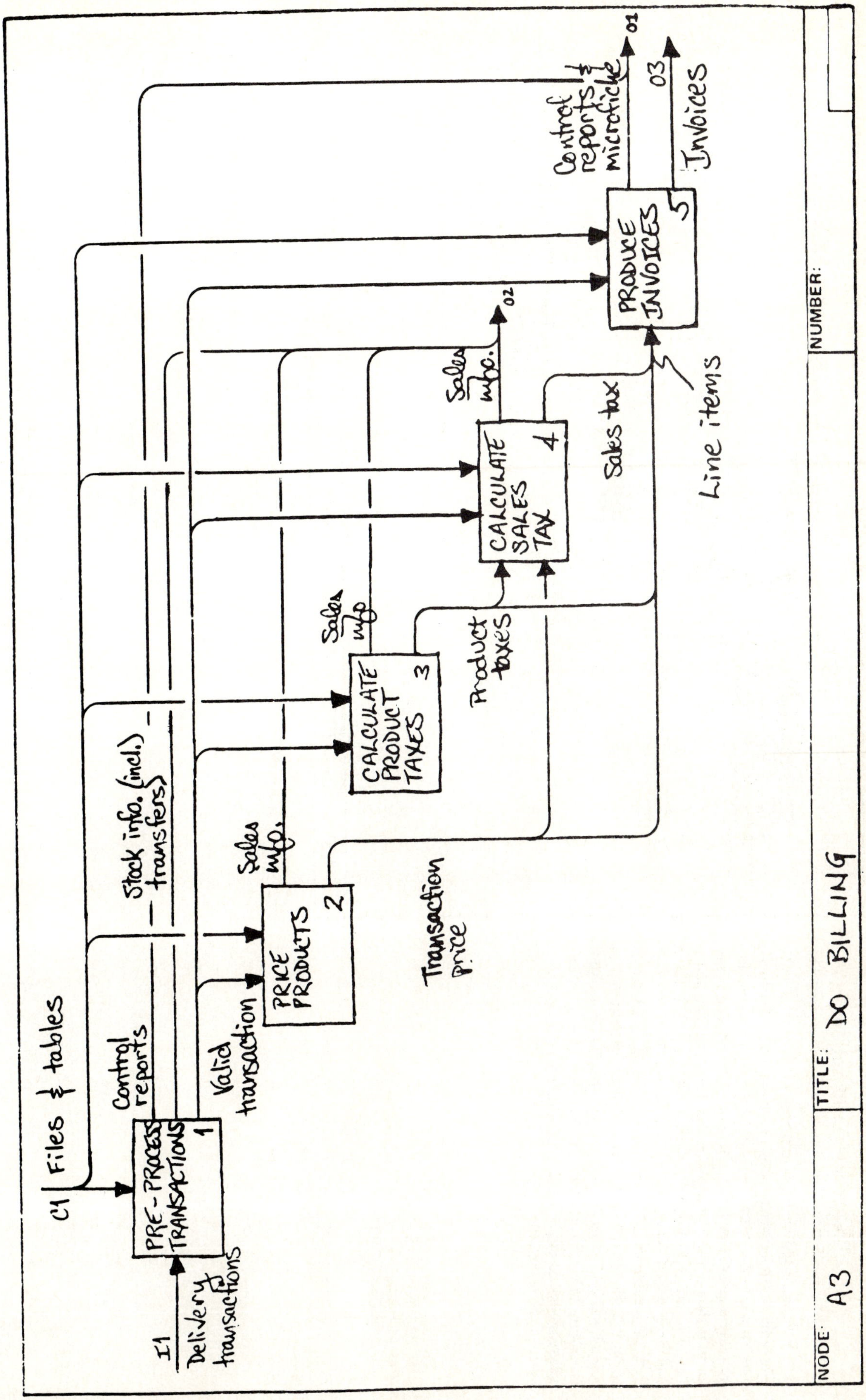

C1 Files & tables
I1 Delivery transactions
PRE-PROCESS TRANSACTIONS 1
Control reports
Valid transaction
Stock info. (incl.) transfers)
PRICE PRODUCTS 2
Sales info.
Transaction price
CALCULATE PRODUCT TAXES 3
Sales info.
Product taxes
CALCULATE SALES TAX 4
Sales info.
O2
Sales tax
Line items
PRODUCE INVOICES 5
Control reports & microfiche
O1
O3
Invoices
NODE: A3
TITLE: DO BILLING
NUMBER:

A31 Preprocess Transactions

Before pricing and taxing transactions, certain overhead functions must be performed, including more validation (box 2). As the note shows, this area was unclear at the time the diagram was drawn. But remember that the A2 diagram asked if the data were proper, and did not test against the files to insure that the transactions could be priced, taxes, and invoiced.

When transactions are sorted (box 1) and validated (box 2), an interface to the stock system is created (output 1 on boxes 1 and 2). Certain control reports are also written (box 3), mainly having to do with the number of batches and tickets processed by the system. (Additional control reports are created in the A35 diagram.)

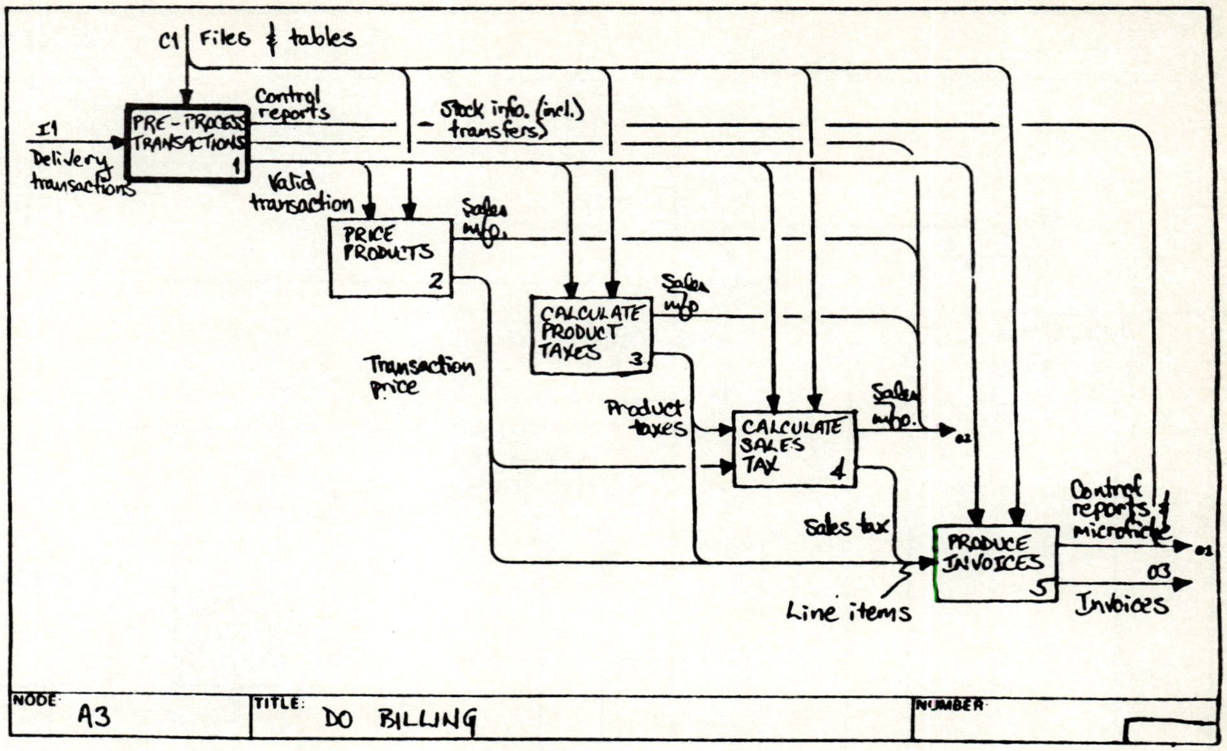

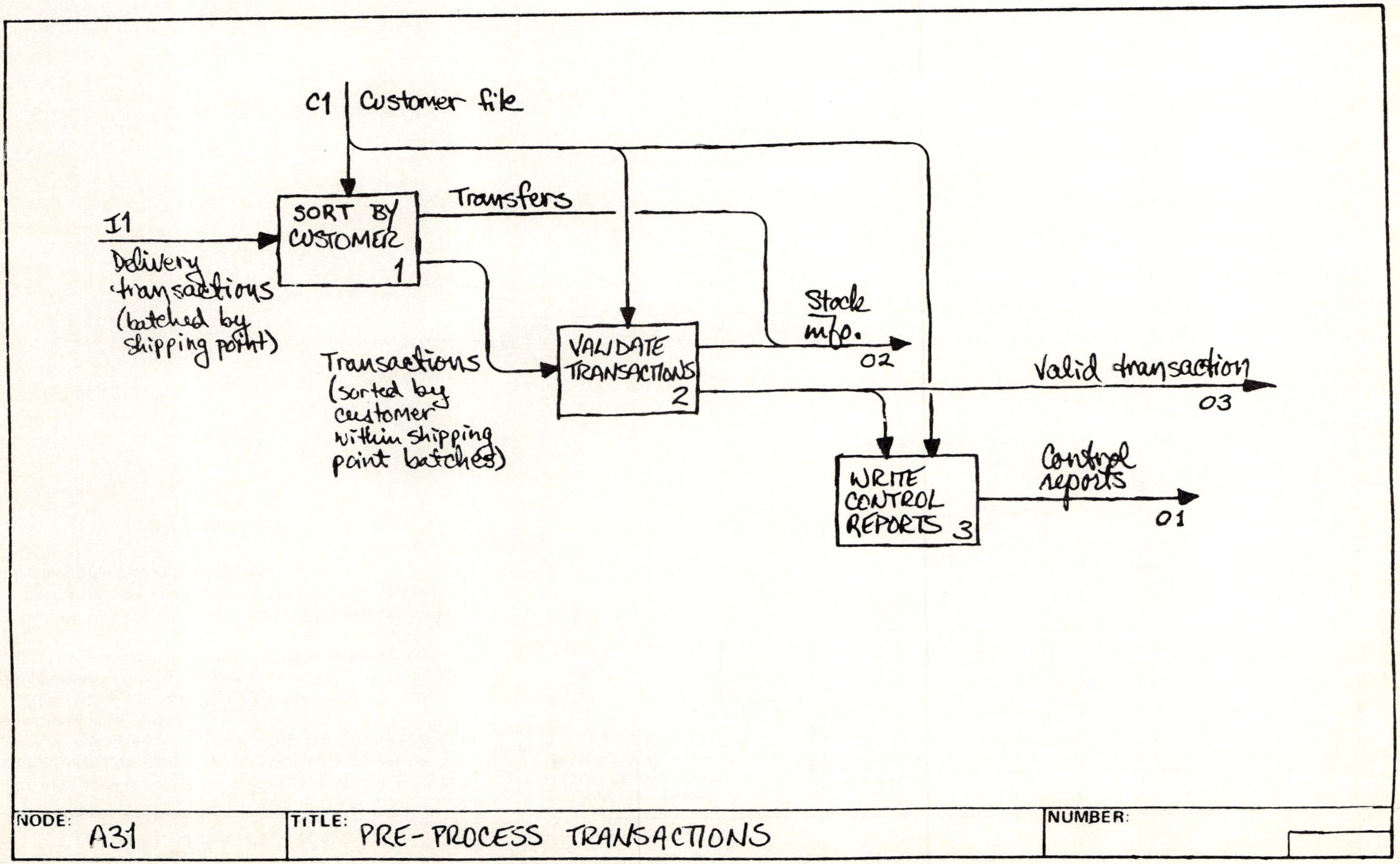
C1 Customer file
I1
Delivery transactions (batched by shipping point)
SORT BY CUSTOMER 1
Transfers
Transactions (sorted by customer within shipping point batches)
VALIDATE TRANSACTIONS 2
Stock info.
O2
Valid transaction
O3
WRITE CONTROL REPORTS 3
Control reports
O1
NODE: A31
TITLE: PRE-PROCESS TRANSACTIONS
NUMBER:

A32 Price Products

The most important part of pricing is to determine if a customer has been granted an exception to the common—or "posted"—price (box 2). If not, then the common price must be applied (box 3).

When a base price has been determined (box 4, control 1), adjustments may be applied from the Price File (box 4, input 1) to produce a price for the transaction (box 4, output 2).

Because the searches for individual price and price adjustments are based on the customer's identity, all necessary customer information must be obtained (box 1).

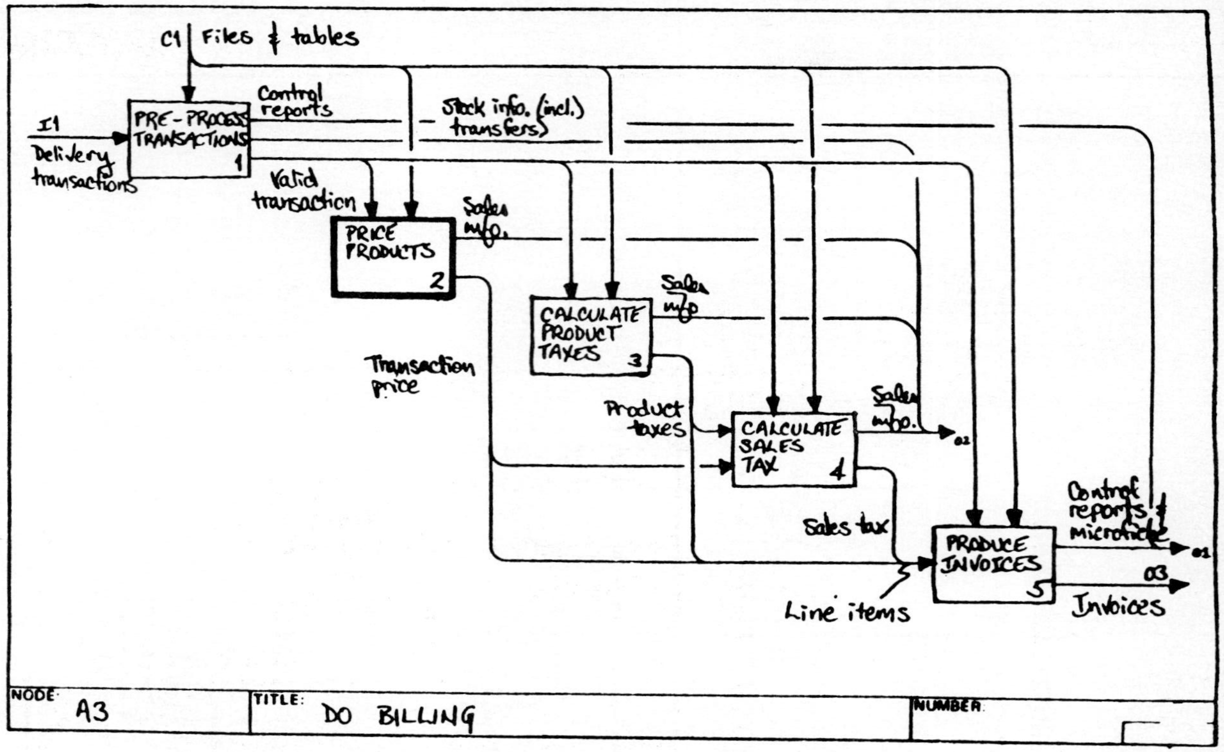

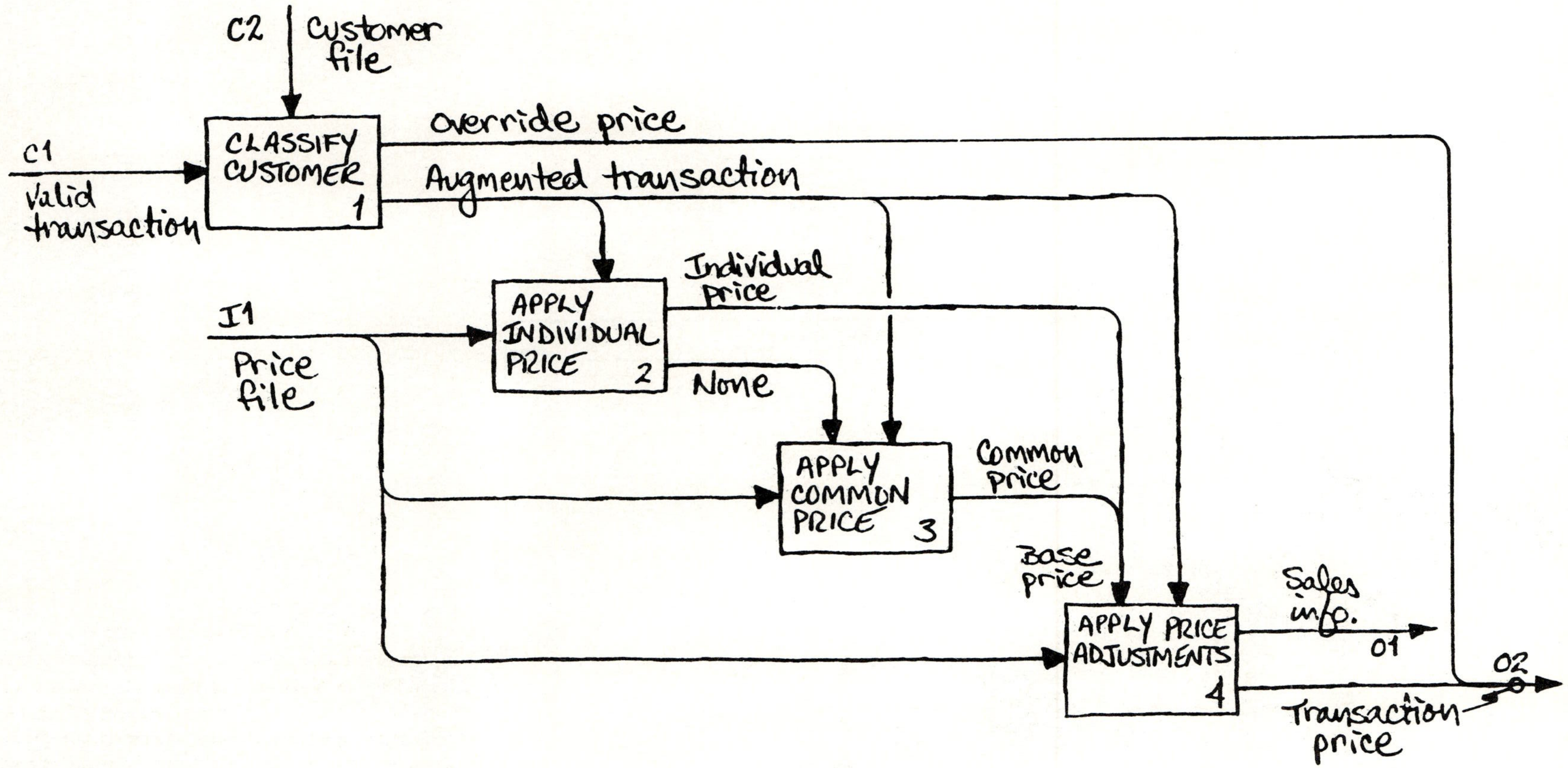

NODE: A32	TITLE: PRICE PRODUCTS	NUMBER:

A33 Calculate Product Taxes

The product involved in the delivery must be classified into a product tax group (box 1). Then we must determine which taxes apply (box 2), based on the type of product (box 2, control 1) and the customer's location (box 2, control 2). The customer may have an exemption (box 2, input 1).

The indication of applicable taxes (box 2, output 1) is used to search tax tables (box 3, input 1) to find the appropriate rates for each taxing municipality (box 3, output 1). These rates are then applied (box 4, input 1) to the quantity involved (box 4, control 2) to determine product taxes (box 4, output 2).

Each product tax produces a separate line item on the invoice.

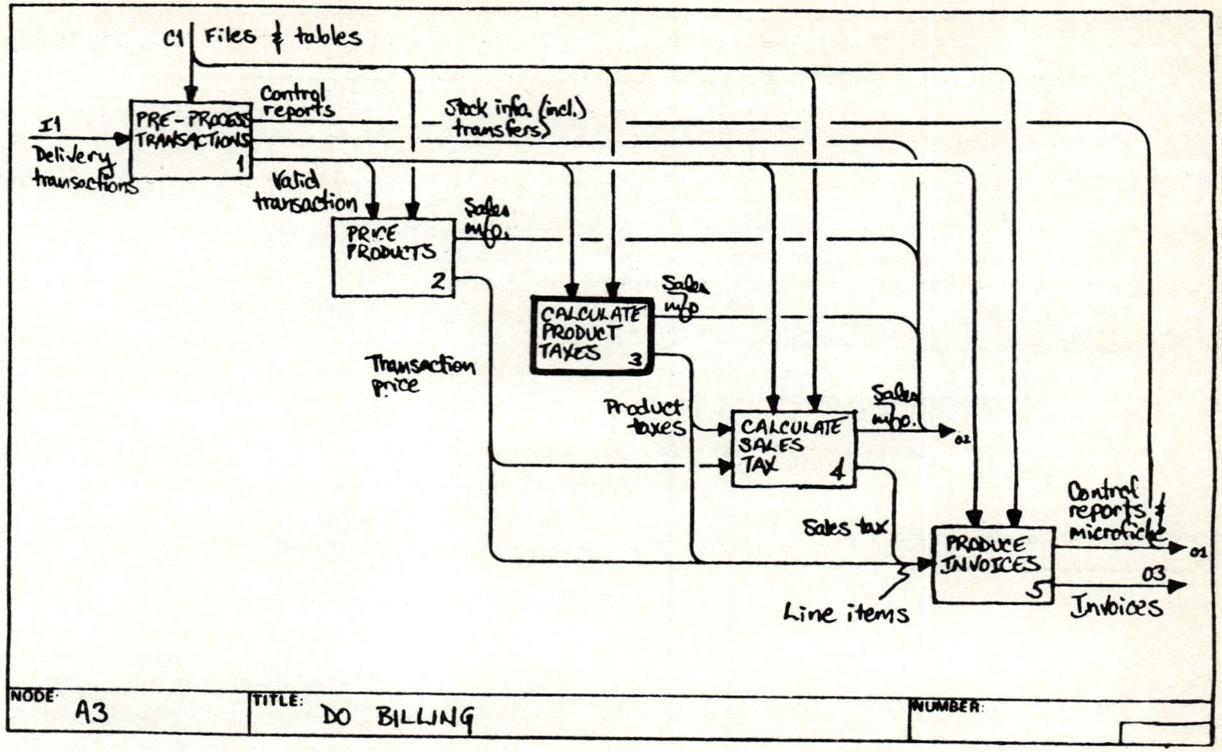

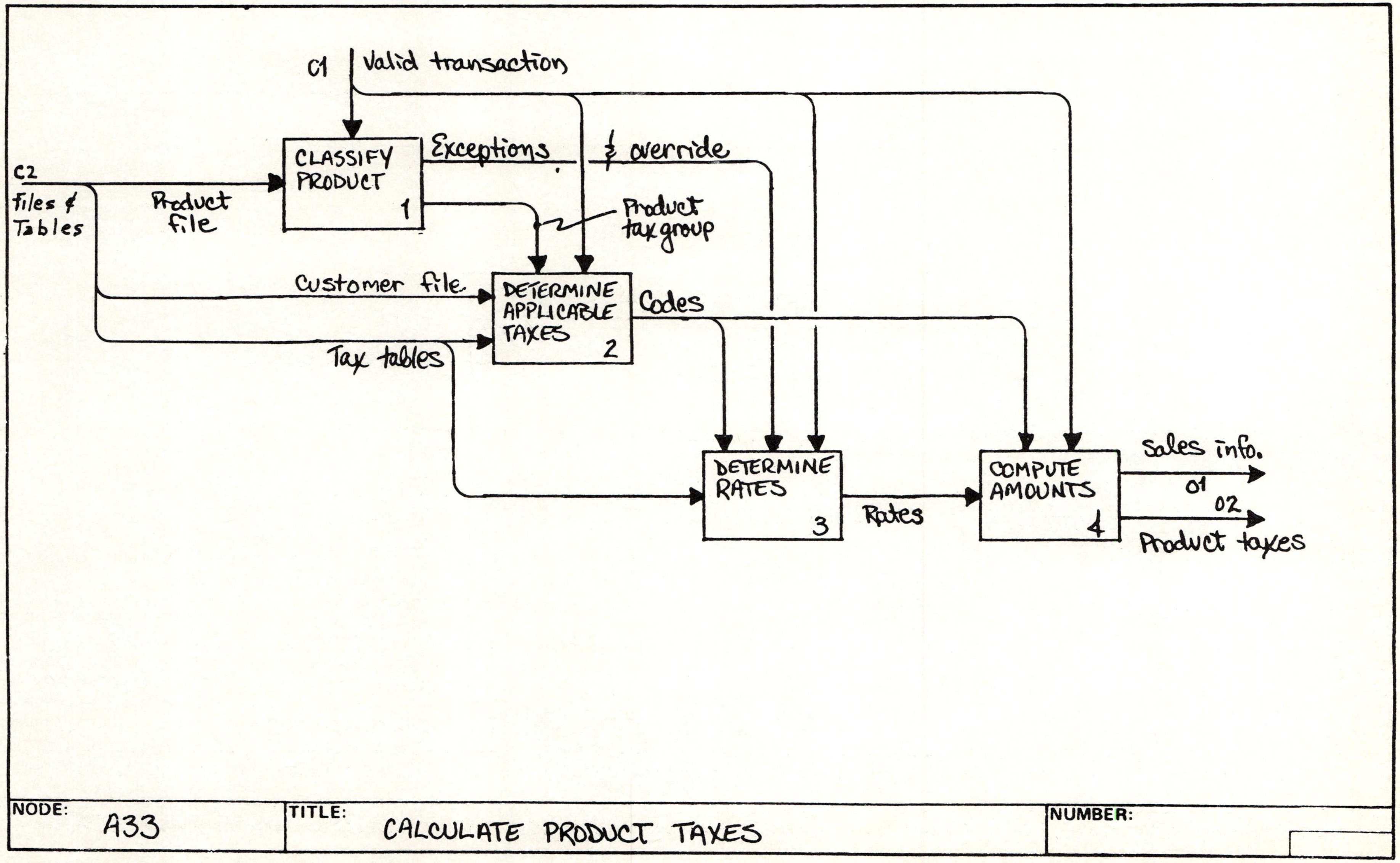

NODE: A33 | TITLE: CALCULATE PRODUCT TAXES | NUMBER:

A34 Calculate Sales Tax

Processing here is very much like application of product taxes, with two differences. First, all sales tax rates are accumulated (box 3, output 1) to produce one line item on the invoice (box 4, output 2).

Second, sales tax may be applied to the transaction base price (box 4, control 2), or to the base price and product taxes (box 4, control 1). The taxing municipality (box 2, output 1) determines how it is to be done.

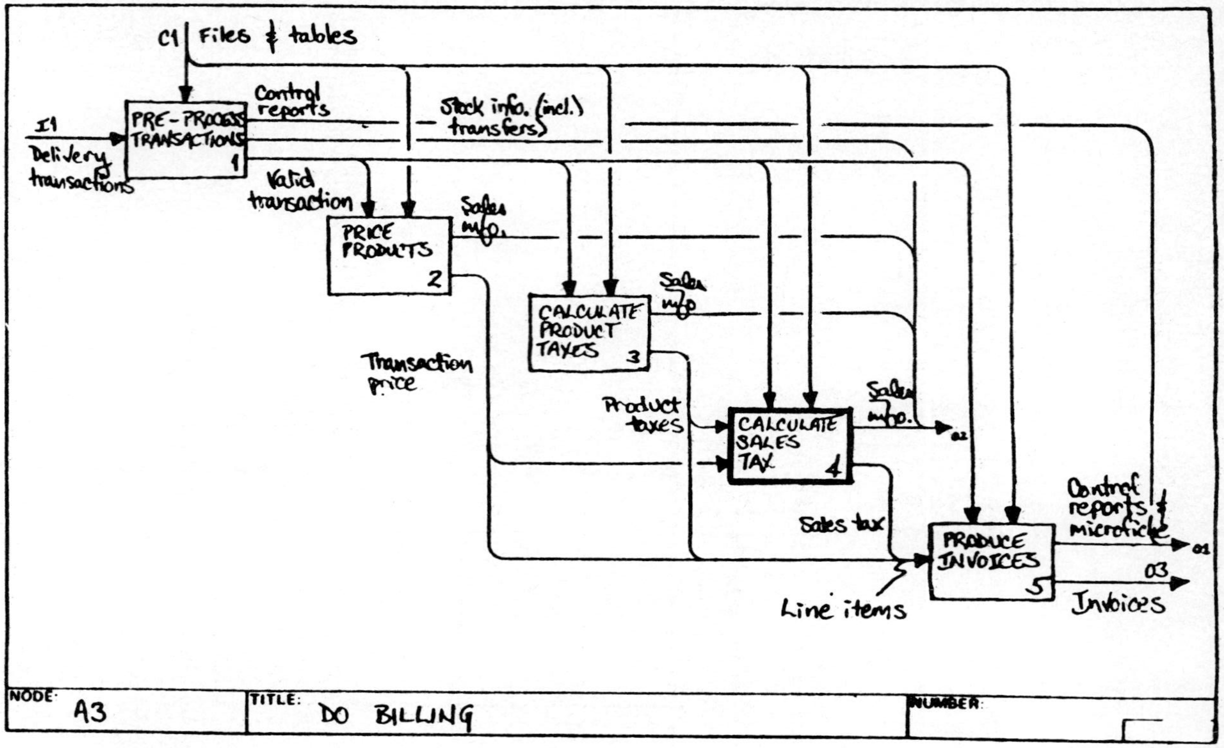

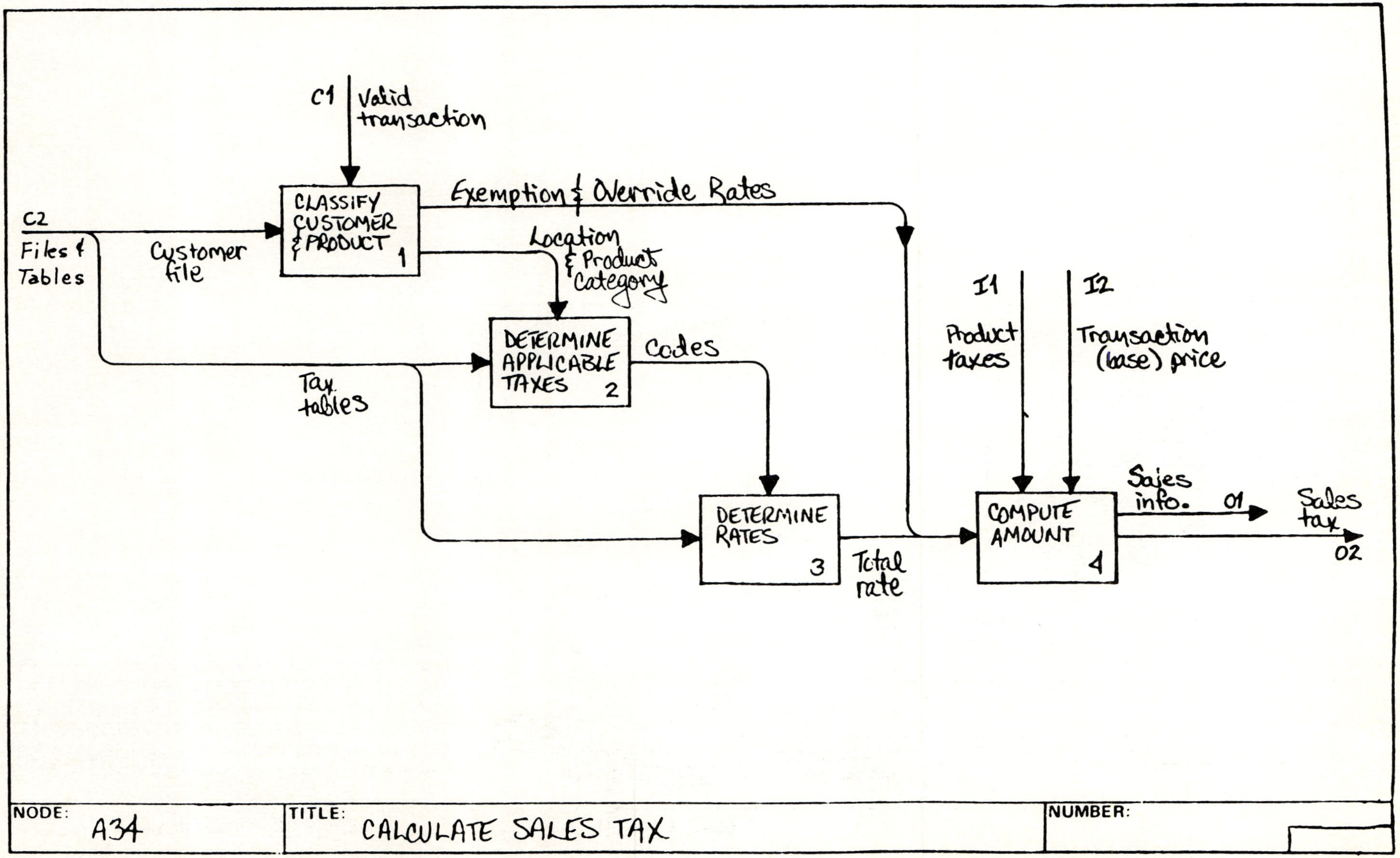
C1
Valid transaction
C2
Files & Tables
Customer file
CLASSIFY CUSTOMER & PRODUCT 1
Exemption & Override Rates
Location & Product category
DETERMINE APPLICABLE TAXES 2
Tax tables
Codes
DETERMINE RATES 3
Total rate
I1
Product taxes
I2
Transaction (base) price
COMPUTE AMOUNT 4
Sales info.
O1
Sales tax
O2
NODE: A34
TITLE: CALCULATE SALES TAX
NUMBER:

A35 Produce Invoices

Before invoices can actually be printed (box 2), certain miscellaneous line item charges must be calculated (box 1). These include fixed charges, variable charges and municipal charges recorded in the Price File (input 1, box 1).

After invoices are produced, they must be sorted into those that are finished and may be sent to customers (box 5, output 1), and those that require special handling (box 5, output 2).

Microfiche copies are made (box 3) and a control report of total dollar volume is written (box 4).

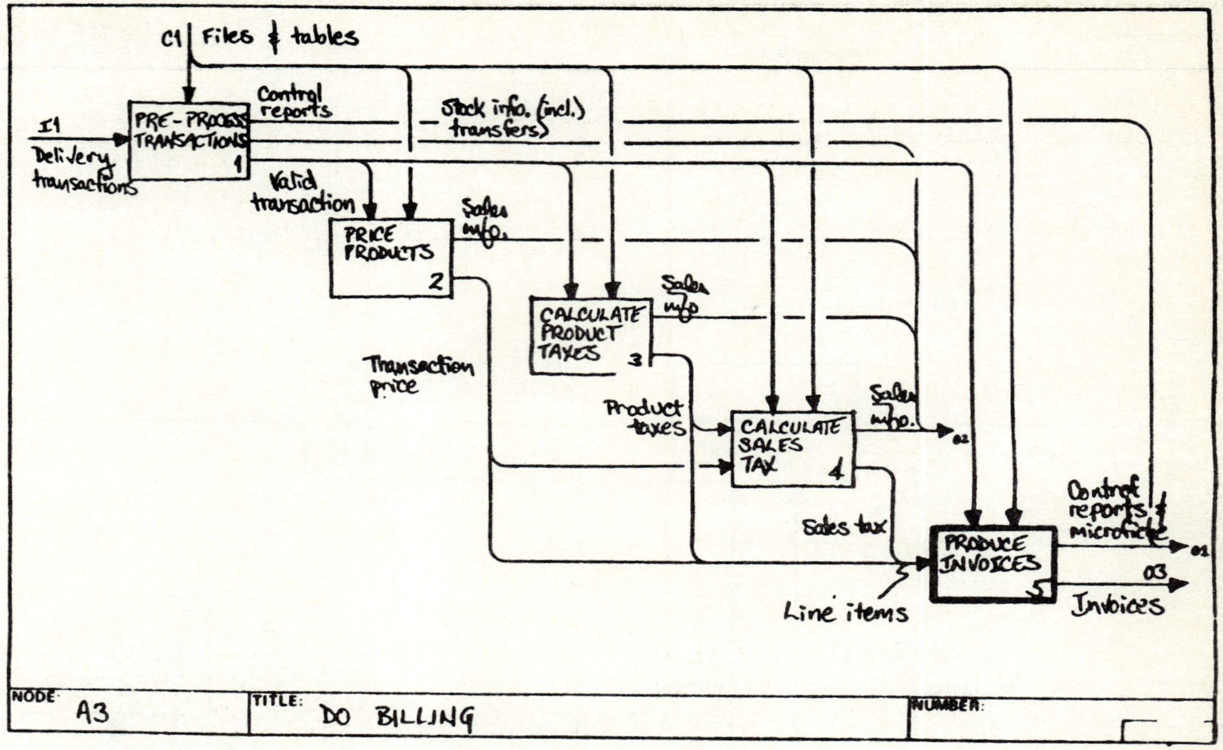

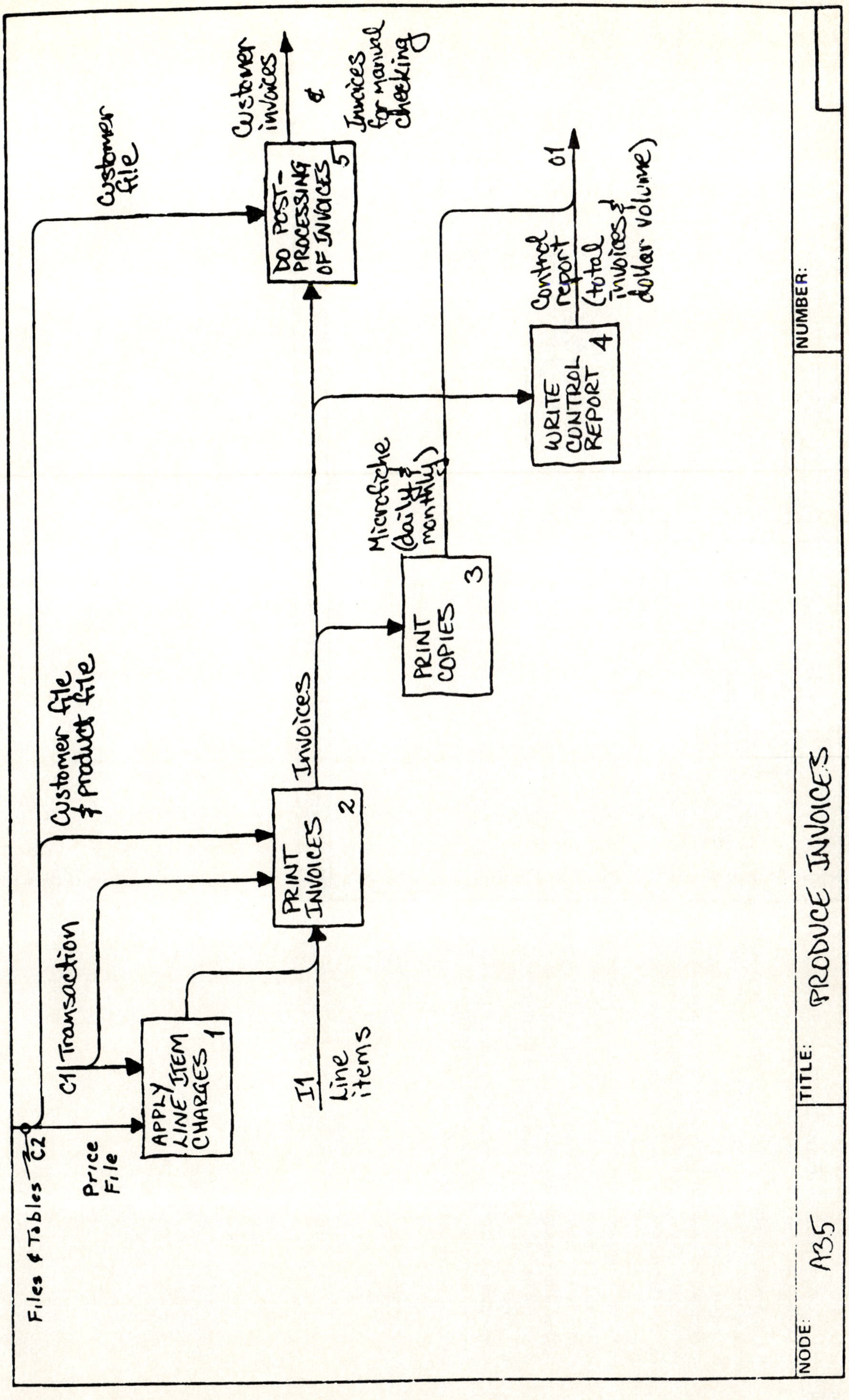
Files & Tables
C2
Price File
C1 Transaction
I1 Line items
APPLY LINE ITEM CHARGES 1
Customer file & product file
PRINT INVOICES 2
Invoices
PRINT COPIES 3
Microfiche (daily & monthly)
WRITE CONTROL REPORT 4
Control report (total invoices & dollar volume)
O1
Customer file
DO POST-PROCESSING OF INVOICES 5
Customer invoices & Invoices for manual checking
NODE: A35
TITLE: PRODUCE INVOICES
NUMBER:

A352 Print Invoices

We first must determine what kind of an invoice is to be printed (box 1). This may modify paging, or items included, or special messages.

Then header information is printed (box 2), depending on the customer (box 2, control 2), the shipper (box 2, input 1), and whether this is a continuation page (box 2, control 3).

All line items are printed (box 3), obtaining product descriptions as necessary (box 3, control 1).

Finally, last page information is printed (box 4), including keyed-in messages (box 4, input 2). If the current page is not the last page, a continuation page is called for (box 4, output 1).

Specifications for the Customer File request a last invoice date (box 4, output 3). This has not been referenced in the available literature on invoicing.

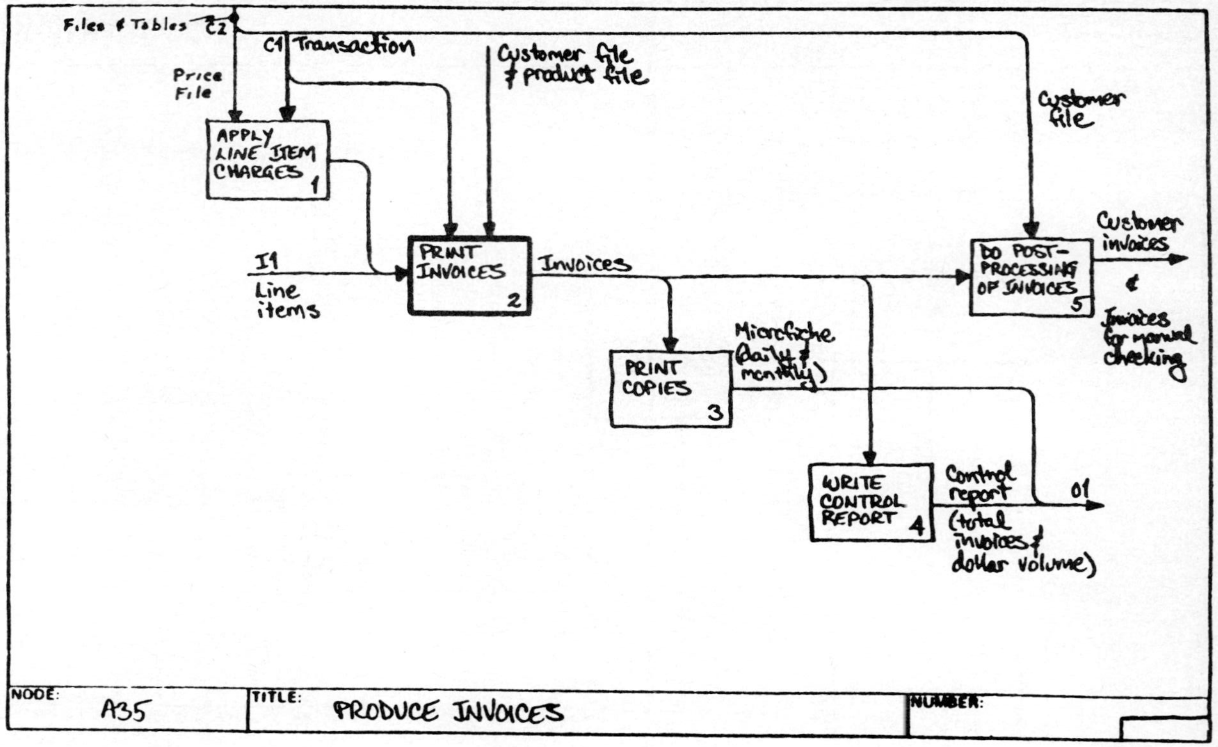

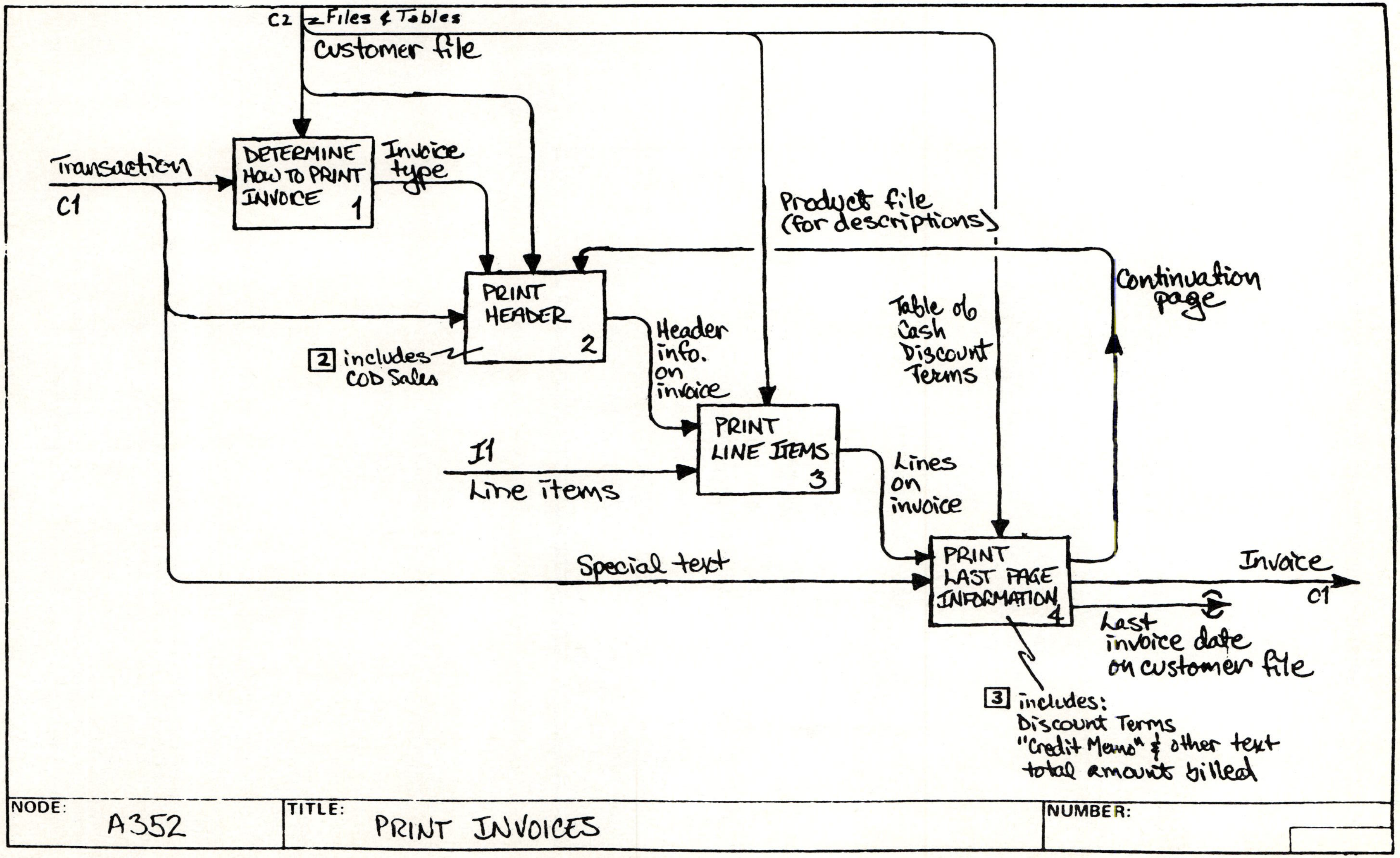
C2
Files & Tables
Customer file
Transaction
C1
DETERMINE HOW TO PRINT INVOICE 1
Invoice type
PRINT HEADER 2
2 includes COD Sales
Header info. on invoice
Product file (for descriptions)
Continuation page
Table of Cash Discount Terms
I1
Line items
PRINT LINE ITEMS 3
Lines on invoice
Special text
PRINT LAST PAGE INFORMATION 4
Invoice
C1
Last invoice date on customer file
3 includes:
Discount Terms
"Credit Memo" & other text
total amounts billed
NODE:
A352
TITLE:
PRINT INVOICES
NUMBER:

Diagram Index for the Central Billing Model

- **A-1** Do Company Billing
  - **A-0** Do Central Billing
    - **A0** Do Central Billing
      - **A1** Manage Database
        - **A12** Maintain Customer Records
        - **A13** Maintain Price Records
        - **A14** Handle Inquiries
      - **A2** Check Transactions
        - **A21** Pass Batches with Valid Controls
        - **A22** Check Delivery Data
        - **A23** Check Controls and Handle Errors
      - **A3** Do Billing
        - **A31** Preprocess Transactions
        - **A32** Price Products
        - **A33** Calculate Product Taxes
        - **A34** Calculate Sales Tax
        - **A35** Produce Invoices
          - **A352** Print Invoices

Chapter 31

Training

31.1 Application Domain

One stumbling block to developing effective training is visualizing how a training group should be organized in order to effectively teach the right kind and amount of information to the right people. For example, when teaching teamwork skills, one works with small groups and not with individuals nor with very large groups. Thus, and especially for large organizations, effective training takes good organization and planning—and this requires a good road map of how training is done.

31.2 Project Background

The U.S. Army Training and Doctrine Command (TRADOC) has the job of maintaining combat readiness and improving the combat effectiveness of the U.S. Army. At one time, TRADOC headquarters needed to: (1) identify changes in Army training required to increase combat effectiveness; (2) describe how future Army training, testing, and evaluation programs would operate after the changes; and (3) plan how the changes would occur and be monitored. An SADT model was developed to satisfy these needs.

31.3 Project Overview

The project was jointly funded by TRADOC and the Defense Advanced Research Projects Agency (DARPA) in order to demonstrate how SADT could be used to improve Army training. An important project objective was to assess whether better systems analysis methods could facilitate the definition of all Army training. Such a definition would then provide the basis for solving smaller training problems with limited resources and tight deadlines. A composite SADT model of U.S. Army training was therefore developed.

31.4 Resources Used to Make the Model

Two SADT authors interviewed about 30 Army personnel during the modeling effort. Another 10 people, some from the military and some from universities, reviewed the model as it was being developed. With this staff, it took 1½ years to complete the model shown here.

31.5 Notes about the Model

Initially, a functional decomposition strategy produced the top diagram. Then, a decomposition-by-type strategy produced first-level and some second-level diagrams. Last, a sequential decomposition strategy was used to describe the details of the training process. The model uses two-way arrows to describe mutually constraining activities. This advanced technique is useful when describing systems in which people interact heavily with each other. In such cases, two-way arrows reduce the clutter caused by large numbers of control or input feedbacks. Also, notice how the judicious use of SADT square notes helps to annotate some important aspects of that feedback information.

A-0 Provide Trained Units

The purpose of this model is to describe the training system as it should be, from the point of view of those who provide trained personnel to the forces. From this viewpoint, the development of advanced weapon designs, tactics, techniques, and organizations are the driving forces behind the development and administration of training which satisfies the needs of the forces. Unimportant from this viewpoint are such areas as general DA guidance and regulations (they affect all activities), current organizational budgets (they affect all activities to the extent that budgets control the quantity and quality of the outputs), and logistic support for all the activities in the model.

As a system, "Provide Trained Units" converts Army missions and objectives (Input 1) into a detailed set of courses and programs (both institutional and non-resident), and trains incoming personnel (Input 2) using whatever material (Input 3) is necessary for training. This training reflects an understanding of the threat (Control 1) and internally attempts to adjust for inadequacies in combat readiness. Training also considers unit manpower requirements and manpower availability, as well as the availability of training facilities such as ranges. The training system also makes adjustments in training techniques and priorities to satisfy current and future funding restraints, which are treated as limitations on resources (Control 2).

The purpose of including combat developments and personnel management within the scope of the model is to describe the extensive interaction between these activities and training, with an eye toward understanding how training and evaluation should influence combat developments and personnel management.

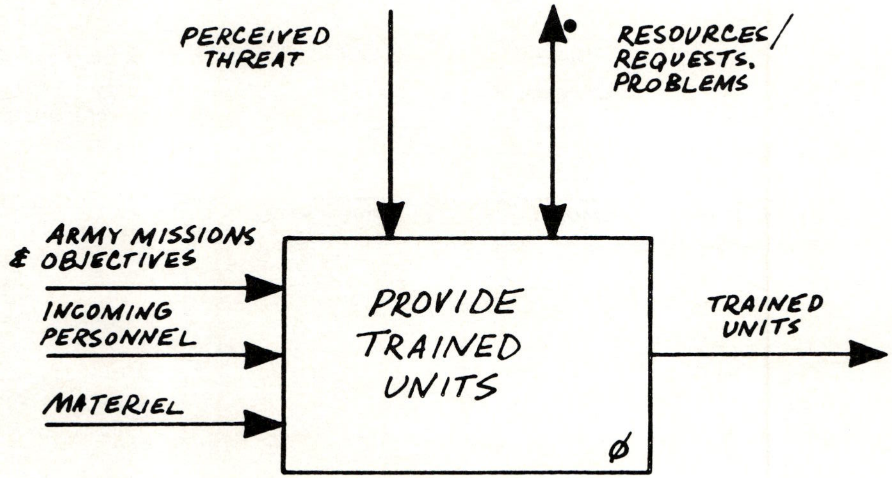

NODE: TRAIN / A-Ø

TITLE: PROVIDE TRAINED UNITS

NUMBER:

A0 Provide Trained Units

In order to fully understand how the Army should provide trained units, it is necessary to describe how fighting systems are designed, implemented, and evaluated. Providing trained units begins with the development of weapons and methods necessary to support the missions and objectives (box 1 input 1) imposed on the Army. The design of this fighting system is then implemented by managing the assembly of manpower and materiel (box 2) and by training the manpower in tactics and techniques (box 3). Finally, the design and implementation of the fighting system is evaluated, providing feedback to stimulate further improvements (box 4).

Box 1 is concerned with developing weapons which are effective in light of the threat perceived by the Army (box 1, control 1). Techniques and tactics as well as organizations are then developed which make the most effective use of those weapons. The resulting output of box 1 describes what hardware/personnel systems the Army will fight with and how the Army will fight.

The function of box 2 is to organize the Army according to the organizational guidelines (box 2 control 1), and to screen and allocate manpower to jobs in units, or assign manpower to institutional training. Box 2 is therefore responsible for the implementation of the hardware/personnel fighting system. It has not yet been further detailed.

The activity of training the Army (box 3) is primarily driven by materiel design, tactics, and techniques (box 3, control 2) in that they yield the training objectives. Institutional and non-resident job training programs are then developed, distributed, and conducted.

Two important outputs of the training activity are the performance test results for use in evaluation (box 3 output 3), and the recommendations concerning materiel designs, tactics and techniques, and organizations (box 3 control 2). These recommendations are the result of difficulties encountered in teaching a particular task or in the conduct of a training exercise.

The performance test results are evaluated by box 4 in light of the perceived threat, and any problems which become obvious are identified as being associated with training, personnel management, or materiel design, tactics, and organization. The resulting recommendations are then fed back to training, personnel management, or combat development. Box 4 has not yet been further detailed.

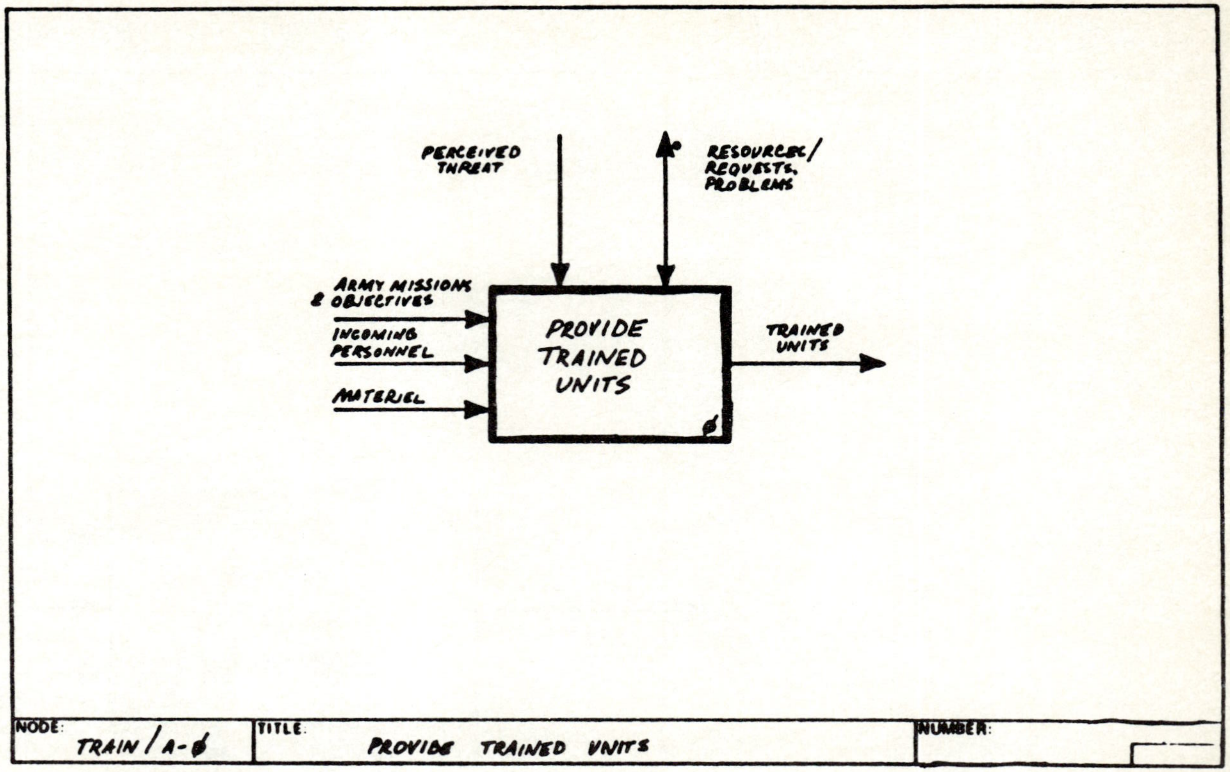

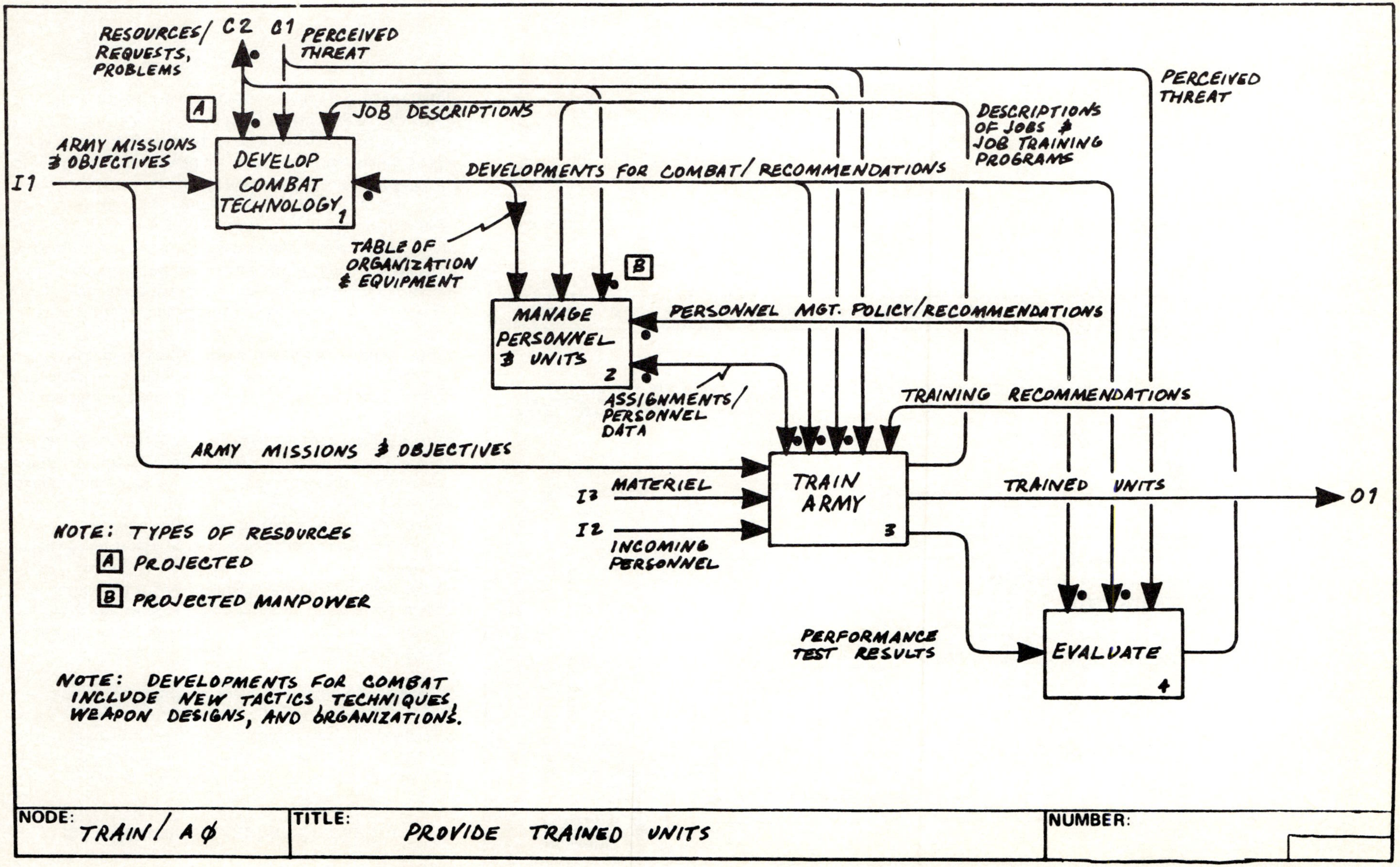
RESOURCES/ REQUESTS, PROBLEMS
C2
C1
PERCEIVED THREAT
A
JOB DESCRIPTIONS
ARMY MISSIONS $ OBJECTIVES
I1
DEVELOP COMBAT TECHNOLOGY 1
DEVELOPMENTS FOR COMBAT/ RECOMMENDATIONS
DESCRIPTIONS OF JOBS $ JOB TRAINING PROGRAMS
PERCEIVED THREAT
TABLE OF ORGANIZATION & EQUIPMENT
B
MANAGE PERSONNEL $ UNITS 2
PERSONNEL MGT. POLICY/RECOMMENDATIONS
ASSIGNMENTS/ PERSONNEL DATA
TRAINING RECOMMENDATIONS
ARMY MISSIONS $ OBJECTIVES
I3
MATERIEL
I2
INCOMING PERSONNEL
TRAIN ARMY 3
TRAINED UNITS
O1
PERFORMANCE TEST RESULTS
EVALUATE 4
NOTE: TYPES OF RESOURCES
A PROJECTED
B PROJECTED MANPOWER
NOTE: DEVELOPMENTS FOR COMBAT INCLUDE NEW TACTICS TECHNIQUES, WEAPON DESIGNS, AND ORGANIZATIONS.
NODE: TRAIN/ A0
TITLE: PROVIDE TRAINED UNITS
NUMBER:

A1 Develop Combat Technology

The function of "Develop Combat Technology" is to, first, develop tactics and techniques which will be effective against the perceived threat, second, develop materiel which will support those tactics, and finally, to develop organization/equipment structure which best suits the tactics and the materiel.

The development of tactics and techniques (box 1) considers the Army missions and objectives (box 1 input 1), and, with knowledge of the weapons, tactics, and organization of the enemy (box 1 control 1), determines how to fight (box 1 output 1). Also produced are requirements for materiel which will be needed to support these tactics (box 1 output 2). Adjustments to the tactics and techniques developed (box 1 output 1) are then made based on the materiel designed by box 2 (box 2 output 1 and box 2 control 1 to box 1 output 2).

Once the materiel has been designed (box 3 control 1) and the tactics for their use (box 3 control 3) have been developed, it then remains the function of box 3 to determine the organization of materiel and manpower (box 3 output 1) which will most easily enable the implementation of the tactics.

An important control on all of the activities shown is the recommendations which are made as a result of evaluating performance test results or as a result of difficulties encountered during the conduct of training. These recommendations can include recommendations to modify tactics (e.g., as a result of field training exercises), (box 1 control 2), recommendations to modify materiel design (e.g., as a result of inability to train soldiers how to use the sights on the LAW), (box 2 control 3), or recommendations to modify organization (e.g., as a result of difficulties to effectively control or display weapons during a field training exercise), (box 3 control 4).

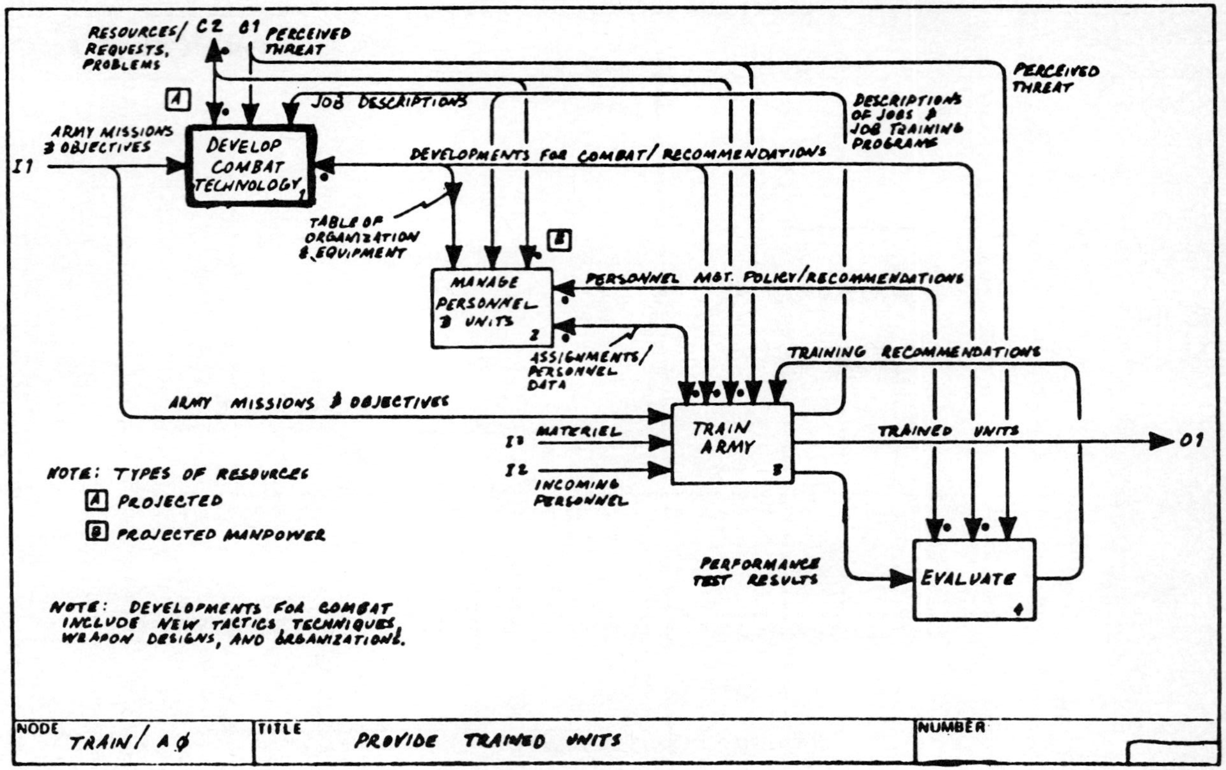

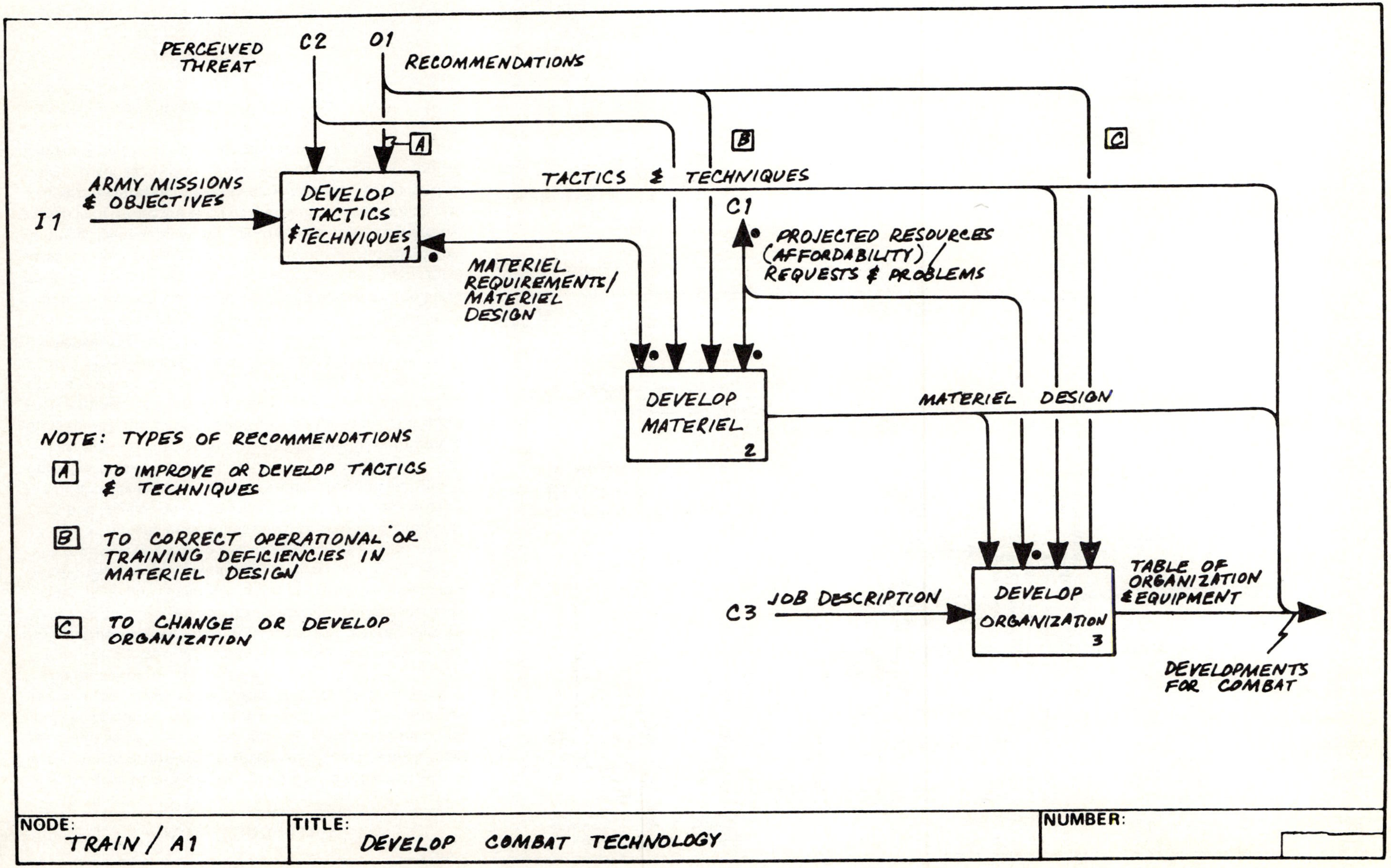
PERCEIVED THREAT
C2
O1
RECOMMENDATIONS
A
B
C
ARMY MISSIONS & OBJECTIVES
I1
DEVELOP TACTICS & TECHNIQUES 1
TACTICS & TECHNIQUES
C1
PROJECTED RESOURCES (AFFORDABILITY)/ REQUESTS & PROBLEMS
MATERIEL REQUIREMENTS/ MATERIEL DESIGN
DEVELOP MATERIEL 2
MATERIEL DESIGN
NOTE: TYPES OF RECOMMENDATIONS
A TO IMPROVE OR DEVELOP TACTICS & TECHNIQUES
B TO CORRECT OPERATIONAL OR TRAINING DEFICIENCIES IN MATERIEL DESIGN
C TO CHANGE OR DEVELOP ORGANIZATION
C3
JOB DESCRIPTION
DEVELOP ORGANIZATION 3
TABLE OF ORGANIZATION & EQUIPMENT
DEVELOPMENTS FOR COMBAT
NODE: TRAIN / A1
TITLE: DEVELOP COMBAT TECHNOLOGY
NUMBER:

A3 Train Army

The purpose of this diagram is to describe the interaction between the component activities of "Train Army". Once training has been developed, the objectives (box 1 output 2) and the training programs (box 1 output 1) are distributed to units (box 3) and institutions (box 2).

The function of "Develop Training" is to first, identify the different jobs that exist in the Army (box 1 output 1); then define what the training objectives for those jobs are (box 1 output 2) including critical tasks, soldier's manuals, "How to Fight" manuals, and performance tests; and finally, develop job training programs (box 1 output 2), which completely define the training for a job/skill level. The development of training must be responsive to problems and suggestions (box 1 control 3) which may arise as a result of evaluation of test results (diagram A0 box 4 output 1 to box 3 control 5).

The activities of training in institutions (box 2) and in units (box 3) are distinguished only by the kinds of training they perform, as limited by the facilities available. Except for trained units (box 3 output 3) incoming personnel (box 2 input 1) and the two classes of trained personnel circulating between boxes 2 and 3, there are no differences in the inputs, controls, and outputs. Specifically, both kinds of training attempt to satisfy the training objectives produced by box 1 (output 2). These objectives include the critical tasks as described by the soldier's manuals and the "How to Fight" manuals, as well as the tests that measure the mastery of those tasks. The particular tasks which are taught in institutions or units may, however, be selected to satisfy some facility or other resource constraint (box 2 control 4 and box 3 control 5).

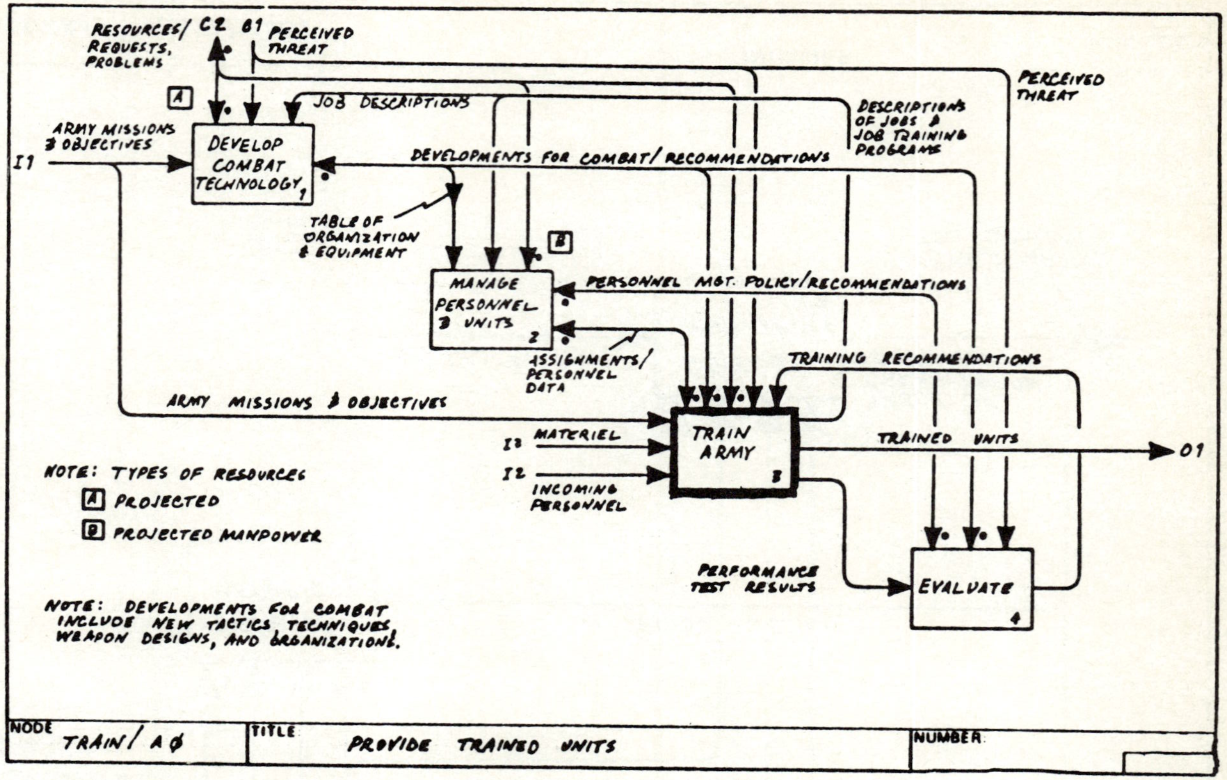

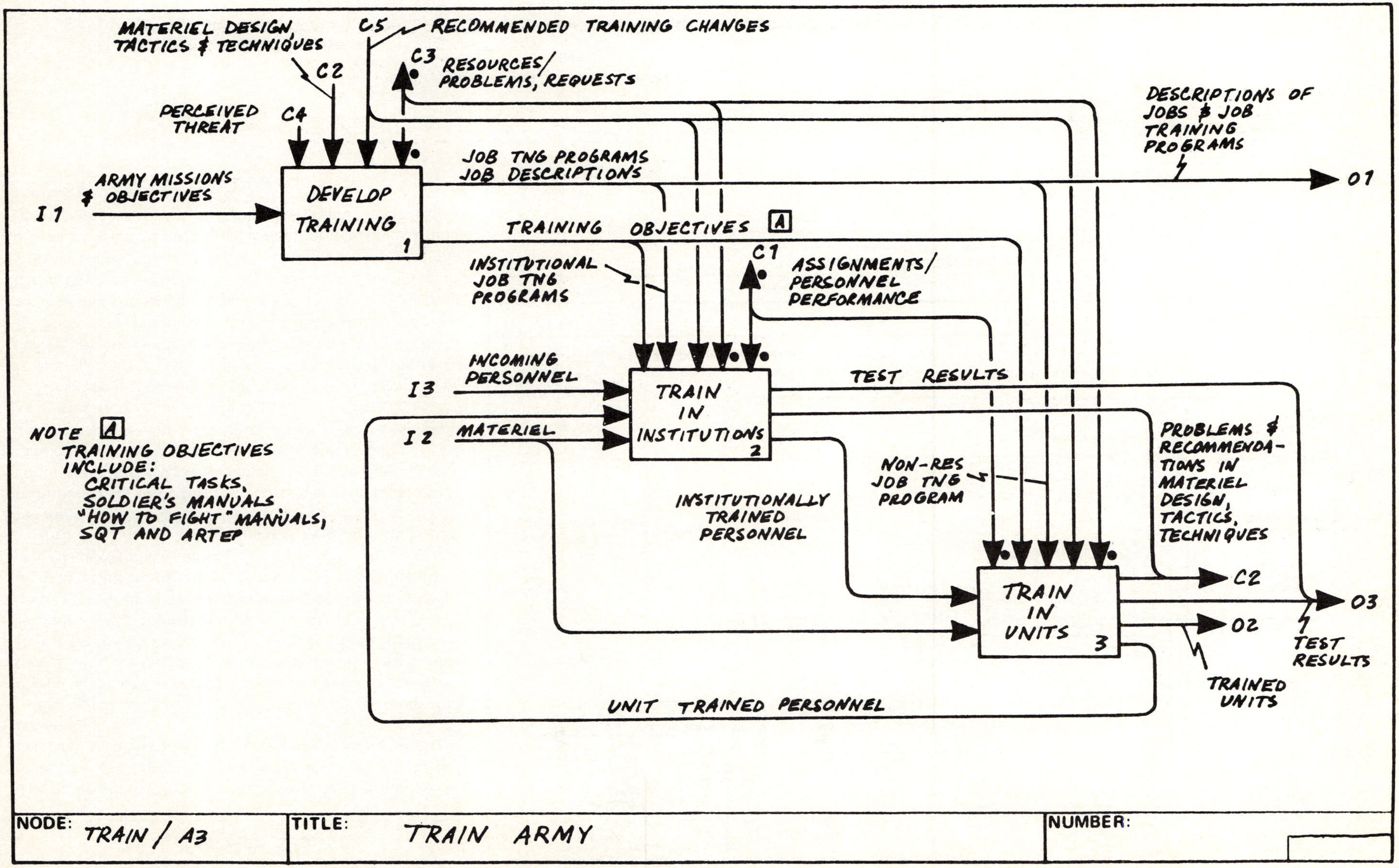

NODE: TRAIN / A3	TITLE: TRAIN ARMY	NUMBER:

A31 Develop Training

The process of developing training consists of an analysis to determine the lowest level objective of training, i.e., the task, and then a synthesis to develop the training programs for jobs.

The function of box 1 is to consider the missions and objectives of the Army (box 1 input 1), the perceived threat (box 1 control 1), and the weapons and methods available (box 1 control 2), and to, first, define the job positions, and second, determine task and job training objectives. Feedback is provided to this activity in that evaluation of training may result in suggestions to modify the resulting list of critical tasks (box 1 control 3).

With the critical tasks defined (box 2 control 1), the function of the second box is to develop courses which will train personnel to perform the particular individual tasks. Along with these specific courses (box 2 output 1), the development of task training specifies required aids, devices, and materiels, and may also specify a required facility, such as a firing range (box 2 output 1).

The process of developing job training (box 3) consists of collecting courses (box 3 input 1) which teach the tasks defined for each job in the soldier's manual (box 3 control 2). These courses are packaged into institutional and non-resident job training programs (box 3 output 1) such as officer basic courses, TEC courses, etc. The decision as to whether training for a particular task belongs in institutions or non-resident training is dependent on job training time constraints (box 3 control 3) and the resources available to institutions (box 3 control 1).

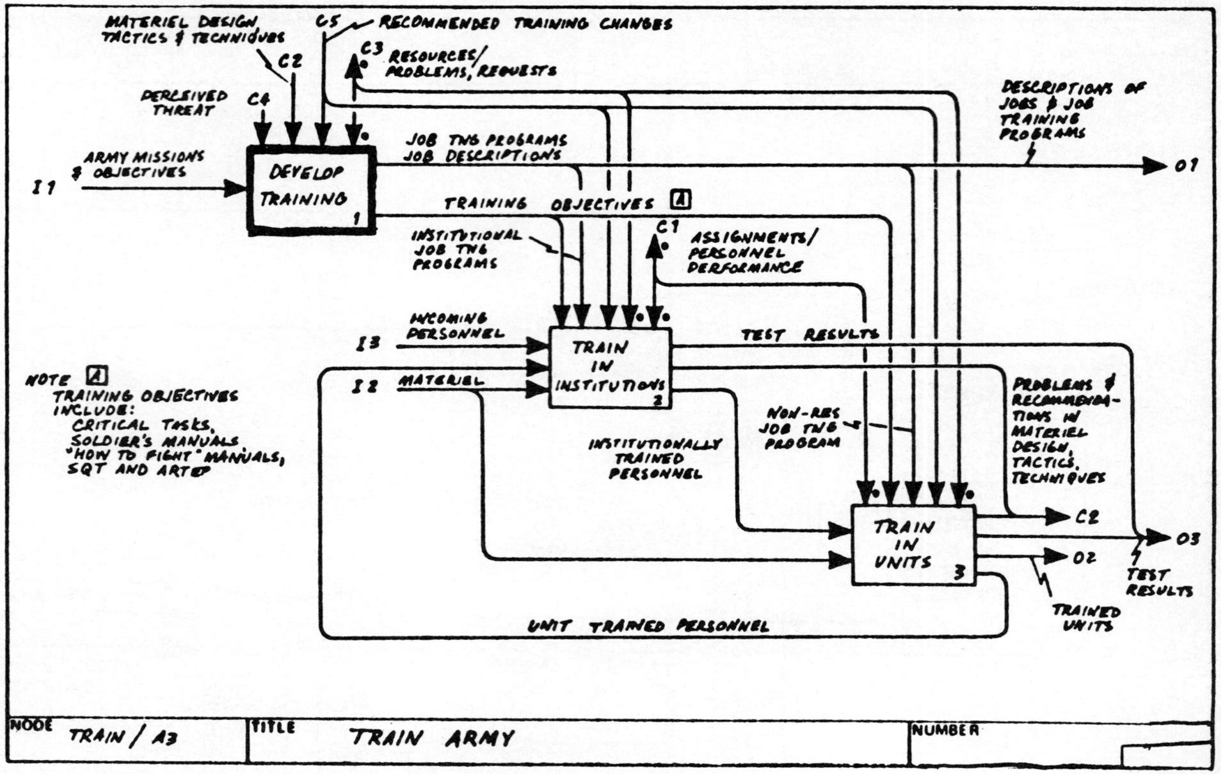

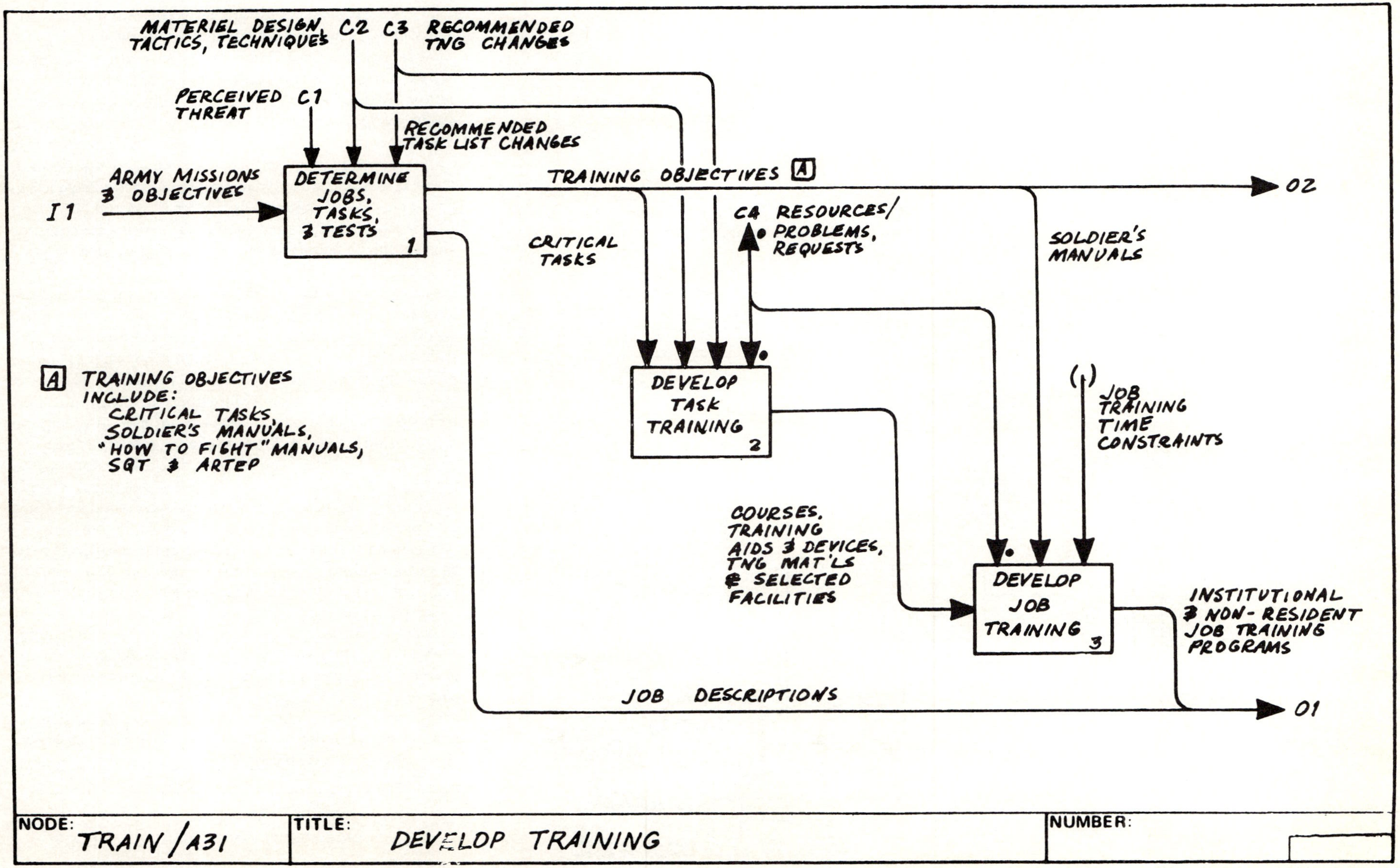
MATERIEL DESIGN, TACTICS, TECHNIQUES
C2
C3
RECOMMENDED TNG CHANGES
PERCEIVED THREAT
C1
RECOMMENDED TASK LIST CHANGES
ARMY MISSIONS & OBJECTIVES
I1
DETERMINE JOBS, TASKS, & TESTS
1
TRAINING OBJECTIVES A
O2
C4 RESOURCES/ PROBLEMS, REQUESTS
CRITICAL TASKS
SOLDIER'S MANUALS
DEVELOP TASK TRAINING
2
( ) JOB TRAINING TIME CONSTRAINTS
A TRAINING OBJECTIVES INCLUDE:
CRITICAL TASKS,
SOLDIER'S MANUALS,
"HOW TO FIGHT" MANUALS,
SQT & ARTEP
COURSES, TRAINING AIDS & DEVICES, TNG MAT'LS & SELECTED FACILITIES
DEVELOP JOB TRAINING
3
INSTITUTIONAL & NON-RESIDENT JOB TRAINING PROGRAMS
JOB DESCRIPTIONS
O1
NODE:
TRAIN/A31
TITLE:
DEVELOP TRAINING
NUMBER:

A311 Determining Jobs, Tasks, and Tests

The determination of training objectives is an analysis process which breaks critical missions down to the level of critical individual tasks and develops the necessary tests. The analysis begins with the identification of critical missions to be performed (box 1 output 1), as determined from the missions and objectives imposed upon the Army (box 1 input 1) and the perceived threat (box 1 control 1). With an understanding of the enemy's capabilities and weaknesses (box 2 control 2), the collective critical tasks as described in the "How to Fight" manuals (box 2 output 1) can be developed from these critical missions (box 2 control 1).

The analysis then continues with the definition of the jobs (duty positions) which must be performed (box 3 output 1) and the identification of the critical tasks to be performed by the individual (box 4 output 1).

Finally, the skill qualification test (box 5 output 1) is developed to test for mastery of the required individual skills, and the Army Training and Evaluation Program is developed to test for mastery of collective skills (box 6 output 1). This latter test may use portions of the appropriate SQT (box 6 input 1) where performance of a collective critical task requires performance of a critical individual task, e.g., radio telephone procedure in a company sized event.

An important point in the understanding of this process concerns responsiveness to the training needs of the Army. First, the process responds to changes in the perceived threat (diagram boundary arrow C1) and changes resulting from combat developments (diagram boundary arrow C2). More important, however, the process compensates for training deficiencies. This is accomplished as follows: In the event that evaluation of performance (diagram A0 box 4) determines that performance is deficient because the critical tasks chosen for training are inappropriate, that activity may recommend that either the critical collective tasks or the critical individual tasks are to be changed.

Another important point to note is that the sum of all the critical individual tasks is less than the critical collective tasks, i.e., there are some collective tasks which cannot be broken down into individual tasks which can be performed alone, such as operation of a crew served weapon. The training objectives developed in the form of critical individual tasks thus serve as the basis for training of the collective critical tasks.

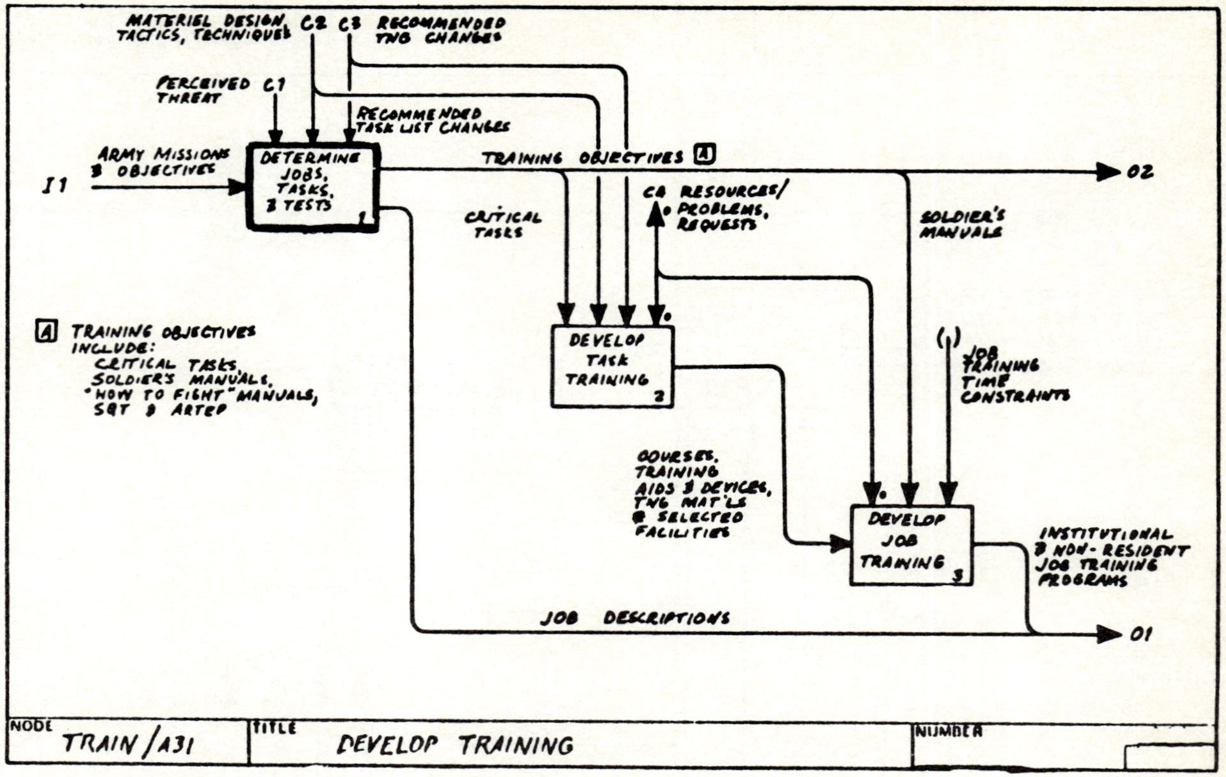

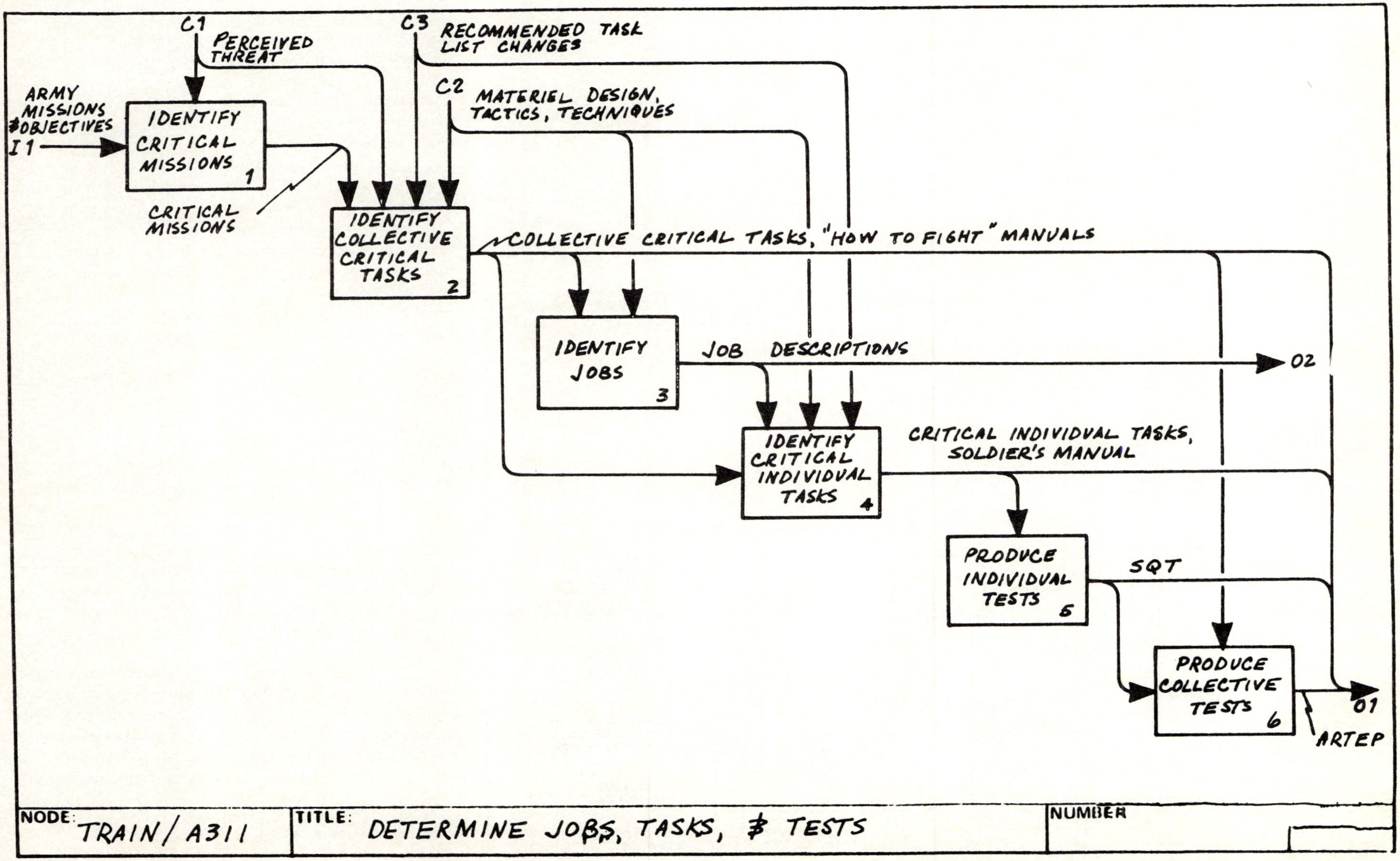
C1
PERCEIVED THREAT
C3
RECOMMENDED TASK LIST CHANGES
C2
MATERIEL DESIGN, TACTICS, TECHNIQUES
ARMY MISSIONS & OBJECTIVES
I1
IDENTIFY CRITICAL MISSIONS 1
CRITICAL MISSIONS
IDENTIFY COLLECTIVE CRITICAL TASKS 2
COLLECTIVE CRITICAL TASKS, "HOW TO FIGHT" MANUALS
IDENTIFY JOBS 3
JOB DESCRIPTIONS
O2
IDENTIFY CRITICAL INDIVIDUAL TASKS 4
CRITICAL INDIVIDUAL TASKS, SOLDIER'S MANUAL
PRODUCE INDIVIDUAL TESTS 5
SQT
PRODUCE COLLECTIVE TESTS 6
O1
ARTEP
NODE: TRAIN/A311
TITLE: DETERMINE JOBS, TASKS, & TESTS
NUMBER

A312 Develop Task Training

Having developed the training objectives (diagram boundary arrow C1), the remainder of the training development process is a synthesis, where training for specific tasks is developed (task training courses) and these task training segments are combined according to the job definitions in the soldier's manual, thus forming job training (diagram A313).

The development of task training begins by grouping tasks which are similar, forming course topics (box 1 output 1). The topic "Automotive Electrical Trouble-Shooting", for instance, could be a grouping of several tasks including "Trouble-Shoot Starter Circuit", "Trouble-Shoot Generating Circuit", "Trouble-Shoot Electrical Accessory Circuits".

The next step is to identify all the teaching methods which could be used to teach that task, e.g., lecture, self-paced instruction, formal OJT (box 2 output 1). The function of box 3 is to determine the training support requirements (box 3 output 1) needed for those possible training methods, e.g., firing ranges, simulators, video recording facilities.

The critical step in task training development is the selection of the presentation method (box 4). This activity must be particularly responsive to the training environment. The training methods (box 4 input 1) and the associated support requirements (box 4 input 2) are evaluated in terms of cost (box 4 control 2) and field effectiveness (box 4 control 1). Also important are two national policies (box 4 control 3): First, use of certain materiel may be restricted (POL, NBC), requiring selection of training methods consistent with this policy; Second, the presence or absence of the draft will significantly affect the quality of the personnel to be trained, and hence, will require adjustment in the presentation method selected. The presentation method selected (box 4 output 3) will also depend on the available training aids (box 6 control 2 to box 4 output 2). Also selected is the type of training facility required, or the specific facility required if such facilities are severely limited (diagram boundary arrow C4).

The remaining two activities organize the task training materiel (box 5 output 1), and develop the training aids (box 6) required to present the materiel.

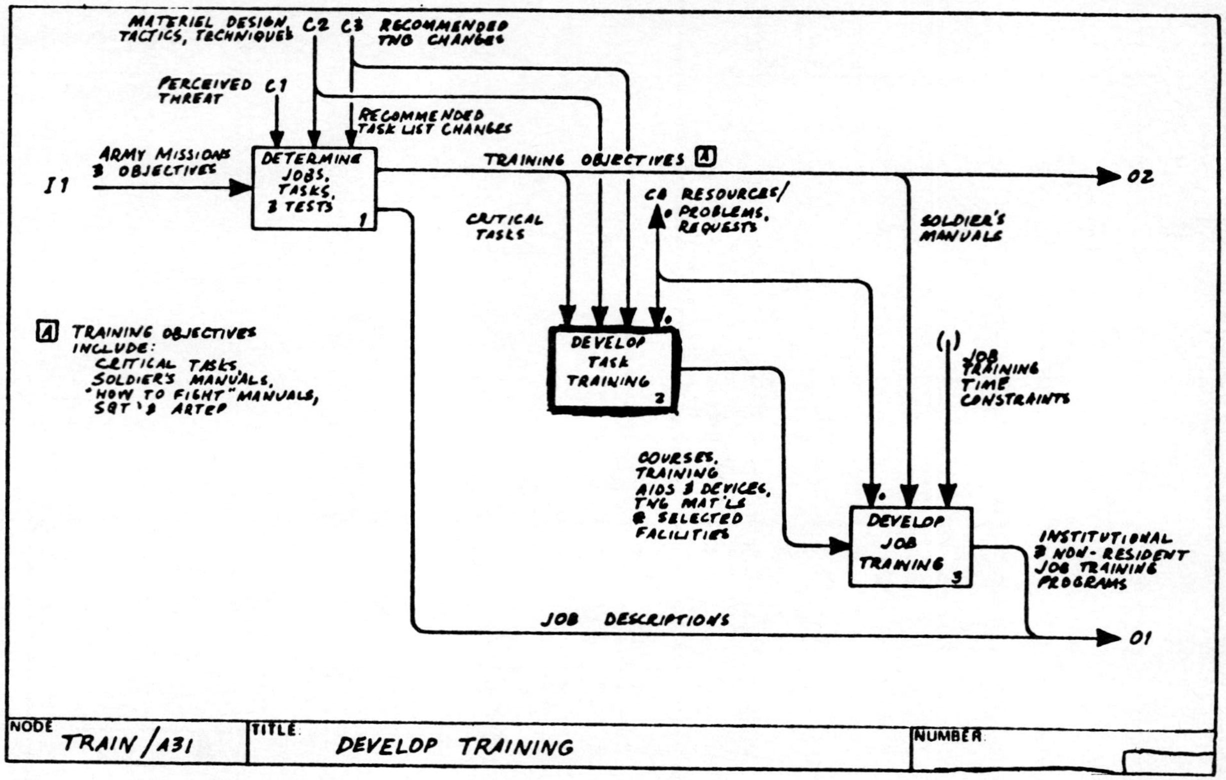

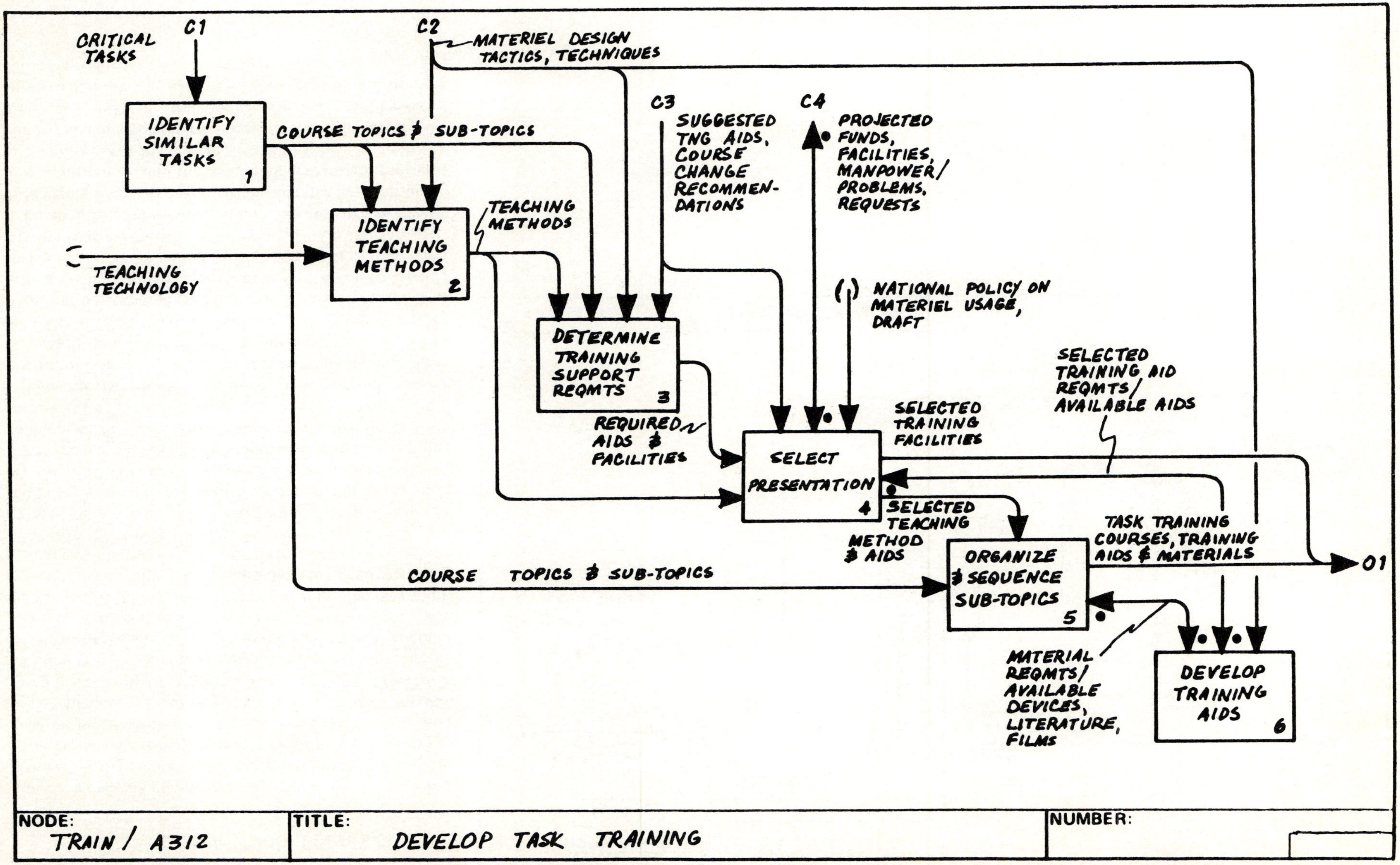
CRITICAL TASKS
C1
C2
MATERIEL DESIGN TACTICS, TECHNIQUES
C3
SUGGESTED TNG AIDS, COURSE CHANGE RECOMMENDATIONS
C4
PROJECTED FUNDS, FACILITIES, MANPOWER/ PROBLEMS, REQUESTS
IDENTIFY SIMILAR TASKS 1
COURSE TOPICS & SUB-TOPICS
IDENTIFY TEACHING METHODS 2
TEACHING METHODS
TEACHING TECHNOLOGY
DETERMINE TRAINING SUPPORT REQMTS 3
REQUIRED AIDS & FACILITIES
( ) NATIONAL POLICY ON MATERIEL USAGE, DRAFT
SELECT PRESENTATION 4
SELECTED TRAINING FACILITIES
SELECTED TRAINING AID REQMTS/ AVAILABLE AIDS
SELECTED TEACHING METHOD & AIDS
ORGANIZE & SEQUENCE SUB-TOPICS 5
TASK TRAINING COURSES, TRAINING AIDS & MATERIALS
O1
COURSE TOPICS & SUB-TOPICS
MATERIAL REQMTS/ AVAILABLE DEVICES, LITERATURE, FILMS
DEVELOP TRAINING AIDS 6
NODE: TRAIN / A312
TITLE: DEVELOP TASK TRAINING
NUMBER:

A313 Develop Job Training

The development of job training continues the synthesis process begun in diagram A312. Task training courses (diagram boundary arrow I1) are combined in various ways to produce job training programs.

The function of box 1 is to select those task training courses which will teach the set of tasks prescribed for a job/skill level in the soldier's manual (box 1 control 1). The set of courses so chosen make up the total training requirements for the job (box 1 output 1). Two operations are then performed on this set of requirements. First, the task training courses are clustered by specific site requirements, if any, (box 2 output 1). Some of these courses may require institutional training, while others may have facility requirements which can be satisfied in units as well. Second, the tasks detailed in the soldier's manual (box 3 control 2) are classified by whether or not they are required for job entry, or required by many jobholders. The task training courses which teach these fundamental critical tasks are then selected for institutional training (box 3 output 1).

Thus far, two classes of courses have been selected for institutional training, those requiring special facilities (box 4 control 2), and those teaching fundamental critical task (box 4 control 1). Courses are then selected from those not requiring institutional training (box 4 control 2) such that institutional training is cost effective (box 4 control 3), and such institutional training satisfies job training time constraints (box 4 control 4).

Finally, the courses selected for institutional training are packaged, producing institutional programs for job training (box 5, output 1), and the remaining courses are packaged into non-resident job training programs (box 6 output 1). Note that non-resident packages also include major portions of institutional training packages so that personnel who enter a job but never took the institutional training could be trained.

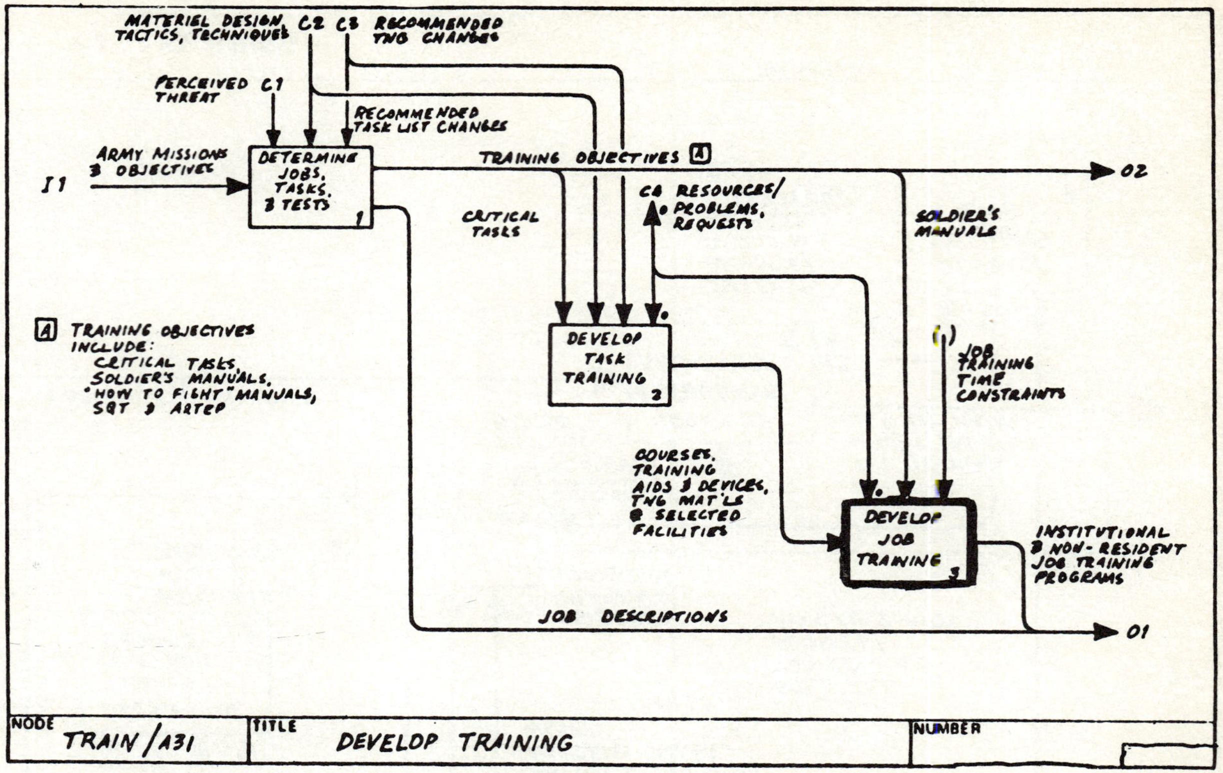

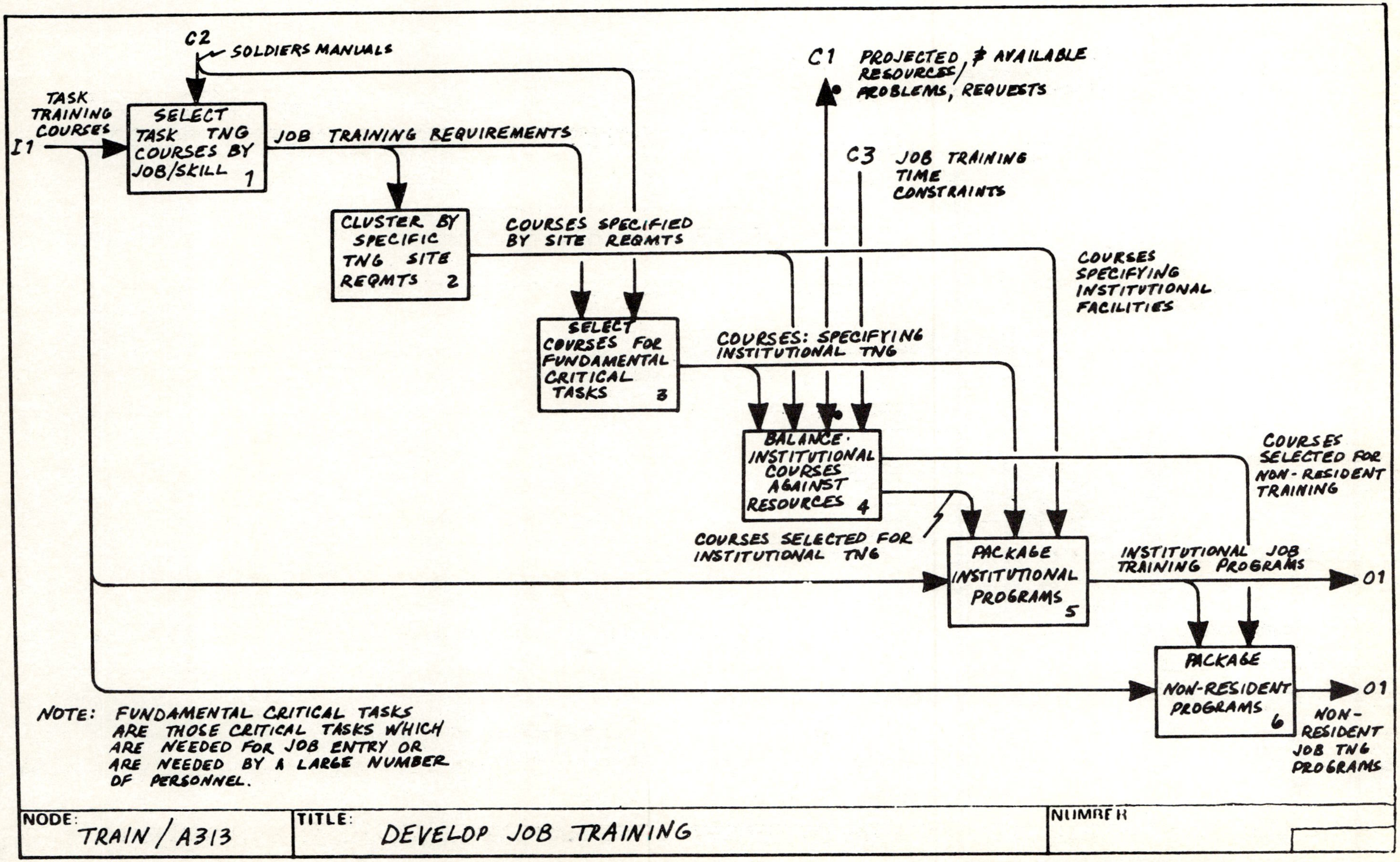
C2
SOLDIERS MANUALS
TASK TRAINING COURSES
I1
SELECT TASK TNG COURSES BY JOB/SKILL 1
JOB TRAINING REQUIREMENTS
CLUSTER BY SPECIFIC TNG SITE REQMTS 2
COURSES SPECIFIED BY SITE REQMTS
SELECT COURSES FOR FUNDAMENTAL CRITICAL TASKS 3
C1
PROJECTED $ AVAILABLE RESOURCES/PROBLEMS, REQUESTS
C3
JOB TRAINING TIME CONSTRAINTS
COURSES SPECIFYING INSTITUTIONAL FACILITIES
COURSES: SPECIFYING INSTITUTIONAL TNG
BALANCE INSTITUTIONAL COURSES AGAINST RESOURCES 4
COURSES SELECTED FOR NON-RESIDENT TRAINING
COURSES SELECTED FOR INSTITUTIONAL TNG
PACKAGE INSTITUTIONAL PROGRAMS 5
INSTITUTIONAL JOB TRAINING PROGRAMS
O1
PACKAGE NON-RESIDENT PROGRAMS 6
O1
NON-RESIDENT JOB TNG PROGRAMS
NOTE: FUNDAMENTAL CRITICAL TASKS ARE THOSE CRITICAL TASKS WHICH ARE NEEDED FOR JOB ENTRY OR ARE NEEDED BY A LARGE NUMBER OF PERSONNEL.
NODE: TRAIN / A313
TITLE: DEVELOP JOB TRAINING
NUMBER

Diagram Index for the Provide Trained Units Model

- **A-0** Provide Trained Units
  - **A0** Provide Trained Units
    - **A1** Develop Combat Technology
    - **A3** Train Army
      - **A31** Develop Training
        - **A311** Determine Jobs, Tasks, and Tests
        - **A312** Develop Task Training
        - **A313** Develop Job Training

Appendix

A

The Prototype Machine Shop Problem

Systems analysis work is as varied as its problem domains. Analysts are often faced with difficult technical problems, even though a system may appear simple on the surface. Also, some feel that only current technological problems are worthy of testing the skill of an analyst. Our experience with teaching SADT in many commercial firms around the world and in many domestic military organizations has shown otherwise. The following problem has been shown to test the expertise of any analyst.

> *Problem statement:* Most aerospace companies have a machine shop for doing prototype jobs. Prototype jobs require the hand machining of individual parts built for special test purposes. The way a prototype machine shop does these unique jobs is described in the paragraphs that follow. Create an SADT model of this process and use it to write a training manual for new personnel.

The problem asks for a narrative description—a specification—of how a prototype machine shop operates. The problem is a realistic one: There is an ongoing process, which must be learned by new members of the organization, and some kind of document is required to teach the process. Such systems analysis problems are encountered every day in the working world. Here is a summary of the work process:

1. An expediter brings work to the shop in the form of a work kit and puts it on the incoming bench. The work kit consists of a job shop order, a blueprint, a time card, raw materials, and unfinished parts.
2. Once a job is received, it may stay on the bench for several days before it is assigned to a machinist. A typical shop has around 30 machinists, 1 supervisor *(foreman)* for every 10 machinists, and a manager who oversees the supervisors. When a machinist is ready for a new job, the machinist and the supervisor look over the next work kit. They first compare the sequence of operations on the job order with the blueprint. Each operation is briefly described, including the tools it requires. Often, a job has a number of machining steps that can be done in one of several ways. The machinist makes notes on a piece of scrap paper about the machining details the machinist and supervisor decide are best. When the detailed process has been worked out, both estimate the time to do the job.
3. After going over the job with the supervisor, the machinist takes the work kit to the machine tool required during the first operation. He or she then goes to the tool crib and takes whatever cutters, clamps, and hand tools are needed to start the operation.
4. Some jobs take as little as ½ hour to do, and others may take several weeks. As a job is being done, the supervisor monitors the work and updates the estimated time for the job on her or his copy of the job order, so that new estimates can be reported to the planning department at any time. Estimates for small jobs are not recorded, but large jobs require new estimates twice a day.
5. Each job is inspected according to the quality level set for it. Inspection instructions for each job are written on the blueprint and may refer to the quality standards manual. Jobs requiring stringent quality levels require inspections after every operation. However, many jobs are partially machined, inspected, and returned to the shop for completion. Jobs rejected during final inspection are returned to the shop for correction. Accepted jobs are given a part tag and an approval seal. Only approved jobs with part tags can be referred to as "parts."
6. When a completed job is tagged, it is returned to the responsible supervisor, who then signs off the work. The tagged part is then put on the outgoing bench, and

any raw material scraps are put in the scrap bin. At this point, the job is considered done, and the machinist can start a new job.

Enough information has been presented above to get you started. Once you have drawn at least the A-0 and A0 diagrams, you should then read the following details about inspections. This information will help you decompose your model to a level of detail sufficient to write the specification:

7. Inspection happens either when the last operation has been completed or when the quality level requires an inspection after all or particular operations. Inspection information on the blueprint indicates the required stringency level. Inspection standards appear on the blueprint in the form of tolerances (i.e., units or percentages that specify the range of acceptable deviation).

8. An inspection ensures that the finished dimensions fall within the given tolerance written on the blueprint. The critical factors which determine inspection standards are the number of operations, the order of operations, and the required measurements with their tolerances. Usually, a supervisor within the shop inspects most jobs, but high-quality jobs are inspected by inspectors from outside the shop.

9. Inspection begins when the inspector identifies the operations that have been thus far completed. "How many?" is the critical question to be asked at this stage. If not all required operations have been carried out, the job is sent back for further work.

10. Jobs requiring appraisal are taken to a designated inspection room where the accuracy of their dimensions are checked. The job is put on a flat surface, special equipment (e.g., calipers, scales, gauges) is used to measure the job, and the measurements are compared with the tolerance specifications written on the blueprint.

11. The job is then appraised. "Are all measurements within standard tolerance?" is the second critical question. If any dimension is beyond tolerance, the job is either returned to the shop for rework or scrapped (if there is insufficient material left for reprocessing). Completed jobs that pass inspection are tagged. Incomplete jobs that pass inspection are returned for completion.

Appendix

B

Diagrams for the Prototype Machine Shop Model

This appendix contains the complete set of diagrams for the Prototype Machine Shop model. These diagrams were put into an appendix both for completeness and to demonstrate how big the Prototype Machine Shop model becomes when it is decomposed to a level adequate for writing the specification (i.e., the training manual for new personnel). Notice that it was not necessary to decompose the model down to an even level to detail to write the specification. For example, some boxes on first-level decompositions (e.g., ***estimate job,*** box 3 on diagram A1) describe a very detailed activity and require no additional detailing. The text of the training manual was not included because of space considerations. See Chapter 18 for how to write the training manual from the diagrams. Figure 18-3 presents the text for 1 page of that manual.

The diagrams are given in SADT node number order (i.e., "depth-first tree-walk" order). This means that the most general diagrams of feeding the family appear immediately before those that are more detailed (see the diagram index on the last page of Appendix B). You can therefore read through the model in two ways: for overview or for detail. To read for overview, read only the high-level diagrams (i.e., A-0, A0, and A1 to A3). To read for detail, read all the second-level diagrams for one particular high-level diagram (e.g., A23 and A24).

A couple of points from our analysis and usage of SADT will help you read these diagrams better. First, we used tunneled arrows to reduce the amount of unimportant detail on the high-level diagrams. For example, we felt ***ready mechanic*** (see diagram A1) and ***current backlog*** (see diagram A11) were such details. Second, because our goal was to describe how everyone in the machine shop contributes to building special parts for experimental aircraft, we used an initial decomposition strategy that centered on the roles people played in the shop (i.e., ***foreman, machinist, inspector***). Then, for each role, we concentrated on the functions each person does, regardless of how those activities are sequenced. This led to a model with several feedback arrows, but one from which very focused text could be written for particular personnel.

USED AT:	AUTHOR: Marca PROJECT: PMS NOTES: 1 2 3 4 5 6 7 8 9 10	DATE: 4/14/86 REV:	■ WORKING DRAFT RECOMMENDED PUBLICATION	READER DATE	CONTEXT: None

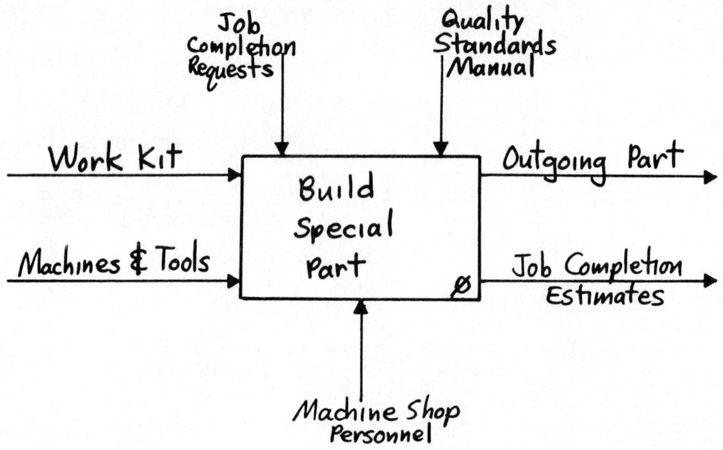

Purpose: Understand the tasks involved in building a special part and how those tasks interrelate, so a training manual for Machine Shop personnel can be written.

Viewpoint: Prototype Machine Shop manager.

NODE: PMS/A-Ø	TITLE: Build Special Part	NUMBER: DAMØØ5 (DAMØØ4)

USED AT:	AUTHOR: Marca PROJECT: PMS NOTES: 1 2 3 4 5 6 7 8 9 10	DATE: 4/15/86 REV:	■ WORKING DRAFT RECOMMENDED PUBLICATION	READER DATE	CONTEXT: Top

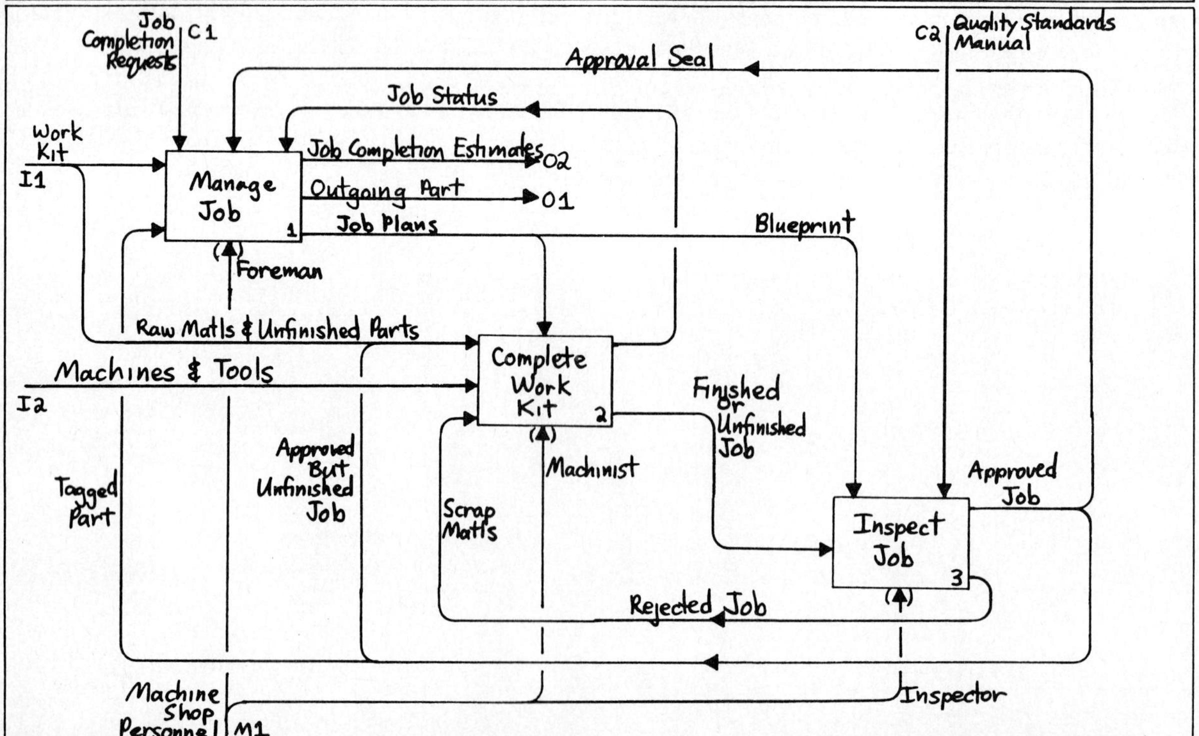

NODE: PMS/AØ	TITLE: Build Special Part	NUMBER: DAMØØ6 (DAMØØ3)

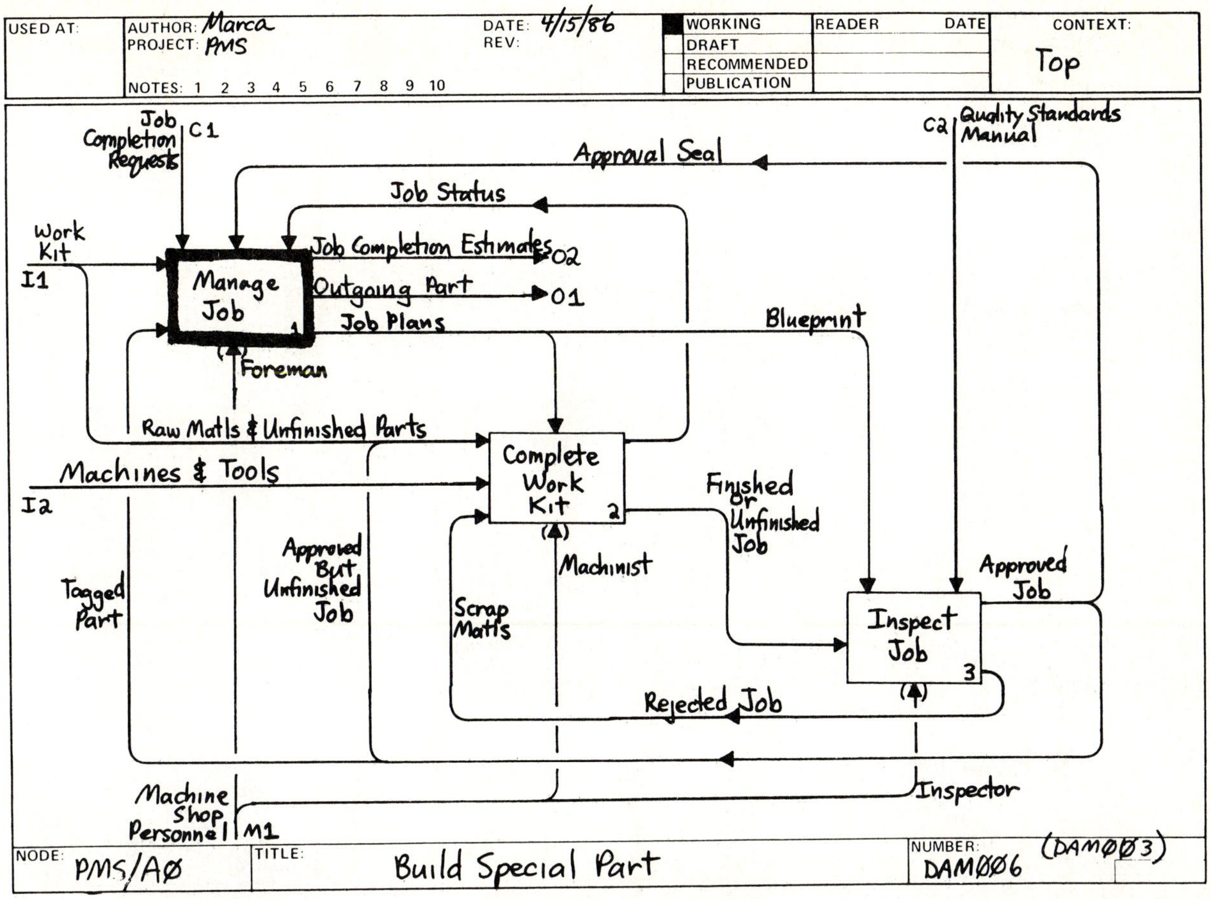
USED AT:
AUTHOR: Marca
PROJECT: PMS
DATE: 4/15/86
REV:
WORKING
DRAFT
RECOMMENDED
PUBLICATION
READER
DATE
CONTEXT:
Top
NOTES: 1 2 3 4 5 6 7 8 9 10
Job Completion Requests
C1
C2
Quality Standards Manual
Approval Seal
Job Status
Work Kit
I1
Job Completion Estimates
O2
Manage Job
1
Outgoing Part
O1
Job Plans
Blueprint
Foreman
Raw Matls & Unfinished Parts
Machines & Tools
I2
Complete Work Kit
2
Finished or Unfinished Job
Approved But Unfinished Job
Tagged Part
Machinist
Scrap Matls
Approved Job
Inspect Job
3
Rejected Job
Machine Shop Personnel
M1
Inspector
NODE:
PMS/A0
TITLE:
Build Special Part
NUMBER:
DAM006
(DAM003)

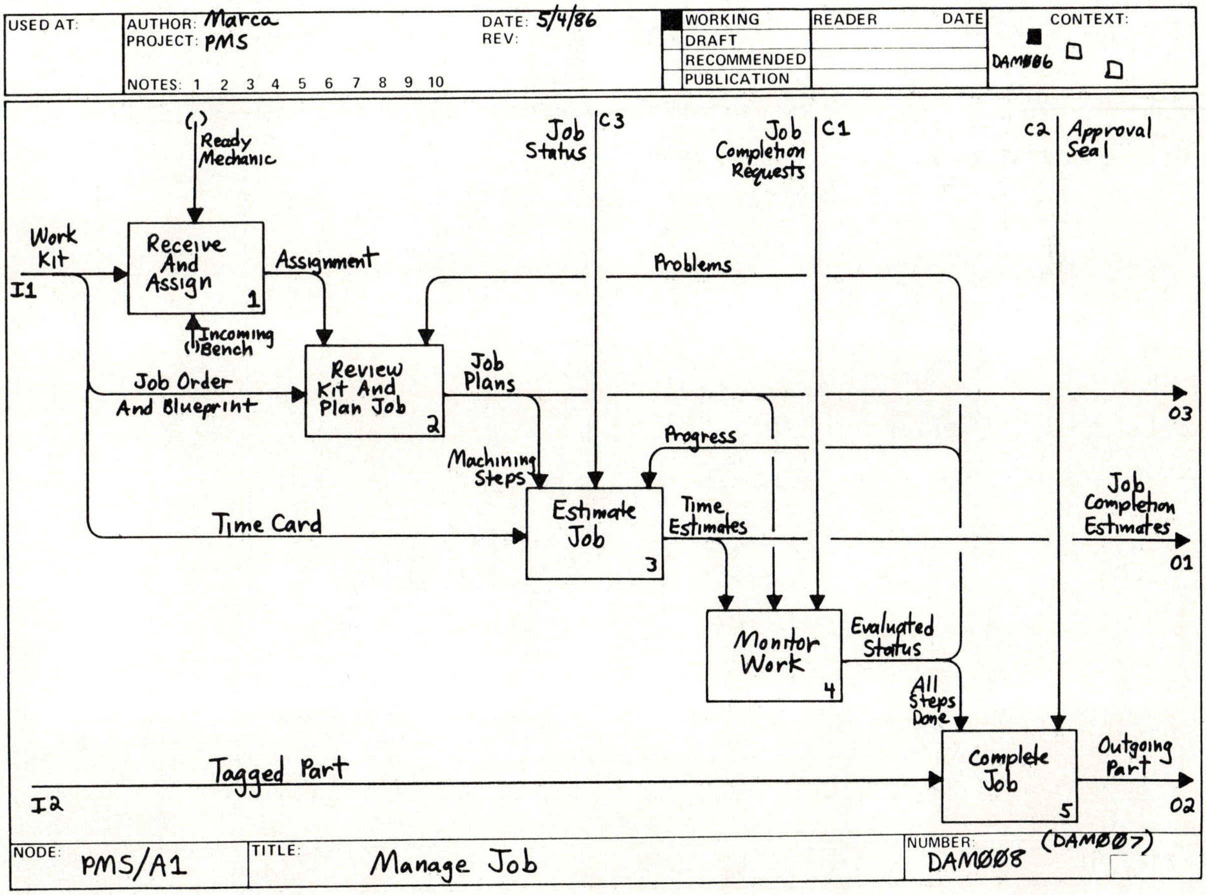
USED AT:
AUTHOR: Marca
PROJECT: PMS
DATE: 5/4/86
REV:
WORKING
DRAFT
RECOMMENDED
PUBLICATION
READER
DATE
CONTEXT:
DAM006
NOTES: 1 2 3 4 5 6 7 8 9 10
Ready Mechanic
Job Status
C3
Job Completion Requests
C1
C2
Approval Seal
Work Kit
I1
Receive And Assign
1
Assignment
Problems
Incoming Bench
Job Order And Blueprint
Review Kit And Plan Job
2
Job Plans
O3
Progress
Machining Steps
Time Card
Estimate Job
3
Time Estimates
Job Completion Estimates
O1
Monitor Work
4
Evaluated Status
All Steps Done
Tagged Part
I2
Complete Job
5
Outgoing Part
O2
NODE:
PMS/A1
TITLE:
Manage Job
NUMBER:
DAM008
(DAM007)

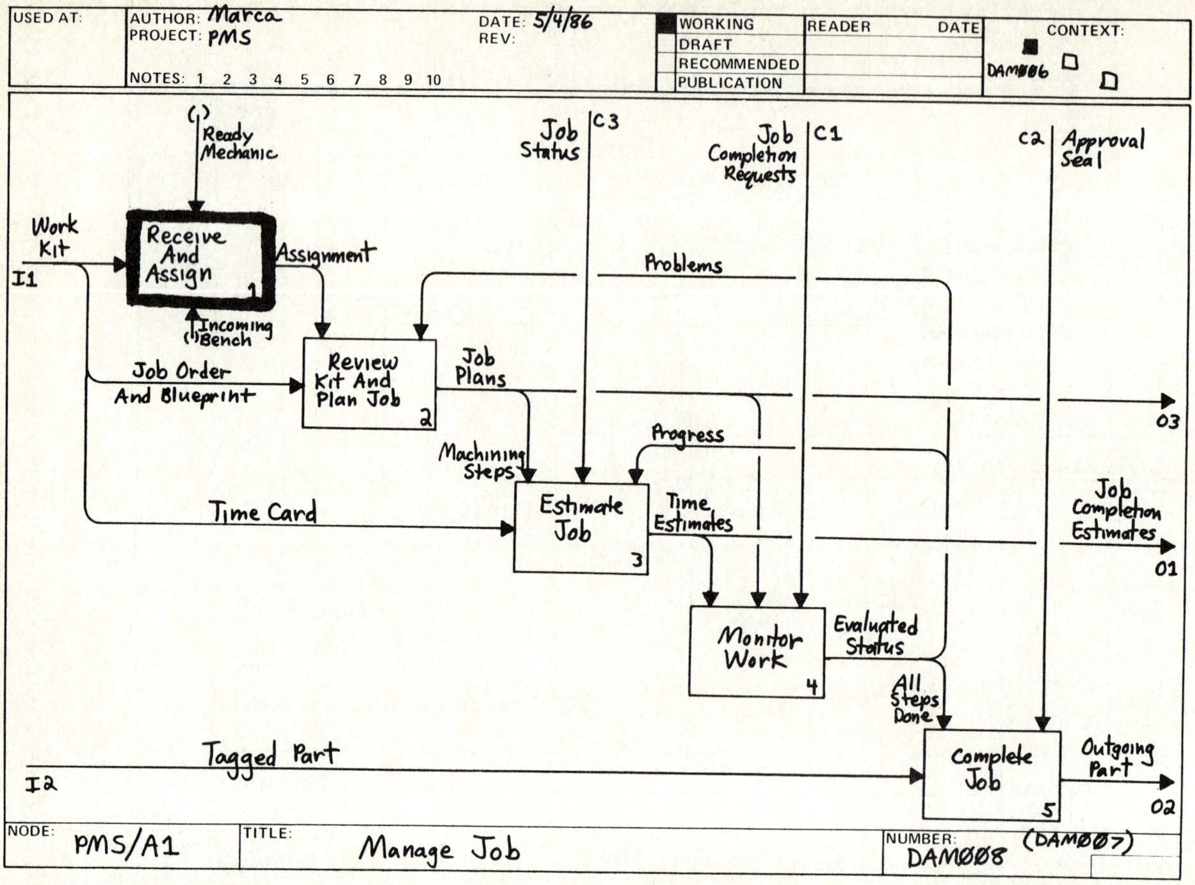
USED AT:
AUTHOR: Marca
PROJECT: PMS
DATE: 5/4/86
REV:
NOTES: 1 2 3 4 5 6 7 8 9 10
WORKING
DRAFT
RECOMMENDED
PUBLICATION
READER
DATE
CONTEXT:
DAM006
Ready Mechanic
Job Status
C3
Job Completion Requests
C1
C2
Approval Seal
Work Kit
I1
Receive And Assign
1
Assignment
Problems
Incoming Bench
Review Kit And Plan Job
2
Job Plans
Job Order And Blueprint
O3
Progress
Machining Steps
Estimate Job
3
Time Card
Time Estimates
Job Completion Estimates
O1
Monitor Work
4
Evaluated Status
All Steps Done
Tagged Part
I2
Complete Job
5
Outgoing Part
O2
NODE: PMS/A1
TITLE: Manage Job
NUMBER: DAM008
(DAM007)

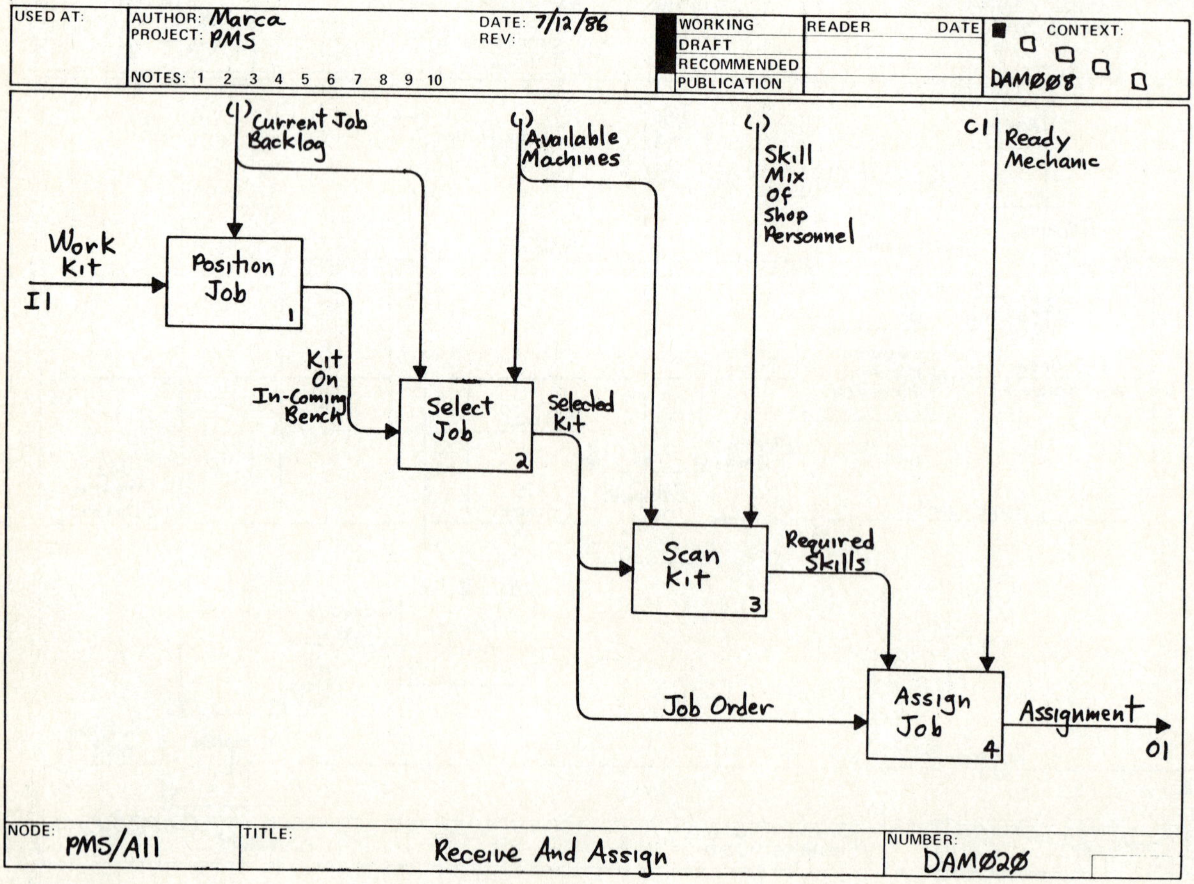
USED AT:
AUTHOR: Marca
PROJECT: PMS
DATE: 7/12/86
REV:
NOTES: 1 2 3 4 5 6 7 8 9 10
WORKING
DRAFT
RECOMMENDED
PUBLICATION
READER
DATE
CONTEXT:
DAM008
Current Job Backlog
Available Machines
Skill Mix Of Shop Personnel
C1
Ready Mechanic
Work Kit
I1
Position Job
1
Kit On In-Coming Bench
Select Job
2
Selected Kit
Scan Kit
3
Required Skills
Job Order
Assign Job
4
Assignment
O1
NODE: PMS/A11
TITLE: Receive And Assign
NUMBER: DAM020

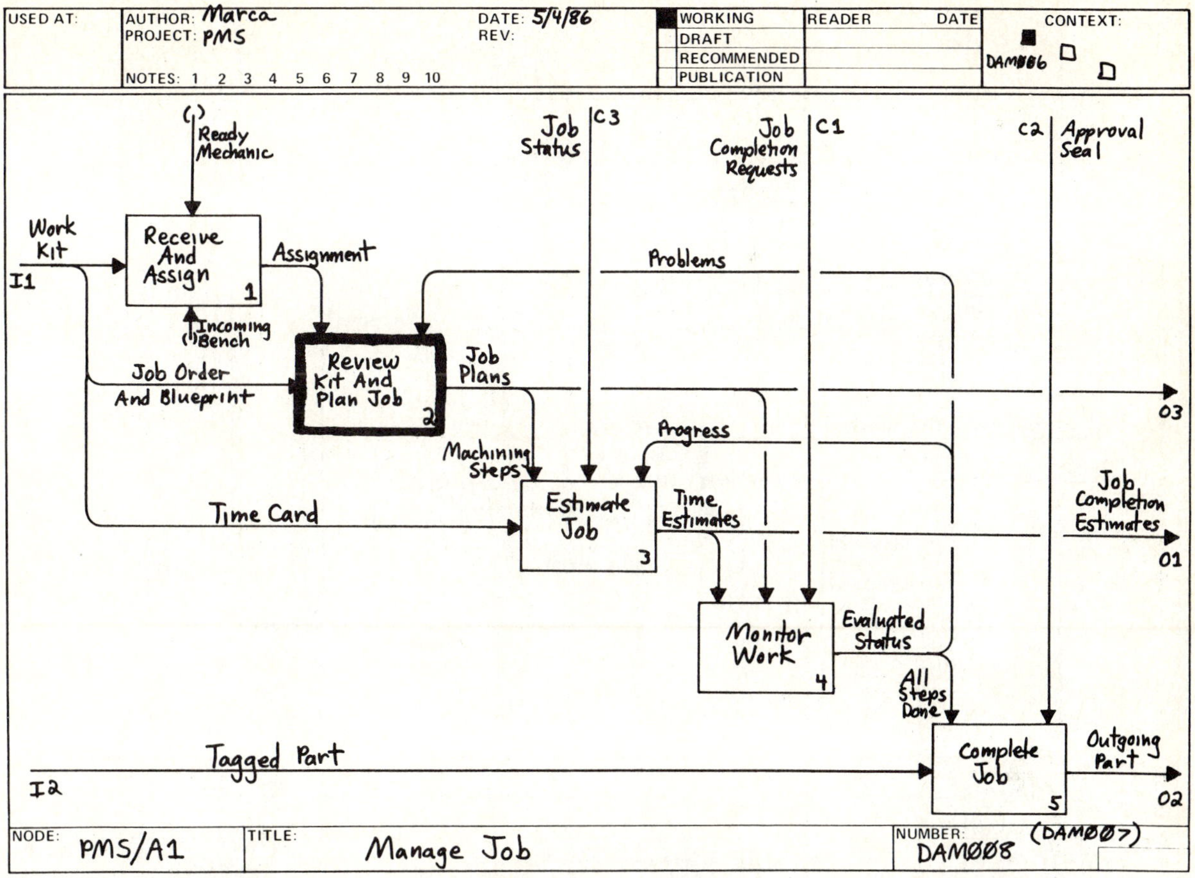
USED AT:
AUTHOR: Marca
PROJECT: PMS
NOTES: 1 2 3 4 5 6 7 8 9 10
DATE: 5/4/86
REV:
WORKING
DRAFT
RECOMMENDED
PUBLICATION
READER
DATE
CONTEXT:
DAM006
Ready Mechanic
Job Status
C3
Job Completion Requests
C1
C2
Approval Seal
Work Kit
I1
Receive And Assign
1
Assignment
Problems
Incoming Bench
Job Order And Blueprint
Review Kit And Plan Job
2
Job Plans
O3
Progress
Machining Steps
Estimate Job
3
Time Card
Time Estimates
Job Completion Estimates
O1
Monitor Work
4
Evaluated Status
All Steps Done
Tagged Part
I2
Complete Job
5
Outgoing Part
O2
NODE:
PMS/A1
TITLE:
Manage Job
NUMBER:
DAM008
(DAM007)

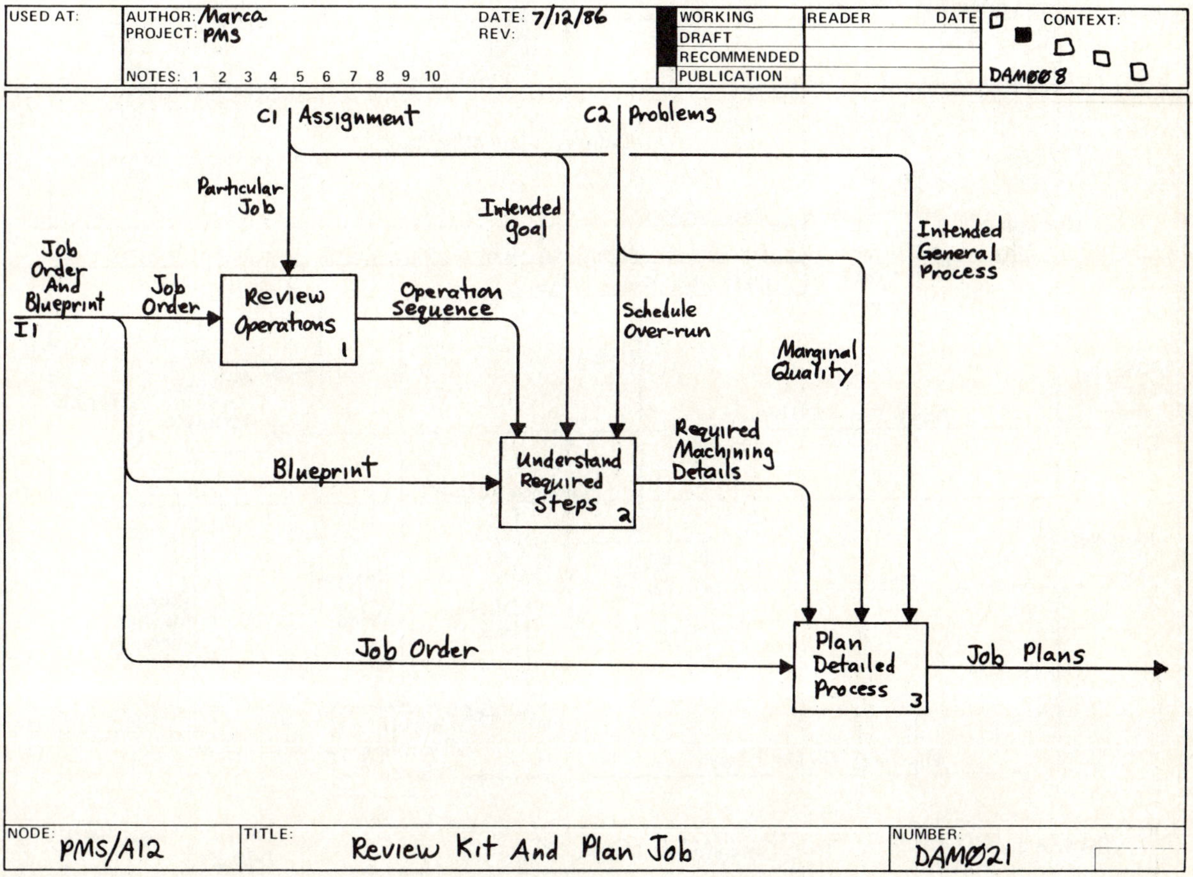
USED AT:
AUTHOR: Marca
PROJECT: PMS
NOTES: 1 2 3 4 5 6 7 8 9 10
DATE: 7/12/86
REV:
WORKING
DRAFT
RECOMMENDED
PUBLICATION
READER
DATE
CONTEXT:
DAM008
C1
Assignment
C2
Problems
Particular Job
Intended Goal
Intended General Process
Job Order And Blueprint
I1
Job Order
Review Operations
1
Operation Sequence
Schedule Over-run
Marginal Quality
Blueprint
Understand Required Steps
2
Required Machining Details
Job Order
Plan Detailed Process
3
Job Plans
NODE:
PMS/A12
TITLE:
Review Kit And Plan Job
NUMBER:
DAM021

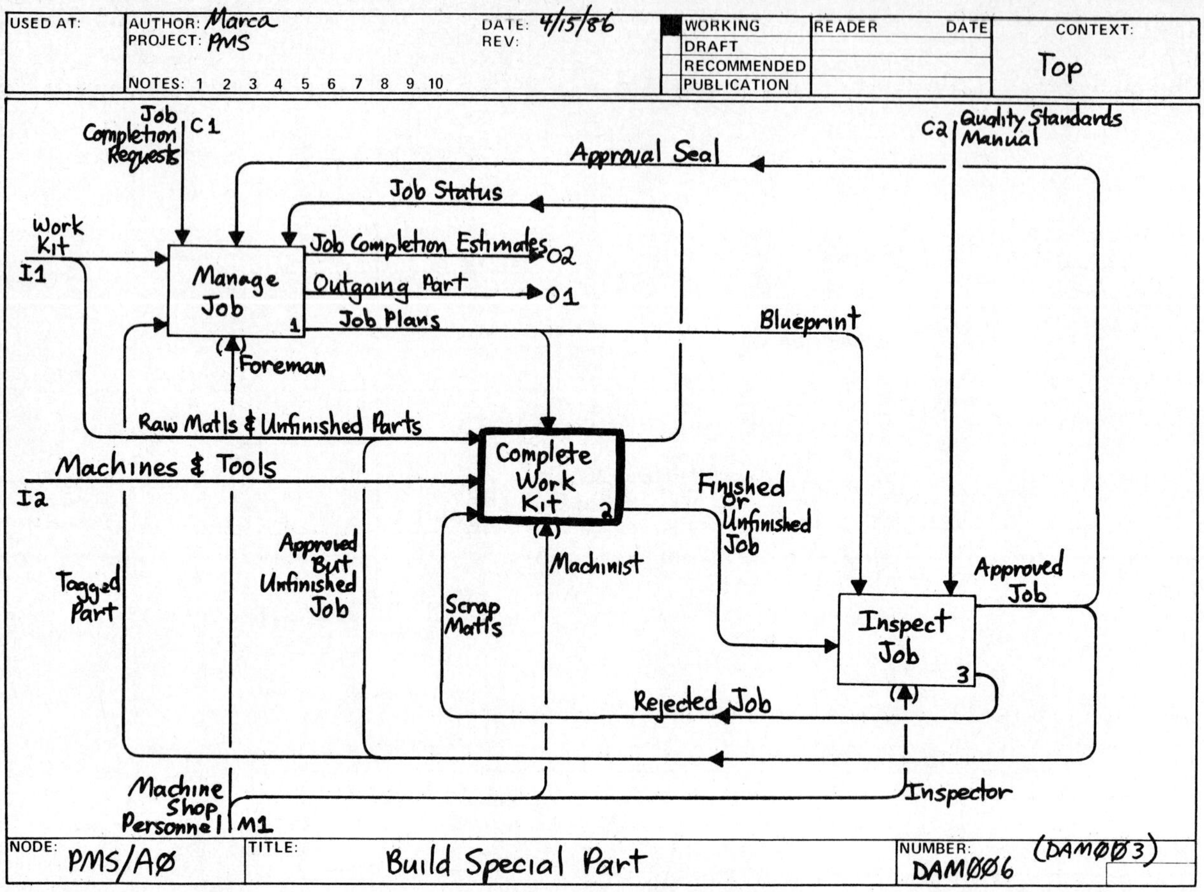
USED AT:
AUTHOR: Marca
PROJECT: PMS
DATE: 4/15/86
REV:
NOTES: 1 2 3 4 5 6 7 8 9 10
WORKING
DRAFT
RECOMMENDED
PUBLICATION
READER
DATE
CONTEXT:
Top
Job Completion Requests
C1
C2
Quality Standards Manual
Approval Seal
Job Status
Work Kit
I1
Manage Job
1
Job Completion Estimates
O2
Outgoing Part
O1
Job Plans
Blueprint
Foreman
Raw Matls & Unfinished Parts
Complete Work Kit
2
Machines & Tools
I2
Finished or Unfinished Job
Approved But Unfinished Job
Machinist
Tagged Part
Scrap Matls
Approved Job
Inspect Job
3
Rejected Job
Machine Shop Personnel
M1
Inspector
NODE:
PMS/AØ
TITLE:
Build Special Part
NUMBER:
DAMØØ6
(DAMØØ3)

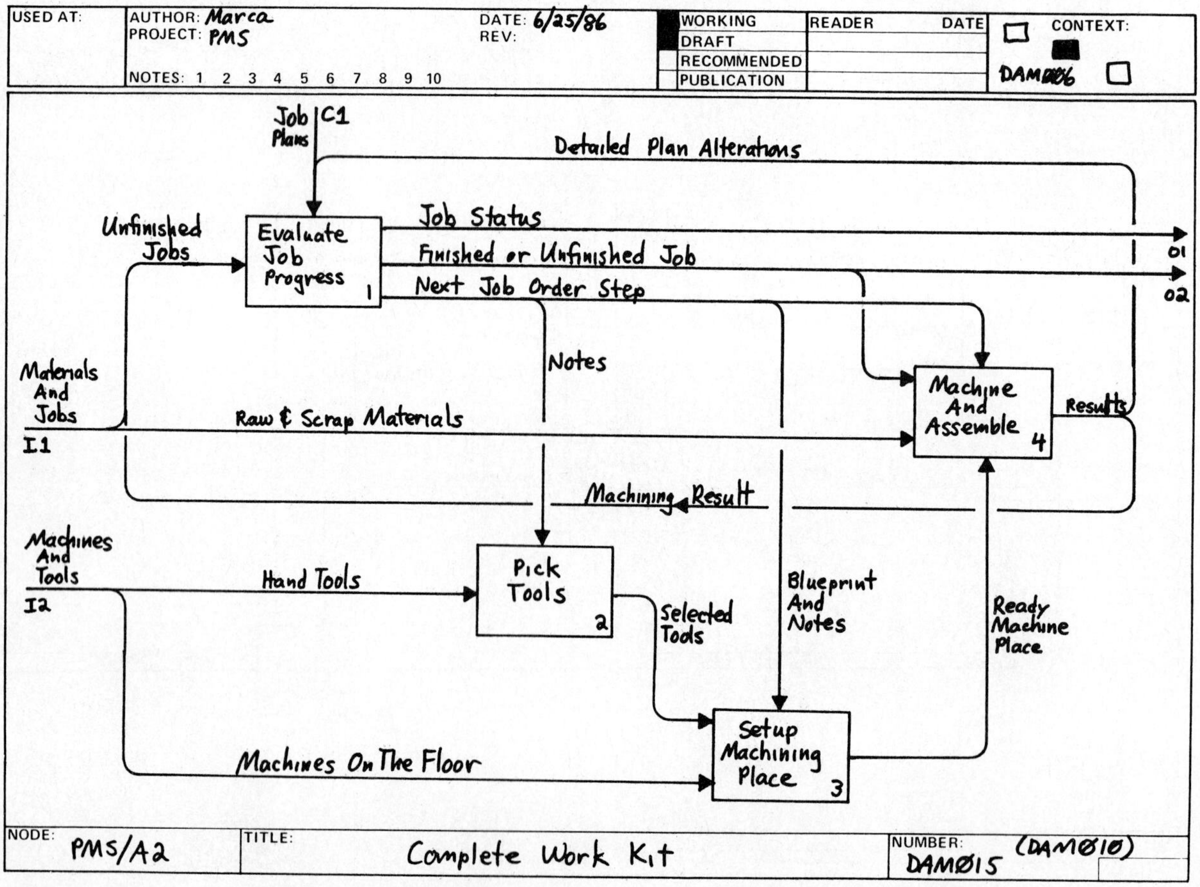
USED AT:
AUTHOR: Marca
PROJECT: PMS
DATE: 6/25/86
REV:
NOTES: 1 2 3 4 5 6 7 8 9 10
WORKING
DRAFT
RECOMMENDED
PUBLICATION
READER
DATE
CONTEXT:
DAMØØ6
Job Plans
C1
Detailed Plan Alterations
Unfinished Jobs
Evaluate Job Progress
1
Job Status
Finished or Unfinished Job
O1
O2
Next Job Order Step
Notes
Materials And Jobs
I1
Raw & Scrap Materials
Machine And Assemble
4
Results
Machining Result
Machines And Tools
I2
Hand Tools
Pick Tools
2
Selected Tools
Blueprint And Notes
Ready Machine Place
Machines On The Floor
Setup Machining Place
3
NODE:
PMS/A2
TITLE:
Complete Work Kit
NUMBER:
DAMØ15
(DAMØ1Ø)

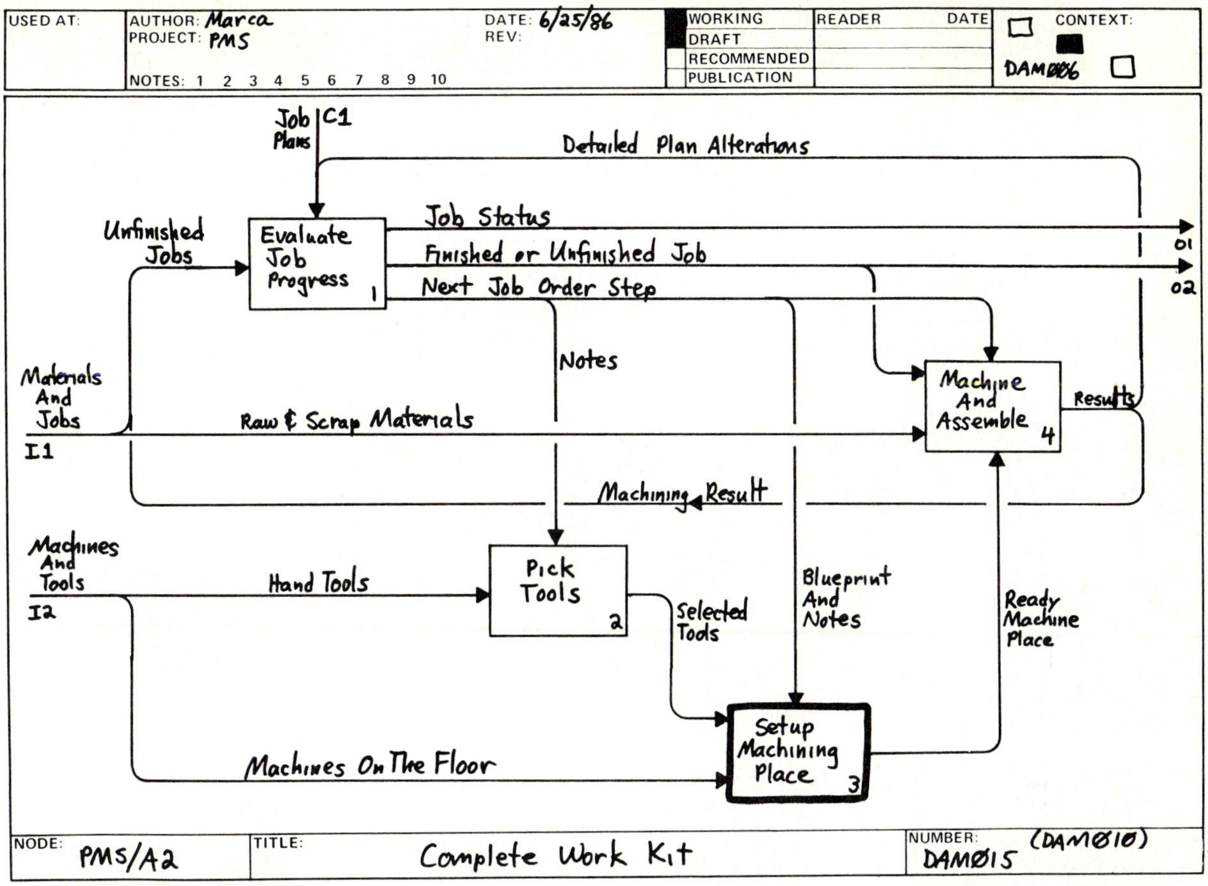
USED AT:
AUTHOR: Marca
PROJECT: PMS
DATE: 6/25/86
REV:
NOTES: 1 2 3 4 5 6 7 8 9 10
WORKING
DRAFT
RECOMMENDED
PUBLICATION
READER
DATE
CONTEXT:
DAMØ06
Job Plans
C1
Detailed Plan Alterations
Unfinished Jobs
Evaluate Job Progress
1
Job Status
Finished or Unfinished Job
Next Job Order Step
O1
O2
Notes
Materials And Jobs
I1
Raw & Scrap Materials
Machine And Assemble
4
Results
Machining Result
Machines And Tools
I2
Hand Tools
Pick Tools
2
Selected Tools
Blueprint And Notes
Ready Machine Place
Setup Machining Place
3
Machines On The Floor
NODE: PMS/A2
TITLE: Complete Work Kit
NUMBER: DAMØ15
(DAMØ10)

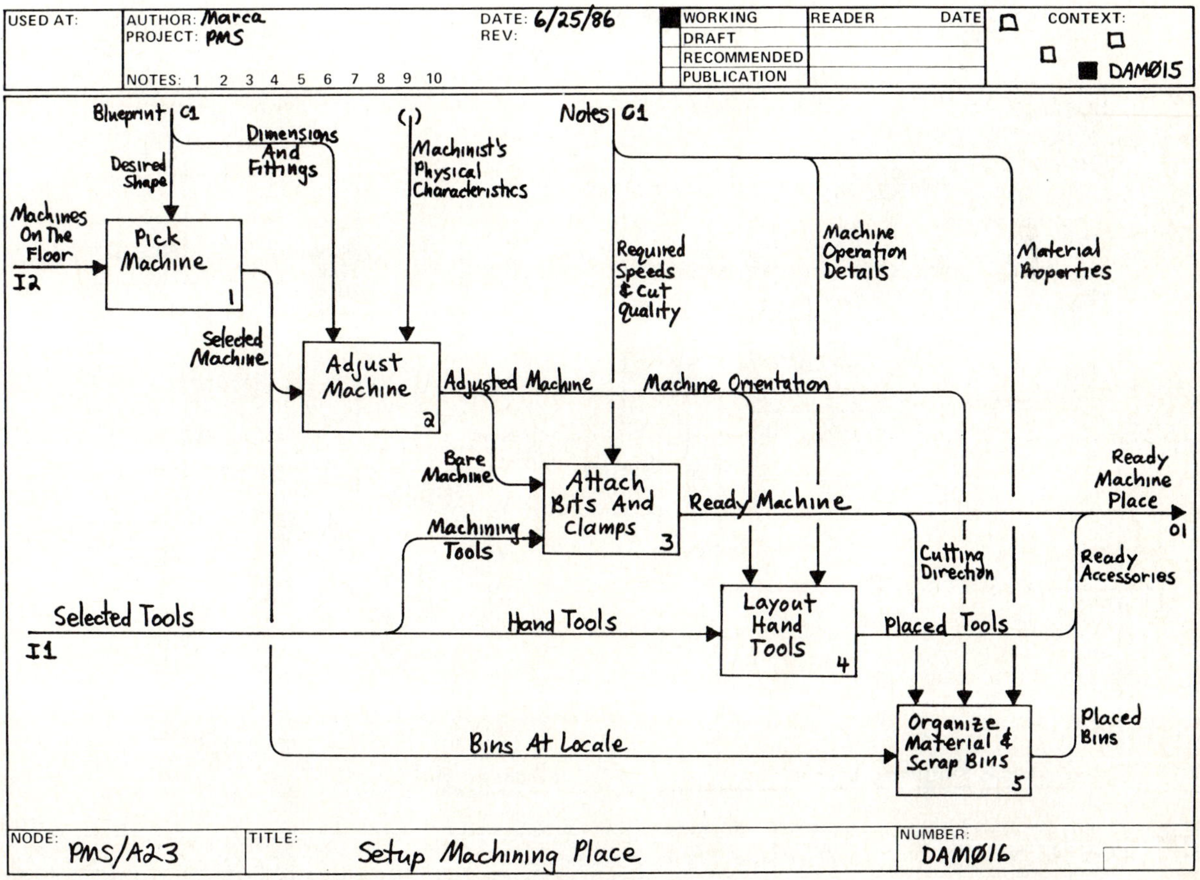
USED AT:
AUTHOR: Marca
PROJECT: PMS
DATE: 6/25/86
REV:
NOTES: 1 2 3 4 5 6 7 8 9 10
WORKING
DRAFT
RECOMMENDED
PUBLICATION
READER
DATE
CONTEXT:
DAMØ15
Blueprint
C1
Desired Shape
Dimensions And Fittings
( )
Machinist's Physical Characteristics
Notes
C1
Machines On The Floor
I2
Pick Machine
1
Selected Machine
Adjust Machine
2
Required Speeds & Cut Quality
Machine Operation Details
Material Properties
Adjusted Machine
Machine Orientation
Bare Machine
Attach Bits And Clamps
3
Machining Tools
Ready Machine
Ready Machine Place
O1
Cutting Direction
Ready Accessories
Selected Tools
I1
Hand Tools
Layout Hand Tools
4
Placed Tools
Placed Bins
Bins At Locale
Organize Material & Scrap Bins
5
NODE: PMS/A23
TITLE: Setup Machining Place
NUMBER: DAMØ16

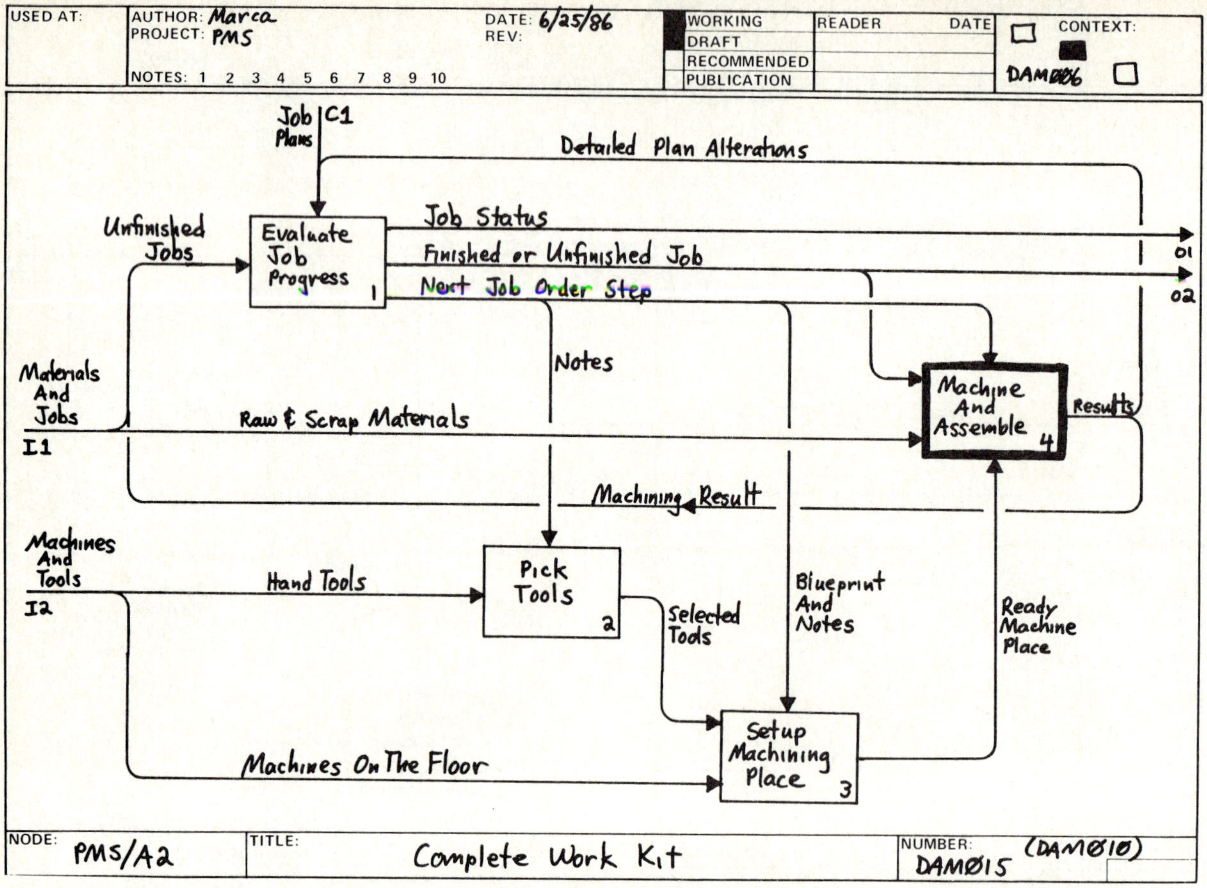
USED AT:
AUTHOR: Marca
PROJECT: PMS
DATE: 6/25/86
REV:
NOTES: 1 2 3 4 5 6 7 8 9 10
WORKING
DRAFT
RECOMMENDED
PUBLICATION
READER
DATE
CONTEXT:
DAM006
Job Plans
C1
Detailed Plan Alterations
Unfinished Jobs
Evaluate Job Progress
1
Job Status
Finished or Unfinished Job
Next Job Order Step
O1
O2
Notes
Materials And Jobs
I1
Raw & Scrap Materials
Machine And Assemble
4
Results
Machining Result
Machines And Tools
I2
Hand Tools
Pick Tools
2
Selected Tools
Blueprint And Notes
Ready Machine Place
Setup Machining Place
3
Machines On The Floor
NODE:
PMS/A2
TITLE:
Complete Work Kit
NUMBER:
DAM015
(DAM010)

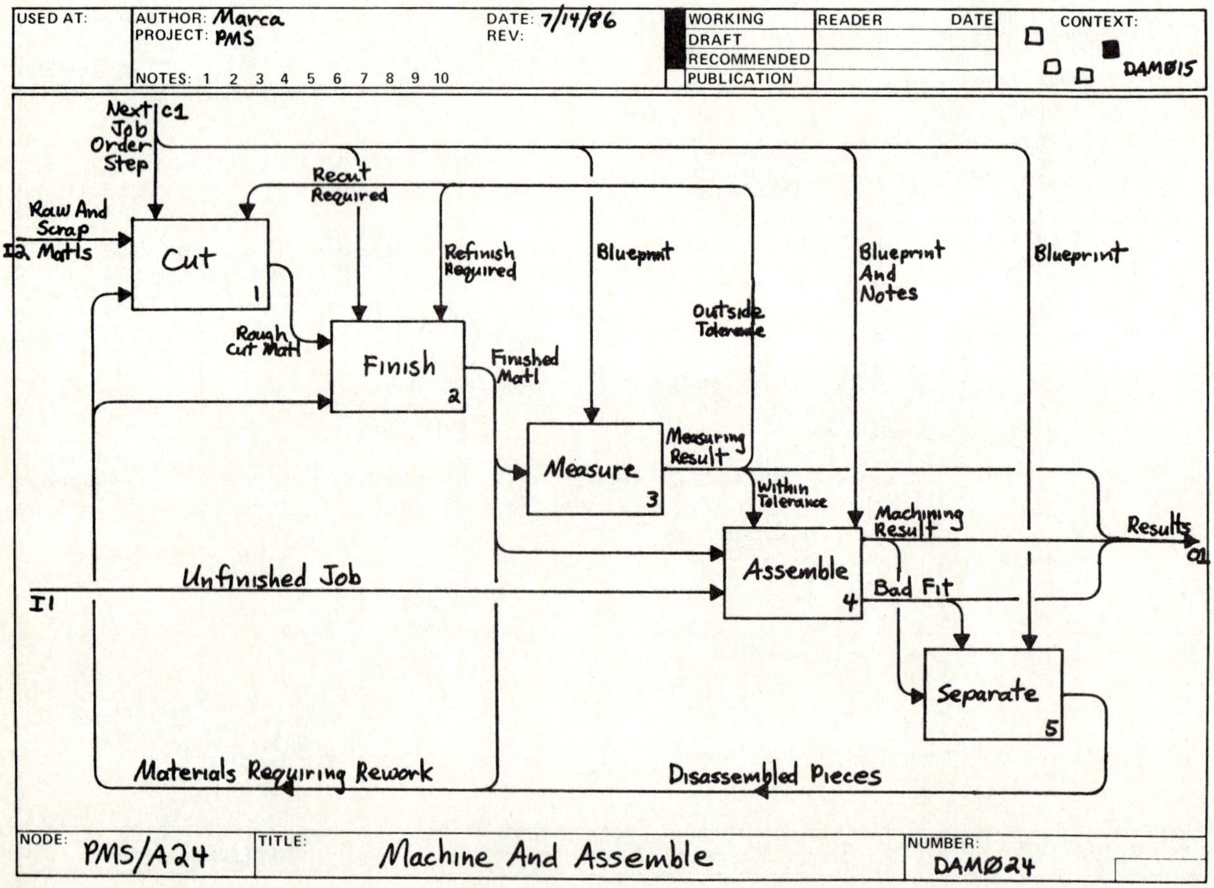
USED AT:
AUTHOR: Marca
PROJECT: PMS
DATE: 7/14/86
REV:
NOTES: 1 2 3 4 5 6 7 8 9 10
WORKING
DRAFT
RECOMMENDED
PUBLICATION
READER
DATE
CONTEXT:
DAM015
Next Job Order Step
C1
Recut Required
Raw And Scrap Matls
I2
Cut
1
Refinish Required
Blueprint
Outside Tolerance
Blueprint And Notes
Blueprint
Rough Cut Matl
Finish
2
Finished Matl
Measure
3
Measuring Result
Within Tolerance
Machining Result
Results
O1
Unfinished Job
I1
Assemble
4
Bad Fit
Separate
5
Materials Requiring Rework
Disassembled Pieces
NODE:
PMS/A24
TITLE:
Machine And Assemble
NUMBER:
DAM024

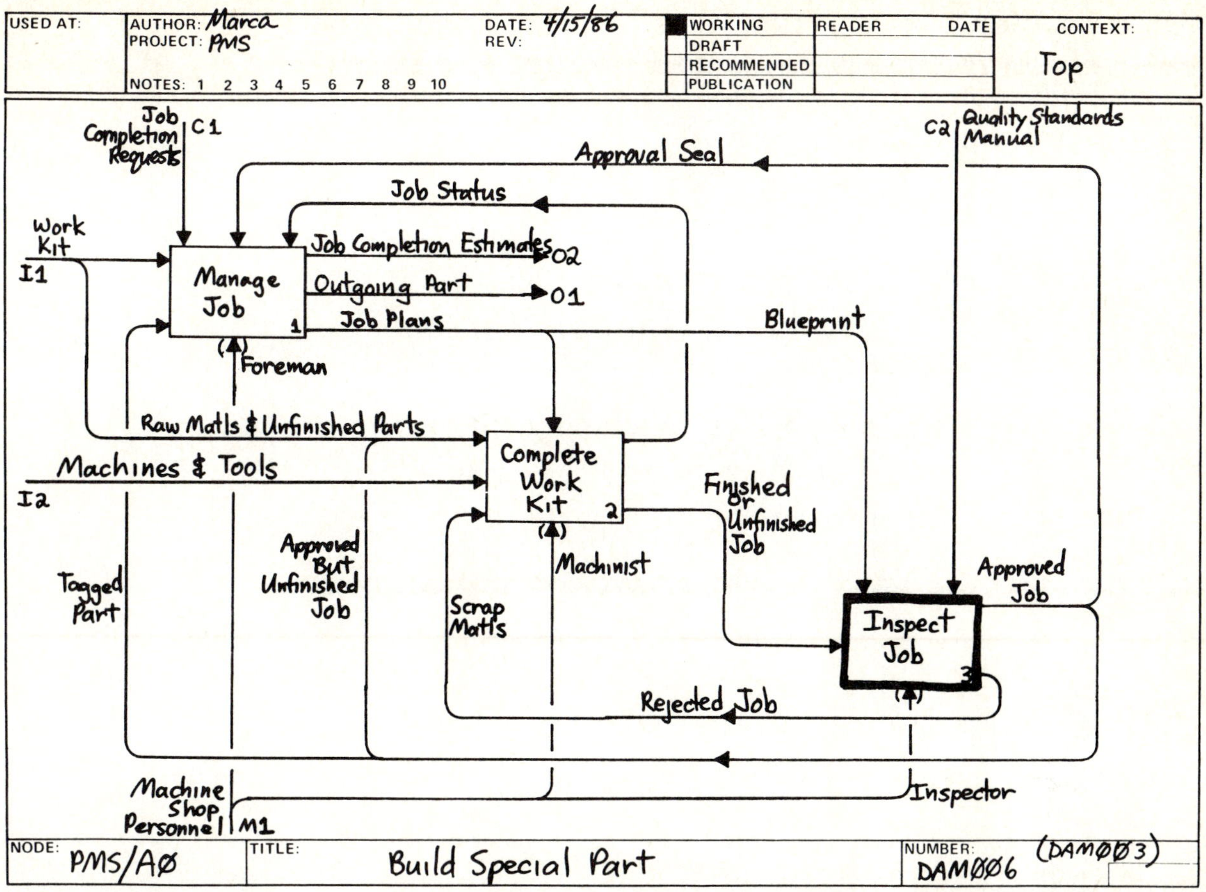
USED AT:
AUTHOR: Marca
PROJECT: PMS
DATE: 4/15/86
REV:
NOTES: 1 2 3 4 5 6 7 8 9 10
WORKING
DRAFT
RECOMMENDED
PUBLICATION
READER
DATE
CONTEXT:
Top
Job Completion Requests
C1
C2
Quality Standards Manual
Approval Seal
Job Status
Work Kit
I1
Manage Job
1
Job Completion Estimates
O2
Outgoing Part
O1
Job Plans
Blueprint
Foreman
Raw Matls & Unfinished Parts
Machines & Tools
I2
Complete Work Kit
2
Finished or Unfinished Job
Approved But Unfinished Job
Tagged Part
Machinist
Scrap Matls
Approved Job
Inspect Job
3
Rejected Job
Machine Shop Personnel
M1
Inspector
NODE: PMS/AØ
TITLE: Build Special Part
NUMBER: DAMØØ6
(DAMØØ3)

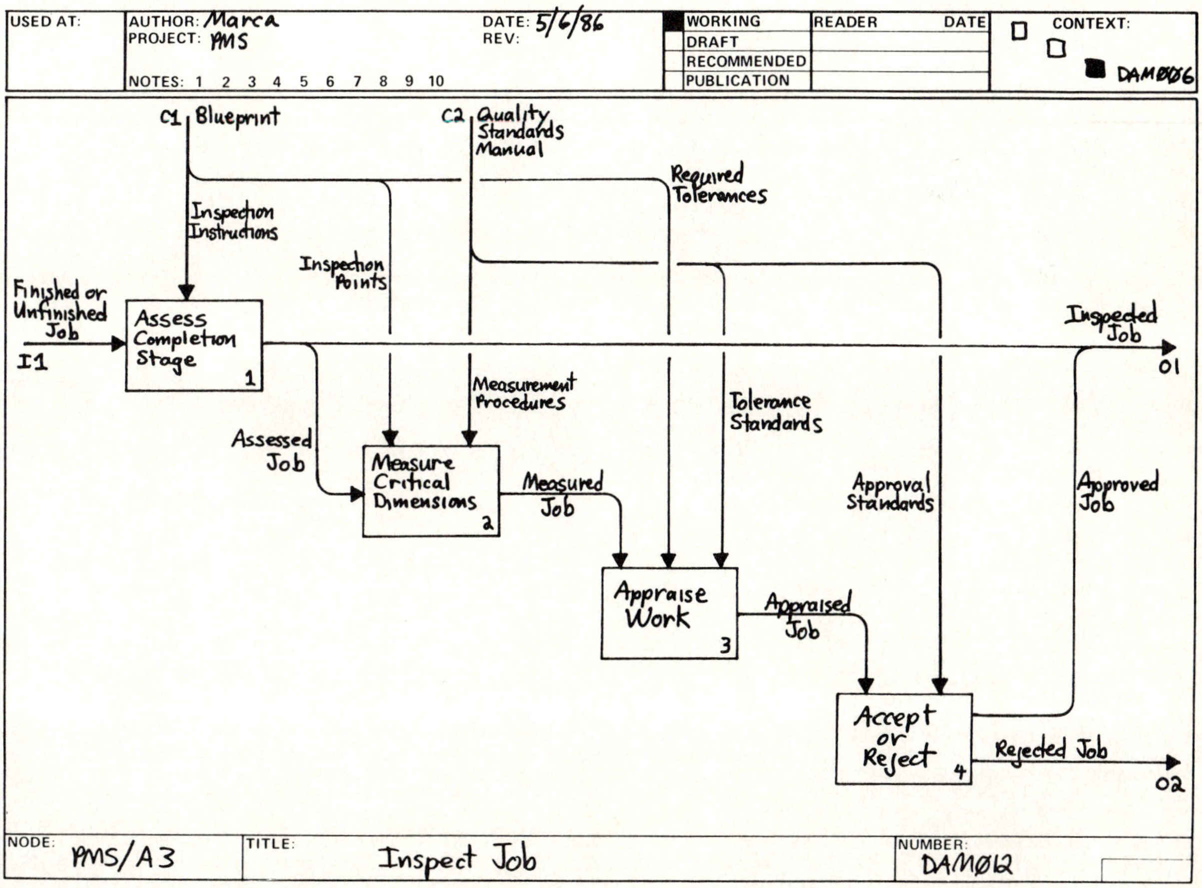
USED AT:
AUTHOR: Marca
PROJECT: PMS
DATE: 5/6/86
REV:
NOTES: 1 2 3 4 5 6 7 8 9 10
WORKING
DRAFT
RECOMMENDED
PUBLICATION
READER
DATE
CONTEXT:
DAMØØ6
C1
Blueprint
C2
Quality Standards Manual
Required Tolerances
Inspection Instructions
Inspection Points
Finished or Unfinished Job
I1
Assess Completion Stage
1
Inspected Job
O1
Measurement Procedures
Tolerance Standards
Assessed Job
Measure Critical Dimensions
2
Measured Job
Approval Standards
Approved Job
Appraise Work
3
Appraised Job
Accept or Reject
4
Rejected Job
O2
NODE: PMS/A3
TITLE: Inspect Job
NUMBER: DAMØ12

Diagram Index for the "Prototype Machine Shop" Model

- **A-0** Build Special Part
  - **A0** Build Special Part
    - **A1** Manage Job
      - **A11** Receive and Assign
      - **A12** Review Kit and Plan Job
    - **A2** Complete Work Kit
      - **A23** Set up Machining Place
      - **A24** Machine and Assemble
    - **A3** Inspect Job

Appendix

C

Diagrams for the Feed Family Model

This appendix contains the complete set of diagrams for the Feed Family model. These diagrams were put into an appendix both for completeness and to demonstrate how big the Feed Family model becomes when it is decomposed to a level adequate for writing the specification. The associated text was not given here, because we felt Workshop 25 to be an adequate example of how to write the specification from the diagrams.

While one can write a complete specification from these diagrams, do not think the diagrams are "the" answer to the workshops. These diagrams are simply one class's solution. Thus, they are an example, for two reasons: First, because each family feeds itself just a bit differently from the way every other family feeds itself. For example, ***unpack*** is part of ***prepare*** instead of being part of ***maintain stock.*** You may decide to do the reverse. Second, cultural and regional differences always cause wide variety in emphasis. For example, Americans are concerned about maintaining inventories, Italians with selecting the right wine, and Japanese with the presentation of the food. This is why we chose the Feed Family problem—because no two models ever come out the same.

We also chose the Feed Family problem because it is *not* a toy problem. The complexities of feeding a family are rich enough to stress any systems analysis methodology. For example, there are mutually constraining activities, details that require hiding, exception handling, and error conditions. In fact, if you study the diagrams carefully, you will find similarities between them and work-related situations. For example:

- ***Prepare kitchen*** is like a shop floor layout problem.
- ***Combine & cook*** is like a coal liquefication process.
- ***Maintain stock*** is like an inventory control problem.
- ***Plan*** and ***maintain stock*** together are like a real-time scheduling problem.

The diagrams are in SADT node number order (i.e., "depth-first tree-walk" order). This means that the most general diagrams of feeding the family appear immediately before those that are more detailed (see the diagram index on the last page of Appendix C). You can therefore read through the model in two ways: for overview or for detail. To read for overview, read only the high-level diagrams (i.e., A-0, A0, and A1 to A5). To read for detail, read all the second-level diagrams for one particular high-level diagram (e.g., A12 to A14).

A couple of points about our usage of SADT will help you read these diagrams better. First, we used tunneled arrows to reduce the amount of unimportant detail on the high-level diagrams (i.e., A0 and A1 to A5). For example, we felt ***store locations & offerings*** was such a detail (see diagram A22). Second, we used tunneled arrows to reduce the complexity of certain diagrams. For example, we tunneled ***dirty kitchen,*** the output of ***combine & cook*** (see diagram A3), to eliminate the need to draw a large number of unnecessary arrows on the decomposition of that box.

Because our goal was to describe how to feed a family, we used a decomposition strategy that centered on the state of food (e.g., bought, stored, cooked, eaten) throughout the development of the model. This resulted in a predominately sequentially flowing model. (In fact, experiments with other decomposition strategies have consistently shown that sequential presentations of feeding a family are more easily followed by most teenagers.) Keeping to such a strategy produced a series of diagrams that describe continually more detailed procedures, which greatly simplified the writing of the specification.

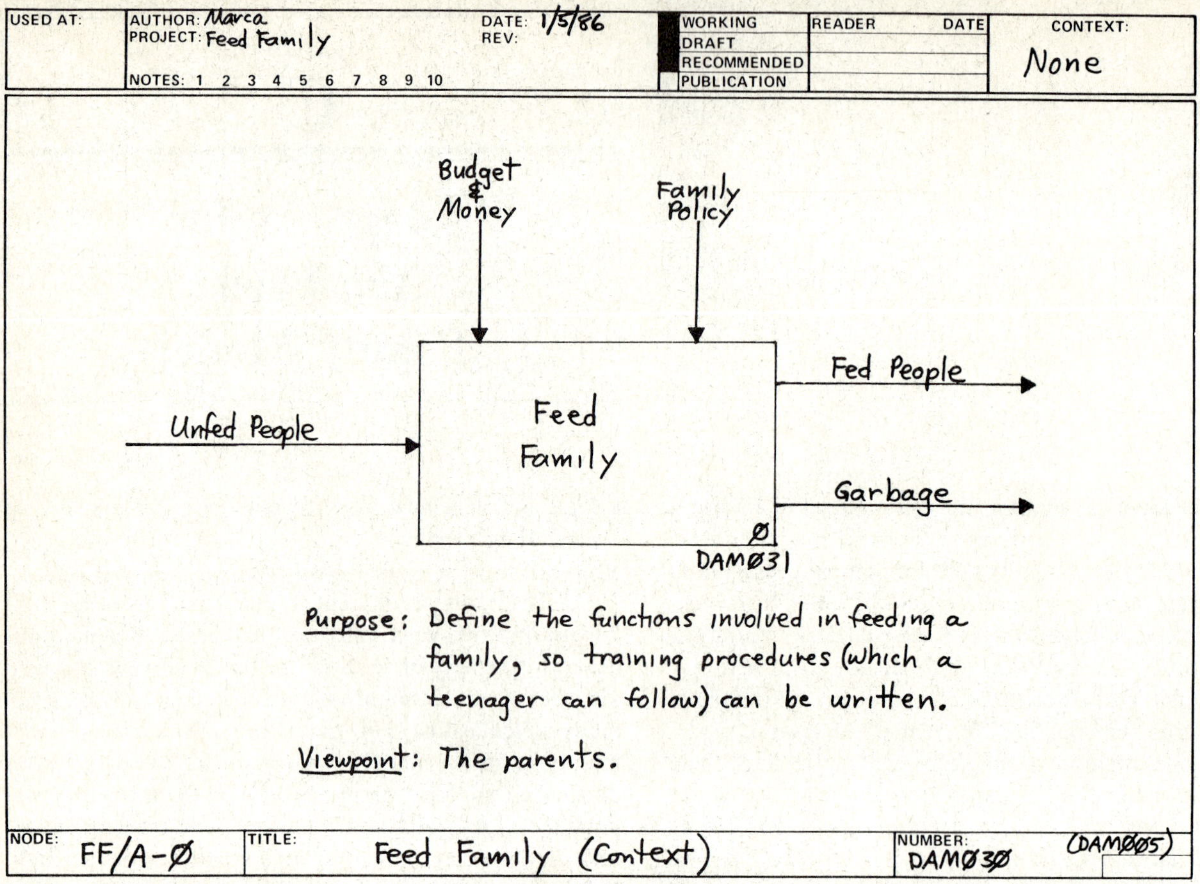
USED AT:
AUTHOR: Marca
PROJECT: Feed Family
NOTES: 1 2 3 4 5 6 7 8 9 10
DATE: 1/5/86
REV:
WORKING
DRAFT
RECOMMENDED
PUBLICATION
READER
DATE
CONTEXT:
None
Budget & Money
Family Policy
Fed People
Unfed People
Feed Family
Garbage
Ø
DAMØ31
Purpose: Define the functions involved in feeding a family, so training procedures (which a teenager can follow) can be written.
Viewpoint: The parents.
NODE:
FF/A-Ø
TITLE:
Feed Family (Context)
NUMBER:
DAMØ3Ø
(DAMØØ5)

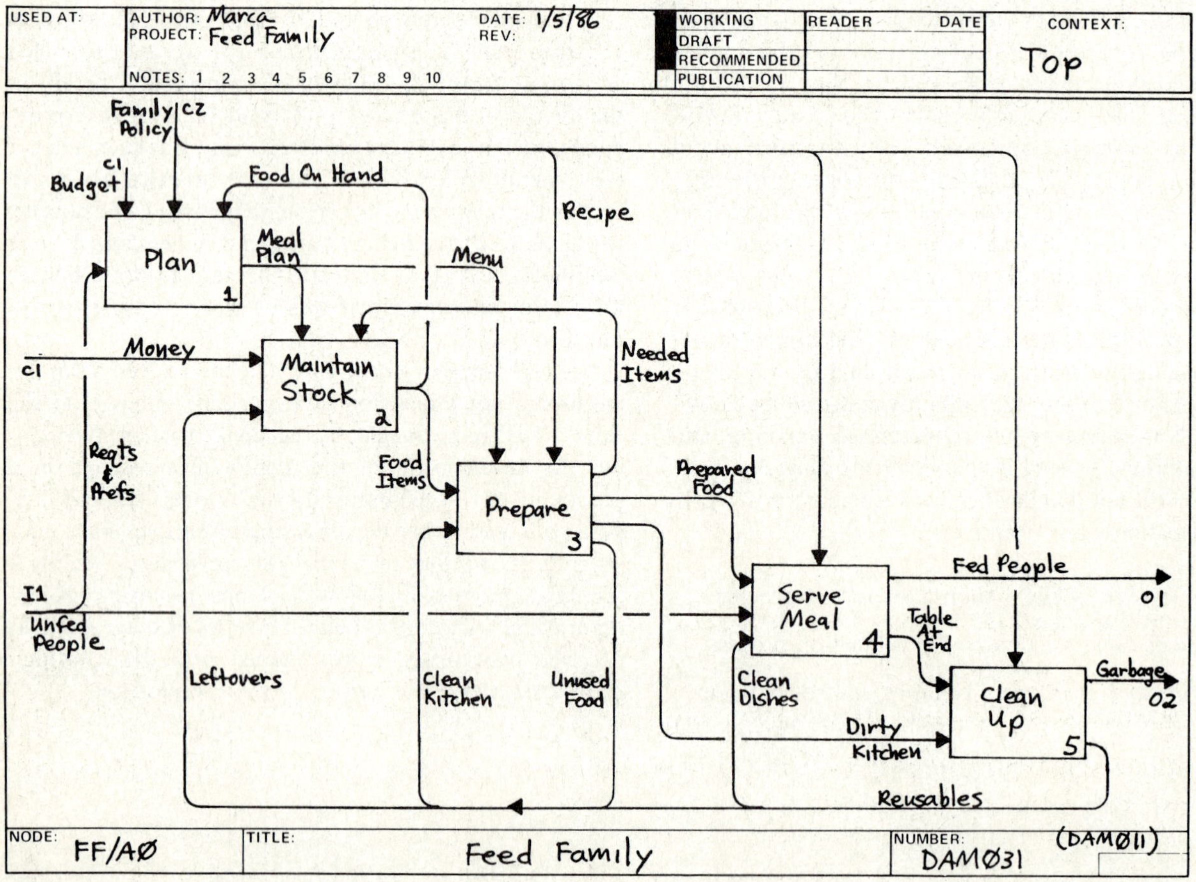
USED AT:
AUTHOR: Marca
PROJECT: Feed Family
NOTES: 1 2 3 4 5 6 7 8 9 10
DATE: 1/5/86
REV:
WORKING
DRAFT
RECOMMENDED
PUBLICATION
READER
DATE
CONTEXT:
Top
Family Policy
C2
C1
Budget
Food On Hand
Recipe
Plan
1
Meal Plan
Menu
Money
C1
Maintain Stock
2
Needed Items
Reqts & Prefs
Food Items
Prepare
3
Prepared Food
I1
Unfed People
Serve Meal
4
Fed People
O1
Table At End
Leftovers
Clean Kitchen
Unused Food
Clean Dishes
Clean Up
5
Garbage
O2
Dirty Kitchen
Reusables
NODE:
FF/AØ
TITLE:
Feed Family
NUMBER:
DAMØ31
(DAMØ11)

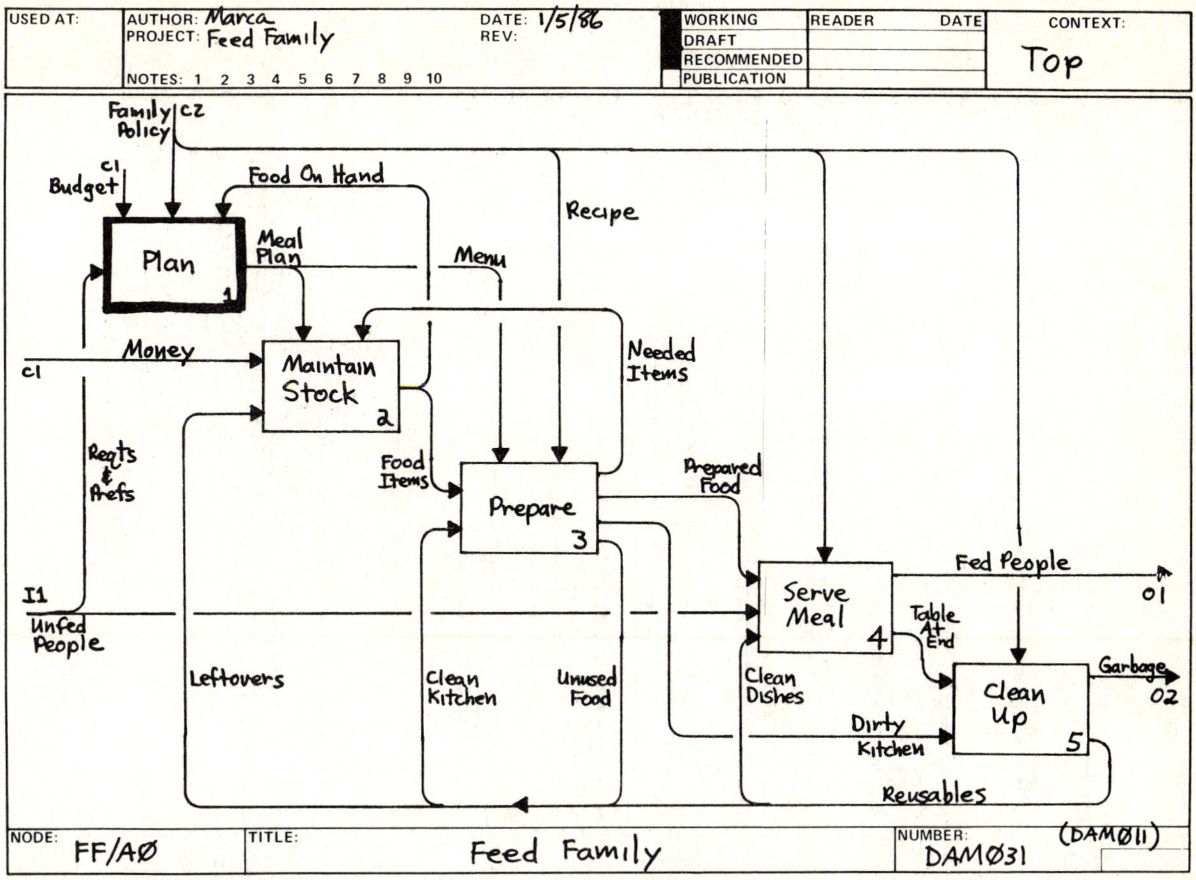
USED AT:
AUTHOR: Marca
PROJECT: Feed Family
DATE: 1/5/86
REV:
NOTES: 1 2 3 4 5 6 7 8 9 10
WORKING
DRAFT
RECOMMENDED
PUBLICATION
READER
DATE
CONTEXT:
Top
Family Policy
C2
C1
Budget
Food On Hand
Plan
1
Meal Plan
Menu
Recipe
Needed Items
C1
Money
Maintain Stock
2
Reqts & Prefs
Food Items
Prepare
3
Prepared Food
I1
Unfed People
Serve Meal
4
Fed People
O1
Table At End
Leftovers
Clean Kitchen
Unused Food
Clean Dishes
Clean Up
5
Garbage
O2
Dirty Kitchen
Reusables
NODE:
FF/AØ
TITLE:
Feed Family
NUMBER:
DAMØ31
(DAMØ11)

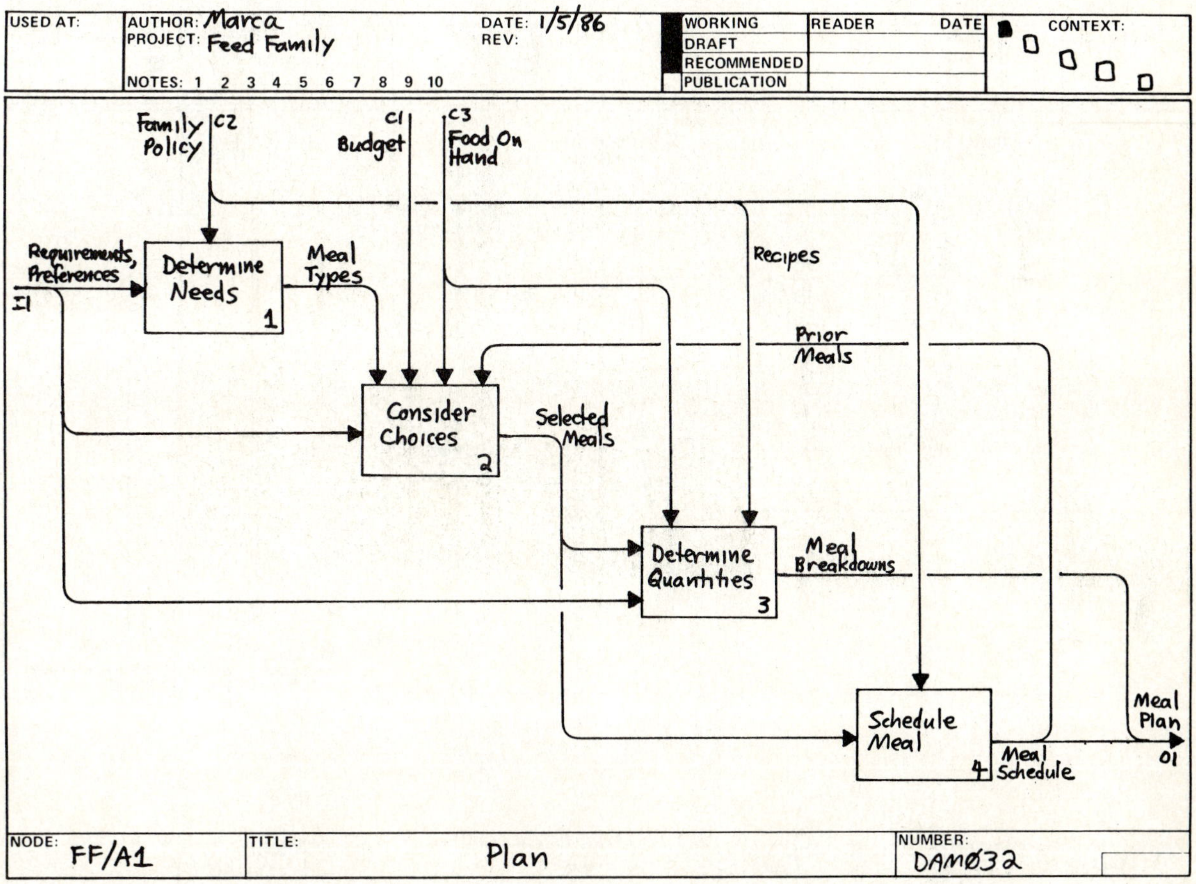
USED AT:
AUTHOR: Marca
PROJECT: Feed Family
DATE: 1/5/86
REV:
NOTES: 1 2 3 4 5 6 7 8 9 10
WORKING
DRAFT
RECOMMENDED
PUBLICATION
READER
DATE
CONTEXT:
Family Policy
C2
C1
Budget
C3
Food On Hand
Requirements, Preferences
I1
Determine Needs
1
Meal Types
Recipes
Prior Meals
Consider Choices
2
Selected Meals
Determine Quantities
3
Meal Breakdowns
Schedule Meal
4
Meal Schedule
Meal Plan
O1
NODE:
FF/A1
TITLE:
Plan
NUMBER:
DAMØ32

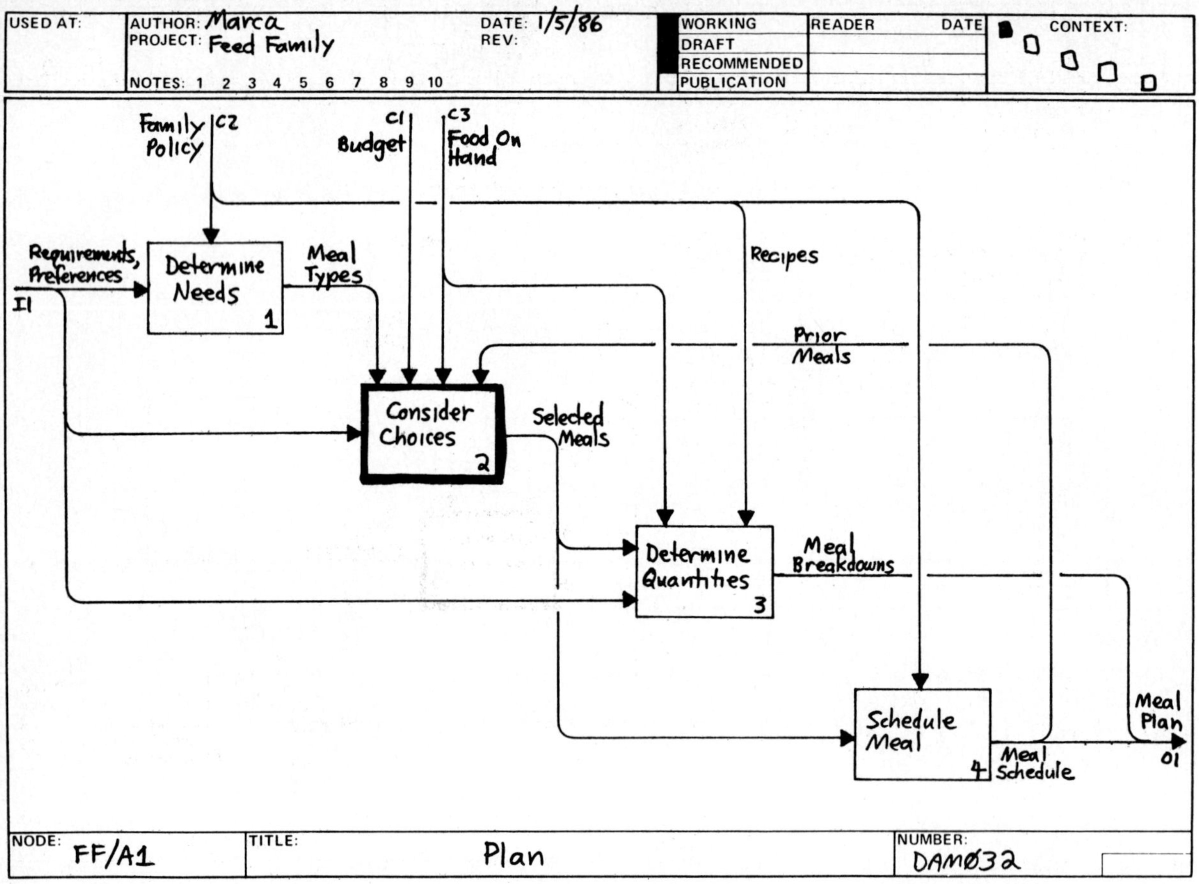
USED AT:
AUTHOR: Marca
PROJECT: Feed Family
DATE: 1/5/86
REV:
NOTES: 1 2 3 4 5 6 7 8 9 10
WORKING
DRAFT
RECOMMENDED
PUBLICATION
READER
DATE
CONTEXT:
Family Policy
C2
C1
Budget
C3
Food On Hand
Requirements, Preferences
I1
Determine Needs
1
Meal Types
Recipes
Prior Meals
Consider Choices
2
Selected Meals
Determine Quantities
3
Meal Breakdowns
Schedule Meal
4
Meal Schedule
Meal Plan
O1
NODE: FF/A1
TITLE: Plan
NUMBER: DAMØ32

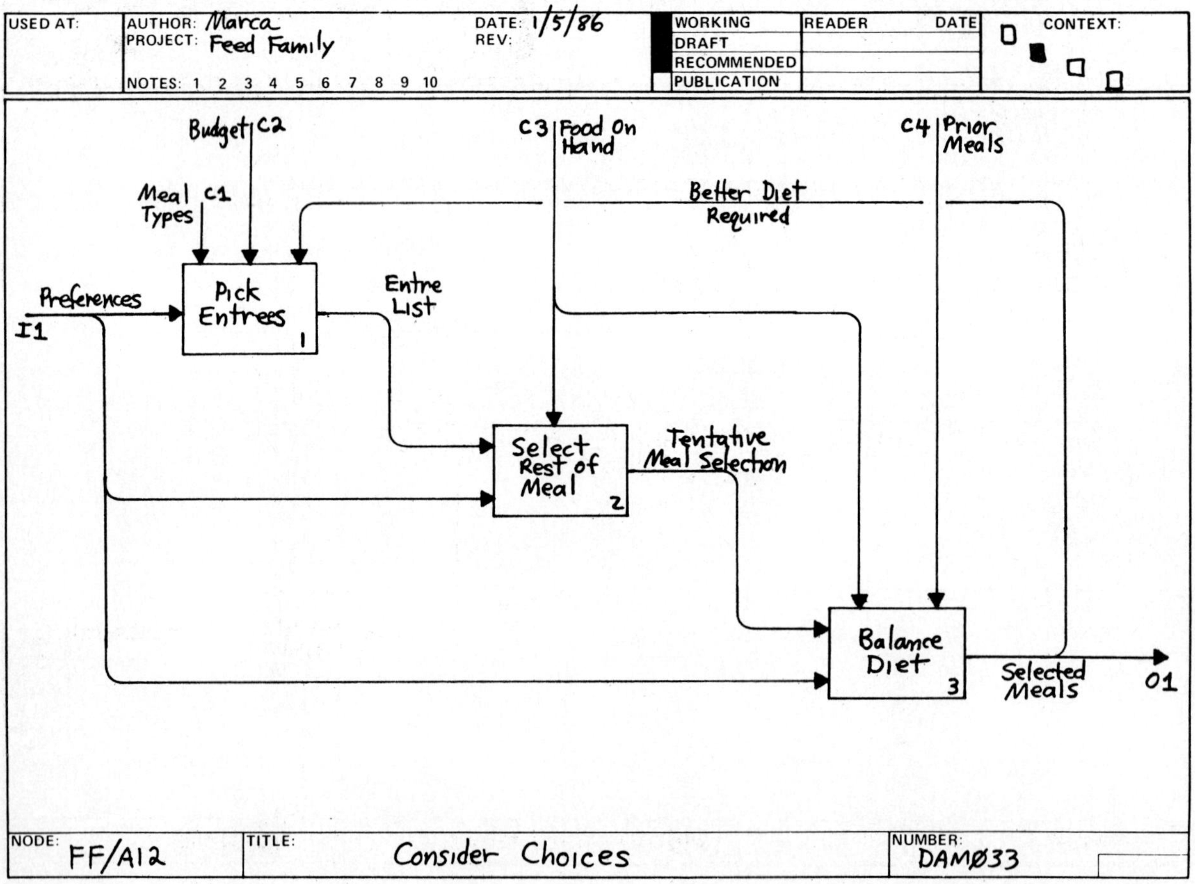
USED AT:
AUTHOR: Marca
PROJECT: Feed Family
DATE: 1/5/86
REV:
NOTES: 1 2 3 4 5 6 7 8 9 10
WORKING
DRAFT
RECOMMENDED
PUBLICATION
READER
DATE
CONTEXT:
Budget
C2
Meal Types
C1
C3
Food On Hand
C4
Prior Meals
Better Diet Required
Preferences
I1
Pick Entrees
1
Entre List
Select Rest of Meal
2
Tentative Meal Selection
Balance Diet
3
Selected Meals
O1
NODE: FF/A12
TITLE: Consider Choices
NUMBER: DAMØ33

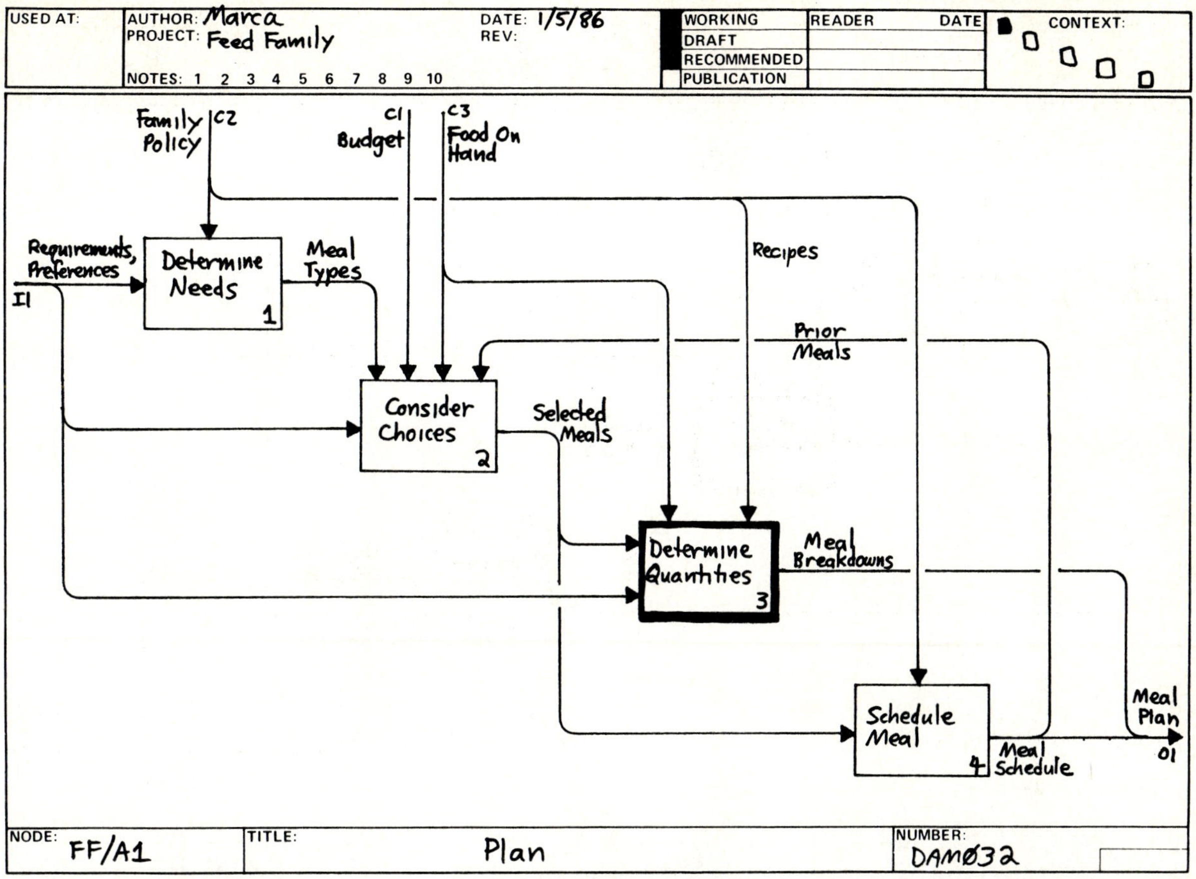
USED AT:
AUTHOR: Marca
PROJECT: Feed Family
DATE: 1/5/86
REV:
NOTES: 1 2 3 4 5 6 7 8 9 10
WORKING
DRAFT
RECOMMENDED
PUBLICATION
READER
DATE
CONTEXT:
Family Policy
C2
C1
Budget
C3
Food On Hand
Requirements, Preferences
I1
Determine Needs
1
Meal Types
Recipes
Prior Meals
Consider Choices
2
Selected Meals
Determine Quantities
3
Meal Breakdowns
Schedule Meal
4
Meal Schedule
Meal Plan
O1
NODE: FF/A1
TITLE: Plan
NUMBER: DAMØ32

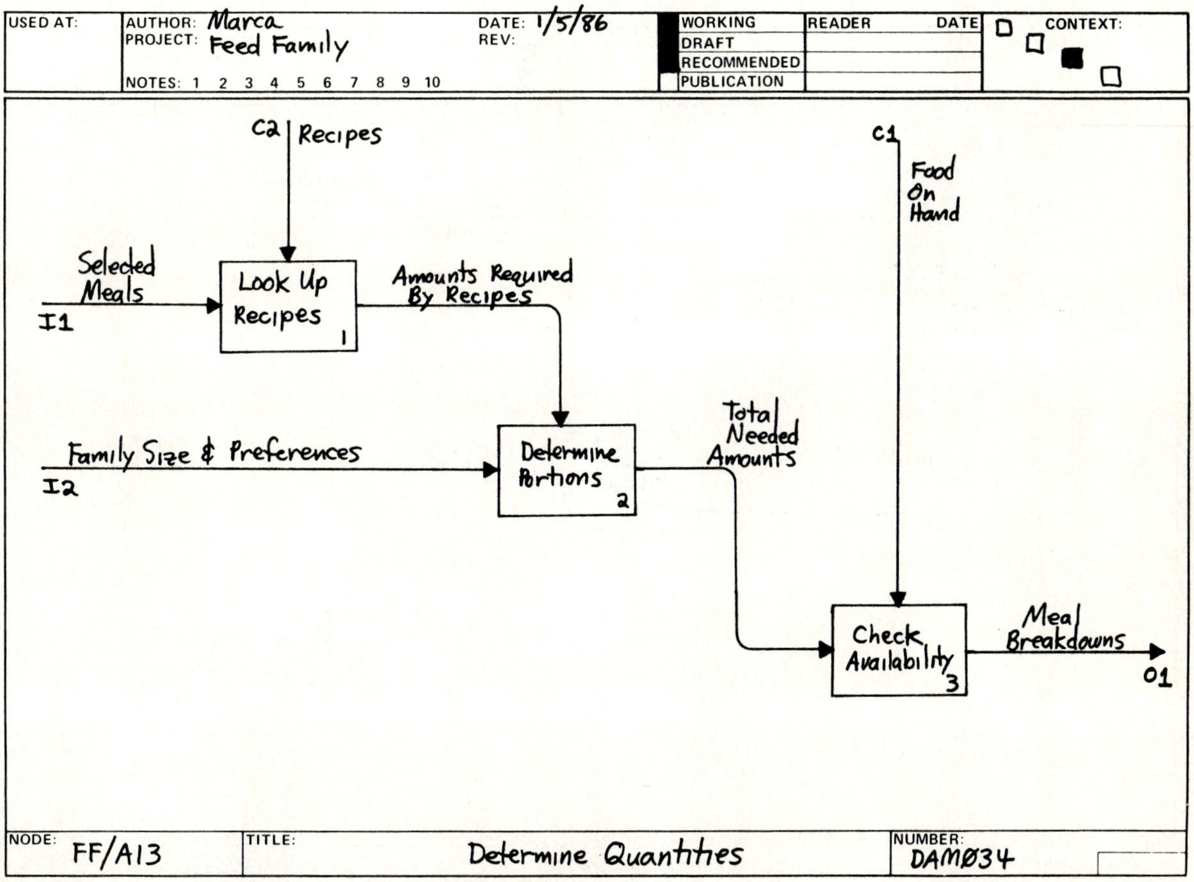
USED AT:
AUTHOR: Marca
PROJECT: Feed Family
DATE: 1/5/86
REV:
NOTES: 1 2 3 4 5 6 7 8 9 10
WORKING
DRAFT
RECOMMENDED
PUBLICATION
READER
DATE
CONTEXT:
C2
Recipes
C1
Food On Hand
Selected Meals
I1
Look Up Recipes
1
Amounts Required By Recipes
Family Size & Preferences
I2
Determine Portions
2
Total Needed Amounts
Check Availability
3
Meal Breakdowns
O1
NODE: FF/A13
TITLE: Determine Quantities
NUMBER: DAMØ34

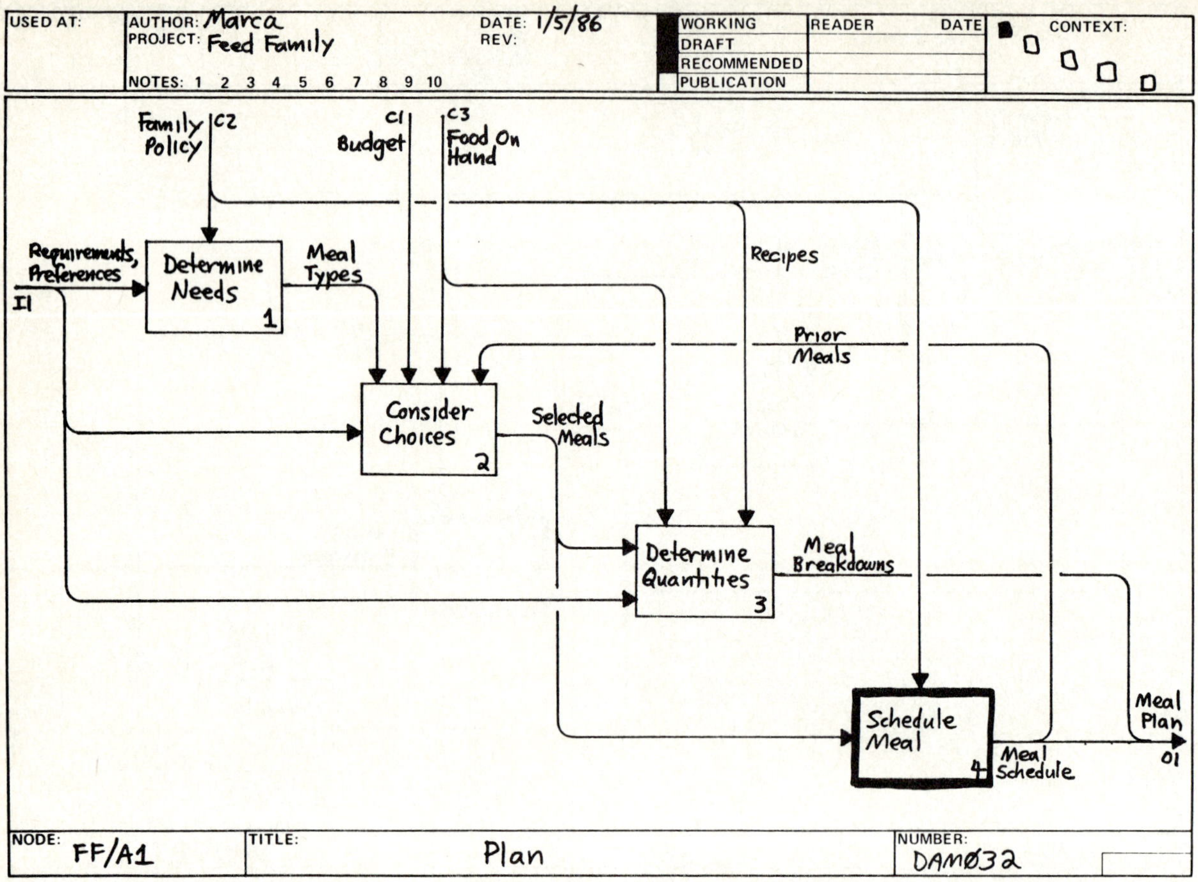
USED AT:
AUTHOR: Marca
PROJECT: Feed Family
DATE: 1/5/86
REV:
NOTES: 1 2 3 4 5 6 7 8 9 10
WORKING
DRAFT
RECOMMENDED
PUBLICATION
READER
DATE
CONTEXT:
Family Policy
C2
C1
Budget
C3
Food On Hand
Requirements, Preferences
I1
Determine Needs
1
Meal Types
Recipes
Prior Meals
Consider Choices
2
Selected Meals
Determine Quantities
3
Meal Breakdowns
Schedule Meal
4
Meal Schedule
Meal Plan
O1
NODE: FF/A1
TITLE: Plan
NUMBER: DAMØ32

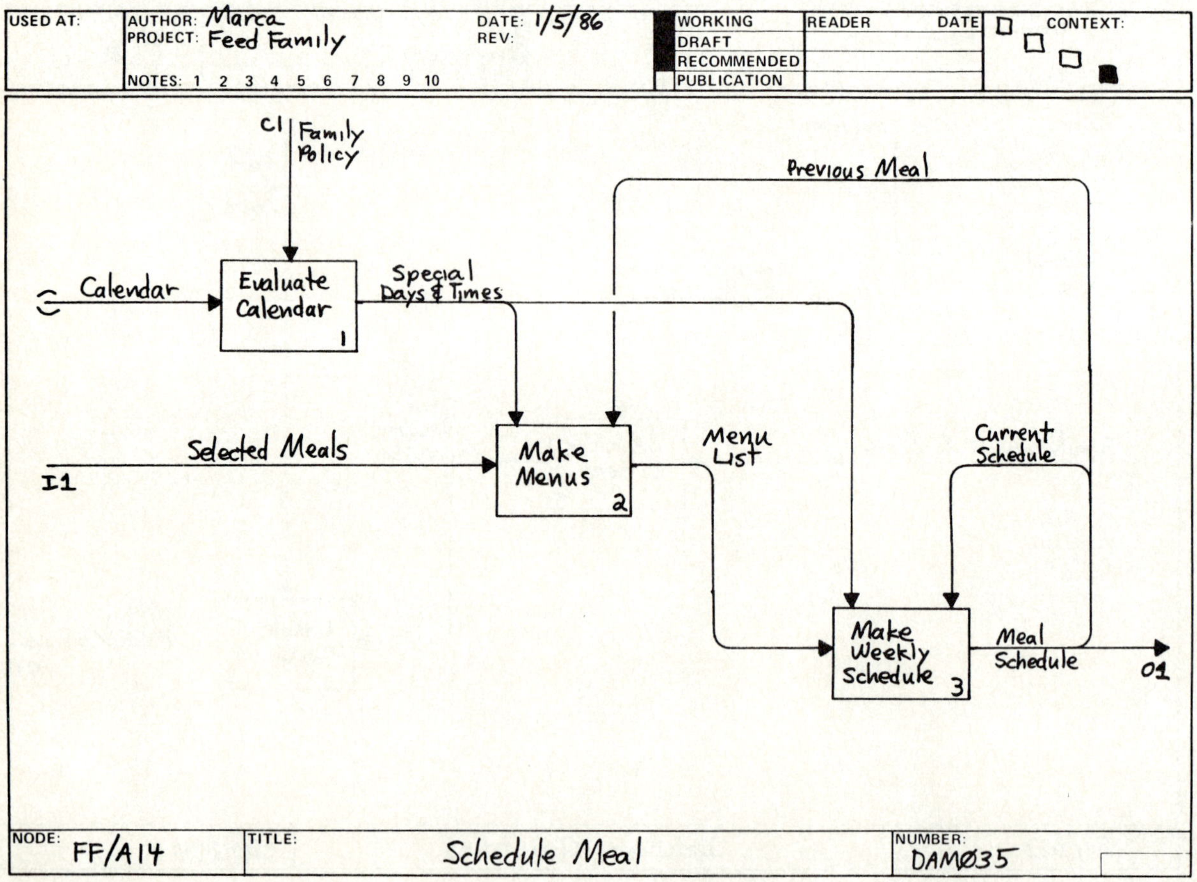
USED AT:
AUTHOR: Marca
PROJECT: Feed Family
DATE: 1/5/86
REV:
NOTES: 1 2 3 4 5 6 7 8 9 10
WORKING
DRAFT
RECOMMENDED
PUBLICATION
READER
DATE
CONTEXT:
C1
Family Policy
Previous Meal
Calendar
Evaluate Calendar
1
Special Days & Times
Selected Meals
I1
Make Menus
2
Menu List
Current Schedule
Make Weekly Schedule
3
Meal Schedule
O1
NODE: FF/A14
TITLE: Schedule Meal
NUMBER: DAMØ35

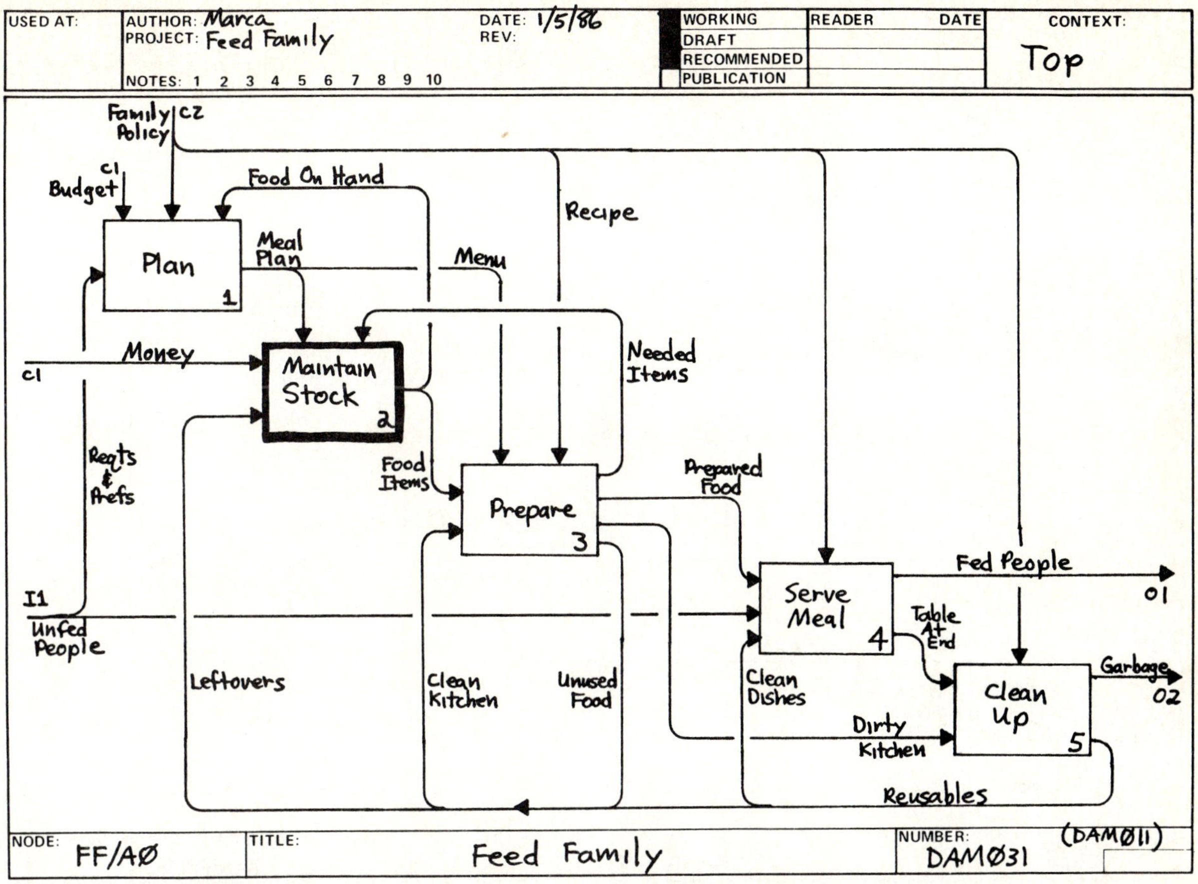
USED AT:
AUTHOR: Marca
PROJECT: Feed Family
DATE: 1/5/86
REV:
NOTES: 1 2 3 4 5 6 7 8 9 10
WORKING
DRAFT
RECOMMENDED
PUBLICATION
READER
DATE
CONTEXT:
Top
Family Policy
C2
Budget
C1
Food On Hand
Recipe
Plan
1
Meal Plan
Menu
Money
C1
Maintain Stock
2
Needed Items
Reqts & Prefs
Food Items
Prepared Food
Prepare
3
I1
Unfed People
Serve Meal
4
Fed People
O1
Table At End
Leftovers
Clean Kitchen
Unused Food
Clean Dishes
Clean Up
5
Garbage
O2
Dirty Kitchen
Reusables
NODE:
FF/AØ
TITLE:
Feed Family
NUMBER:
DAMØ31
(DAMØ11)

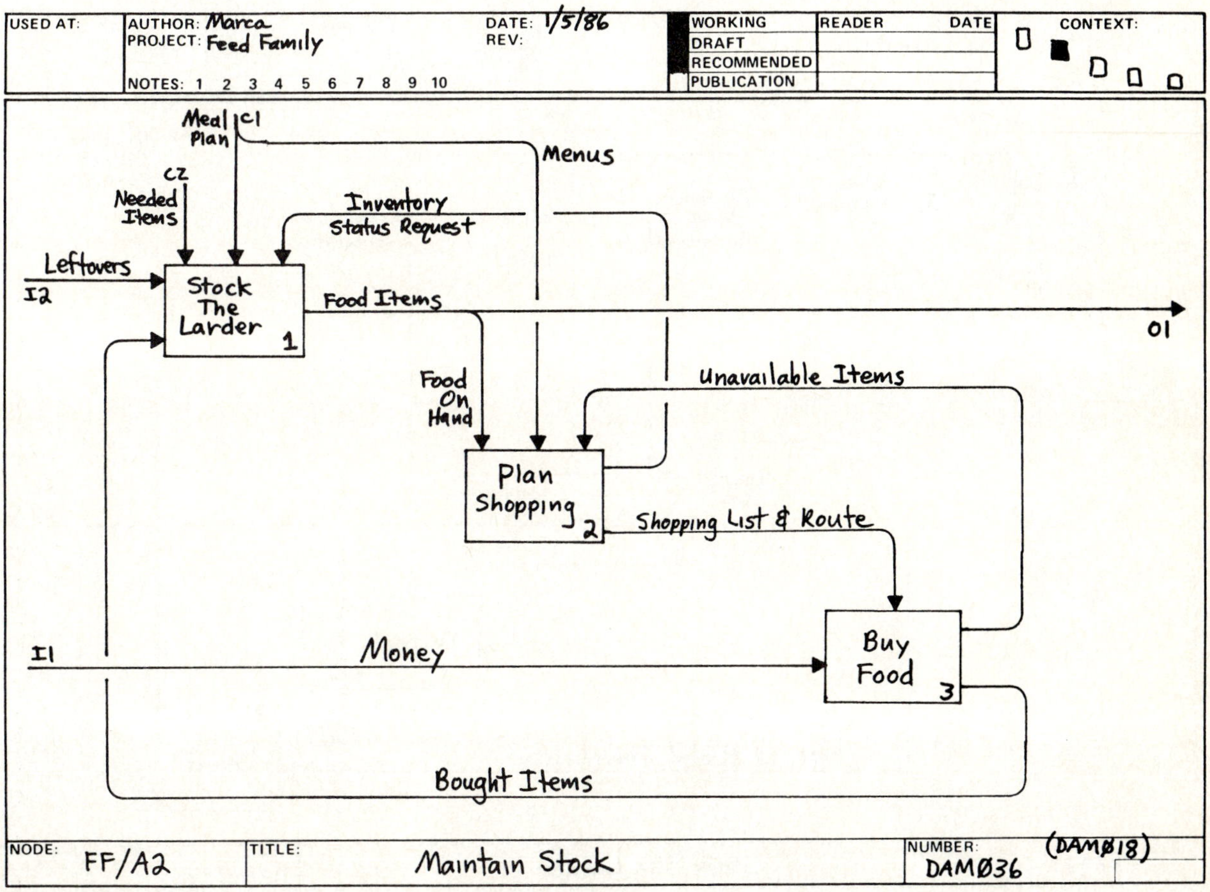
USED AT:
AUTHOR: Marca
PROJECT: Feed Family
DATE: 1/5/86
REV:
NOTES: 1 2 3 4 5 6 7 8 9 10
WORKING
DRAFT
RECOMMENDED
PUBLICATION
READER
DATE
CONTEXT:
Meal Plan
C1
Menus
C2
Needed Items
Inventory Status Request
Leftovers
I2
Stock The Larder
1
Food Items
O1
Unavailable Items
Food On Hand
Plan Shopping
2
Shopping List & Route
I1
Money
Buy Food
3
Bought Items
NODE:
FF/A2
TITLE:
Maintain Stock
NUMBER:
DAMØ36
(DAMØ18)

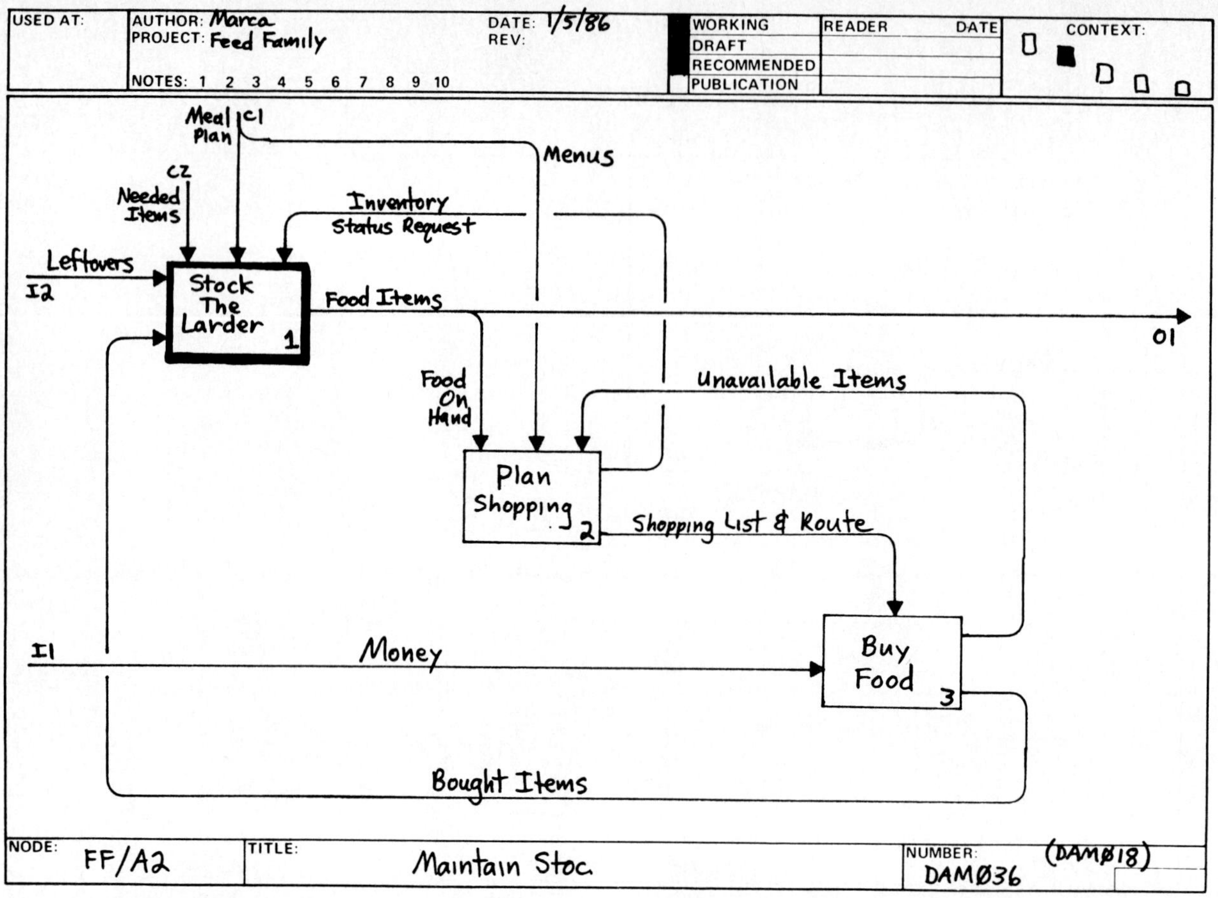
USED AT:
AUTHOR: Marca
PROJECT: Feed Family
DATE: 1/5/86
REV:
NOTES: 1 2 3 4 5 6 7 8 9 10
WORKING
DRAFT
RECOMMENDED
PUBLICATION
READER
DATE
CONTEXT:
Meal Plan
C1
C2
Needed Items
Menus
Inventory Status Request
Leftovers
I2
Stock The Larder 1
Food Items
O1
Food On Hand
Unavailable Items
Plan Shopping 2
Shopping List & Route
Buy Food 3
I1
Money
Bought Items
NODE: FF/A2
TITLE: Maintain Stoc.
NUMBER: DAMØ36
(DAMØ18)

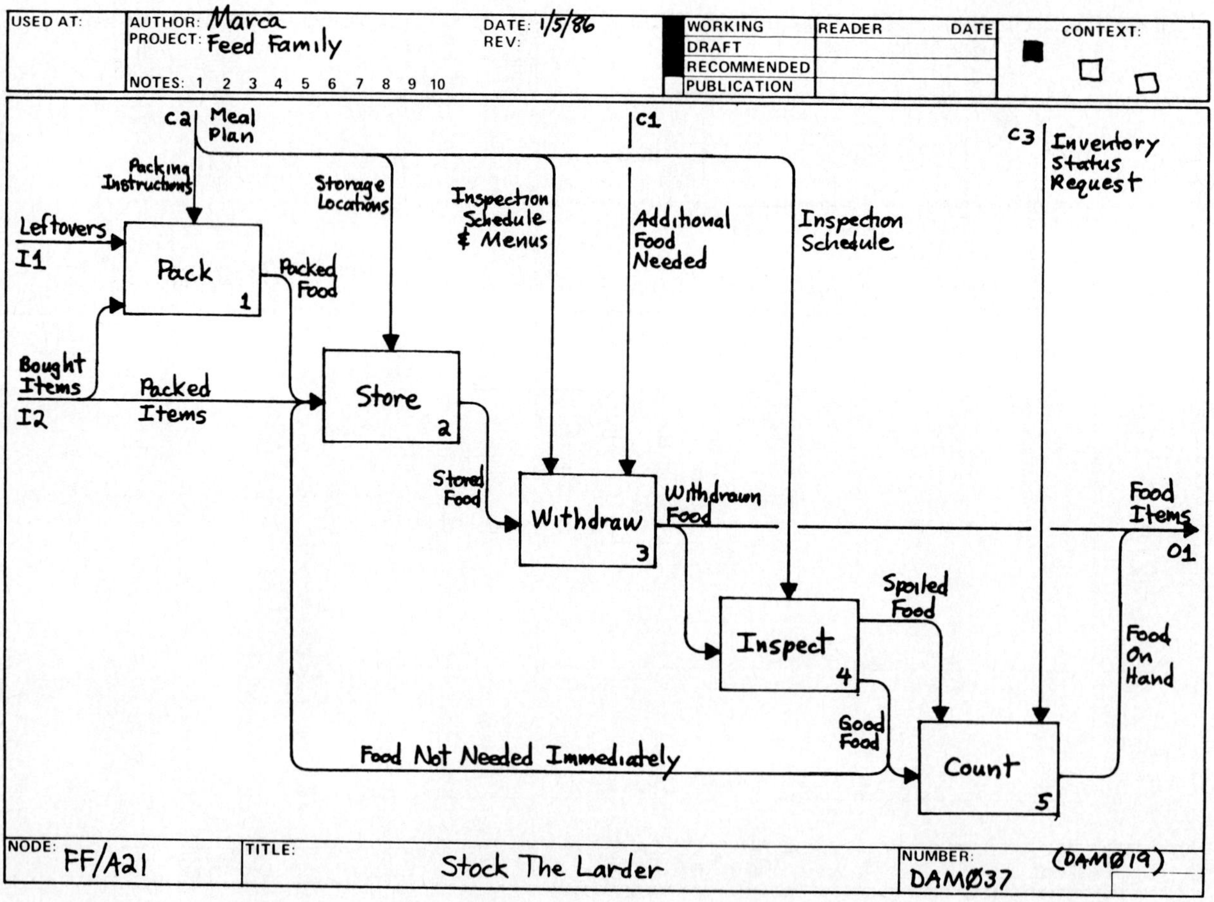
USED AT:
AUTHOR: Marca
PROJECT: Feed Family
DATE: 1/5/86
REV:
NOTES: 1 2 3 4 5 6 7 8 9 10
WORKING
DRAFT
RECOMMENDED
PUBLICATION
READER
DATE
CONTEXT:
C2
Meal Plan
C1
C3
Inventory Status Request
Packing Instructions
Storage Locations
Inspection Schedule & Menus
Additional Food Needed
Inspection Schedule
Leftovers
I1
Pack 1
Packed Food
Bought Items
Packed Items
I2
Store 2
Stored Food
Withdraw 3
Withdrawn Food
Food Items
O1
Inspect 4
Spoiled Food
Food On Hand
Good Food
Food Not Needed Immediately
Count 5
NODE: FF/A21
TITLE: Stock The Larder
NUMBER: DAMØ37
(DAMØ19)

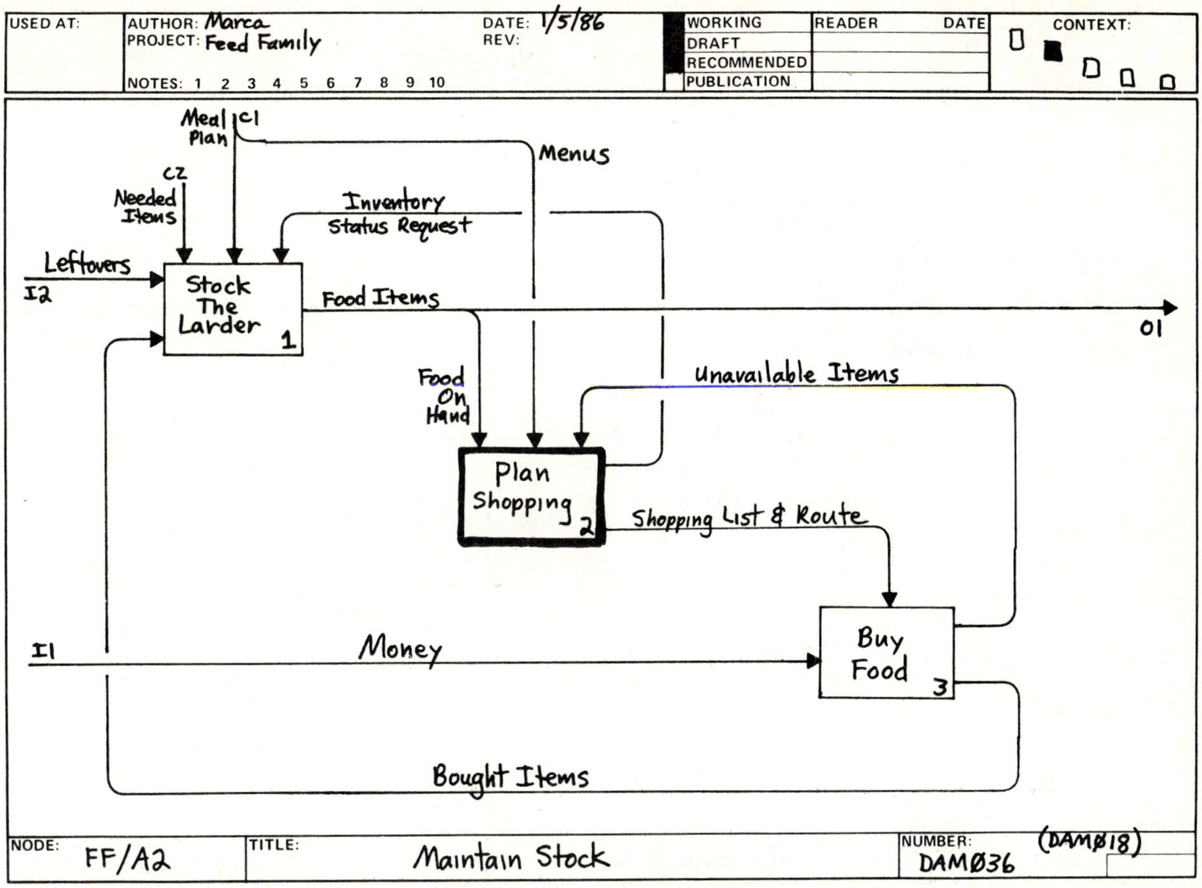
USED AT:
AUTHOR: Marca
PROJECT: Feed Family
DATE: 1/5/86
REV:
NOTES: 1 2 3 4 5 6 7 8 9 10
WORKING
DRAFT
RECOMMENDED
PUBLICATION
READER
DATE
CONTEXT:
Meal Plan
C1
Menus
C2
Needed Items
Inventory Status Request
Leftovers
I2
Stock The Larder 1
Food Items
O1
Food On Hand
Unavailable Items
Plan Shopping 2
Shopping List & Route
I1
Money
Buy Food 3
Bought Items
NODE: FF/A2
TITLE: Maintain Stock
NUMBER: DAMØ36
(DAMØ18)

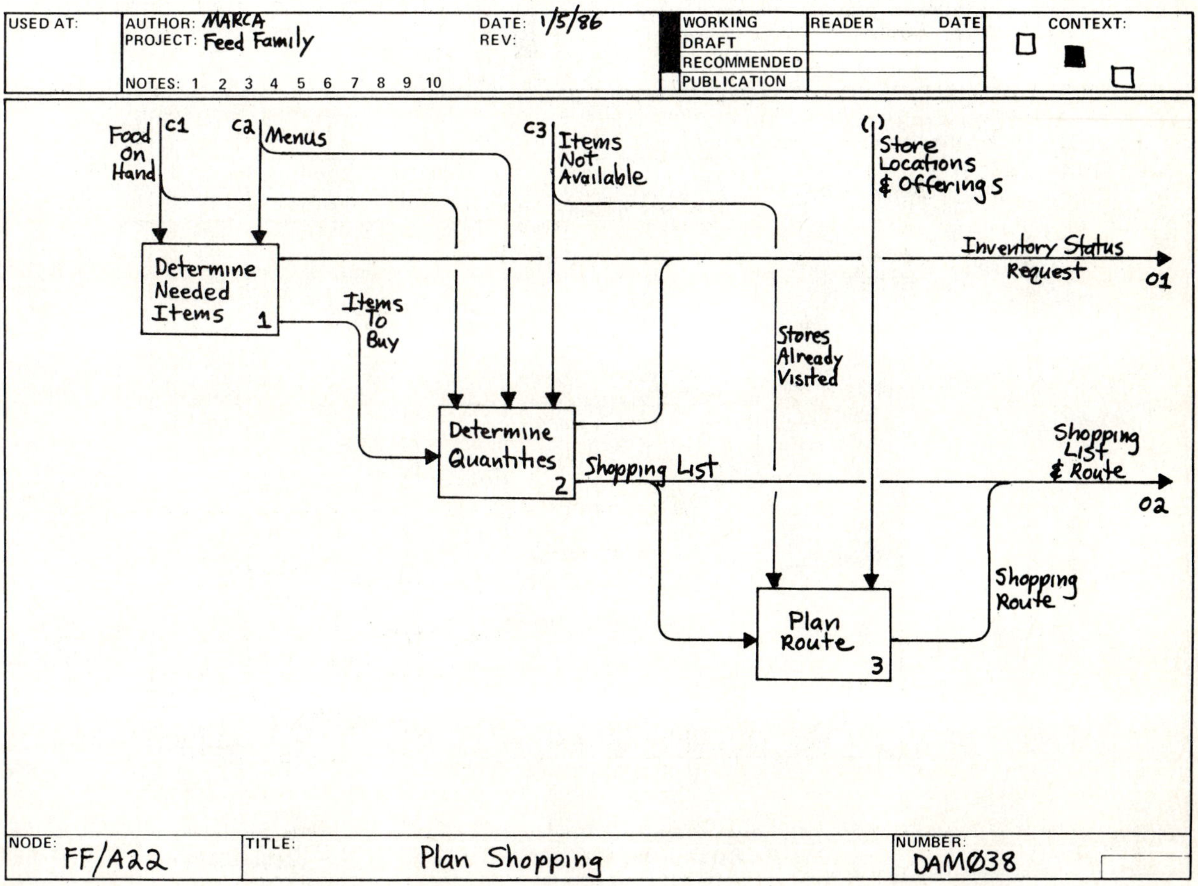
USED AT:
AUTHOR: MARCA
PROJECT: Feed Family
DATE: 1/5/86
REV:
NOTES: 1 2 3 4 5 6 7 8 9 10
WORKING
DRAFT
RECOMMENDED
PUBLICATION
READER
DATE
CONTEXT:
Food On Hand
C1
C2
Menus
C3
Items Not Available
(1)
Store Locations & Offerings
Inventory Status Request
O1
Determine Needed Items 1
Items To Buy
Stores Already Visited
Determine Quantities 2
Shopping List
Shopping List & Route
O2
Shopping Route
Plan Route 3
NODE: FF/A22
TITLE: Plan Shopping
NUMBER: DAMØ38

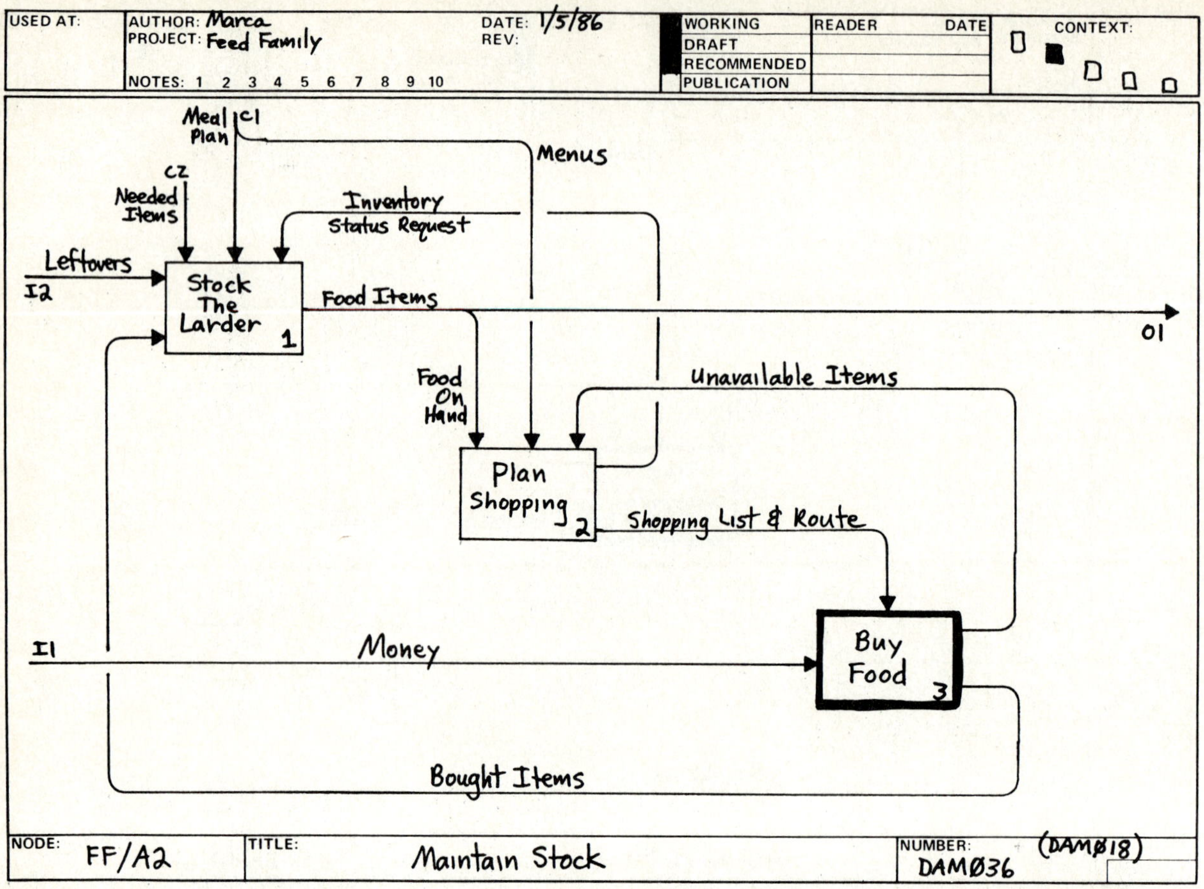
USED AT:
AUTHOR: Marca
PROJECT: Feed Family
DATE: 1/5/86
REV:
NOTES: 1 2 3 4 5 6 7 8 9 10
WORKING
DRAFT
RECOMMENDED
PUBLICATION
READER
DATE
CONTEXT:
Meal Plan
C1
Menus
C2
Needed Items
Inventory Status Request
Leftovers
I2
Stock The Larder 1
Food Items
O1
Food On Hand
Unavailable Items
Plan Shopping 2
Shopping List & Route
I1
Money
Buy Food 3
Bought Items
NODE: FF/A2
TITLE: Maintain Stock
NUMBER: DAMØ36
(DAMØ18)

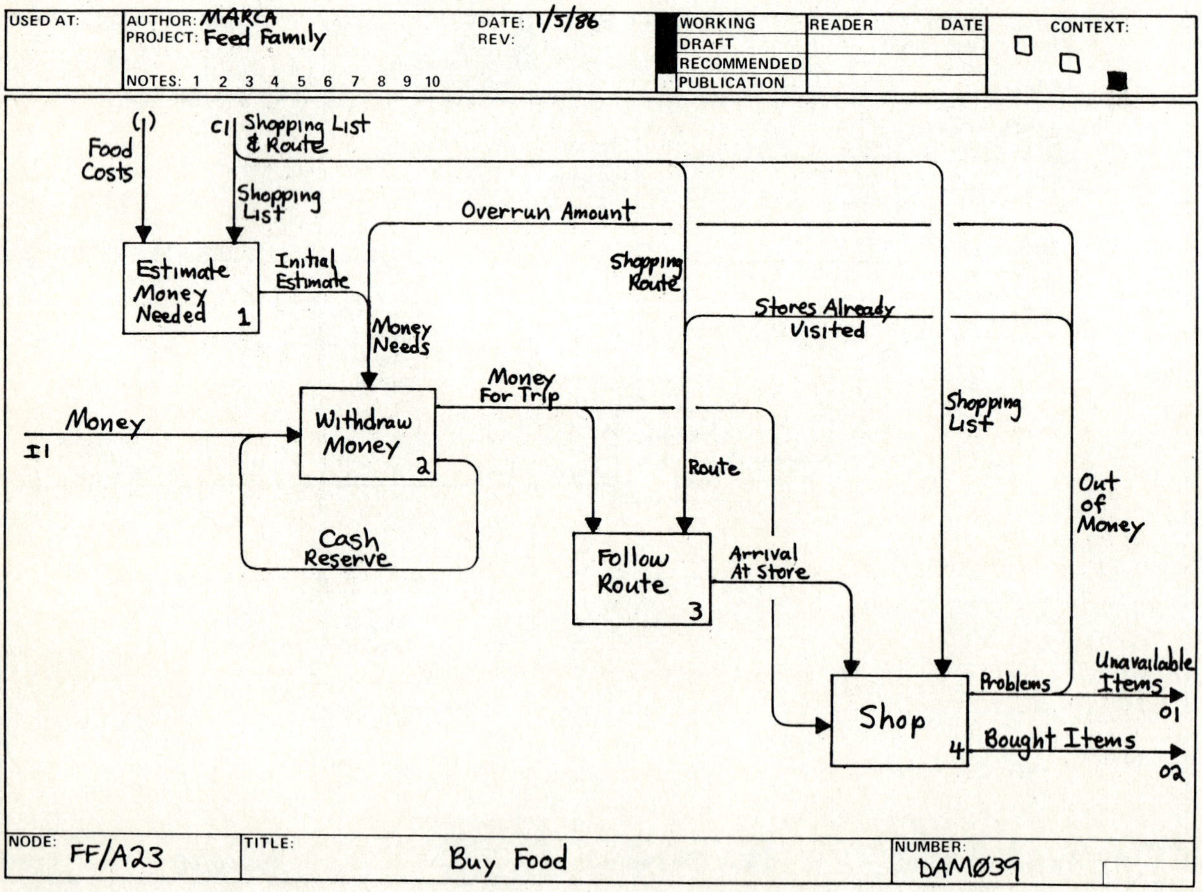
USED AT:
AUTHOR: MARCA
PROJECT: Feed Family
DATE: 1/5/86
REV:
NOTES: 1 2 3 4 5 6 7 8 9 10
WORKING
DRAFT
RECOMMENDED
PUBLICATION
READER
DATE
CONTEXT:
(1)
Food Costs
C1
Shopping List & Route
Shopping List
Overrun Amount
Estimate Money Needed 1
Initial Estimate
Shopping Route
Money Needs
Stores Already Visited
Money For Trip
Money
I1
Withdraw Money 2
Shopping List
Route
Out of Money
Cash Reserve
Follow Route 3
Arrival At Store
Problems
Unavailable Items
O1
Shop 4
Bought Items
O2
NODE: FF/A23
TITLE: Buy Food
NUMBER: DAMØ39

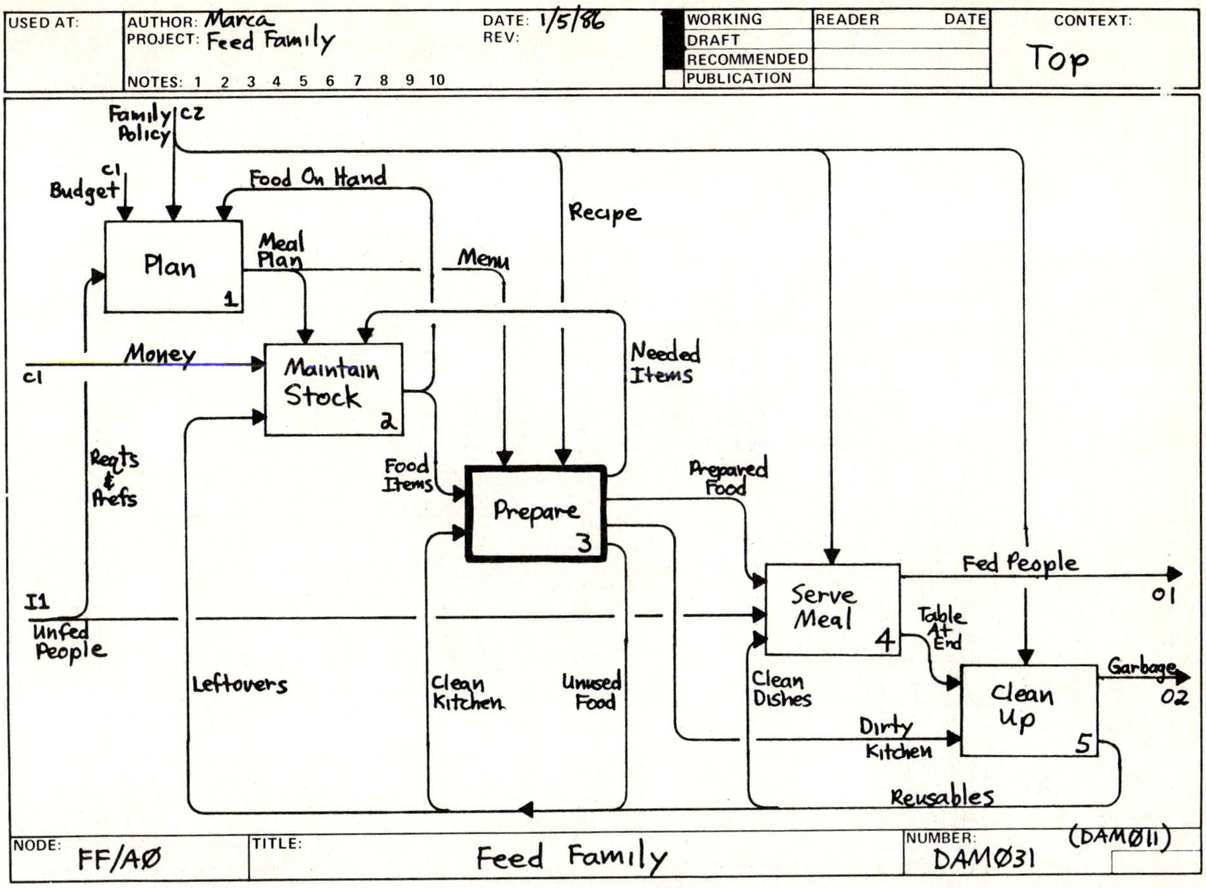
USED AT:
AUTHOR: Marca
PROJECT: Feed Family
DATE: 1/5/86
REV:
NOTES: 1 2 3 4 5 6 7 8 9 10
WORKING
DRAFT
RECOMMENDED
PUBLICATION
READER
DATE
CONTEXT:
Top
Family Policy C2
C1 Budget
Food On Hand
Plan 1
Meal Plan
Menu
Recipe
Money
C1
Maintain Stock 2
Needed Items
Reqts & Prefs
Food Items
Prepare 3
Prepared Food
I1
Unfed People
Serve Meal 4
Fed People
O1
Table At End
Clean Up 5
Garbage
O2
Leftovers
Clean Kitchen
Unused Food
Clean Dishes
Dirty Kitchen
Reusables
NODE: FF/AØ
TITLE: Feed Family
NUMBER: DAMØ31
(DAMØ11)

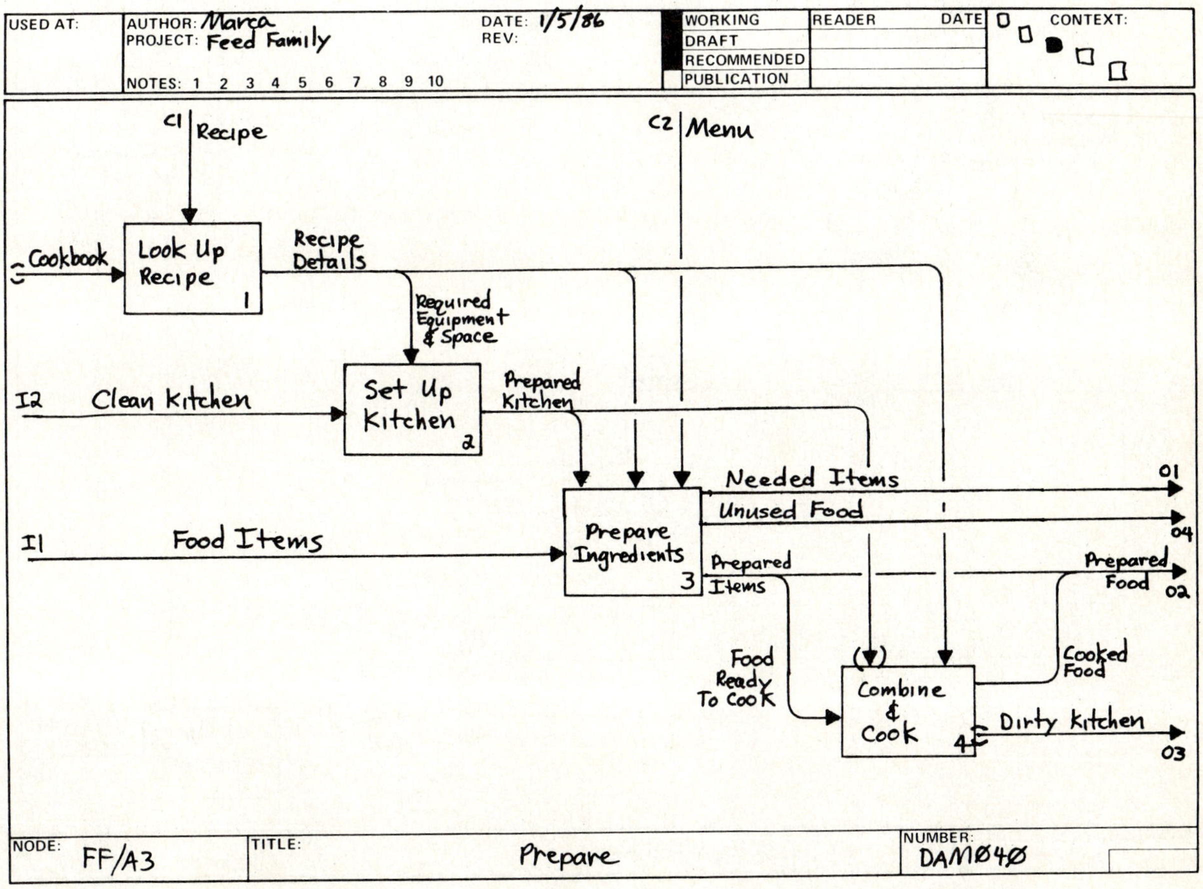
USED AT:
AUTHOR: Marca
PROJECT: Feed Family
DATE: 1/5/86
REV:
NOTES: 1 2 3 4 5 6 7 8 9 10
WORKING
DRAFT
RECOMMENDED
PUBLICATION
READER
DATE
CONTEXT:
C1 Recipe
C2 Menu
Cookbook
Look Up Recipe 1
Recipe Details
Required Equipment & Space
I2 Clean Kitchen
Set Up Kitchen 2
Prepared Kitchen
Needed Items
O1
Unused Food
O4
I1 Food Items
Prepare Ingredients 3
Prepared Items
Prepared Food
O2
Food Ready To Cook
Cooked Food
Combine & Cook 4
Dirty Kitchen
O3
NODE: FF/A3
TITLE: Prepare
NUMBER: DAMØ4Ø

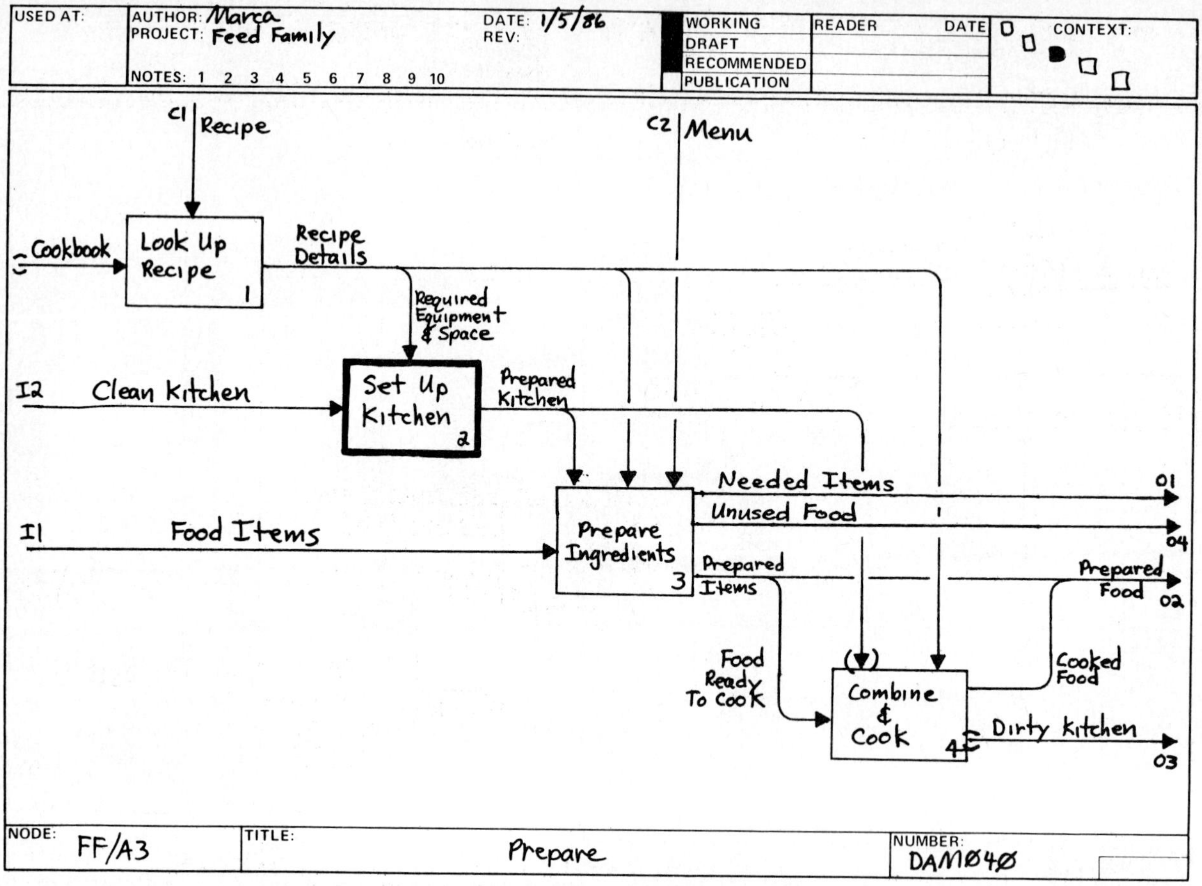
USED AT:
AUTHOR: Marca
PROJECT: Feed Family
DATE: 1/5/86
REV:
NOTES: 1 2 3 4 5 6 7 8 9 10
WORKING
DRAFT
RECOMMENDED
PUBLICATION
READER
DATE
CONTEXT:
C1 Recipe
C2 Menu
Cookbook
Look Up Recipe 1
Recipe Details
Required Equipment & Space
I2 Clean Kitchen
Set Up Kitchen 2
Prepared Kitchen
Needed Items O1
Unused Food O4
I1 Food Items
Prepare Ingredients 3
Prepared Items
Prepared Food O2
Food Ready To Cook
Combine & Cook 4
Cooked Food
Dirty Kitchen O3
NODE: FF/A3
TITLE: Prepare
NUMBER: DAMØ4Ø

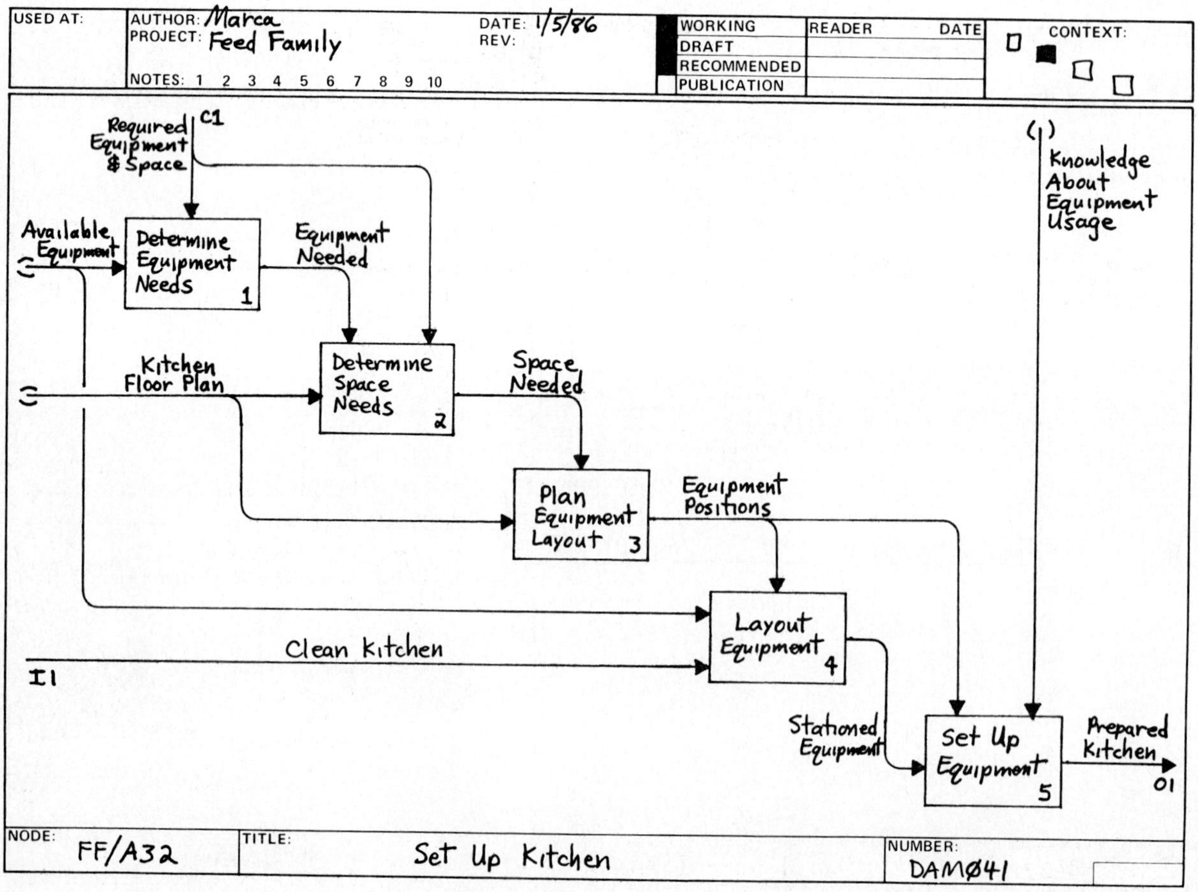
USED AT:
AUTHOR: Marca
PROJECT: Feed Family
DATE: 1/5/86
REV:
NOTES: 1 2 3 4 5 6 7 8 9 10
WORKING
DRAFT
RECOMMENDED
PUBLICATION
READER
DATE
CONTEXT:
Required Equipment & Space C1
Knowledge About Equipment Usage
Available Equipment
Determine Equipment Needs 1
Equipment Needed
Kitchen Floor Plan
Determine Space Needs 2
Space Needed
Plan Equipment Layout 3
Equipment Positions
Layout Equipment 4
Clean Kitchen
I1
Stationed Equipment
Set Up Equipment 5
Prepared Kitchen O1
NODE: FF/A32
TITLE: Set Up Kitchen
NUMBER: DAMØ41

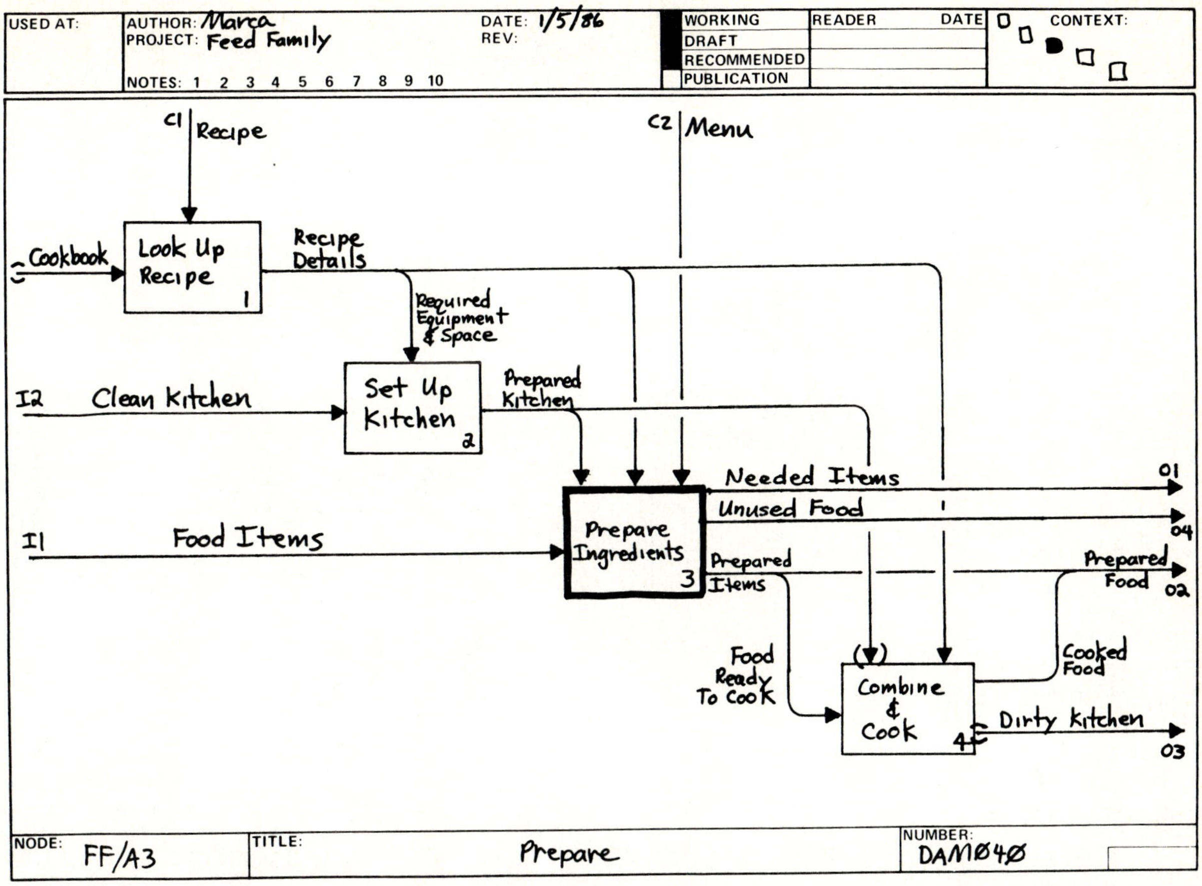
USED AT:
AUTHOR: Marca
PROJECT: Feed Family
DATE: 1/5/86
REV:
NOTES: 1 2 3 4 5 6 7 8 9 10
WORKING
DRAFT
RECOMMENDED
PUBLICATION
READER
DATE
CONTEXT:
C1 Recipe
C2 Menu
Cookbook
Look Up Recipe 1
Recipe Details
Required Equipment & Space
I2 Clean Kitchen
Set Up Kitchen 2
Prepared Kitchen
Needed Items O1
Unused Food O4
I1 Food Items
Prepare Ingredients 3
Prepared Items
Prepared Food O2
Food Ready To Cook
Cooked Food
Combine & Cook 4
Dirty kitchen O3
NODE: FF/A3
TITLE: Prepare
NUMBER: DAMØ4Ø

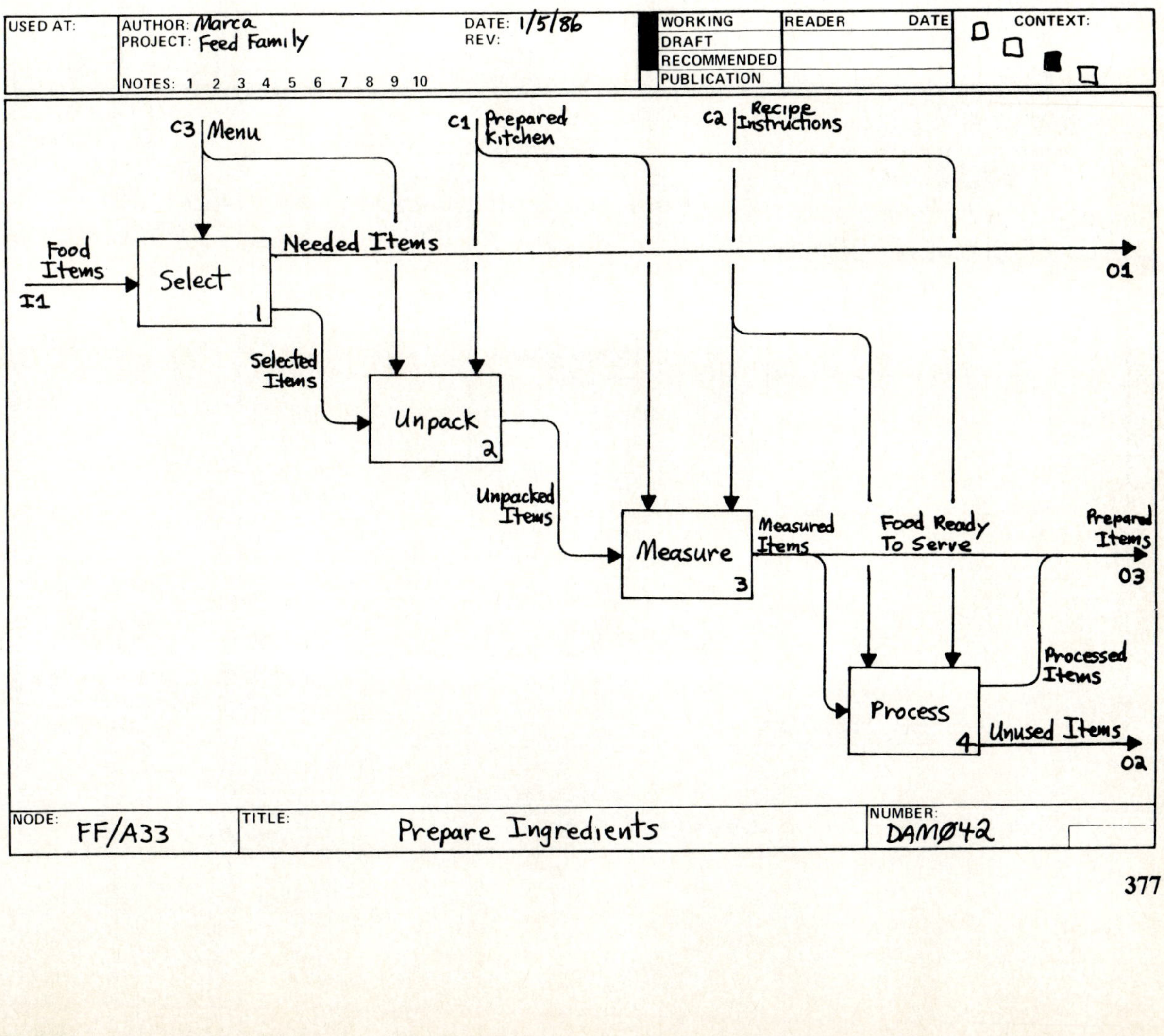
USED AT:
AUTHOR: Marca
PROJECT: Feed Family
DATE: 1/5/86
REV:
NOTES: 1 2 3 4 5 6 7 8 9 10
WORKING
DRAFT
RECOMMENDED
PUBLICATION
READER
DATE
CONTEXT:
C3 Menu
C1 Prepared Kitchen
C2 Recipe Instructions
Food Items
I1
Select 1
Needed Items O1
Selected Items
Unpack 2
Unpacked Items
Measure 3
Measured Items
Food Ready To Serve
Prepared Items O3
Processed Items
Process 4
Unused Items O2
NODE: FF/A33
TITLE: Prepare Ingredients
NUMBER: DAMØ42

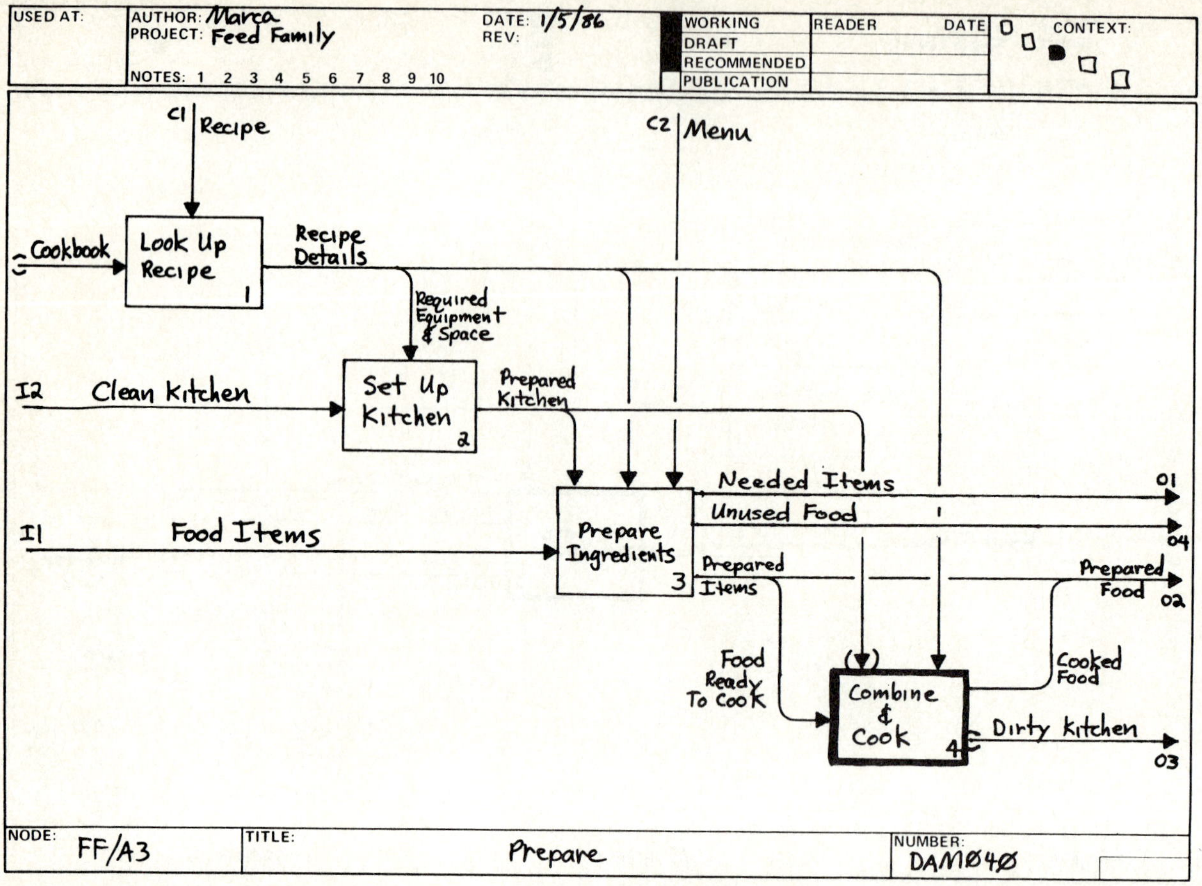
USED AT:
AUTHOR: Marca
PROJECT: Feed Family
DATE: 1/5/86
REV:
NOTES: 1 2 3 4 5 6 7 8 9 10
WORKING
DRAFT
RECOMMENDED
PUBLICATION
READER
DATE
CONTEXT:
C1 Recipe
C2 Menu
Cookbook
Look Up Recipe 1
Recipe Details
Required Equipment & Space
I2 Clean Kitchen
Set Up Kitchen 2
Prepared Kitchen
Needed Items
O1
Unused Food
O4
I1 Food Items
Prepare Ingredients 3
Prepared Items
Prepared Food
O2
Food Ready To Cook
( )
Combine & Cook 4
Cooked Food
Dirty Kitchen
O3
NODE: FF/A3
TITLE: Prepare
NUMBER: DAMØ4Ø

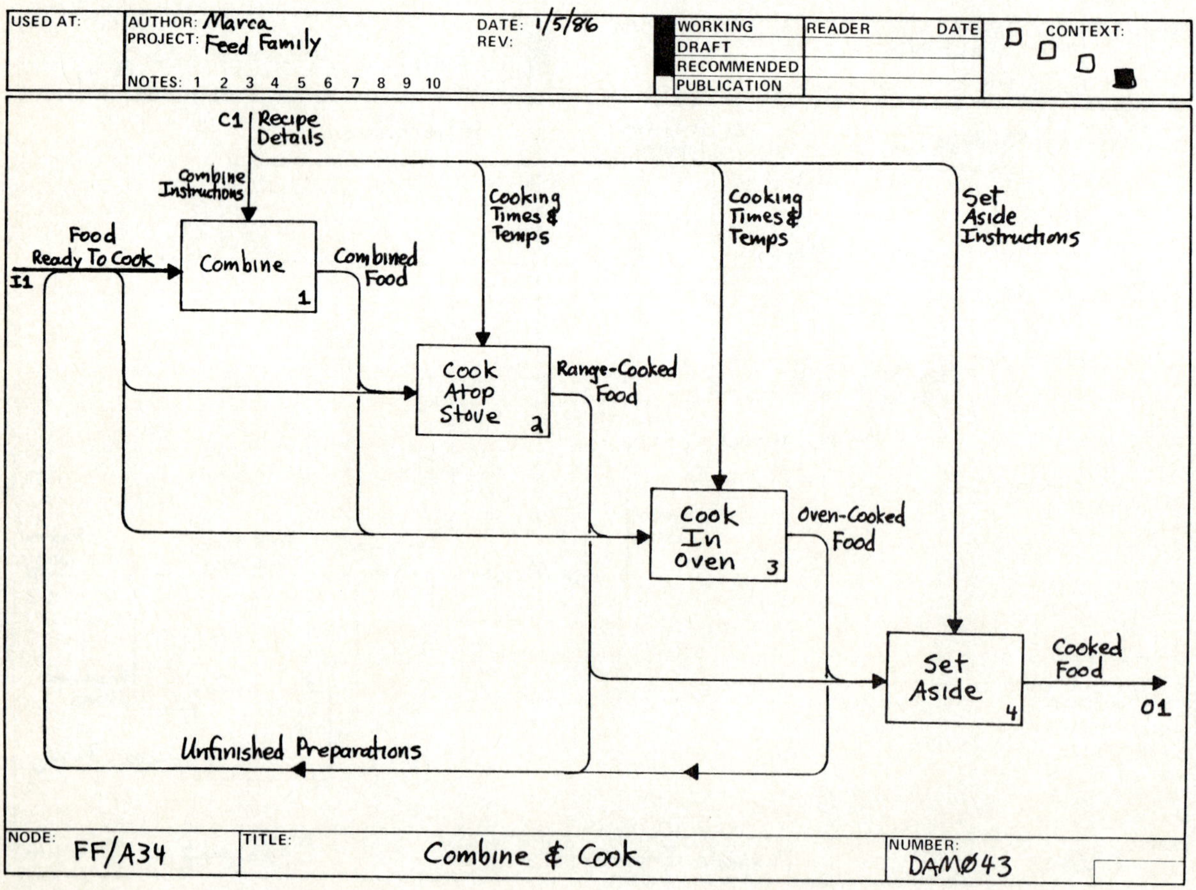
USED AT:
AUTHOR: Marca
PROJECT: Feed Family
DATE: 1/5/86
REV:
NOTES: 1 2 3 4 5 6 7 8 9 10
WORKING
DRAFT
RECOMMENDED
PUBLICATION
READER
DATE
CONTEXT:
C1 Recipe Details
Combine Instructions
Cooking Times & Temps
Cooking Times & Temps
Set Aside Instructions
Food Ready To Cook
I1
Combine 1
Combined Food
Cook Atop Stove 2
Range-Cooked Food
Cook In Oven 3
Oven-Cooked Food
Set Aside 4
Cooked Food
O1
Unfinished Preparations
NODE: FF/A34
TITLE: Combine & Cook
NUMBER: DAMØ43

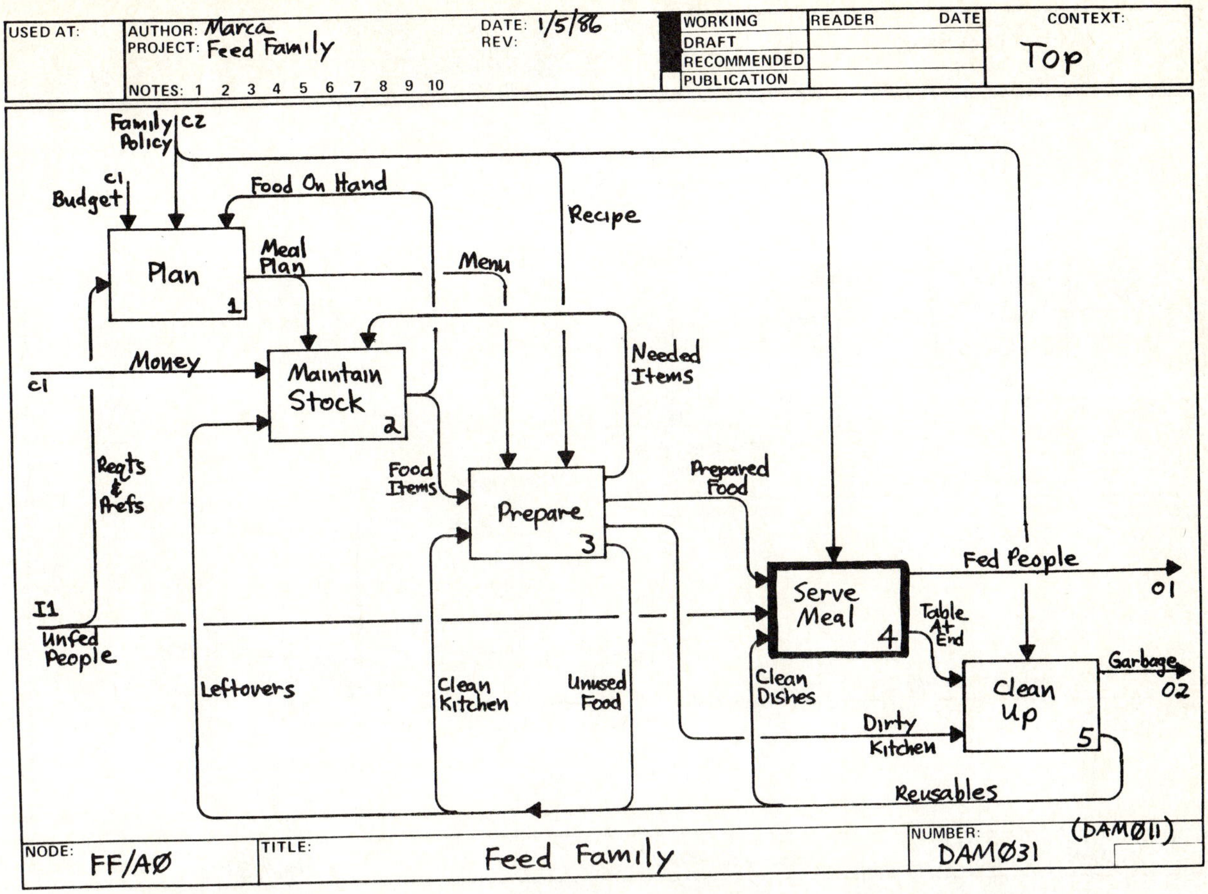
USED AT:
AUTHOR: Marca
PROJECT: Feed Family
DATE: 1/5/86
REV:
NOTES: 1 2 3 4 5 6 7 8 9 10
WORKING
DRAFT
RECOMMENDED
PUBLICATION
READER
DATE
CONTEXT:
Top
Family Policy C2
C1 Budget
Food On Hand
Recipe
Plan 1
Meal Plan
Menu
Money
C1
Maintain Stock 2
Needed Items
Reqts & Prefs
Food Items
Prepare 3
Prepared Food
Serve Meal 4
Fed People
O1
I1
Unfed People
Table At End
Garbage
O2
Clean Up 5
Leftovers
Clean Kitchen
Unused Food
Clean Dishes
Dirty Kitchen
Reusables
NODE: FF/AØ
TITLE: Feed Family
NUMBER: DAMØ31
(DAMØ11)

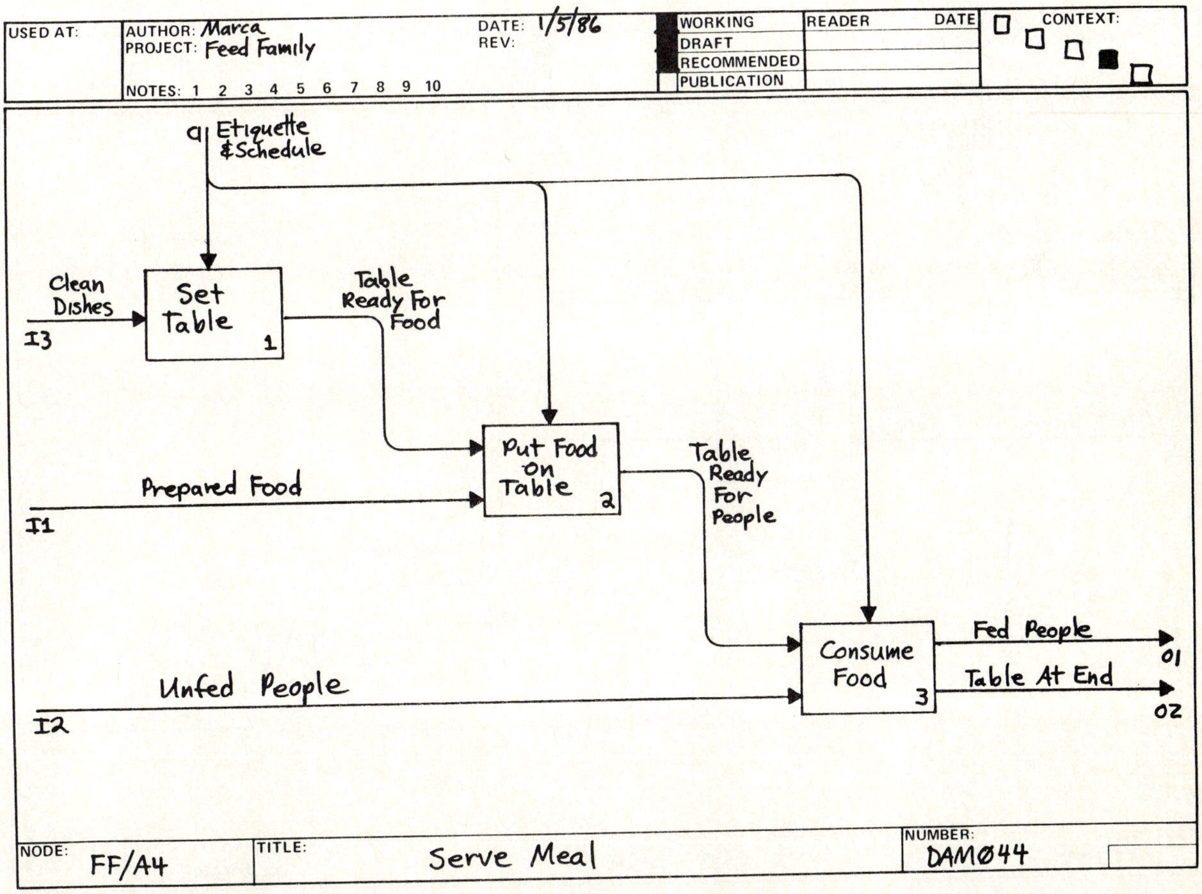
USED AT:
AUTHOR: Marca
PROJECT: Feed Family
DATE: 1/5/86
REV:
NOTES: 1 2 3 4 5 6 7 8 9 10
WORKING
DRAFT
RECOMMENDED
PUBLICATION
READER
DATE
CONTEXT:
C1 Etiquette & Schedule
Clean Dishes
I3
Set Table 1
Table Ready For Food
Put Food On Table 2
Prepared Food
I1
Table Ready For People
Consume Food 3
Fed People
O1
Table At End
O2
Unfed People
I2
NODE: FF/A4
TITLE: Serve Meal
NUMBER: DAMØ44

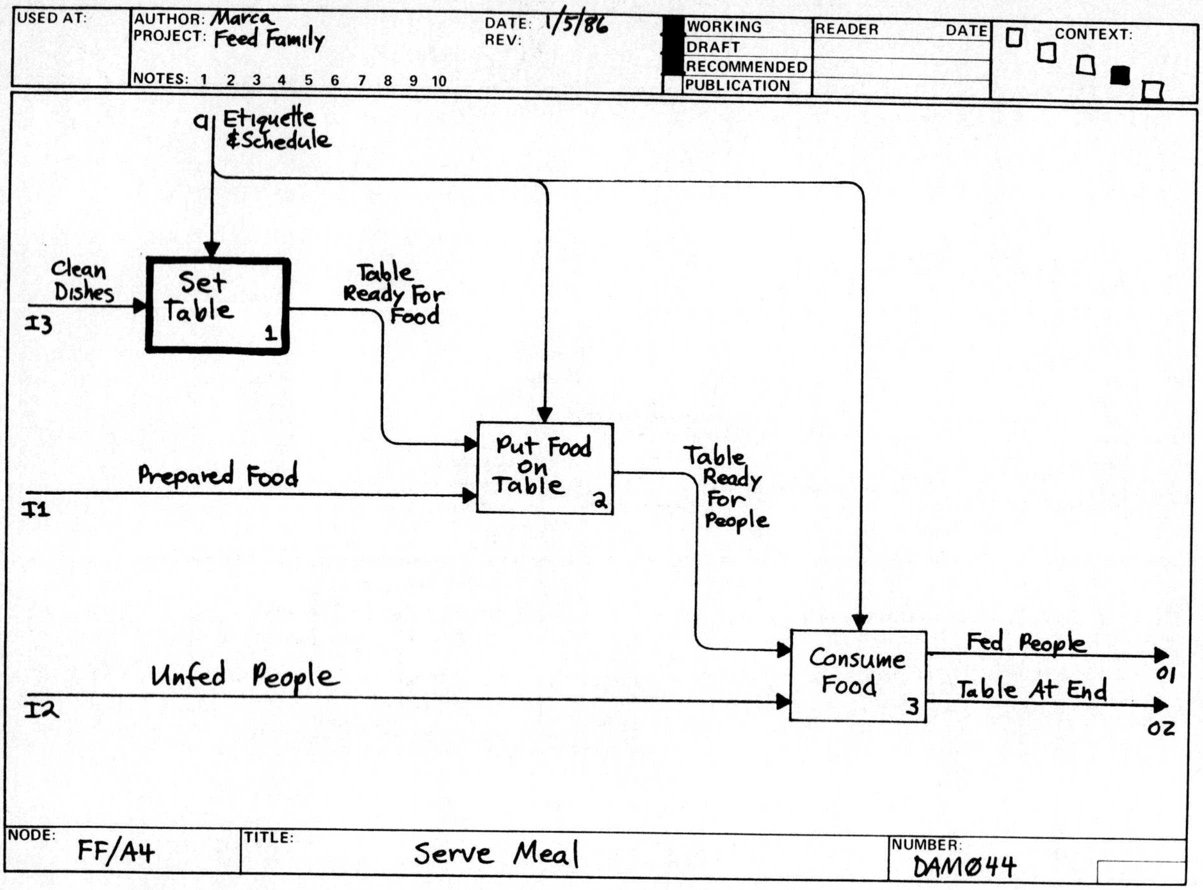
USED AT:
AUTHOR: Marca
PROJECT: Feed Family
DATE: 1/5/86
REV:
NOTES: 1 2 3 4 5 6 7 8 9 10
WORKING
DRAFT
RECOMMENDED
PUBLICATION
READER
DATE
CONTEXT:
C1 Etiquette & Schedule
Clean Dishes
I3
Set Table 1
Table Ready For Food
Put Food on Table 2
Prepared Food
I1
Table Ready For People
Consume Food 3
Unfed People
I2
Fed People
O1
Table At End
O2
NODE:
FF/A4
TITLE:
Serve Meal
NUMBER:
DAMØ44

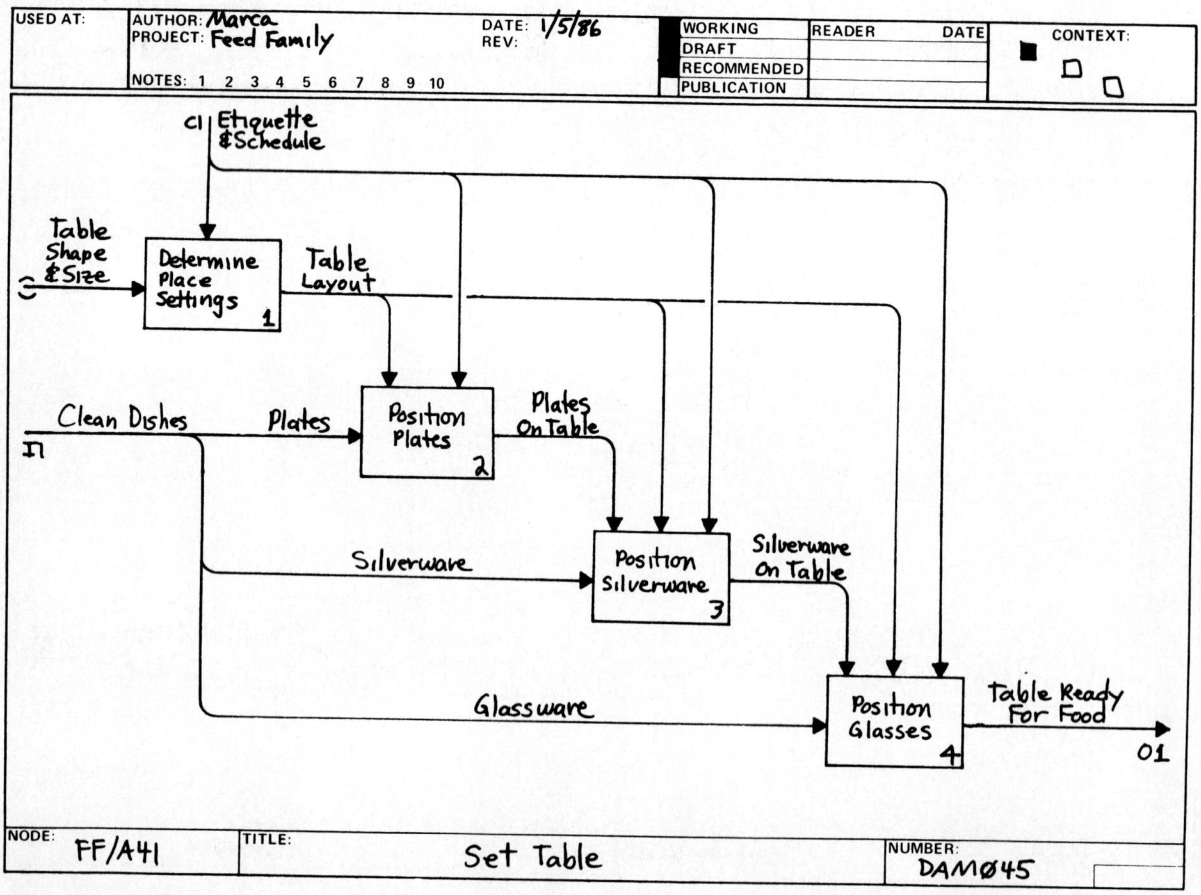
USED AT:
AUTHOR: Marca
PROJECT: Feed Family
DATE: 1/5/86
REV:
NOTES: 1 2 3 4 5 6 7 8 9 10
WORKING
DRAFT
RECOMMENDED
PUBLICATION
READER
DATE
CONTEXT:
C1 Etiquette & Schedule
Table Shape & Size
Determine Place Settings 1
Table Layout
Clean Dishes
I1
Plates
Position Plates 2
Plates On Table
Silverware
Position Silverware 3
Silverware On Table
Glassware
Position Glasses 4
Table Ready For Food
O1
NODE:
FF/A41
TITLE:
Set Table
NUMBER:
DAMØ45

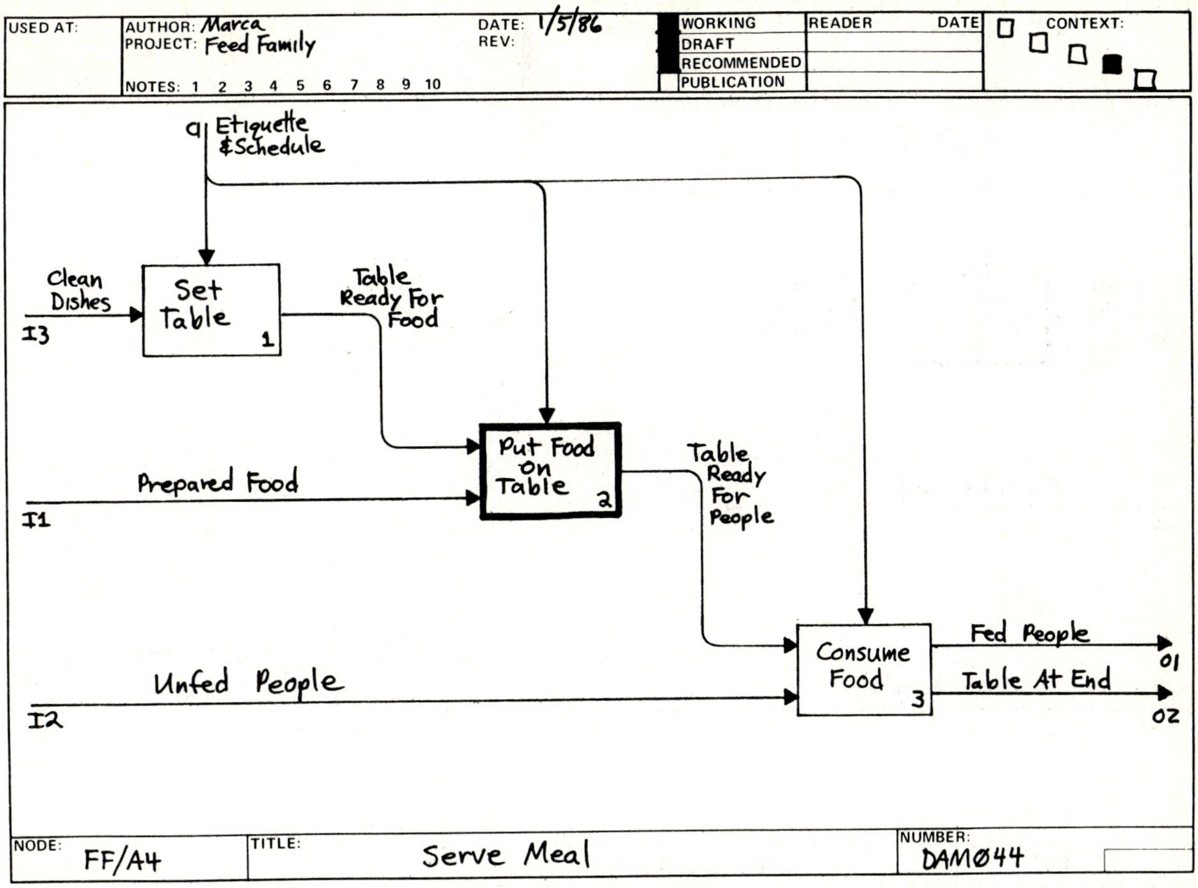
USED AT:
AUTHOR: Marca
PROJECT: Feed Family
DATE: 1/5/86
REV:
NOTES: 1 2 3 4 5 6 7 8 9 10
WORKING
DRAFT
RECOMMENDED
PUBLICATION
READER
DATE
CONTEXT:
C1 Etiquette & Schedule
Clean Dishes
I3
Set Table 1
Table Ready For Food
Put Food On Table 2
Prepared Food
I1
Table Ready For People
Consume Food 3
Fed People
O1
Table At End
O2
Unfed People
I2
NODE: FF/A4
TITLE: Serve Meal
NUMBER: DAMØ44

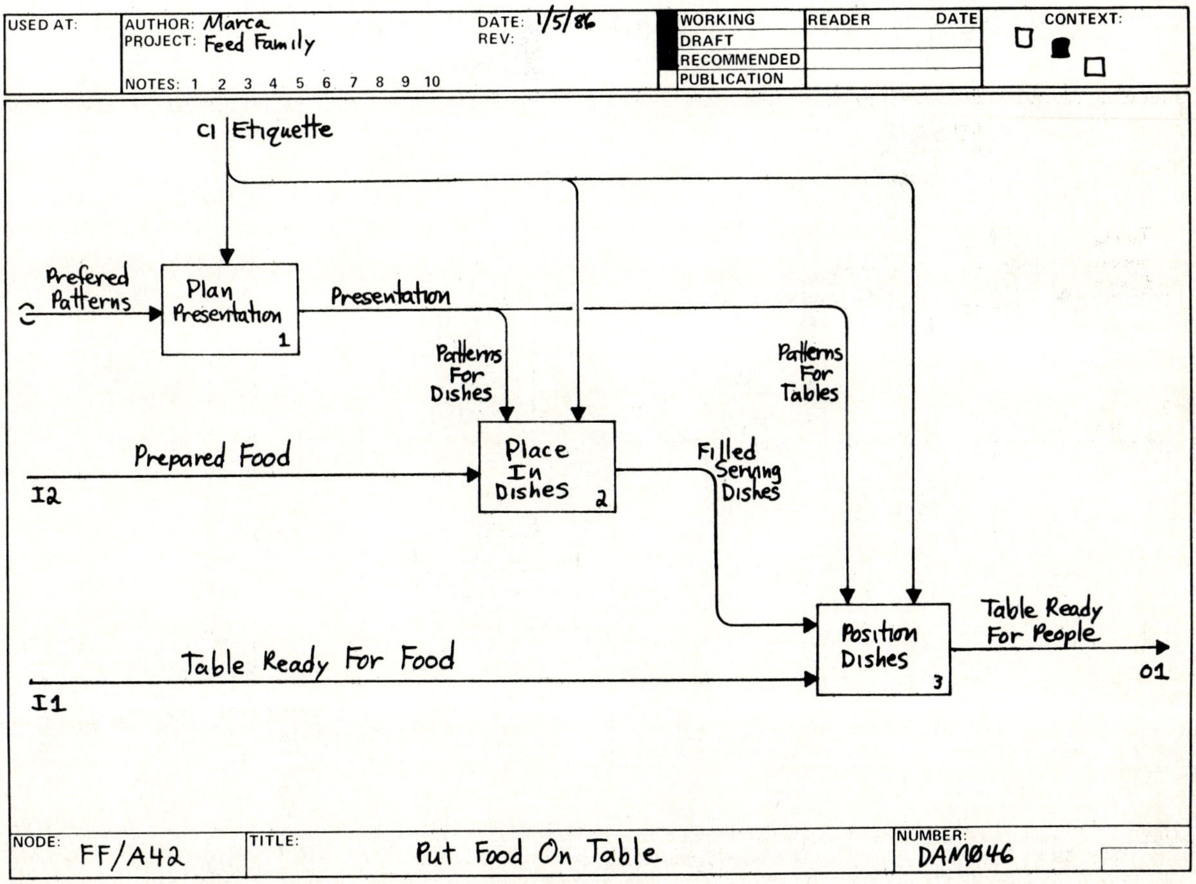
USED AT:
AUTHOR: Marca
PROJECT: Feed Family
DATE: 1/5/86
REV:
NOTES: 1 2 3 4 5 6 7 8 9 10
WORKING
DRAFT
RECOMMENDED
PUBLICATION
READER
DATE
CONTEXT:
C1 Etiquette
Prefered Patterns
Plan Presentation 1
Presentation
Patterns For Dishes
Patterns For Tables
Prepared Food
I2
Place In Dishes 2
Filled Serving Dishes
Position Dishes 3
Table Ready For People
O1
Table Ready For Food
I1
NODE: FF/A42
TITLE: Put Food On Table
NUMBER: DAMØ46

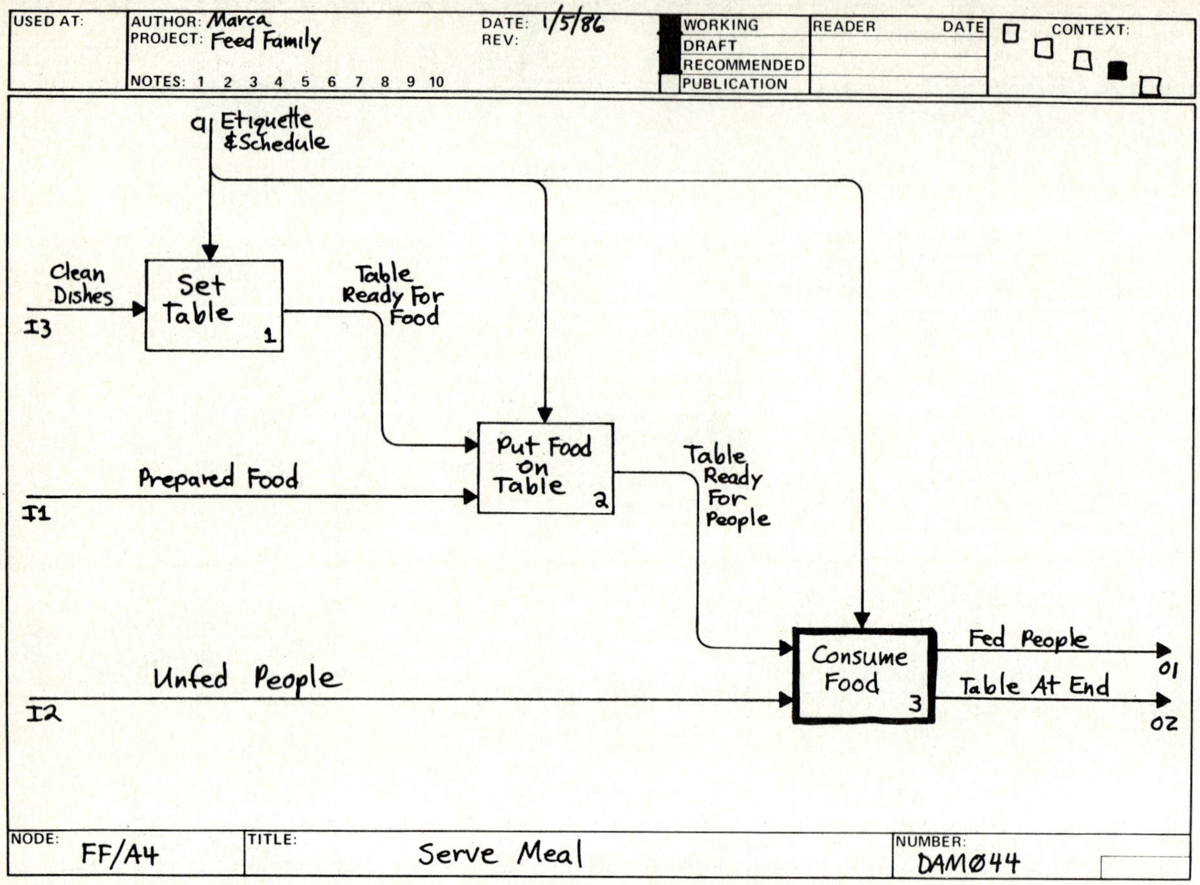
USED AT:
AUTHOR: Marca
PROJECT: Feed Family
DATE: 1/5/86
REV:
NOTES: 1 2 3 4 5 6 7 8 9 10
WORKING
DRAFT
RECOMMENDED
PUBLICATION
READER
DATE
CONTEXT:
C1 Etiquette & Schedule
Clean Dishes
I3
Set Table 1
Table Ready For Food
Put Food on Table 2
Prepared Food
I1
Table Ready For People
Consume Food 3
Fed People
O1
Table At End
O2
Unfed People
I2
NODE: FF/A4
TITLE: Serve Meal
NUMBER: DAMØ44

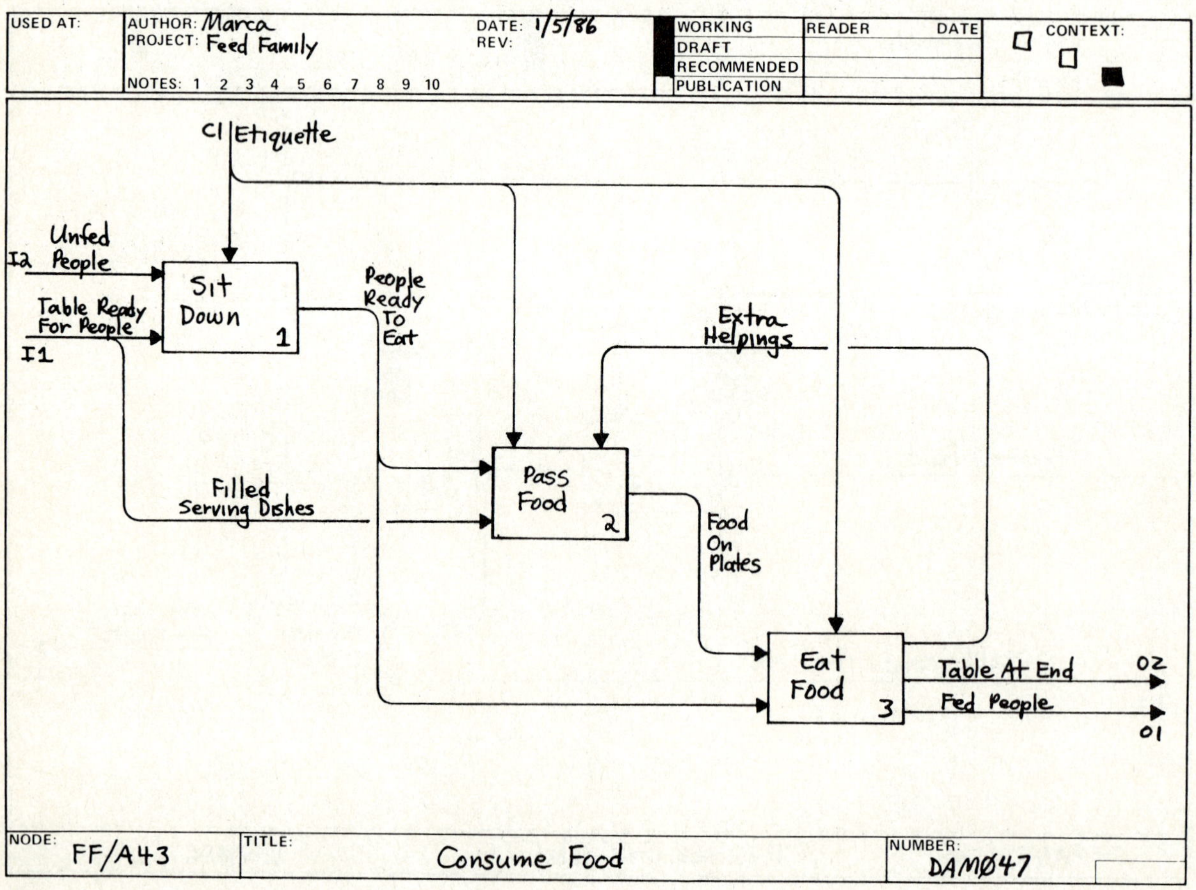
USED AT:
AUTHOR: Marca
PROJECT: Feed Family
DATE: 1/5/86
REV:
NOTES: 1 2 3 4 5 6 7 8 9 10
WORKING
DRAFT
RECOMMENDED
PUBLICATION
READER
DATE
CONTEXT:
C1 Etiquette
Unfed People
I2
Table Ready For People
I1
Sit Down 1
People Ready To Eat
Extra Helpings
Pass Food 2
Filled Serving Dishes
Food On Plates
Eat Food 3
Table At End
O2
Fed People
O1
NODE: FF/A43
TITLE: Consume Food
NUMBER: DAMØ47

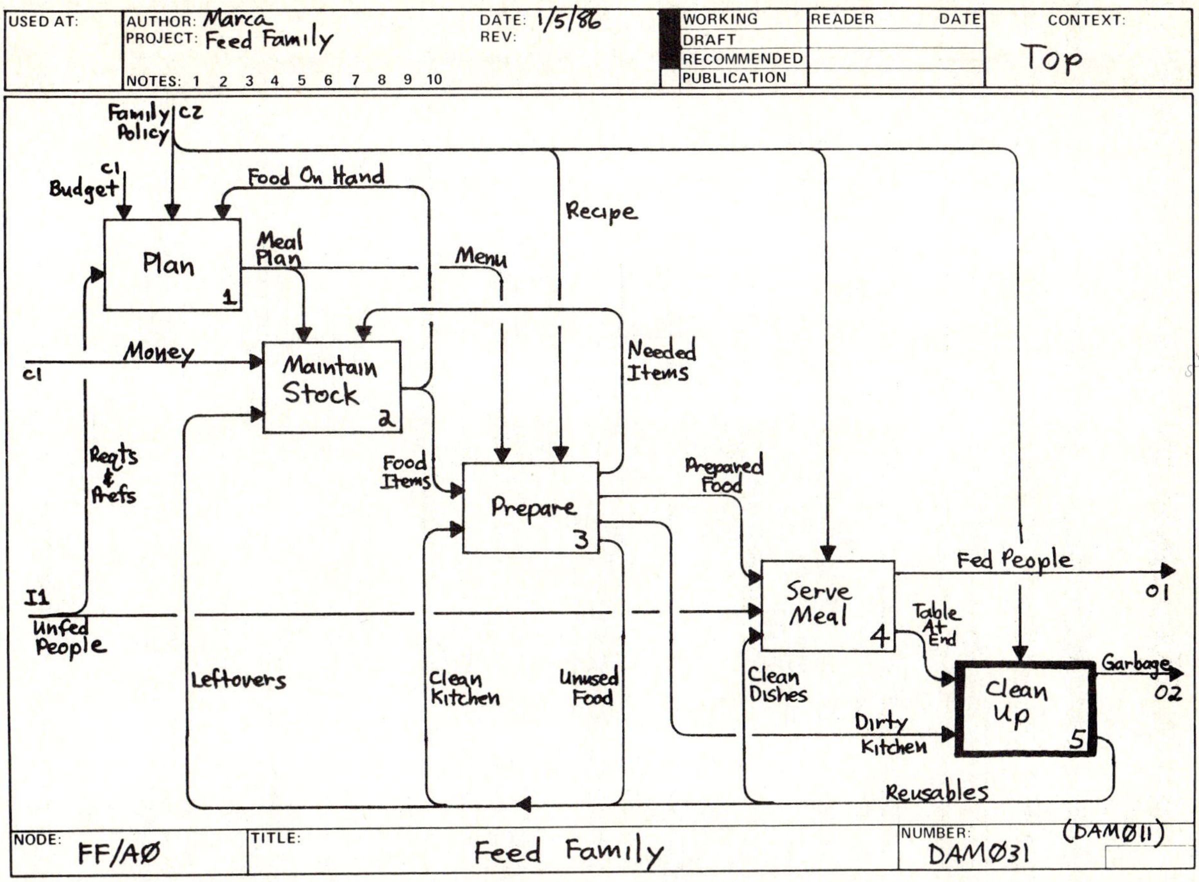
USED AT:
AUTHOR: Marca
PROJECT: Feed Family
DATE: 1/5/86
REV:
NOTES: 1 2 3 4 5 6 7 8 9 10
WORKING
DRAFT
RECOMMENDED
PUBLICATION
READER
DATE
CONTEXT:
Top
Family Policy
C2
C1
Budget
Food On Hand
Recipe
Plan
1
Meal Plan
Menu
Needed Items
C1
Money
Maintain Stock
2
Reqts & Prefs
Food Items
Prepared Food
Prepare
3
Fed People
O1
Serve Meal
4
Table At End
I1
Unfed People
Leftovers
Clean Kitchen
Unused Food
Clean Dishes
Garbage
O2
Clean Up
5
Dirty Kitchen
Reusables
NODE:
FF/AØ
TITLE:
Feed Family
NUMBER:
DAMØ31
(DAMØ11)

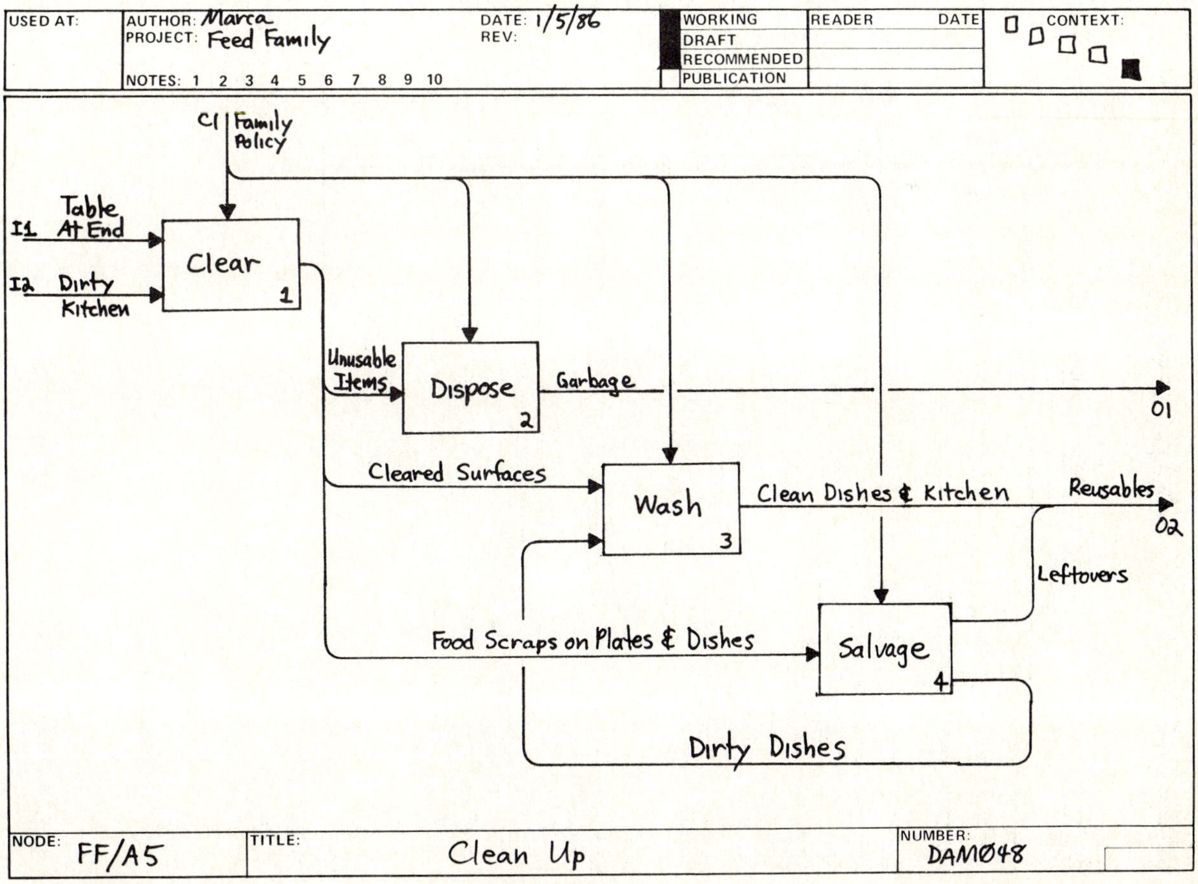
USED AT:
AUTHOR: Marca
PROJECT: Feed Family
DATE: 1/5/86
REV:
NOTES: 1 2 3 4 5 6 7 8 9 10
WORKING
DRAFT
RECOMMENDED
PUBLICATION
READER
DATE
CONTEXT:
C1
Family Policy
I1
Table At End
Clear
1
I2
Dirty Kitchen
Unusable Items
Dispose
2
Garbage
O1
Cleared Surfaces
Wash
3
Clean Dishes & Kitchen
Reusables
O2
Leftovers
Food Scraps on Plates & Dishes
Salvage
4
Dirty Dishes
NODE:
FF/A5
TITLE:
Clean Up
NUMBER:
DAMØ48

Diagram Index for the "Feed Family" Model

A-0 Feed Family
- **A0** Feed Family
  - **A1** Plan
    - **A12** Consider Choices
    - **A13** Determine Quantities
    - **A14** Schedule Meal
  - **A2** Maintain Stock
    - **A21** Stock the Larder
    - **A22** Plan Shopping
    - **A23** Buy Food
  - **A3** Prepare
    - **A32** Set up Kitchen
    - **A33** Prepare Ingredients
    - **A34** Combine and Cook
  - **A4** Serve Meal
    - **A41** Set Table
    - **A42** Put Food on Table
    - **A43** Consume Food
  - **A5** Clean Up

Appendix

D

Forms for Making SADT Activity Diagrams and Models

This appendix includes the three forms you will need to develop SADT activity diagrams and models. The first is the Diagram Form, used to draw diagrams and make the diagram supplements of Glossary, text, and figures. The second is the Kit Cover Page, used to collect diagrams and supplements into reviewable packets of material. The third is the C-Number Log Form, used to record the identity of each diagram made during a modeling effort. Parts I to IV of this book describe in detail how these forms should be used.

These forms have been included here to give you an opportunity to start making SADT activity diagrams and models on your own. These are the same forms used in the IDEF0 methodology—that subset of SADT adopted for use by the Integrated Computer-Aided Manufacturing effort initially sponsored by the U.S. Air Force—which has been in the public domain since the late 1970s. We hope you make use of the forms until you can acquire training and automated support tools.

USED AT:	AUTHOR: DATE: PROJECT: REV: NOTES: 1 2 3 4 5 6 7 8 9 10	WORKING DRAFT RECOMMENDED PUBLICATION	READER DATE	CONTEXT:

NODE:	TITLE:	NUMBER:

AUTHOR:
PROJECT:

DATE:
REV:

NOTES: 1 2 3 4 5 6 7 8 9 10

☐ WORKING
☐ DRAFT
☐ RECOMMENDED
☐ PUBLICATION

READER	DATE

☐ LOG
☐ FILE
☐ AUTHOR

DOCUMENT NUMBER

Received ____________

Completed ____________

COPYING INSTRUCTIONS:

_______ copies of _______ pages = _______ total

☐ as soon as possible ☐ by ____________

READERS:	New Kit To Reader	Due Back	Comments To Author	Due Back	Response To Reader

RESPONSE REQUIRED: ☐ Fast ☐ Normal ☐ Slow ☐ None

CONTENTS:

Pg.	Node	Title	C-Number	Status
A		COVER SHEET		
B				
C				
D				
E				
F				
G				
H				
I				
J				
K				
L				
M				
N				

COMMENTS:

☐ UPDATE
☐ REPLACE
} Model File ____________ with this kit.

SPECIAL INSTRUCTIONS:

☐ no author copy

_____ extra author copies

NODE: TITLE: NUMBER:

AUTHOR:
PROJECT:
DATE:

NOTES: 1 2 3 4 5 6 7 8 9 10

C-NUMBERS STARTING AT____00

00	C-NUMBER LOG	25		50		75	
01		26		51		76	
02		27		52		77	
03		28		53		78	
04		29		54		79	
05		30		55		80	
06		31		56		81	
07		32		57		82	
08		33		58		83	
09		34		59		84	
10		35		60		85	
11		36		61		86	
12		37		62		87	
13		38		63		88	
14		39		64		89	
15		40		65		90	
16		41		66		91	
17		42		67		92	
18		43		68		93	
19		44		69		94	
20		45		70		95	
21		46		71		96	
22		47		72		97	
23		48		73		98	
24		49		74		99	

NODE:
TITLE: C-NUMBER LOG
NUMBER: ______00

Selected Bibliography

Bachert, R. F., K. H. Evers, and **P. R. Santucci:** "SADT/SAINT Simulation Technique," *Proceedings of the National Aerospace and Electronics Conference,* 1981, pp. 4–9.

Borgida, A., S. Greenspan, and **J. Mylopoulos:** "Knowledge Representation as the Basis for Requirements Specifications," *Computer,* vol. 18, no. 4, April 1985, pp. 82–91.

Brackett, J. W., and **C. L. McGowan:** "Applying SADT to Large System Problems," *Proceedings of the Conference on Life Cycle Management,* August 1977, pp. 539–551.

Combelic, D.: "User Experience with New Software Methods (SADT)," *Proceedings of the NCC,* vol. 47, 1978, pp. 631–633.

Connor, M.: "Structured Analysis and Design Technique—SADT," Auerbach Portfolio 32-04-02, 1979.

Dickover, M. E., C. L. McGowan, and **D. T. Ross:** "Software Design Using SADT," *Proceedings of the ACM Annual Conference,* October 1977, pp. 125–133.

Evers, K. H., R. F. Bachert, and **P. R. Santucci:** "SADT/SAINT: Large Scale Analysis Simulation Methodology," *Proceedings of the 1981 Winter Simulation Conference,* 1981.

Greenspan, S.: *Requirements Modeling: A Knowledge Representation Approach to Software Requirements Definition,* University of Toronto Technical Report CSRG-155 (Ph.D. thesis), March 1984.

Hori, S.: *Human-Directed Activity Cell Model,* CAM-I Long Range Planning Final Report, CAM-I Inc., 1972.

LeClair, S. R., et al.: "Integrated Computer-Aided Manufacturing (ICAM) Architecture Part II," vol. I–VI, NTIS B062454–B052459, June 1981.

Marca, D., and **C. McGowan:** "Static and Dynamic Data Modeling for Information System Design," *Proceedings of the 6th International Conference on Software Engineering,* September 1982, pp. 137–146.

Pederson, J., and **J. Buckle:** "Kongsberg's Road to an Industrial Software Methodology," *IEEE Transactions on Software Engineering,* vol. 4, no. 4, July 1978.

Ross, D. T.: "Structured Analysis (SA): A Language for Communicating Ideas," *IEEE Transactions on Software Engineering,* vol. 3, no. 1, January 1977, pp. 16–34.

———: "Applications and Extensions of SADT," *Computer,* vol. 18, no. 4, April 1985, pp. 25–35.

——— and **J. Brackett:** "An Approach to Structured Analysis," *Computer Decisions,* vol. 8, no. 9, September 1976, pp. 40–44.

——— and **K. E. Schoman:** "Structured Analysis for Requirements Definition," *IEEE Transactions on Software Engineering,* vol. 3, no. 1, January 1977, pp. 6–15.

SofTech, Inc.: "IDEF0 Author's Guide to Creating Activity Diagrams," SofTech Deliverable no. 7500–13, September 1979.

———: "Integrated Computer-Aided Manufacturing (ICAM) Final Report: IDEF0 Functional Modeling Manual," SofTech Deliverable for Contract no. F33612-78-C-5158, January 1981.

———: "DWS/CS Emergency Preset Extensions to the A-7E Methodology," SofTech TP 150, March 1983.

Thomas, M.: "Function Decomposition: SADT," *Proceedings of the Infotech Conference on Structured Design,* 1976, pp. 115–135.

Wallace, R, J. Stockenburg, and **R. Charrette:** *A Unified Methodology for Developing Systems,* McGraw-Hill, New York, 1987.

Index

ABOUT THE AUTHORS

DAVID MARCA is a principal software engineer in the Intelligent Systems Technologies Group at Digital Equipment Corporation. He is currently applying software engineering and artificial intelligence technologies to computer integrated manufacturing (CIM) problems. Previously, Mr. Marca was with SofTech for seven years, where he used SADT on a variety of problem domains and taught it to commercial firms and government agencies. He is the author of the book *Applying Software Engineering Principles.*

CLEMENT MCGOWAN is manager of the Computer and Information Systems Department at GTE Laboratories. He presently directs a research staff in the areas of software to define telephone network services, executable specifications, information retrieval, and business modeling. Previously, Dr. McGowan worked at SofTech, where he used and taught SADT on a variety of systems analysis and design projects. He also served on the faculty of Brown University, where he coauthored the book *Top-Down Structured Programming Techniques.*